KEY ELEMENTS IN POLYMERS FOR ENGINEERS AND CHEMISTS
From Data to Applications

KEY ELEMENTS IN POLYMERS FOR ENGINEERS AND CHEMISTS

From Data to Applications

Edited by
**Alexandr A. Berlin, DSc, Viktor F. Kablov, DSc,
Andrey A. Pimerzin, DSc, and Simon S. Zlotsky, PhD**

Gennady E. Zaikov, DSc, and A. K. Haghi, PhD
Reviewers and Advisory Board Members

Apple Academic Press

TORONTO   NEW JERSEY

Apple Academic Press Inc.
3333 Mistwell Crescent
Oakville, ON L6L 0A2
Canada

Apple Academic Press Inc.
9 Spinnaker Way
Waretown, NJ 08758
USA

©2014 by Apple Academic Press, Inc.
Exclusive worldwide distribution by CRC Press, a member of Taylor & Francis Group

No claim to original U.S. Government works

International Standard Book Number-13: 978-1-926895-80-2 (Hardcover)

This book contains information obtained from authentic and highly regarded sources. Reprinted material is quoted with permission and sources are indicated. Copyright for individual articles remains with the authors as indicated. A wide variety of references are listed. Reasonable efforts have been made to publish reliable data and information, but the authors, editors, and the publisher cannot assume responsibility for the validity of all materials or the consequences of their use. The authors, editors, and the publisher have attempted to trace the copyright holders of all material reproduced in this publication and apologize to copyright holders if permission to publish in this form has not been obtained. If any copyright material has not been acknowledged, please write and let us know so we may rectify in any future reprint.

Trademark Notice: Registered trademark of products or corporate names are used only for explanation and identification without intent to infringe.

Library of Congress Control Number: 2014937165

Library and Archives Canada Cataloguing in Publication

Key elements in polymers for engineers and chemists: from data to applications/edited by Alexandr A. Berlin, DSc, Viktor F. Kablov, DSc, Andrey A. Pimerzin, DSc, and Simon S. Zlotsky, PhD; Gennady E. Zaikov, DSc, and A. K. Haghi, PhD, Reviewers and Advisory Board Members.

Includes bibliographical references and index.
ISBN 978-1-926895-80-2 (bound)
1. Polymers. 2. Polymerization. I. Kablov, Viktor F. (Viktor Fedorovich), editor II. Berlin, Alexandr A., editor III. Pimerzin, Andrey A., editor IV. Zlotsky, Simon S., editor

QD381.K49 2014   547'.7   C2014-902294-8

Apple Academic Press also publishes its books in a variety of electronic formats. Some content that appears in print may not be available in electronic format. For information about Apple Academic Press products, visit our website at **www.appleacademicpress.com** and the CRC Press website at **www.crcpress.com**

ABOUT THE EDITORS

Alexandr A. Berlin, DSc
Professor Alexandr A. Berlin, DSc, is Director of the N. N. Semenov Institute of Chemical Physics at the Russian Academy of Sciences, Moscow, Russia. He is a member of the Russian Academy of Sciences and many national and international associations. Dr. Berlin is world-renowned scientist in the field of chemical kinetics (combustion and flame), chemical physics (thermodynamics), chemistry and physics of oligomers, polymers, and composites and nanocomposites. He is the contributor of 100 books and volumes and 1000 original papers and reviews.

Viktor F. Kablov, DSc
Viktor F. Kablov, DSc, was appointed Director of VolzhskyPolytechnical Institute (branch) of VSTU–VPI (branch) of VSTU in 2000, and in 2002, he was elected Head of the Department of Polymer Chemical Technology and Industrial Ecology at the Institute. Professor Kablov has organized new training courses in modeling, computer-assisted methods and information systems in polymer engineering, engineering creativity methods, polymer chemistry and physics, and biotechnology. In 2010, he opened a Master's degree program in elastomer processing technology. He has also authored over 150 inventions, more than 300 research publications, 6 textbooks and 3 monographs.

Prof. Kablov's area of research relates to developing a scientific basis for obtaining elastomer materials operating in extreme conditions (heat-, fire-, and corrosion-resistant materials, antifriction composite materials); developing nano- and microheterogenic processes; modifying and operational additives to improve adhesive; thermal oxidation; and processing properties of polymer materials, as well as increasing the fire resistance of materials; developing materials for medical applications and fundamentally new hybrid polymers based on biopolymers and synthetic polymers; obtaining high selectivity sorbents based on template synthesis; developing waste bio recycling methods, as well as research in other areas of technology.

Andrey Pimerzin, DSc
AndreyPimerzin, DSc, was appointed Vice Rector of the Samara State Technical University, Samara, Russia, in 2007. He was also elected as Head of the Department of Chemical Technology Oil Gas Refining at the same university in 2001. In 2009, Professor Pinerzin was invited as a researcher at the Thermodynamics

Research Center NIST in the USA, and in 2011, he was one of the organizers of the international conference RCCT-2011. Under Andrey Pimerzin's supervision, eight PhD and one Doctor of Science theses have been defended. He is the author of more than 250 science publications, including 11 textbooks and 9 patents. His research interests include chemical and statistical thermodynamics of organic compounds, thermochemistry, kinetics, heterogeneous catalysis, and refining technology.

Simon S. Zlotsky, PhD
Professor Simon Solomonovich Zlotsky received his BS in chemistry from Chemistry and Technological Department of Ufa Petroleum Institute in 1968. From 1970 to 1973 he studied and gained his PhD from the Moscow Institute of the Petrochemical and Gas Industry. Dr. Zlotsky worked as a research fellow at the Ufa State Petroleum Technological University (1973–77) and researched his doctoral thesis on chemistry of cyclic acetals. From 1982 to 2010 he was a Professor in the Department of General Chemistry in the Ufa State Petroleum Technological University, and he has headed the department since 2010.

REVIEWERS AND ADVISORY BOARD MEMBERS

Gennady E. Zaikov, DSc

Gennady E. Zaikov, DSc, is Head of the Polymer Division at the N. M. Emanuel Institute of Biochemical Physics, Russian Academy of Sciences, Moscow, Russia, and Professor at Moscow State Academy of Fine Chemical Technology, Russia, as well as Professor at Kazan National Research Technological University, Kazan, Russia. He is also a prolific author, researcher, and lecturer. He has received several awards for his work, including the the Russian Federation Scholarship for Outstanding Scientists. He has been a member of many professional organizations and on the editorial boards of many international science journals.

A. K. Haghi, PhD

A. K. Haghi, PhD, holds a BSc in urban and environmental engineering from University of North Carolina (USA); a MSc in mechanical engineering from North Carolina A&T State University (USA); a DEA in applied mechanics, acoustics and materials from Université de Technologie de Compiègne (France); and a PhD in engineering sciences from Université de Franche-Comté (France). He is the author and editor of 65 books as well as 1000 published papers in various journals and conference proceedings. Dr. Haghi has received several grants, consulted for a number of major corporations, and is a frequent speaker to national and international audiences. Since 1983, he served as a professor at several universities. He is currently Editor-in-Chief of the *International Journal of Chemoinformatics and Chemical Engineering* and *Polymers Research Journal* and on the editorial boards of many international journals. He is also a member of the Canadian Research and Development Center of Sciences and Cultures (CRDCSC), Montreal, Quebec, Canada.

CONTENTS

List of Contributors .... *xi*
List of Abbreviations .... *xv*
Preface ....*xvii*

1. **RAFT-Polymerization of Styrene—Kinetics and Mechanism** 1
   Nikolai V. Ulitin, Aleksey V. Oparkin, Rustam Ya. Deberdeev, Evgenii B. Shirokih, and Gennady E. Zaikov

2. **A Detailed Review on Pore Structure Analysis of Electrospun Porous Membranes** 29
   Bentolhoda Hadavi Moghadam, Vahid Mottaghitalab, Mahdi Hasanzadeh, and Akbar Khodaparast Haghi

3. **Experimental Techniques for Application of Recycled Polymers in Construction Industries** 51
   O. Emamgholipour and A. K. Haghi

4. **Nanocomposites Polyethylene/Organoclay on Suprasegmental Level** 107
   G. V. Kozlov, K. S. Dibirova, G. M. Magomedov, and G. E. Zaikov

5. **Quantum and Wave Characteristics of Spatial-Energy Interactions** 115
   G. A. Korablev, and G. E. Zaikov

6. **Key Elements on Nanopolymers—From Nanotubes to Nanofibers** ... 129
   A. K. Haghi and G. E. Zaikov

7. **Polyacetylene** 251
   V. A. Babkin, G. E. Zaikov, M. Hasanzadeh, and A. K. Haghi

8. **Features of Macromolecule Formation by RAFT-Polymerization of Styrene in the Presence of Trithiocarbonates** 313
   Nikolai V. Ulitin, Timur R. Deberdeev, Aleksey V. Oparkin, and Gennady E. Zaikov

9. **A Study on Physical Properties of Composites Based on Epoxy Resin** 333
   J. Aneli, O. Mukbaniani, E. Markarashvili, G. E. Zaikov, and E. Klodzinska

10. **Synthesis, Structural Properties, Development and Applications of Metal-Organic Frameworks in Textile** 343
    M. Hasanzadeh and B. Hadavi Moghadam

11. **Thermal Behavior and Ionic Conductivity of the PEO/PAc and PEO/PAc Blends** 359
    Amirah Hashifudin, Sim Lai Har, Chan Chin Han, Hans Werner Kammer, and Siti Nor Hafiza Mohd Yusoff

12. **Correlation Between the Storage Time of the NRL and the Efficiency of PMMA Grafting to NR** 371
    Yoga Sugama Salim, Nur Aziemah Zainudin, Chin Han Chan, and Kai Weng Chan

13. **A Study on Composite Polymer Electrolyte** 385
    Tan Winie, N.H.A. Rosli, M.R. Ahmad, R.H.Y. Subban, and C.H. Chan

14. **A Study on Solid Polymer Electrolytes** 399
    Siti Nor Hafiza Mohd Yusoff, Sim Lai Har, Chan Chin Han, Amirah Hashifudin, and Hans–Werner Kammer

15. **Modification of PC-PHBH Blend Monolith** 411
    Yuanrong Xin and Hiroshi Uyama

    Index *423*

LIST OF CONTRIBUTORS

M. R. Ahmad
Faculty of Applied Sciences, Universiti Teknologi MARA, 40450 Shah Alam, Malaysia

J. Aneli
R. Dvali Institute of Machine Mechanis, email: JimAneli@yahoo.com

V. A. Babkin
Volgograd State Architect-build University, Sebrykov Departament, Russia

C. H. Chan
Faculty of Applied Sciences, Universiti Teknologi MARA, 40450 Shah Alam, Malaysia

Rustam Ya. Deberdeev
Kazan National Research Technological University, 68 Karl Marx Street, 420015 Kazan, Republic of Tatarstan, Russian Federation, Fax: +7 (843) 231-41-56; E-mail: n.v.ulitin@mail.ru

K. S Dibirova
Dagestan State Pedagogical University, Makhachkala 367003, Yaragskii st., 57, Russian Federation

O. Emamgholipour
University of Guilan, Iran

Akbar Khodaparast Haghi
Department of Textile Engineering, University of Guilan, Rasht, Iran, E-mail: Haghi@Guilan.ac.ir

Sim Lai Har
Centre of Foundation Studies, Universiti Teknologi MARA, 42300 Puncak Alam, Selangor.Malysia

Mahdi Hasanzadeh
Department of Textile Engineering, University of Guilan, Rasht, Iran, Email: m_hasanzadeh@aut.ac.ir, Tel.: +98-21-33516875; fax: +98-182-3228375

Amirah Hashifudin
Faculty of Applied Sciences, Universiti Teknologi MARA, 40450 Shah Alam, Selangor. Malysia

E. Markarashvili
Javakhishvili Tbilisi State University Faculty of Exact and Natural Sciences, Department of Macromolecular Chemistry. I. Chavchavadze Ave., 3, Tbilisi 0128, Republic of Georgia

O. Mukbaniani
Javakhishvili Tbilisi State University Faculty of Exact and Natural Sciences, Department of Macromolecular Chemistry. I. Chavchavadze Ave., 3, Tbilisi 0128, Republic of Georgia, email: OmariMui@yahoo.com

Hans-Werner Kammer
University of Halle, Mansfelder Str. 28, D-01309 Dresden, Germany

E. Klodzinska
Institute for Engineering of Polymer Materials and Dyes, 55 M. Sklodowskiej-Curie str., 87-100 Torun, Poland, E-mail: S.Kubica@impib.pl

G. A. Korablev
[1]Izhevsk State Agricultural Academy, Basic Research and Educational Center of Chemical Physics and Mesoscopy, Udmurt Scientific Center, Ural Division, Russian Academy of Science, Russia, Izhevsk, 426000, e-mail: korablev@udm.net, biakaa@mail.ru

G. V. Kozlov
Dagestan State Pedagogical University, Makhachkala 367003, Yaragskii st., 57, Russian Federation

G. M. Magomedov
Dagestan State Pedagogical University, Makhachkala 367003, Yaragskii st., 57, Russian Federation

Bentolhoda Hadavi Moghadam
Department of Textile Engineering, University of Guilan, Rasht, Iran, E-mail address: motaghitalab@guilan.ac.ir

Siti Nor Hafiza Mohd Yusoff
Faculty of Applied Sciences, Universiti Teknologi MARA, 40450 Shah Alam, Selangor. Malysia

Vahid Mottaghitalab
Department of Textile Engineering, University of Guilan, Rasht, Iran, E-mail address: motaghitalab@guilan.ac.ir

V Nikolai
Kazan National Research Technological University, 68 Karl Marx Street, 420015 Kazan, Republic of Tatarstan, Russian Federation, Fax: +7 (843) 231-41-56, E-mail: n.v.ulitin@mail.ru

Aleksey V. Oparkin
Kazan National Research Technological University, 68 Karl Marx Street, 420015 Kazan, Republic of Tatarstan, Russian FederationFax: +7 (843) 231-41-56; E-mail: n.v.ulitin@mail.ru

N. H. A. Rosli
Faculty of Applied Sciences, Universiti Teknologi MARA, 40450 Shah Alam, Malaysia

Yoga Sugama Salim
Department of Chemistry, University of Malaya, 50603 Kuala Lumpur, Malaysia

Evgenii B. Shirokih
Kazan National Research Technological University, 68 Karl Marx Street, 420015 Kazan, Republic of Tatarstan, Russian Federation, Fax: +7 (843) 231-41-56; E-mail: n.v.ulitin@mail.ru

R. H. Y. Subban
Faculty of Applied Sciences, Universiti Teknologi MARA, 40450 Shah Alam, Malaysia

Ulitin, Timur
Kazan National Research Technological University, 68 Karl Marx Street, 420015 Kazan, Republic of Tatarstan, Russian Federation, Fax: +7 (843) 231-41-56; E-mail: n.v.ulitin@mail.ru

Nikolai V. Ulitin
Kazan National Research Technological University, 68 Karl Marx Street, 420015 Kazan, Republic of Tatarstan, Russian Federation Fax: +7 (843) 231-41-56; E-mail: n.v.ulitin@mail.ru

Hiroshi Uyama
Department of Applied Chemistry, Graduate School of Engineering, Osaka University, Suita, Osaka 565-0871, Japan

List of Contributors xiii

Kai Weng Chan
Synthomer (M) Sdn. Bhd., 86000 Kluang, Malaysia

Tan Winie
Faculty of Applied Sciences, Universiti Teknologi MARA, 40450 Shah Alam, Malaysia*. Tel.: +60-3-55435738; fax: +60-3-55444562, E-mail address: tanwinie@salam.uitm.edu.my

Yuanrong Xin
Department of Applied Chemistry, Graduate School of Engineering, Osaka University, Suita, Osaka 565-0871, Japan

Siti Nor Hafiza Mohd Yusoff
Faculty of Applied Sciences, Universiti Teknologi MARA, 40450 Shah Alam, Selangor, Malaysia

Gennady E. Zaikov
N. M. Emanuel Institute of Biochemical Physics, Russian Academy of Sciences, 4 Kosygina st., Moscow, Russian Federation, 119334, Fax: +7(499)137-41-01; E-mail: chembio@sky.chph.ras.ru

Nur Aziemah Zainudin
Faculty of Applied Sciences, Universiti Teknologi MARA, 40500 Shah Alam, Malaysia Chin Han Chan Faculty of Applied Sciences, Universiti Teknologi MARA, 40500 Shah Alam, Malaysia

LIST OF ABBREVIATIONS

AFM	atomic force microscopy
AIBN	azobisisobutyronitrile
ATR	attenuated total reflection
CPE	composite polymer electrolyte
CTMP	chemithermomechanical pulp
DBTC	dibenzyltritiocarbonate
DCAA	dichloroacetic acid
DPNR	deproteinized natural rubber
DSC	differential scanning calorimetry
ENR	epoxidized natural rubber
eV	electron-volts
IS	impedance spectroscopy
ITS	indirect tensile strength
LB	Lattice–Boltzmann
LLDPE	linear low density polyethylene
M&S	modeling and simulation
MDD	maximum dry density
MEMS	microelectromechanical systems
MMA	methyl methacrylate
MMD	molecular-mass distribution
MMD	molecular-mass distribution
MMT	montmorillonite
MNR	modified natural rubber
MOFs	metal-organic frameworks
MWNT	multi-walled nanotubes
NIPS	non-solvent induced phase separation
NR	natural rubber
MST	mechanical stability time
NRL	natural rubber latex
ODE's	ordinary differential equations
OSA	objective-based simulated Annealing
OWC	optimum water content
Pac	polyacrylate
PAN	polyacrylonitrile
PANi	polyaniline

PC	polycarbonate
PD	polydispersity
PEI	polyethylenimine
PEO	poly ethylene oxide
PPy	polypyrrole
PSM	post-synthetic modification
PSW	plastic solid waste
RAFT	reversible addition-fragmentation chain transfer
RDP	radial density profile
RH	rice husk ash
RVP	radial velocity profile
SBUs	secondary building units
SC	clayey sand
SEM	scanning electron microscopy
SEP	spatial-energy parameter
SF	silica fume
SPE	solid polymer electrolyte
SRNF	resistant nanofiltration
SWNT	single-wall nanotubes
TEM	transmission electron microscopy
TEOS	tetraethoxysilane
TEPA	tetraethylenepentamine
THF	tetrahydrofuran
TMTD	thiuramdisulphide
TSC	total solid content

PREFACE

Polymers have played a significant part in the existence of humans. They have a role in every aspect of modern life, including health care, food, information technology, transportation, energy industries, etc. The speed of developments within the polymer sector is phenomenal and, at the same time, crucial to meet demands of today's and future life. Specific applications for polymers range from using them in adhesives, coatings, painting, foams, and packaging to structural materials, composites, textiles, electronic and optical devices, biomaterials and many other uses in industries and daily life. Polymers are the basis of natural and synthetic materials. They are macromolecules, and in nature are the raw material for proteins and nucleic acids, which are essential for human bodies.

Cellulose, wool, natural rubber and synthetic rubber, plastics are well-known examples of natural and synthetic types. Natural and synthetic polymers play a massive role in everyday life, and a life without polymers really does not exist. A correct understanding of polymers did not exist until 1920s. In 1922, Staudinger published his idea that polymers were long chain molecules with normal chemical bonds holding them together. But for nearly 10 years this idea did not attract much attention. Around this period other researchers like Carothers who tended towards Staudinger's idea, discovered a type of synthetic material, which could be produced by its constituent monomers. Later on it was shown that as well as addition reaction, polymers could be prepared through condensation mechanism.

Previously it was believed that polymers could only be prepared through addition polymerization. The mechanism of the addition reaction was also unknown and hence there was no sound basis of proposing a structure for the polymers. This lack of information was the main controversy existed between Staudinger and his critics. The studies by Carothers and other researchers resulted in theorizing the condensation polymerization. It became clear that difunctional molecules like dihydric alcohols and dicarboxylic acids could react repeatedly with the release of water to form polyesters of high molecular mass. This mechanism became well understood and the structure of the resultant polyester could be specified with greater confidence.

In 1941/42 the world witnessed the infancy of polyethylene terephthalate or better known as the polyester. A decade later for the first time polyester/cotton blends introduced. In those days Terylene and Dacron (commercial names for polyester fibers) were miracle fibers but still overshadowed by nylon. Not many would have predicted those decades later, polyester would have become the world's inexpensive, general purpose fibers as well as becoming a premium fiber for special functions in engineering textiles, fashion and many other technical end-uses. From the time nylon and polyester were first used there have been an amazing technological advances which have made them so cheap to manufacture and widely available.

These developments have made the polymers such as polyesters to contribute enormously in today's modern life. One of the most important applications is the furnishing sector (home, office, cars, aviation industry, etc.), which benefits hugely from the advances in technology. There are a number of requirements for a fabric to function in its chosen end use, for example, resistance to pilling and abrasion, as well as, dimensional stability. Polyester is now an important part of the upholstery fabrics. The shortcomings attributed to the fiber in its early days have mostly been overcome. Now it plays a significant part in improving the life span of a fabric, as well as its dimensional stability, which is due to its heat-setting properties.

About half century has passed since synthetic leather a composite material completely different from conventional ones came to the market. Synthetic leather was originally developed for end-uses such as, the upper of shoes. Gradually other uses like clothing steadily increased the production of synthetic leather and suede. Synthetic leathers and suede have a continuous ultrafine porous structure comprising a three-dimensional entangled nonwoven fabric and an elastic material principally made of polyurethane. Polymeric materials consisting of the synthetic leathers are polyamide and polyethylene terephthalate for the fiber and polyurethanes with various soft segments, such as aliphatic polyesters, polyethers and polycarbonates for the matrix.

The introduction of plastics is associated with the twentieth century but the first plastic material, celluloid, were made in 1865. During the 1970s, clothes of polyester became fashionable but by the 1980s synthetics lost the popularity in favor of natural materials. Although people were less enthusiastic about synthetic fabrics for everyday wear, Gore–Tex and other synthetics became popular for outdoors and workout clothing. At the same time as the use of synthetic materials in clothing declined, alternative uses were found. One great example is the use of polyester for making beverage bottles where it replaced glass with its shatterproof properties as a significant property.

In general it can be said that plastics enhance and even preserve life. Kevlar, for instance, when it is used in making canoes for recreation or when used to make a bulletproof vest. Polyester enhances life, when this highly nonreactive material is used to make replacement human blood vessels or even replacement skin for burn victims. With all the benefits attributed to plastics, they have their negative side. A genuine environmental problem exists due to the fact that the synthetic polymers do not break down easily compared with the natural polymers. Hence the need not only to develop biodegradable plastics, but also to work on more effective means of recycling. A lot of research needed to study the methods of degradation and stabilization of polymers in order to design polymers according to the end-use.

Among the most important and versatile of the hundreds of commercial plastics is polyethylene. Polyethylene is used in a wide variety of applications because it can be produced in many different forms. The first type to be commercially exploited was called low-density polyethylene (LDPE). This polymer is characterized by a large degree of branching, forcing the molecules to pack together rather than loosely forming a low-density material. LDPE is soft and pliable and has applications ranging from plastic bags, containers, textiles, and electrical insulation, to coatings for packaging materials.

Another form of polyethylene differing from LDPE in structure is high-density polyethylene (HDPE). HDPE demonstrates little or no branching, resulting in the molecules to be tightly packed. HDPE is much more rigid than LDPE and is used in applications where rigidity is important. Major uses of HDPE are plastic tubing, bottles, and bottle caps. Other variations of polyethylene include high and ultra-high molecular mass ones. These types are used in applications where extremely tough and resilient materials are needed.

Natural polymers unlike the synthetic ones do possess very complex structure. Natural polymers such as cellulose, wool, and natural rubber are used in many products in large proportions. Cellulose derivatives are one of the most versatile groups of regenerated materials with various fields of application. Cellulose is found in nature in all forms of plant life, particularly in wood and cotton. The purest form of cellulose is obtained from the seed hairs of the cotton plant, which contain up to 95% cellulose. The first cellulose derivatives came to stage around 1845 when the nitration of starch and paper led to discovery of cellulose nitrate. In 1865 for the first time a moldable thermoplastic made of cellulose nitrate and castor oil.

In 1865 the first acetylation of cellulose was carried out but the first acetylation process for use in industry was announced in 1894. In 1905 an acetylation process was introduced which yielded a cellulose acetate soluble in the cheap solvent, acetone. It was during the First World War when cellulose acetate dope found importance for weather proofing and stiffening the fabric of aircraft wings. There was a large surplus production capacity after the war, which led to civilian end uses such as the production of cellulose acetate fibers by 1920's. Cellulose acetate became the main thermoplastic molding material when the first modern injection molding machines were designed. Among the cellulose derivatives, cellulose acetates are produced in the largest volume. Cellulose acetate can be made into fibers, transparent films and the less substituted derivatives are true thermoplastics. Cellulose acetates are moldable and can be fabricated by the conventional processes. They have toughness, good appearance, capable of many color variations including white transparency.

New applications are being developed for polymers at a very fast rate all over the world at various research centers. Examples of these include electro active polymers, nanoproducts, robotics, etc. Electro active polymers are special types of materials, which can be used for example as artificial muscles and facial parts of robots or even in nanorobots. These polymers change their shape when activated by electricity or even by chemicals. They are lightweight but can bear a large force, which is very useful when being utilized for artificial muscles. Electro active polymers together with nanotubes can produce very strong actuators. Currently research works are carried out to combine various types of electro active polymers with carbon nanotubes to make the optimal actuator. Carbon nanotubes are very strong, elastic, and conduct electricity. When they are used as an actuator, in combination with an electro active polymer the contractions of the artificial muscle can be controlled by electricity. Already works are under way to use electro active polymers in space. Various space agencies are investigating the possibility of using these polymers in space. This technology has a lot to offer for the future, and with the ever-increasing work on nanotechnology, electro active materials will play very important part in modern life.

— **Alexandr A. Berlin, DSc, Viktor F. Kablov, DSc, Andrey A. Pimerzin, DSc, and Simon S. Zlotsky, PhD**

CHAPTER 1

RAFT-POLYMERIZATION OF STYRENE—KINETICS AND MECHANISM

NIKOLAI V. ULITIN, ALEKSEY V. OPARKIN,
RUSTAM YA. DEBERDEEV, EVGENII B. SHIROKIH,
and GENNADY E. ZAIKOV

CONTENTS

1.1 Introduction 2
1.2 Experimental Part 2
      1.2.1 Mathematical Modeling of Polymerization Process 3
      1.2.2 Rate Constants 9
      1.2.3 Model's Adequacy 18
      1.2.4 Numerical approach 21
1.3 Conclusion 26
Keywords 27
References 27

1.1 INTRODUCTION

The kinetic modeling of styrene controlled radical polymerization, initiated by 2,2'-asobis(isobutirnitrile) and proceeding by a reversible chain transfer mechanism was carried out and accompanied by "addition-fragmentation" in the presence dibenzyltritiocarbonate. An inverse problem of determination of the unknown temperature dependences of single elementary reaction rate constants of kinetic scheme was solved. The adequacy of the model was revealed by comparison of theoretical and experimental values of polystyrene molecular-mass properties. The influence of process controlling factors on polystyrene molecular-mass properties was studied using the model

The controlled radical polymerization is one of the most developing synthesis methods of narrowly dispersed polymers nowadays [1–3]. Most considerations were given to researches on controlled radical polymerization, proceeding by a reversible chain transfer mechanism and accompanied by "addition-fragmentation" (RAFT – reversible addition-fragmentation chain transfer) [3]. It should be noted that for classical RAFT-polymerization (proceeding in the presence of sulphur-containing compounds, which formula is Z–C(=S)–S–R', where Z – stabilizing group, R' – outgoing group), valuable progress was obtained in the field of synthesis of new controlling agents (RAFT-agents), as well as in the field of research of kinetics and mathematical modeling; and for RAFT-polymerization in symmetrical RAFT-agents' presence, particularly, tritiocarbonates of formula R'–S–C(=S)–S–R', it came to naught in practice: kinetics was studied in extremely general form [4] and mathematical modeling of process hasn't been carried out at all. Thus, the aim of this research is the kinetic modeling of polystyrene controlled radical polymerization initiated by 2,2'-asobis(isobutirnitrile) (AIBN), proceeding by reversible chain transfer mechanism and accompanied by "addition-fragmentation" in the presence of dibenzyltritiocarbonate (DBTC), and also the research of influence of the controlling factors (temperature, initial concentrations of monomer, AIBN and DBTC) on molecular-mass properties of polymer.

1.2 EXPERIMENTAL PART

Prior using of styrene (Aldrich, 99%), it was purified of aldehydes and inhibitors at triple cleaning in a separatory funnel by 10%-th (mass) solution of NaOH (styrene to solution ratio is 1:1), then it was scoured by distilled water to neutral reaction and after that it was dehumidified over $CaCl_2$ and rectified in vacuo.

AIBN (Aldrich, 99%) was purified of methanol by re-crystallization.

DBTC was obtained by the method presented in research [4]. Masses of initial substances are the same as in Ref. [4]. Emission of DBTC was 81%. NMR ^{13}C (CCl_3D) δ, ppm: 41.37, 127.60, 128.52, 129.08, 134.75, and 222.35.

Examples of polymerization were obtained by dissolution of estimated quantity of AIBN and DBTC in monomer. Solutions were filled in tubes, 100 mm long, and having internal diameter of 3 mm, and after degassing in the mode of "freezing-defrosting" to residual pressure 0.01-mmHg column, the tubes were unsoldered. Polymerization was carried out at 60°C.

Research of polymerization's kinetics was made with application of the calorimetric method on Calvet type differential automatic microcalorimeter DAK-1–1 in the mode of immediate record of heat emission rate in isothermal conditions at 60°C. Kinetic parameters of polymerization were calculated basing on the calorimetric data as in the work [5]. The value of polymerization enthalpy $\Delta H = -73.8$ kJ × mol^{-1} [5] was applied in processing of the data in the calculations.

Molecular-mass properties of polymeric samples were determined by gel-penetrating chromatography in tetrahydrofuran at 35°C on chromatograph GPCV 2000 "Waters". Dissection was performed on two successive banisters PLgel MIXED–C 300×7.5 mm, filled by stir gel with 5 μm vesicles. Elution rate – 0.1 mL × min^{-1}. Chromatograms were processed in programme "Empower Pro" with use of calibration by polystyrene standards.

1.2.1 MATHEMATICAL MODELING OF POLYMERIZATION PROCESS

Kinetic scheme, introduced for description of styrene controlled radical polymerization process in the presence of trithiocarbonates, includes the following phases.

1. Real initiation

$$I \xrightarrow{k_d} 2R(0)^\cdot$$

2. Thermal initiation [6]. It should be noted that polymer participation in thermal initiation reactions must reduce the influence thereof on molecular-mass distribution (MMD). However, since final mechanism of these reactions has not been ascertained in recording of balance differential equations for polymeric products so far, we will ignore this fact.

$$3M \xrightarrow{k_{i1}} 2R(1),$$

$$2M+P \xrightarrow{k_{i2}} R(1)+R(i),$$

$$2P \xrightarrow{k_{i3}} 2R(i)\cdot$$

In these three reactions summary concentration of polymer is recorded as P.

3. Chain growth

$$R(0)+M \xrightarrow{k_p} R(1),$$

$$R'+M \xrightarrow{k_p} R(1),$$

$$R(i)+M \xrightarrow{k_p} R(i+1).$$

4. Chain transfer to monomer

$$R(i)+M \xrightarrow{k_{tr}} P(i, 0, 0, 0) + R(1).$$

5. Reversible chain transfer [4]. As a broadly used assumption lately, we shall take that intermediates fragmentation rate constant doesn't depend on leaving radical's length [7].

$$R(i)+RAFT(0, 0) \underset{k_f}{\overset{k_{a1}}{\rightleftarrows}} Int(i, 0, 0) \underset{k_{a2}}{\overset{k_f}{\rightleftarrows}} RAFT(i, 0)+R' \quad (I)$$

$$R(j)+RAFT(i, 0) \underset{k_f}{\overset{k_{a2}}{\rightleftarrows}} Int(i, j, 0) \underset{k_{a2}}{\overset{k_f}{\rightleftarrows}} RAFT(i, j)+R' \quad (II)$$

$$R(k)+RAFT(i, j) \underset{k_f}{\overset{k_{a2}}{\rightleftarrows}} Int(i, j, k) \quad (III)$$

6. Chain termination [4]. For styrene's RAFT-polymerization in the trithiocarbonates presence, besides reactions of radicals quadratic termination

$$R(0)+R(0) \xrightarrow{k_{t1}} R(0)\text{-}R(0),$$

$$R(0)+R' \xrightarrow{k_{t1}} R(0)\text{-}R',$$

$$R'+R' \xrightarrow{k_{t1}} R'\text{-}R',$$

$$R(0)+R(i) \xrightarrow{k_{t1}} P(i, 0, 0, 0),$$

$$R'+R(i) \xrightarrow{k_{t1}} P(i, 0, 0, 0),$$

$$R(j)+R(i\text{-}j) \xrightarrow{k_{t1}} P(i, 0, 0, 0)$$

are character reactions of radicals and intermediates cross termination.

$$R(0) + Int(i, 0, 0) \xrightarrow{k_{t2}} P(i, 0, 0, 0),$$

$$R(0) + Int(i, j, 0) \xrightarrow{k_{t2}} P(i, j, 0, 0),$$

$$R(0) + Int(i, j, k) \xrightarrow{k_{t2}} P(i, j, k, 0),$$

$$R' + Int(i, 0, 0) \xrightarrow{k_{t2}} P(i, 0, 0, 0),$$

$$R' + Int(i, j, 0) \xrightarrow{k_{t2}} P(i, j, 0, 0),$$

$$R' + Int(i, j, k) \xrightarrow{k_{t2}} P(i, j, k, 0),$$

$$R(j) + Int(i, 0, 0) \xrightarrow{k_{t2}} P(i, j, 0, 0),$$

$$R(k) + Int(i, j, 0) \xrightarrow{k_{t2}} P(i, j, k, 0),$$

$$R(m) + Int(i, j, k) \xrightarrow{k_{t2}} P(i, j, k, m).$$

In the introduced kinetic scheme: I, R(0), R(i), R', M, RAFT(i, j), Int(i, j, k), P(i, j, k, m) – reaction system's components (refer to Table 1); i, j, k, m – a number of monomer links in the chain; kd – a real rate constant of the initiation reaction; ki1, ki2, ki3, – thermal rate constants of the initiation reaction's; kp, ktr, ka1, ka2, kf, kt1, kt2 are the values of chain growth, chain transfer to monomer, radicals addition to low-molecular RAFT-agent, radicals addition to macromolecular RAFT-agent, intermediates fragmentation, radicals quadratic termination and radicals and intermediates cross termination reaction rate constants, respectively.

TABLE 1 Signs of components in a kinetic scheme.

Symbol	Structure	Symbol	Structure
I	$NC-C(CH_3)_2-N=N-C(CH_3)_2-CN$	Int(i, 0, 0)	PhCH$_2$-S-C(•)(SR(i))-S-CH$_2$Ph
R(0)	$NC-C^{\bullet}(CH_3)_2$	Int(i, j, 0)	PhCH$_2$-S-C(•)(SR(i))(SR(j))
R'	Ph$^{\bullet}$	Int(i, j, k)	R(k)S-C(•)(SR(i))(SR(j))
M	CH$_2$=CH-Ph	P(i, 0, 0, 0)	–CH(Ph)–CH(Ph)–; R(0)-branch with PhCH$_2$-S-C(SR(i))-S-CH$_2$Ph; R'-branch with PhCH$_2$-S-C(SR(i))-S-CH$_2$Ph
R(i)	~CH$_2$-C$^{\bullet}$H-Ph	P(i, j, 0, 0)	R(0)-C(S-CH$_2$Ph)(SR(i))(SR(j)); R'-C(S-CH$_2$Ph)(SR(i))(SR(j)); R(j)-S-C(SR(i))-S-CH$_2$Ph
RAFT(0,0)	PhCH$_2$-S-C(=S)-S-CH$_2$Ph	P(i, j, k, 0)	R(0)-C(SR(k))(SR(i))(SR(j)); R'-C(SR(k))(SR(i))(SR(j))

RAFT(i, 0)

P(i, j, k, m)

RAFT(i, j)

The differential equations system describing this kinetic scheme, is as follows:

$$d[I] / dt = -k_d[I];$$

$$d[R(0)]/dt = 2f\, k_d[I] - [R(0)](k_p[M] + k_{t1}(2[R(0)] + [R'] + [R]) + k_{t2}(\sum_{i=1}^{\infty}[Int(i, 0, 0)] +$$

$$+ \sum_{i=1}^{\infty}\sum_{j=1}^{\infty}[Int(i, j, 0)] + \sum_{i=1}^{\infty}\sum_{j=1}^{\infty}\sum_{k=1}^{\infty}[Int(i, j, k)]));$$

$$d[M] / dt = -(k_p([R(0)] + [R'] + [R]) + k_{tr}[R])[M] - 3k_{i1}[M]^3 - 2k_{i2}[M]^2([M]_0 - [M]);$$

$$d[R']/dt = -k_p[R'][M] + 2k_f \sum_{i=1}^{\infty}[Int(i, 0, 0)] - k_{a2}[R']\sum_{i=1}^{\infty}[RAFT(i, 0)] +$$

$$+ k_f \sum_{i=1}^{\infty}\sum_{j=1}^{\infty}[Int(i, j, 0)] - k_{a2}[R']\sum_{i=1}^{\infty}\sum_{j=1}^{\infty}[RAFT(i, j)] - [R'](k_{t1}([R(0)] + 2[R'] + [R]) +$$

$$+ k_{t2}(\sum_{i=1}^{\infty}[Int(i, 0, 0)] + \sum_{i=1}^{\infty}\sum_{j=1}^{\infty}[Int(i, j, 0)] + \sum_{i=1}^{\infty}\sum_{j=1}^{\infty}\sum_{k=1}^{\infty}[Int(i, j, k)]));$$

$$d[RAFT(0,0)] / dt = -k_{a1}[RAFT(0,0)][R] + k_f\sum_{i=1}^{\infty}[Int(i, 0, 0)];$$

$$d[R(1)]/dt = 2k_{i1}[M]^3 + 2k_{i2}[M]^2([M]_0-[M]) + 2k_{i3}([M]_0-[M])^3 + k_p[M]([R(0)]+[R']$$

$$-[R(1)]) + k_{tr}[R(i)][M] - k_{a1}[R(1)][RAFT(0,0)] + k_f[Int(1,0,0)]-$$

$$-k_{a2}[R(1)]\sum_{i=1}^{\infty}[RAFT(i,0)] + 2k_f[Int(1,1,0)] - k_{a2}[R(1)]\sum_{i=1}^{\infty}\sum_{j=1}^{\infty}[RAFT(i,j)]+$$

$$+3k_f[Int(1,1,1)] - [R(1)](k_{t1}([R(0)]+[R']+[R]) + k_{t2}(\sum_{i=1}^{\infty}[Int(i,0,0)]+$$

$$+\sum_{i=1}^{\infty}\sum_{j=1}^{\infty}[Int(i,j,0)] + \sum_{i=1}^{\infty}\sum_{j=1}^{\infty}\sum_{k=1}^{\infty}[Int(i,j,k)])), \; i=2,\ldots;$$

$$d[R(i)]/dt = k_p[M]([R(i-1)]-[R(i)]) - k_{tr}[R(i)][M] - k_{a1}[R(i)][RAFT(0,0)] + k_f[Int(i,0,0)]-$$

$$-k_{a2}[R(i)]\sum_{i=1}^{\infty}[RAFT(i,0)] + 2k_f[Int(i,j,0)] - k_{a2}[R(i)]\sum_{i=1}^{\infty}\sum_{j=1}^{\infty}[RAFT(i,j)] + 3k_f[Int(i,j,k)]-$$

$$-[R(i)](k_{t1}([R(0)]+[R']+[R]) + k_{t2}(\sum_{i=1}^{\infty}[Int(i,0,0)] + \sum_{i=1}^{\infty}\sum_{j=1}^{\infty}[Int(i,j,0)]+$$

$$+\sum_{i=1}^{\infty}\sum_{j=1}^{\infty}\sum_{k=1}^{\infty}[Int(i,j,k)])), \; i=2,\ldots;$$

$$d[Int(i,0,0)]/dt = k_{a1}[RAFT(0,0)][R(i)] - 3k_f[Int(i,0,0)] + k_{a2}[R'][RAFT(i,0)]-$$

$$-k_{t2}[Int(i,0,0)]([R(0)]+[R']+[R]);$$

$$d[Int(i,j,0)]/dt = k_{a2}[RAFT(i,0)][R(j)] - 3k_f[Int(i,j,0)] + k_{a2}[R'][RAFT(i,j)]-$$

$$-k_{t2}[Int(i,j,0)]([R(0)]+[R']+[R]);$$

$$d[Int(i,j,k)]/dt = k_{a2}[RAFT(i,j)][R(k)] - 3k_f[Int(i,j,k)] - k_{t2}[Int(i,j,k)]([R(0)]+[R']+[R]);$$

$d[RAFT(i, 0)]/dt = 2k_f[Int(i, 0, 0)] - k_{a2}[R'][RAFT(i, 0)] - k_{a2}[RAFT(i, 0)][R] + 2k_f[Int(i, j, 0)];$

$d[RAFT(i, j)]/dt = k_f[Int(i, j, 0)] - k_{a2}[R'][RAFT(i, j)] - k_{a2}[RAFT(i, j)][R] + 3k_f[Int(i, j, k)];$

$d[P(i, 0, 0, 0)]/dt = [R(i)](k_{t1}([R(0)]+[R']) + k_{tr}[M]) + \dfrac{k_{t1}}{2}\sum_{j=1}^{i-1}[R(j)][R(i-j)] +$

$+ k_{t2}[Int(i, 0, 0)]([R(0)]+[R']);$

$d[P(i, j, 0, 0)]/dt = k_{t2}([Int(i, j, 0)]([R(0)]+[R']) + \sum_{i+j=2}^{\infty}[R(j)][Int(i, 0, 0)]);$

$d[P(i, j, k, 0)]/dt = k_{t2}([Int(i, j, k)]([R(0)]+[R']) + \sum_{i+j+k=3}^{\infty}[R(k)][Int(i, j, 0)]);$

$d[P(i, j, k, m)]/dt = k_{t2}\sum_{i+j+k+m=4}^{\infty}[R(m)][Int(i, j, k)].$

where f – initiator's efficiency; $[R] = \sum_{i=1}^{\infty}[R(i)]$ – summary concentration of macro-radicals; t – time.

A method of generating functions was used for transition from this equation system to the equation system related to the unknown MMD moments [8].

Number-average molecular mass (Mn), polydispersity index (PD) and weight-average molecular mass (Mw) are linked to MMD moments by the following expressions:

$$M_n = (\Sigma\mu_1 / \Sigma\mu_0)M_{ST}, \quad PD = \Sigma\mu_2\Sigma\mu_0/(\Sigma\mu_1)^2, \quad M_w = PD \cdot M_n,$$

where $\Sigma\mu_0$, $\Sigma\mu_1$, $\Sigma\mu_2$ – sums of all zero, first and second MMD moments; $M_{ST} = 104$ g/mol – styrene's molecular mass.

1.2.2 RATE CONSTANTS

1.2.2.1 REAL AND THERMAL INITIATION

The efficiency of initiation and temperature dependence of polymerization real initiation reaction rate constant by AIBN initiator are determined basing on the data in this research, which have established a good reputation for mathematical modeling of leaving in mass styrene radical polymerization [6]:

$$f = 0.5, \quad k_d = 1.58 \cdot 10^{15} e^{-15501/T}, \text{ s}^{-1},$$

where T – temperature, K.

As it was established in the research, thermal initiation reactions' rates constants depend on the chain growth reactions rate constants, the radicals' quadratic termination and the monomer initial concentration:

$$k_{i1}=1.95\cdot 10^{13}\frac{k_{t1}}{k_p^2 M_0^3}e^{-20293/T}, \text{ L}^2\cdot \text{mol}^{-2}\cdot \text{s}^{-1};$$

$$k_{i2}=4.30\cdot 10^{17}\frac{k_{t1}}{k_p^2 M_0^3}e^{-23878/T}, \text{ L}^2\cdot \text{mol}^{-2}\cdot \text{s}^{-1};$$

$$k_{i3}=1.02\cdot 10^{8}\frac{k_{t1}}{k_p^2 M_0^2}e^{-14807/T}, \text{ L}\cdot \text{mol}^{-1}\cdot \text{s}^{-1}. \qquad (6).$$

1.2.2.2 CHAIN TRANSFER TO MONOMER REACTION'S RATE CONSTANT

On the basis of the data in research [6]:

$$k_{tr}=2.31\cdot 10^{6}e^{-6376/T}, \text{ L·mol}^{-1}\text{·s}^{-1}.$$

1.2.2.3 RATE CONSTANTS FOR THE ADDITION OF RADICALS TO LOW–MOLECULAR AND MACROMOLECULAR RAFT–AGENTS

In research [9], it was shown by the example of dithiobenzoates at first that chain transfer to low- and macromolecular RAFT-agents of rate constants are functions of respective elementary constants. Let us demonstrate this for our process. For this record, the change of concentrations [Int(i, 0, 0)], [Int(i, j, 0)], [RAFT(0,0)] and [RAFT(i, 0)] in quasistationary approximation for the initial phase of polymerization is as follows:

$$d[\text{Int}(i, 0, 0)]/dt=k_{a1}[\text{RAFT}(0,0)][R]-3k_f[\text{Int}(i, 0, 0)] \approx 0, \qquad (1)$$

$$d[\text{Int}(i, j, 0)]/dt=k_{a2}[\text{RAFT}(i, 0)][R]-3k_f[\text{Int}(i, j, 0)] \approx 0, \qquad (2)$$

RAFT-Polymerization of Styrene—Kinetics and Mechanism

$$d[RAFT(0,0)]/dt = -k_{a1}[RAFT(0,0)][R] + k_f[Int(i, 0, 0)], \qquad (3)$$

$$d[RAFT(i, 0)]/dt = 2k_f[Int(i, 0, 0)] - k_{a2}[RAFT(i, 0)][R] + 2k_f[Int(i, j, 0)]. \qquad (4)$$

The Eq. (1) expresses the following concentration $[Int(i, 0, 0)]$:

$$[Int(i, 0, 0)] = \frac{k_{a1}}{3k_f}[RAFT(0,0)][R].$$

Substituting the expansion gives the following $[Int(i, 0, 0)]$ expression to Eq. (3):

$$d[RAFT(0,0)]/dt = -k_{a1}[RAFT(0,0)][R] + k_f \frac{k_{a1}}{3k_f}[RAFT(0,0)][R].$$

After transformation of the last equation, we have:

$$\frac{d[RAFT(0,0)]}{[RAFT(0,0)]} = -\frac{2}{3}k_{a1}[R]dt.$$

Solving this equation (initial conditions: $t = 0$, $[R] = [R]_0 = 0$, $[RAFT(0,0)] = [RAFT(0,0)]_0$), we obtain:

$$\ln\frac{[RAFT(0,0)]}{[RAFT(0,0)]_0} = -\frac{2}{3}k_{a1}[R]t. \qquad (5)$$

To transfer from time t, being a part of Eq. (5), to conversion of monomer C_M, we put down a balance differential equation for monomer concentration, assuming that at the initial phase of polymerization, thermal initiation and chain transfer to monomer are not of importance:

$$d[M]/dt = -k_p[R][M]. \qquad (6)$$

Transforming the Eq. (6) with its consequent solution at initial conditions $t = 0$, $[R] = [R]_0 = 0$, $[M] = [M]_0$:

$$d[M]/[M] = -k_p[R]dt,$$

$$\ln \frac{[M]}{[M]_0} = -k_p[R]t. \tag{7}$$

Link rate [M]/[M]$_0$ with monomer conversion (C_M) in an obvious form like this:

$$C_M = \frac{[M]_0 - [M]}{[M]_0} = 1 - \frac{[M]}{[M]_0},$$

$$\frac{[M]}{[M]_0} = 1 - C_M.$$

We substitute the last ratio to Eq. (7) and express time t:

$$t = \frac{-\ln(1 - C_M)}{k_p[R]}. \tag{8}$$

After substitution of the expression (8) by the Eq. (5), we obtain the next equation:

$$\ln \frac{[RAFT(0,0)]}{[RAFT(0,0)]_0} = \frac{2}{3} \frac{k_{a1}}{k_p} \ln(1 - C_M). \tag{9}$$

By analogy with introduced [M]/[M]$_0$ to monomer conversion, reduce ratio [RAFT(0,0)] / [RAFT(0,0)]$_0$ to conversion of low-molecular RAFT-agent – $C_{RAFT(0,0)}$. As a result, we obtain:

$$\frac{[RAFT(0,0)]}{[RAFT(0,0)]_0} = 1 - C_{RAFT(0,0)}. \tag{10}$$

Substitute the derived expression for [RAFT(0,0)] / [RAFT(0,0)]$_0$ from Eq. (10) to Eq. (9):

$$\ln(1 - C_{RAFT(0,0)}) = \frac{2}{3}\frac{k_{a1}}{k_p}\ln(1 - C_M). \qquad (11)$$

In the research [9], the next dependence of chain transfer to low-molecular RAFT-agent constant C_{trl} is obtained on the monomer and low-molecular RAFT-agent conversions:

$$C_{trl} = \frac{\ln(1 - C_{RAFT(0,0)})}{\ln(1 - C_M)}. \qquad (12)$$

Comparing Eqs. (12) and (11), we obtain dependence of chain transfer to low-molecular RAFT-agent constant C_{trl} on the constant of radicals' addition to macromolecular RAFT-agent and chain growth reaction rate constant:

$$C_{trl} = \frac{2}{3}\frac{k_{a1}}{k_p}. \qquad (13)$$

From Eq. (13), we derive an expression for constant k_{a1}, which will be based on the following calculation:

$$k_{a1} = 1.5 C_{trl} k_p, \; L \cdot mol^{-1} \cdot s^{-1}$$

As a numerical value for C_{trl}, we assume value 53, derived in research [4] on the base of Eq. (12), at immediate experimental measurement of monomer and low-molecular RAFT-agent conversions. Since chain transfer reaction in RAFT-polymerization is usually characterized by low value of activation energy, compared to activation energy of chain growth, it is supposed that constant C_{trl} doesn't depend or slightly depends on temperature. We will propose as an assumption that C_{trl} doesn't depend on temperature [10].

By analogy with k_{a1}, we deduce equation for constant k_{a2}. From Eq. (2) we express such concentration [Int(i, j, 0)]:

$$[Int(i, j, 0)] = \frac{k_{a2}}{3k_f}[RAFT(i, 0)][R].$$

Substitute expressions, derived for [Int(i, 0, 0)] and [Int(i, j, 0)] in Eq. (4):

$$d[RAFT(i, 0)]/dt = \frac{2}{3}k_{a1}[RAFT(0,0)][R] - \frac{1}{3}k_{a2}[RAFT(i, 0)][R]. \qquad (14)$$

Since in the end it was found that constant of chain transfer to low-molecular RAFT-agent C_{tr1} is equal to divided to constant k_p coefficient before expression [RAFT(0,0)][R] in the balance differential equation for [RAFT(0,0)], from Eq. (14) for constant of chain transfer to macromolecular RAFT-agent, we obtain the next expression:

$$C_{tr2} = \frac{1}{3}\frac{k_{a2}}{k_p}.$$

From the last equation we obtain an expression for constant k_{a2}, which based on the following calculation:

$$k_{a2} = 3C_{tr2}k_p, \text{ L·mol}^{-1}\cdot\text{s}^{-1}. \tag{15}$$

In research [4] on the base of styrene and DBTK, macromolecular RAFT-agent was synthesized, thereafter with a view to experimentally determine constant C_{tr2}, polymerization of styrene was performed with the use of the latter. In the course of experiment, it may be supposed that constant C_{tr2} depends on monomer and macromolecular RAFT-agent conversions by analogy with Eq. (12). As a result directly from the experimentally measured monomer and macromolecular RAFT-agent conversions, value C_{tr2} was derived, equal to 1,000. On the ground of the same considerations as for that of C_{tr1}, we assume independence of constant C_{tr2} on temperature.

1.2.2.4 RATE CONSTANTS OF INTERMEDIATES FRAGMENTATION, TERMINATION BETWEEN RADICALS AND TERMINATION BETWEEN RADICALS AND INTERMEDIATES

In research [4] it was shown, that RAFT-polymerization rate is determined by this equation:

$$(W_0/W)^2 = 1 + \frac{k_{t2}}{k_{t1}}K[RAFT(0,0)]_0 + \frac{k_{t3}}{k_{t1}}K^2[RAFT(0,0)]_0^2,$$

where W_0 and W – polymerization rate in the absence and presence of RAFT-agent, respectively, s^{-1}; K – constant of equilibrium (III), L·mol^{-1}; k_{t3} – constant of termination between two intermediates reaction rate, L·mol^{-1}·s^{-1} [11].

For initiated AIBN styrene polymerization in DBTC's presence at 80°C, it was shown that intermediates quadratic termination wouldn't be implemented and RAFT-polymerization rate was determined by equation [4]:

$$(W_0 / W)^2 = 1 + 8[RAFT(0,0)]_0.$$

Since $\frac{k_{t2}}{k_{t1}} \approx 1$, then at 80°C K = 8 L·mol^{-1} [4]. In order to find dependence of constant K on temperature, we made research of polymerization kinetics at 60°C. It was found, (Fig. 1), that the results of kinetic measurements well rectify in coordinates $(W_0 / W)^2 = f([RAFT(0,0)]_0)$. At 60°C, K = 345 L·mol^{-1} was obtained. Finally dependence of equilibrium constant on temperature has been determined in the form of Vant–Goff's equation:

$$K = 4.85 \cdot 10^{-27} e^{22123/T}, \text{L·mol}^{-1}. \tag{16}$$

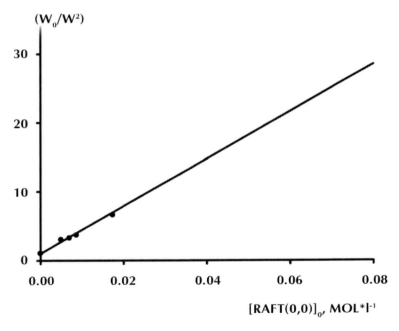

FIGURE 1 Dependence $(W_0 / W)^2$ on DBTC concentration at 60°C.

In compliance with the equilibrium (III), the constant is equal to

$$K = \frac{k_{a2}}{3k_f}, \text{ L·mol}^{-1}.$$

Hence, reactions of intermediates fragmentation rate constant will be as such:

$$k_f = \frac{k_{a2}}{3K}, \text{ s}^{-1}. \qquad (17)$$

The reactions of intermediates fragmentation rate constant was built into the model in the form of dependence (17) considering Eqs. (15) and (16).

As it has been noted above, ratio k_{t2}/k_{t1} equals approximately to one, therefore it will taken, that $k_{t2} \approx k_{t1}$ [4]. For description of gel-effect, dependence as a function of monomer conversion C_M and temperature T (K) [12] was applied:

$$k_{t2} \approx k_{t1} \approx 1.255 \cdot 10^9 e^{-844/T} e^{-2(A_1 C_M + A_2 C_M^2 + A_3 C_M^3)}, \text{ L·mol}^{-1}\cdot\text{s}^{-1},$$

where $A_1 = 2.57 - 5.05 \cdot 10^{-3} T$; $A_2 = 9.56 - 1.76 \cdot 10^{-2} T$; $A_3 = -3.03 + 7.85 \cdot 10^{-3} T$.

1.2.2.5 RATE CONSTANT FOR CHAIN GROWTH

The method of polymerization, being initiated by pulse laser radiation [13] is used for determination of rate constant for chain growth k_p lately. It is anticipated that such an estimation method is more correct, than the traditionally used revolving sector method [12]. We made our choice on temperature dependence of the rate constant for chain growth that was derived on the ground of method of polymerization, being initiated by pulse laser radiation:

$$k_p = 4.27 \cdot 10^7 e^{-3910/T}, \text{ L·mol}^{-1}\cdot\text{s}^{-1}, \qquad (18)$$

since this dependence is more adequately describes the change of polymerization reduced rate with monomer conversion in the network of the developed mathematical model (Fig. 2), than temperature dependence, which is derived by revolving sector method [12]:

$$k_p = 1.057 \cdot 10^7 e^{-3667/T}, \text{ L·mol}^{-1}\cdot\text{s}^{-1}. \qquad (19)$$

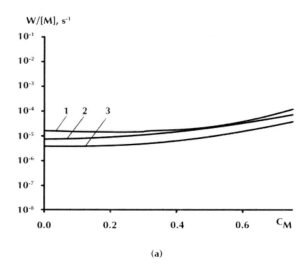

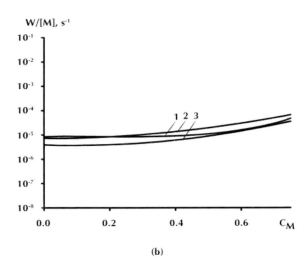

FIGURE 2 Dependence of initiated AIBN ($[I]_0$=0.01 mol·L^{-1}) styrene polymerization reduced rate on monomer conversion at 60°C (1 – experiment; 2 – estimation by introduced in this research mathematical model with temperature dependence of k_p (18); 3 – estimation by introduced in this research mathematical model with temperature dependence of k_p (19): [RAFT(0,0)]$_0$ = 0 mol·L^{-1} (a), 0.007 (b).

1.2.3 MODEL'S ADEQUACY

The results of polystyrene molecular-mass properties calculations by the introduced mathematical model are presented in Figs. 3 and 4. Mathematical model of styrene RAFT-polymerization in the presence of trithiocarbonates, taking into account the radicals and intermediates cross termination, adequately describes the experimental data that prove the process mechanism, built in the model. The essential proof of the mechanism correctness is that in case of conceding the absence of radicals and intermediates cross termination – the experimental data wouldn't substantiate theoretical calculation by the mathematical model, introduced in this assumption (Fig. 5).

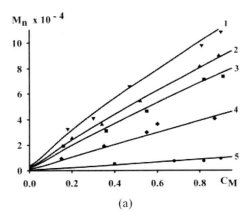

(a)

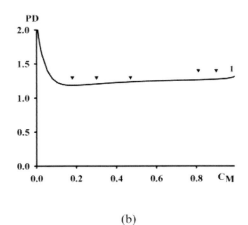

(b)

FIGURE 3 *(Continued)*

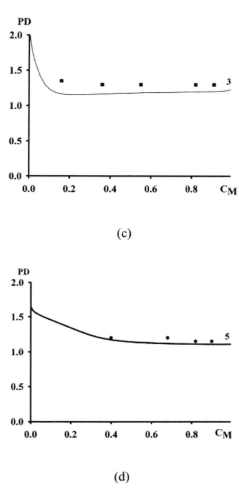

FIGURE 3 Dependence of number-average molecular mass (a) and polydispersity index (b)–(d) on monomer conversion for being initiated by AIBN ($[I]_0$=0.01 mol·L^{-1}) styrene bulk RAFT-polymerization at 60°C in the presence of DBTC (lines – estimation by model; points – experiment): $[RAFT(0,0)]_0$ = 0.005 mol·L^{-1} (1), 0.007 (2), 0.0087 (3), 0.0174 (4), 0.087 (5).

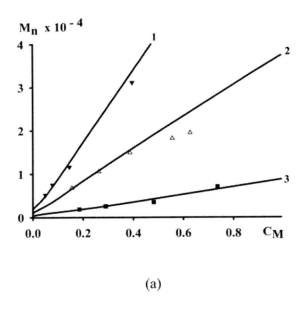

(a)

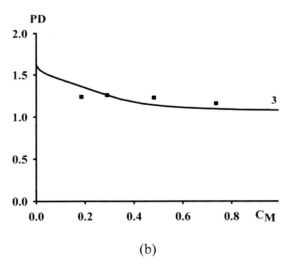

(b)

FIGURE 4 Dependence of number-average molecular mass (a) and polydispersity index (b) on monomer conversion for being initiated by AIBN ($[I]_0$=0.01 mol·L^{-1}) styrene bulk RAFT-polymerization at 80°C in DBTC presence (lines – estimation by model; points – experiment): $[RAFT(0,0)]_0$ = 0.01 mol·L^{-1} (1), 0.02 (2), 0.1 (3) [4].

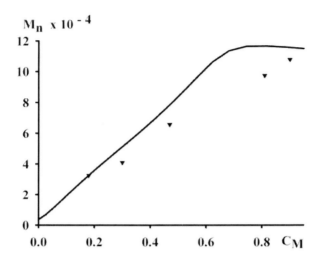

FIGURE 5 Dependence of number-average molecular mass on monomer conversion for initiated AIBN ($[I]_0$=0.01 mol·L^{-1}) styrene bulk RAFT-polymerization at 60 °C in DBTC presence [RAFT(0,0)]0 = 0.005 mol·L^{-1} (lines – estimation by model assuming that radicals and intermediates cross termination are absent; points – experiment).

Due to adequacy of the model realization at numerical experiment it became possible to determine the influence of process controlling factors on polystyrene molecular-mass properties.

1.2.4 NUMERICAL APPROACH

Research of influence of the process controlling factors on molecular-mass properties of polystyrene, synthesized by RAFT-polymerization method in the presence of AIBN and DBTC, was made in the range of initial concentrations of: initiator – 0–0.1 mol·L^{-1}, monomer – 4.35–8.7 mol·L^{-1}, DBTC – 0.001–0.1 mol·L^{-1}; and at temperatures – 60–120°C.

1.2.4.1 THE INFLUENCE OF AIBN INITIAL CONCENTRATION BY NUMERICAL APPROACH

It was set forth that generally in the same other conditions, with increase of AIBN initial concentration number-average, the molecular mass of polystyrene decreases (Fig. 6). At all used RAFT-agent initial concentrations, there is a linear or close to linear growth of number- average molecular mass of polystyrene with monomer conversion. This means that even the lowest RAFT-agent initial concentrations affect the process of radical polymerization. It should be noted that at high RAFT-agent initial concentrations (Fig. 7) the change of AIBN initial concentration practically doesn't have any influence on number-average molecular mass of polystyrene. But at increased temperatures (Fig. 8), in case of high AIBN initial concentration, it is comparable to high RAFT-agent initial concentration; polystyrene molecular mass would be slightly decreased due to thermal initiation.

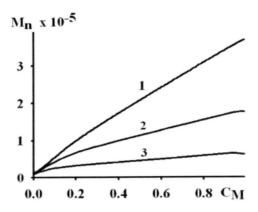

FIGURE 6 Dependence of number-average molecular mass M_n on monomer conversion C_M (60°C) $[M]_0 = 6.1$ mol·L^{-1}, $[RAFT(0, 0)]_0 = 0.001$ mol·L^{-1}, $[I]_0 = 0.001$ mol·L^{-1} (1), 0.01 (2), 0.1 (3).

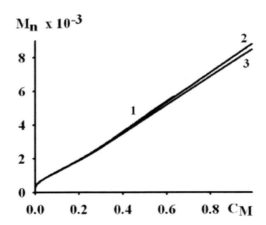

FIGURE 7 Dependence of number-average molecular mass M_n on monomer conversion C_M (60°C) $[M]_0 = 8.7$ mol·L^{-1}, $[RAFT(0, 0)]_0 = 0.1$ mol·L^{-1}, $[I]_0 = 0.001$ mol·L^{-1} (1), 0.01 (2), 0.1 (3).

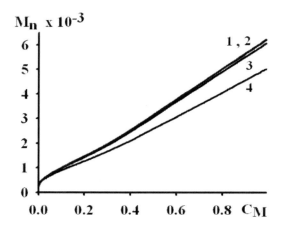

FIGURE 8 Dependence of number-average molecular mass M_n on monomer conversion C_M (120°C) $[M]_0 = 6.1$ mol·L^{-1}, $[RAFT(0, 0)]_0 = 0.1$ mol·L^{-1}, $[I]_0 = $ mol·L^{-1} (1), 0.001 (2), 0.01 (3), 0.1 (4).

Since the main product of styrene RAFT-polymerization process, proceeding in the presence of trithiocarbonates, is a narrow-dispersed high-molecular RAFT-agent (marked in kinetic scheme as RAFT(i, j)), which is formed as a result of reversible chain transfer, and widely-dispersed (minimal polydispersity – 1.5) polymer, forming by the radicals quadratic termination, so common polydispersity index of synthesizing product is their ratio. In a broad sense, with increase of AIBN initial concentration, the part of widely-dispersed polymer, which is formed as a result of the radicals quadratic termination, increase in mixture, thereafter general polydispersity index of synthesizing product increases.

However, at high temperatures this regularity can be discontinued – at low RAFT-agent initial concentrations the increase of AIBN initial concentration leads to a decrease of polydispersity index (Fig. 9, curves 3 and 4). This can be related only thereto that at high temperatures thermal initiation and elementary reactions rate constants play an important role, depending on temperature, chain growth and radicals quadratic termination reaction rate constants, monomer initial concentration in a complicated way [6]. Such complicated dependence makes it difficult to analyze the influence of thermal initiation role in process kinetics, therefore the expected width of MMD of polymer, which is expected to be synthesized at high temperatures, can be estimated in every specific case in the frame of the developed theoretical regularities.

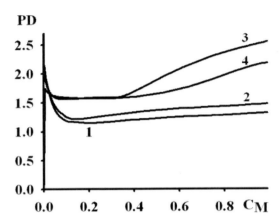

FIGURE 9 Dependence of polydispersity index PD on monomer conversion C_M (120°C) $[M]_0 = 8.7$ mol·L^{-1}, $[RAFT(0, 0)]_0 = 0.001$ mol·L^{-1}, $[I]_0 = 0$ mol·L^{-1} (1), 0.001 (2), 0.01 (3), 0.1 (4).

Special attention shall be drawn to the fact that for practical objectives, realization of RAFT-polymerization process without an initiator is of great concern. In all cases at high temperatures as the result of styrene RAFT-polymerization implementation in the presence of RAFT-agent without AIBN, more high-molecular (Fig. 10) and more narrow-dispersed polymer (Fig. 9, curve 1) is built-up than in the presence of AIBN (Fig. 9, curves 2–4).

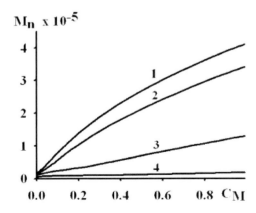

FIGURE 10 Dependence of number-average molecular mass M_n on monomer conversion C_M (120°C) $[M]_0 = 8.7$ mol·L^{-1}, $[RAFT(0, 0)]_0 = 0.001$ mol·L^{-1}, $[I]_0 = 0$ mol·L^{-1} (1), 0.001 (2), 0.01 (3), 0.1 (4).

1.2.4.2 THE INFLUENCE OF MONOMER INITIAL CONCENTRATION BY NUMERICAL EXPERIMENT

In other identical conditions, the decrease of monomer initial concentration reduces the number-average molecular mass of polymer. Polydispersity index doesn't practically depend on monomer initial concentration.

1.2.4.3 INFLUENCE OF RAFT-AGENT INITIAL CONCENTRATION BY NUMERICAL EXPERIMENT

In other identical conditions, increase of RAFT-agent initial concentration reduces the number-average molecular mass and polydispersity index of polymer (Fig. 11).

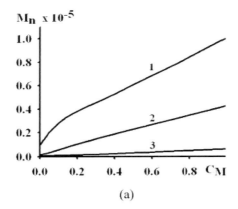

(a)

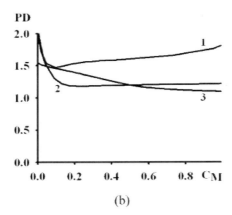

(b)

FIGURE 11 Dependence of number-average molecular mass M_n (a) and polydispersity index PD (b) on monomer conversion C_M (90°C) $[I]_0 = 0.01$ mol·L^{-1}, $[M]_0 = 6.1$ mol·L^{-1}, $[RAFT(0, 0)]_0 = 0.001$ mol·L^{-1} (1), 0.01 (2), 0.1 (3).

1.2.4.4 THE INFLUENCE OF TEMPERATURE BY NUMERICAL EXPERIMENT

Generally, in other identical conditions, the increase of temperature leads to a decrease of number-average molecular mass of polystyrene (Fig. 12 (a)). Thus, polydispersity index increases (Fig. 12 (b)). If RAFT-agent initial concentration greatly exceeds AIBN initial concentration, then the temperature practically doesn't influence the molecular-mass properties of polystyrene.

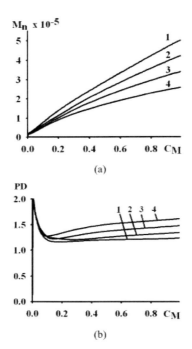

FIGURE 12 Dependence of number-average molecular mass M_n (a) and polydispersity index PD (b) on monomer conversion C_M $[I]_0 = 0.001$ mol·L^{-1}, $[M]_0 = 8.7$ mol·L^{-1}, $[RAFT(0, 0)]_0 = 0.001$ mol·L^{-1}, T = 60°C (1), 90 (2), 120 (3), 150 (4).

1.3 CONCLUSION

The kinetic model developed in this research allows an adequate description of molecular-mass properties of polystyrene, obtained by controlled radical polymerization, which proceeds by reversible chain transfer mechanism and accompanied by "addition-fragmentation." This means, that the model can be used for development of technological applications of styrene RAFT-polymerization in the presence of trithiocarbonates.

Researches were supported by Russian Foundation for Basic Research (project. no. 12–03–97050-r_povolzh'e_a).

KEYWORDS

- controlled radical polymerization
- dibenzyltritiocarbonate
- mathematical modeling
- polystyrene
- reversible addition-fragmentation chain transfer

REFERENCES

1. Matyjaszewski K.: Controlled/Living Radical Polymerization: Progress in ATRP, D.C.: American Chemical Society, Washington (2009).
2. Matyjaszewski K.: Controlled/Living Radical Polymerization: Progress in RAFT, DT, NMP and OMRP, D.C.: American Chemical Society, Washington (2009).
3. Barner-Kowollik C.: Handbook of RAFT Polymerization, Wiley–VCH Verlag GmbH, Weinheim (2008).
4. Chernikova E.V., Terpugova P.S., Garina E.S., Golubev V.B.: Controlled radical polymerization of styrene and n-butyl acrylate mediated by tritiocarbonates. Polymer Science, Vol. 49(A), №2, 108 (2007).
5. Stephen Z.D. Cheng: Handbook of Thermal Analysis and Calorimetry. Volume 3 – Applications to Polymers and Plastics: New York, Elsevier, (2002).
6. LI I., Nordon I., Irzhak V.I., Polymer Science, 47, 1063 (2005).
7. Zetterlund P.B., Perrier S., Macromolecules, 44, 1340 (2011).
8. Biesenberger J.A., Sebastian D.H.: Principles of polymerization engineering, John Wiley & Sons Inc., New York (1983).
9. Chong Y.K., Krstina J., Le T.P.T., Moad G., Postma A., Rizzardo E., Thang S.H., Macromolecules, 36, 2256 (2003).
10. Goto A., Sato K., Tsujii Y., Fukuda T., Moad G., Rizzardo E., Thang S.H., Macromolecules, 34, 402 (2001).
11. Kwak Y., Goto A., Fukuda T., Macromolecules, 37, 1219 (2004).
12. Hui A.W., Hamielec A.E., J. Appl. Polym. Sci, 16, 749 (1972).
13. Li D., Hutchinson R.A., Macromolecular Rapid Communications, 28, 1213 (2007).

CHAPTER 2

A DETAILED REVIEW ON PORE STRUCTURE ANALYSIS OF ELECTROSPUN POROUS MEMBRANES

BENTOLHODA HADAVI MOGHADAM,
VAHID MOTTAGHITALAB, MAHDI HASANZADEH,
and AKBAR KHODAPARAST HAGHI

CONTENTS

Abstract	30
2.1 Introduction	30
2.2 Polymer Membranes	31
2.2.1 Types of Membranes	31
2.3 Nanofibrous Membrane	32
2.3.1 Electrospinning Process	33
2.4 Membrane Porosity	35
2.5 Summary of Literature Review	40
2.6 Conclusion	45
Keywords	45
Reference	45

ABSTRACT

Nanoporous membranes are an important class of nanomaterials that can be used in many applications, especially in micro and nanofiltration. Electrospun nanofibrous membranes have gained increasing attention due to the high porosity, large surface area per mass ratio along with small pore sizes, flexibility, and fine fiber diameter, and their production and application in development of filter media. Image analysis is a direct and accurate technique that can be used for characterization of porous media. This technique, due to its convenience in detecting individual pores in a porous media, has some advantages for pore measurement. The three-dimensional reconstruction of porous media, from the information obtained from a two-dimensional analysis of photomicrographs, is a relatively new research area. In the present paper, we have reviewed the recent progress in pore structure analysis of porous membranes with emphasis in image analysis technique. Pore characterization techniques, properties, and characteristics of nanoporous structures are also discussed in this paper.

2.1 INTRODUCTION

Nanofibrous membranes have received increasing attention in recent years as an important class of nanoporous materials. Although there are various techniques to produce polymer nanofiber mats, electrospinning is considered as one of the most efficient ways to obtain nonwoven nanofiber mats with pore sizes ranging from tens of nanometers to tens of micrometers.[1–3]

In recent years, significant progress has been done in the understanding and modeling of pore-scale processes and phenomena. By using increased computational power, realistic pore-scale modeling in recognition of tomographic and structure of porous membranes can be obtained.[4–6] Information about the pore structure of membranes is often obtained by several methods including mercury intrusion porosimetry,[7–9] liquid extrusion porosimetry,[9–12] flow porosimetry[9–15] and image analysis of thin section images.[16–23] Image analysis is a useful technique that is gaining attention due to its convenience in detecting individual pores in the membrane image.

The three-dimensional reconstruction of porous media, such as nanofiber mats, from the information obtained from a two-dimensional micrograph has attracted considerable interest for many applications. A successful reconstruction procedure leads to significant improvement in predicting the macroscopic properties of porous media.[23–44]

This short review intends to introduce recent progress in pore structure analysis of porous membranes, with emphasis on electrospun polymer nanofiber mats. The paper is organized as follows. Section 2.2 present polymer membranes and their types. Section 2.3 deals with the nanofibrous membranes as one of the most important porous media. Section 2.4 presents the porosity of membranes and the techniques used to evaluate the pore characteristics of porous membranes. Finally, Section 2.5 surveys the most characteristic and important recent examples, which image analysis technique was used for characterization of porous media, especially three-dimensional reconstruction of porous structure. The report ends with a conclusion.

2.2 POLYMER MEMBRANES

Membrane technologies are already serving as a useful tool for industrial processes, such as health sector, food industry, sustainable water treatment and energy conversion and storage. Membrane is a selective barrier between two phases and defined as a very thin layer or cluster of layers that allows one or more selective components to permeate through readily when mixtures of different kinds of components are driven to its surface, thereby producing a purified product. The ability to control the diffusion rate of a chemical species through the membrane is key property of membranes.[45–54] Several membrane processes have been proposed based on the barrier structure, including microfiltration,[54–55] ultrafiltration,[56–57] nanofiltration,[54,58] and reverse osmosis.[58] The membrane processes can be categorized based on the barrier structure by different driving forces as shown in Fig. 1.

FIGURE 1 Size range of particles in various membrane separation processes.

Pore size distribution, specific surface area, outer surface, and cross section morphology are some of the most important characteristics of membranes.

2.2.1 TYPES OF MEMBRANES

According to the morphology, membranes can be classified into two main types: 'isotropic' and 'anisotropic'. Isotropic membranes include microporous, nonporous, and dense membranes and are made of single layer with uniform structure through the depth of the membrane. On the other hand, anisotropic membranes consist of more than one layer supported by a porous substrate.

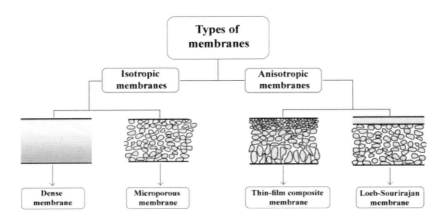

FIGURE 2 Classification of membranes.

2.2.1.1 ISOTROPIC MEMBRANES

Most of the available membranes are porous or consist of a dense top layer on a porous structure. Isotropic dense membranes can be prepared (i) by melt extrusion of polymer or (ii) by solution casting (solvent evaporation).[50–52]

Isotropic microporous membranes have a rigid, highly voided structure, and interconnected pores. These membranes have higher fluxes than isotropic dense membranes. Microporous membranes are prepared by some methods, such as track-etching, expanded-film, and template leaching.[50,59]

2.2.1.2 ANISOTROPIC MEMBRANES

Anisotropic membranes are consists of a very thin (0.1 to 1 pm) selective skin layer on a highly permeable microporous substrate and highly membrane fluxes in which the porosity, pore size, or even membrane composition change from the top to the bottom surface of the membrane. An asymmetrical structure is now produced from a wide variety of polymers, which are currently applied in pressure driven membrane processes, such as reverse osmosis, ultrafiltration, or gas separation.

Anisotropic membranes may be prepared by various techniques, including phase inversion,[50,59] interfacial polymerization,[50] solution coating,[50] plasma deposition,[50] and electrospinning in the laboratory or on a small industrial scale.[60] Nanofibrous membranes, as simple and interesting anisotropic membranes, are described in the following section.

2.3 NANOFIBROUS MEMBRANE

Nanofibrous membranes have high specific surface area, high porosity, small pore size, and flexibility to conform to a wide variety of sizes and shapes. Therefore, they have been suggested as excellent candidate for many applications, especially in micro and nanofiltration. Nanofibrous membrane can be processed by a number of techniques such as drawing,[61] template synthesis,[62] phase separation,[63] self-assembly,[64] and electrospinning.[65] Among them electrospinning has an advantage with its comparative low cost and relatively high production rate.

With regard to the low mechanical properties of nanofibrous membrane, many attempts have been made to improve mechanical properties of nanofibers. In this regard, electrospinning process is the best method to fabricate the mixed membrane of microfiber and nanofibers simultaneously. The good physical and mechanical properties of microfibers (as a substrate) are favorable for enhancing the mechanical performance of nanofibrous membrane.

2.3.1 ELECTROSPINNING PROCESS

Within the past several years, electrospinning process has garnered increasing attention, due to its capability and feasibility to generate large quantities of nanofibers with well-defined structures. Figure 3 shows a schematic illustration of electrospinning setup. In this process, a strong electric field is applied between polymer solution contained in a syringe with a capillary tip and grounded collector. When the electric field overcomes the surface tension force, the charged polymer solution forms a liquid jet and travels towards collection plate. As the jet travels through the air, the solvent evaporates and dry fibers deposits on the surface of a collector.

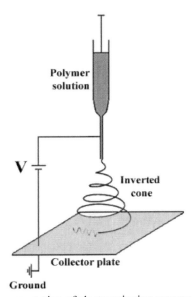

FIGURE 3 Schematic representation of electrospinning process.

The morphology and the structure of the electrospun nanofibrous membrane are dependent upon many parameters which are mainly divided into three categories: solution properties (the concentration, liquid viscosity, surface tension, and dielectric properties of the polymer solution), processing parameters (applied voltage, volume flow rate, tip to collector distance, and the strength of the applied electric field), and ambient conditions (temperature, atmospheric pressure and humidity).[66–80]

2.3.2 POROUS NANOFIBERS

There are various techniques for the fabrication of highly porous nanofiber, including fiber bonding, solvent casting, particle leaching, phase separation, emulsion freeze-drying, gas foaming, and 3–D printing. These porous membranes can also be produced by a combination of electrospinning and phase inversion techniques. Nanofibrous membrane produced by this approach can generate additional space and surface area within as-spun fibrous scaffolds. Due to the fine fiber size and large expected surface area, electrospun nanofibrous membranes have a desirable property for filter media, catalyst immobilization substrates, absorbent media and encapsulated active ingredients, such as activated carbon and various biocides.

Studied showed that the formation of pores on nanofibers during electrospinning process affected by many parameters such as humidity, type of polymer, solvent vapor pressure, electrospinning conditions, etc. Although no generally agreed set of definitions exists, porous materials can be classified in terms of their pore sizes into various categories including capillaries (>200 nm), macropores (50–200 nm), mesopores (2–50 nm) and micropores (0.5–2 nm). According to the literature, the mechanism that forms porous surface on polymer casting film is applicable to the phenomenon on electrospun nanofibers.[81–88] The rapid solvent evaporation and subsequent condensation of moisture into water particles result in the formation of nano or micropores on the fiber surface. When the environment humidity increases, the pore size becomes larger. However, this result was observed only when the solution used a highly volatile organic solvent, such as chloroform, tetrahydrofuran and acetone.

2.3.3 POTENTIAL APPLICATIONS

The research and development of electrospun nanofibrous membrane has evinced more interest and attention in recent years due to the heightened awareness of its potential applications in various fields. The electrospun nanofibrous membrane, due to their high specific surface area, high porosity, flexibility, and small pore size, have been suggested as excellent candidate for many applications including filtration, multifunctional membranes, reinforcements in light weight composites, biomedical agents, tissue engineering scaffolds, wound dressings, full cell and protective clothing.[72,91–92]

A Detailed Review on Pore Structure Analysis

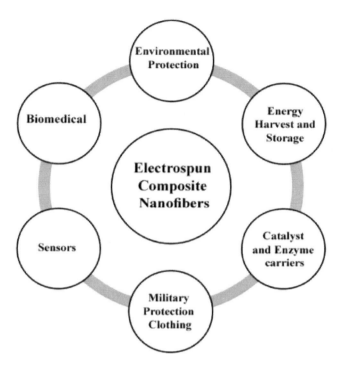

FIGURE 4 Potential applications of electrospun nanofibrous membranes.

Membranes have been widely used for a great variety of applications such as bioseparations, sterile filtration, bioreactors, and so on. Each application depends on the membrane material and structure. For microfiltration and ultrafiltration process, the efficiency was determined by the selectivity and permeability of the membrane.

2.4 MEMBRANE POROSITY

As it has been mentioned earlier, the most important characteristics of membranes are: thickness, pore diameter, solvent permeability and porosity. Moreover, the filtration performance of membranes is also strongly related to their pore structure parameters, that is, percent open area, and pore size distribution. Hence, the porosity and pore structure characteristics play a more important role in membrane process and applications.[93–97]

The porosity, ε_V, is defined as the percentage of the volume of the voids, V_v, to the total volume (voids plus constituent material), V_t, and is given by

$$\varepsilon_V = \frac{V_v}{V_t} \times 100 \tag{1}$$

Similarly, the percent open area, ε_A, that is defined as the percentage of the open area, A_0, to the total area, A_t, is given by

$$\varepsilon_A = \frac{A_0}{A_t} \times 100 \tag{2}$$

Usually, porosity is determined for membranes with a three-dimensional structure (e.g., relatively thick nonwoven fabrics). Nevertheless, for two-dimensional structures such as woven fabrics and relatively thin nonwovens, it is often assumed that porosity and percent open area are equal.[98]

2.4.1 PORE CHARACTERIZATION TECHNIQUES

Development and application of effective procedures for membrane characterization are one of indispensable components of membrane research. Pore structure, as the main characteristics of porous membrane, has significant influence on the performance of membranes. In general, there are three types of pores in membrane: (1) the closed pore that are not accessible; (2) the blind pore that terminate within the material; and (3) the through pore that permit fluid flow through the material and determine the barrier characteristics and permeability of the membrane (*see* Fig. 5).

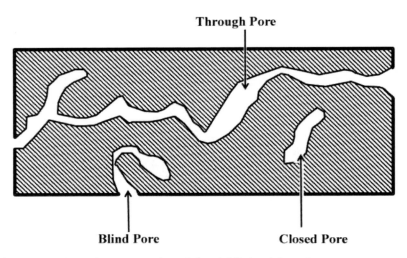

FIGURE 5 Schematic representation of closed, blind and through pores.

A Detailed Review on Pore Structure Analysis

Pore structure characteristics, as one of the main tools for evaluating the performance of any porous membrane, can be performed using microscopic and macroscopic approaches. Microscopic techniques usually consist of high-resolution light microscopy, electron microscopy, and X-ray diffraction. The major disadvantage of this technique is that it cannot determine flow properties and also is time consuming and expensive. Various techniques may be used to evaluate the pore characteristics of porous membranes through macroscopic approach, including mercury intrusion porosimetry, liquid extrusion porosimetry and liquid extrusion flow porometry. These techniques are also used for pore structure characterization of nanofibrous membranes, but the low stiffness and high-pressure sensitivity of nanofiber mats limit application of these techniques.[99–108]

2.4.1.1 MERCURY INTRUSION POROSIMETRY

Mercury intrusion porosimetry is a well-known method, which is often used to evaluate pore characteristics of porous membranes. This technique provides statistical information about various aspects of porous media such as pore size distribution, or the volume distribution.

Due to the fact that mercury, as a non-wetting liquid, does not intrude into pore spaces (except under application of sufficient pressure), a relationship between the size of pores and the pressure applied can be found. In this technique, a porous membrane is completely surrounded by mercury and pressure is applied to force the mercury into pores. As mercury pressure increases the large pores are filled with mercury first and the pore sizes are calculated as the mercury pressure increases. At higher pressures, mercury intrudes into the fine pores and, when the pressure reaches a maximum, total pore volume and porosity are calculated.[98]

According to Jena and Gupta,[9–11] the relationship of pressure and pore size is determined by the Laplace equation:

$$D = -\frac{4\gamma\cos\theta}{p} \quad (1)$$

where D is pore diameter, γ is surface tension of mercury, θ is contact angle of mercury and p is pressure on mercury for intrusion into the pore.

In addition to the pore size and its distribution, the total pore volume and the total pore area can also be determined by mercury intrusion porosimetry method. On the other hand, this method gives no information about the number of pores and is generally applicable to porous membrane with pore sizes ranging from 0.0018 μm to 400 μm. Moreover, mercury intrusion porosimetry does not account for closed pores because mercury does not intrude into them. Due to the application of high pressures, sample collapse and compression is possible, hence it is not suitable for fragile compressible materials such as nanofiber sheets. Other concerns include the fact that it is assumed that the pores are cylindrical, which is not the case in reality.[98]

2.4.1.2 Liquid Extrusion Porosimetry

In liquid extrusion porosimetry, a wetting liquid spontaneously intrudes into the pores and then is extrude from pores by a non-reacting gas. The differential pressure p is related to pore diameter D by

$$D = -\frac{4\gamma \cos \theta}{p} \qquad (2)$$

where γ is surface tension of wetting liquid, and θ is contact angle of wetting liquid. The volume of extruded liquid and the differential pressure is measured by this technique.

In this method, a membrane is placed under the sample such that the largest pore of the membrane is smaller than the smallest pore of interest in the sample. First, the pores of the sample and the membrane are filled with a wetting liquid and then the pressure on gas is increased to displace the liquid from pores of the sample. Because the gas pressure is inadequate to empty the pores of the membrane, the liquid filled pores of the membrane allow the extruded liquid from the pores of the sample to flow out while preventing the gas to escape.[9–10]

Through pore volume and diameter were determined by measuring the volume of the liquid flowing out of the membrane and differential pressure, respectively. It should be noted that liquid extrusion porosimetry measures only the volume and diameters of through pores (blind pore are not measured), whereas mercury intrusion porosimetry measures all pore diameter.[11–12]

2.4.1.3 FLOW POROSIMETRY (BUBBLE POINT METHOD)

Flow porosimetry or bubble point method is based on the principle that a porous membrane will allow a fluid to pass only when the pressure applied exceeds the capillary attraction of the fluid in the largest pore. In this test method, the pore of the membrane is filled with a liquid and continues airflow is used to remove liquid from the pores. At a critical pressure, the first bubble will come through the largest pore in the wetted specimen. As the pressure increases, the smaller pores are emptied of liquid and gas flow increases. Once the flow rate and the applied pressure are known, particle size distribution, the number of pores, and porosity can be derived. In flow porosimetry, the membrane with pore sizes in the range of 0.013–500 μm can be measured.[98]

It is important to note that flow porosimetry measures only the throat diameter of each through pore and cannot measure the blind pore. This technique is based on the assumption that the pores are cylindrical, which is not the case in reality.[15]

2.4.1.4 IMAGE ANALYSIS

Image analysis technique, due to its convenience in detecting individual pores in a nonwoven image, has some advantages for pore measurement. Image analysis technique has been used to measure the pore characteristics of electrospun nanofiberwebs.[98] To measure the pore characteristics of electrospun nanofibrous membranes using image analysis, images (or micrograph) of the nanofiber webs, which are usually obtained by scanning electron microscopy (SEM), transmission electron microscopy (TEM) or atomic force microscopy (AFM), are required. This is highly relevant in that a picture to be used for image analysis must be of high quality and taken under appropriate magnifications.[98] The major advantage of porosity characterization by image analysis technique is that the cross sections provide detailed information about the spatial and size distribution of pores as well as their shape.

In this technique, initial segmentation of the micrographs is required to produce binary images. The typical way of producing a binary image from a greyscale image is by 'global thresholding' in which a single constant threshold is applied to segment the image. All pixels up to and equal to the threshold belong to the object and the remaining belong to the background. Global thresholding is very sensitive to inhomogeneities in the grey-level distributions of object and background pixels. In order to eliminate the effect of inhomogeneities in global thresholding, local thresholding scheme could be used. Firstly, the image is divided into sub-images where the inhomogeneities are negligible. Then the optimal thresholds are found for each sub-image. It can be found that this process is equivalent to segmenting the image with locally varying thresholds.[98] Fig. 6 shows global thresholding and local thresholding of electrospun nanofibrous mat. It is obvious that global thresholding resulted in some broken fiber segments. However, this problem was solved by using local thresholding. It should be mentioned that this process is extremely sensitive to noise contained in the image. So, a procedure to clean the noise and enhance the contrast of the image is necessary before the segmentation.[19,98]

FIGURE 6 (a) SEM image of a real web, (b) global thresholding, (c) local thresholding.

2.4.2 APPROPRIATE TECHNIQUE FOR NANOFIBROUS MEMBRANES

As mentioned above, through pore volume of nanofibrous membranes can be measured by mercury intrusion porosimetry and liquid extrusion porosimetry. Due to the high pressure that is applied to the nanofibrous membranes in mercury intrusion porosimetry method, the pores can get enlarged, which leads to overestimation of porosity values. Blind pores in the nanofiber mat are negligible. Therefore, porosity of the nanofibrous membranes can be obtained from the measured pore volume and bulk density of the material. Liquid extrusion technique can give liquid permeability and surface area of through pores, which could not be measured by mercury intrusion porosimetry. It is important to note that for many applications such as filtration, pore throat diameters of nanofiber mats are required in addition to pore volume. While mercury intrusion and liquid extrusion porosimetry cannot measure pore throat diameter, flow porosimetry can measures pore throat diameters without distorting pore structure. Therefore, flow porosimetry is more suited for pore characterization of nanofibrous membranes.[18,109–110]

2.5 SUMMARY OF LITERATURE REVIEW

Establishing the quantitative relationships between the microstructure of porous media and their properties are an important goal, with a broad relevance to many scientific sectors and engineering applications. Since variations in pore shape and pore space connectivity are intrinsic features of many porous media, a pore structure model must involve both geometric and topological descriptions of their complex microstructure. Nowadays modeling and simulation of nanoporous membrane is of special interest to many researchers.

According to the literature, there are two types of pore-scale modeling; (1) Lattice–Boltzmann (LB) model and (2) pore network model. LB models capable of simulating flow and transport in the actual pore space. Pore network model has been considered as an effective tools used to investigate macroscopic properties from fundamental pore-scale behavior of processes and phenomena based on geometric volume averaging. This model has been used in chemical engineering, petroleum engineering and hydrology fields to study a wide range of single and multiphase flow processes. Pore network model utilizes an idealization of the complex pore space geometry of the porous media. For this purpose the pore space is represented by pore elements having simple geometric shapes such as pore-bodies and pore-throats that have been represented by spheres and cylinders, respectively.[31,111–113]

A Detailed Review on Pore Structure Analysis

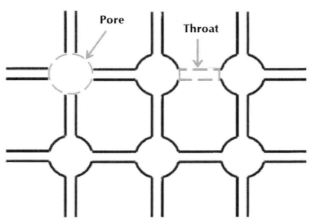

FIGURE 7 Schematic of a pore network illustrating location of pores and throats.

2.5.1 2D IMAGE ANALYSIS OF POROUS MEDIA

Definition of pore network structure, such as pore-body locations, pore-body size distributions, pore-throat size distributions, connectivity, and the spatial correlation between pore-bodies, is an important step towards analysis of porous media. Many valuable attempts have been made for characterization of porous media based on image analysis techniques. For example, Masselin et al.[20] employed image analysis to determine the parameters, such as porosity, pore density, mean pore radius, pore size distribution, and thickness of five asymmetrical ultrafiltration membranes. The results obtained from image analysis for the pore size were found to be in good agreement with rejection data.

Ekneligoda et al.[114] used image analysis technique to extract the area and perimeter of each pore from SEM images of two sandstones, Berea and Fontainebleau. The compressibility of each pore was calculated using boundary elements, and estimated from a perimeter-area scaling law. After the macroscopic bulk modulus of the rock was estimated by the area-weighted mean pore compressibility and the differential effective medium theory, the predicted results were compared with experimental values of the bulk modulus. The resulting predictions are close to the experimental values of the bulk modulus.

A variety of image analysis techniques was used by Lange et al.[19] to characterize the pore structure of cement-based materials, including plain cement paste, pastes with silica fume, and mortars. These techniques include sizing, two-point correlation, and fractal analyses. Backscattered electron images of polished sections were used to observe the pore structure of cement pastes and mortars. They measured pore size distribution of specimen by using image analysis techniques and compared with mercury intrusion porosimetry derived pore size distribution curves. They found that the image-based pore size distribution was able to better describe the large porosity than the mercury intrusion porosimetry.

In the study on pore structure of electrospun nanofibrous membranes by Ziabari et al.[98] a novel image analysis-based method was developed for measuring pore characteristics of electrospun nanofiber webs. Their model was direct, very fast, and presents valuable and comprehensive information regarding pore structure parameters of the webs. In this method, SEM images of nanofiber webs were converted to binary images and used as an input. First, voids connected to the image border are identified and cleared by using morphological reconstruction where the mask image is the input image and marker image is zero everywhere except along the border. Total area, which is the number of pixels in the image, is measured. Then the pores are labelled and each considered as an object. Here the number of pores may be obtained. In the next step, the number of pixels of each object as the area of that object is measured. Having the area of pores, the porosity may be calculated.

They also investigated the effects of web density, fiber diameter and its variation on pore characteristics of the webs by using some simulated images and found that web density and fiber diameter significantly influence the pore characteristics, whereas the effect of fiber diameter variations was insignificant. Furthermore, it seemed that the changes in number of pores were independent of variation of fiber diameter and that this could be attributed to the arrangement of the fibers.

In another study, Ghasemi–Mobarakeh et al.[115] demonstrated the possibility of porosity measurement of various surface layers of nanofibers mat using image analysis. They found that porosity of various surface layers is related to the number of layers of nanofibers mat. This method is not dependent on the magnification and histogram of images. Other methods such as mercury intrusion porosimetry, indirect method, and also calculation of porosity by density measurement cannot be used for porosity measurement of various surface layers and measure the total porosity of nanofibers mat. These methods show high porosity values (higher than 80%) for the nanofibers mat, while the porosity measurements based on thickness and apparent density of nanofibers mat demonstrated the porosity of between 60 and 70%.[116] This value was calculated using the following equation:

$$\varepsilon_V = 1 - \frac{\rho_a}{\rho_b} \times 100 \qquad (1)$$

$$\rho_a = \frac{m}{T \times A} \qquad (2)$$

where ρ_a and ρ_b are apparent density and bulk density of nanofiber mat, m is nanofiber mat mass, A is nanofiber mat area, and T is thickness of nanofiber mat.[116]

2.5.2 3D IMAGE ANALYSIS OF POROUS MEDIA

Several instrumental characterization techniques have been suggested to obtain 3D volume images of pore space, such as X-ray computed micro tomography and magnetic resonance computed micro tomography. However, these techniques may be limited by their resolution. So, the 3D stochastic reconstruction of porous media from statistical information (produced by analysis of 2D photomicrographs) has been suggested. Although pore network models can be two- or three-dimensional, 2D image analysis, due to their restricted information about the whole microstructure, was unable to predict morphological characteristics of porous membrane. Therefore, 3D reconstruction of porous structure will lead to significant improvement in predicting the pore characteristics. Recently research work has focused on the 3D image analysis of porous membranes.

Wiederkehr et al.[117] in their study of three-dimensional reconstruction of pore network, utilized an image morphing technique to construct a three-dimensional multiphase model of the coating from a number of such cross section images. They show that the technique can be successfully applied to light microscopy images to reconstruct 3D pore networks. The reconstructed volume was converted into a tetrahedron-based mesh representation suited for the use in finite element applications using a marching cubes approach. Comparison of the results for three-dimensional data and two-dimensional cross-section data suggested that the 3D-simulation should be more realistic due to the more exacter representation of the real microstructure.

Delerue et al.[118] utilized skeletization method to obtain a reconstructed image of the spatialized pore sizes distribution, i.e. a map of pore sizes, in soil or any porous media. The Voronoi diagram, as an important step towards the calculation of pore size distribution both in2D and 3D media, was employed to determine the pore space skeleton. Each voxel has been assigned a local pore size and a reconstructed image of a spatialized local pore size distribution was created. The reconstructed image not only provides a means for calculating the global volume versus size pore distribution, but also performs fluid invasion simulation, which take into account the connectivity of and constrictions in the pore network. In this case, mercury intrusion in a 3D soil image was simulated.

Al–Raoush et al.[31] employed a series of algorithms, based on the three-dimensional skeletonization of the pore space in the form of nodes connected to paths, to extract pore network structure from high-resolution three-dimensional synchrotron microtomography images of unconsolidated porous media systems. They used dilation algorithms to generate inscribed spheres on the nodes and paths of the medial axis to represent pore-bodies and pore-throats of the network, respectively. The authors have also determined the pore network structure, i.e. three-dimensional spatial distribution (x-, y-, and z-coordinates) of pore-bodies and pore-throats, pore-body and pore-throat sizes, and the connectivity, as well as the porosity, specific surface area, and representative elementary volume analysis on the porosity. They show that X-ray microtomography is an effective tool to non-destructively extract the structure of porous media. They concluded that spatial correlation between pore-bodies in the network is important and controls many processes and phenomena in single and multiphase flow and transport problems. Furthermore, the impact of resolution on the properties of the network structure was also investigated and the results showed that it has a significant impact and can be controlled by two factors: the grain size/resolution ratio and the uniformity of the system.

In another study, Liang et al.[5] proposed a truncated Gaussian method based on Fourier transform to generate 3D pore structure from 2D images of the sample. The major advantage of this method is that the Gaussian field is directly generated from its autocorrelation function and also the use of a linear filter transform is avoided Moreover, it is not required to solve a set of nonlinear equations associated with this transform. They show that the porosity and autocorrelation function of the reconstructed porous media, which are measured from a 2D binarized image of a thin section of the sample, agree with measured values. By truncating the Gaussian distribution, 3D porous media can be generated. The results for a Berea sandstone sample showed that the mean pore size distribution, taken as the result of averaging between several serial cross-sections of the reconstructed 3D representation, is in good agreement with the original thresholded 2D image. It is believed that by 3D reconstruction of porous media, the macroscopic properties of porous structure such as permeability, capillary pressure, and relative permeability curves can be determined.

Diógenes et al.[119] reported the reconstruction of porous bodies from 2D photomicrographic images by using simulated annealing techniques. They proposed the following methods to reconstruct a well-connected pore space: (i) Pixel-based Simulated Annealing (PSA), and (ii) Objective-based Simulated Annealing (OSA). The difference between the present methods and other research studies, which tried to reconstruct porous media using pixel-movement based simulated techniques, is that this method is based in moving the microstructure grains (spheres) instead of the pixels. They applied both methods to reconstruct reservoir rocks microstructures, and compared the 2D and 3D results with microstructures reconstructed by truncated Gaussian methods. They found that PSA method is not able to reconstruct connected porous media in 3D, while the OSA reconstructed microstructures with good pore space connectivity. The OSA method also tended to have better permeability determination results than the other methods. These results indicated that the OSA method can reconstruct better microstructures than the present methods.

In another study, a 3D theoretical model of random fibrous materials was employed by Faessel et al.[120] They used X-ray tomography to find 3D information on real networks. Statistical distributions of fibers morphology properties (observed at microscopic scale) and topological characteristics of networks (derived from mesoscopic observation), is built using mathematical morphology tools. The 3D model of network is assembled to simulate fibrous networks. They used a number of parameter describing a fiber, such as length, thickness, parameters of position, orientation and curvature, which derived from the morphological properties of the real network.

2.6 CONCLUSION

In recent years, great efforts have been devoted to nanoporous membranes. As a conclusion, much progress has been made in the preparation and characterization of porous media. Among several porous membranes, electrospun nanofibrous membranes, due to the high porosity, large surface area-to-volume ratios, small pores, and fine fiber diameter, have gained increasing attention. Useful techniques for evaluation of the pore characteristics of porous membranes are reviewed. Image analysis techniques have been suggested as a useful method for characterization of porous media due to its convenience in detecting individual pores. It is believed that the three-dimensional reconstruction of porous media, from the information obtained from a two-dimensional analysis of photomicrographs, will bring a promising future tonanoporous membranes.

KEYWORDS

- image analysis
- nanofibrous membrane
- porous media
- three-dimensional analysis

REFERENCE

1. Rutledge G.C., Li Y., Fridrikh S., *National Textile Center Research Briefs – Materials Competency*, 2004.
2. Yao C., Li X., Song T., *J. Appl. Polym. Sci.*, **103**, 380 (2007).
3. Kim G., Kim W., J. Biomed. Mater. Res B: *Appl. Bio. Mater.*, **81B**, 104 (2007).
4. Silina D., Patzekb T., *Phys. A: Stat. Mech. Appl.*, **371**, 336 (2006).
5. Liang Z.R., Fernandes C.P., Magnani F.S., Philippi P.C., *J. Petrol. Sci. Eng.*, **21**, 273 (1998).
6. Kim K.J., Fane A.G., R. Ben Aim, Liu M.G., Jonsson G., Tessaro I.C., Broek A.P., D.Bargeman, *J. Membran. Sci.*, **81**, 35(1994).
7. Dullien F.A.L., Dhawan G.K., *J. Colloid. Interf. Sci.*, **47**, 337 (1974).
8. Liabastre A.A., Orr C., *J. Colloid. Interf. Sci.*, **64**, 1 (1978).
9. Jena A., Gupta K., *Fluid/particle Sep. J.*, **14**, 227 (2002).
10. Jena A., Gupta K., *J. Filtration Soc.*, **1**, 23 (2001).
11. Jena A., Gupta K., *Int. Nonwovens. J.*, Fall, 45 (2003).

12. Barrett E.P., Joyner L.G., Halenda P.P., *J. Am. Chem. Soc.* **73**, 373 (1951).
13. Calvo J.I., A. Herna Ndez, P. Pra Danos, Martibnez L., Bowen W.R., *J. Colloid. Interf. Sci.*, **176**, 467 (1995).
14. Jena A.K., Gupta K.M., J. Power. Sources, **80**, 46 (1999).
15. Kraus G., Ross J.W., and Girifalco L.A., *Phys. Chem.*, **57**, 330 (1953).
16. Gribble C.M., Matthews G.P., Laudone G.M., Turner A., C.J.Ridgway, Schoelkopf J., Gane P.A.C., *J. Chem. Eng. Sci.*, **66**, 3701 (2011).
17. Deshpande S., Kulkarni A., Sampath S., Herman H., *Surf. Coat. Tech.*, **187**, 6(2004).
18. Tomba E., Facco P., Roso M., Modesti M., Bezzo F., Barolo M., *Ind. Eng. Chem. Res.*, **49**, 2957(2010).
19. Lange D.A., *Cement. Concrete. Res.*, **24**, 841(1994).
20. Masselin I., L. Durand–Bourlier, Laine J.M., Sizaret P.Y., Chasseray X., Lemordant D., *J. Membrane Sci.*, **186**, 85 (2001).
21. Garboczi E.J., Bentz D.P., Martys N.S., *Exp. Meth. Phys. Sci.*, **35**, 1 (1999).
22. Mickel W., Munster S., Jawerth L.M., Vader D.A., Weitz D.A., Sheppard A.P., K.Mecke, Fabry B., G.E. Schroder–Turk, *Biophys. J.*, **95**, 6072 (2008).
23. Roysam B., Lin G., M. Amri Abdul–Karim, O. Al–Kofahi, K. Al–Kofahi, Shain W., Szarowski D.H., Turner J.N., Handbook of Biological Confocal Microscopy, 3rd edition, Springer, New York, 2006.
24. Quiblier J.A., *J. Colloid. Interf. Sci.*, **98**, 84 (1984).
25. Santos L.O.E., Philippi P.C., Damiani M.C., Fernandes C.P., *J. Petrol. Sci. Eng.,* **35**, 109 (2002).
26. Sambaer W., Zatloukal M., Kimmer D., *Chem. Eng. Sci.*, **66**, 613 (2011).
27. Shin C.H., Seo J.M., Bae J.S., *J. Ind. Eng. Chem.*, **15**, 784 (2009).
28. Ye G., K. van Breugel and Fraaij A.L.A., *Cement. Concrete Res.*, **33**, 215 (2003).
29. Holzer L., Münch B., Rizzi M., Wepf R., Marschall P., Graule T., *Appl. Clay. Sci.*, **47**, 330 (2010).
30. Liang Z., Ioannidis M.A., Chatzis I., *J. Colloid. Interf. Sci.*, **221**, 13(2000).
31. R.I. Al–Raoush and Willson C.S., *J. Hydrol.*, **300**, 44 (2005).
32. Fenwick D.H., Blunt M.J., *Adv. Water. Resour.*, **21**, 143 (1998).
33. Santos L.O.E., Philippi P.C., Damiani M.C., Fernandes C.P., *J. Petrol. Sci. Eng.,* **35**,109(2002).
34. Mendoza F., Verboven P., Mebatsion H.K., Kerckhofs, G. M.Wevers and B. Nicolaï, *Planta*, **226**, 559 (2007).
35. Bakke S. and Øren P., *SPE.J.*, **2**, 136 (1997).
36. Yee Ho A.Y., Gao H., Y. Cheong Lam and Rodrıguez I., *Adv. Funct. Mater*, **18**, 2057 (2008).
37. Sakamoto Y., Kim T.W., Ryoo R., Terasaki O., *Angew.Chem*, **116**, 5343 (2004).
38. R.I. Al–Raoush, PhD Thesis, Louisiana State University, 2002.
39. Boissonnat J.D., *ACGraphic M.T.,* **3**, 266 (1984). P
40. Holzer L., Indutnyi F., Ph. Gasser, Münch B., Wegmann M., *J. Microsc.*, **216**, 84 (2004).
41. Pothuaud L., Porion P., Lespessailles E., Benhamou C.L., Levitz P., *J. Microsc.*, **199**, 149 (2000).
42. Yeong C.L.Y., Torquato S., *Phys. Rev. E*, **58**, 224 (1998).
43. Biswal B., Manwart C., Hilfer R., *Physica. A*, **255**, 221 (1998).

44. Desbois G., Urai J.L., Kukla P.A., J.Konstanty and C.Baerle, *J. Petrol. Sci. Eng.*, **78**, 243 (2011).
45. Ulbricht M., *Polymer*, **47**, 2217 (2006).
46. Amendt M.A., PhD Thesis, University Of Minnesota, (2010).
47. P.Szewczykowski, PhD Thesis, University of Denmark, (2009).
48. T.Gullinkala, PhD Thesis, University of Toledo, (2010).
49. S.Naveed and Bhatti I., *J. Res. Sci.*, **17**, 155 (2006).
50. Baker R.W., John Wiley & Sons, England, (2004).
51. Roychowdhury A., MSc Thesis, Louisiana State University, (2007).
52. W.F. Catherina Kools, PhD Thesis, University of Twente, (1998).
53. M.K. Buckley–Smith, PhD Thesis, University of Waikato, (2006).
54. Nunes S.P., Peinemann K.V., Wiley–VCH, Germany, (2001).
55. Li W., PhD Thesis, University of Cincinnati, 2009.
56. Childress A.E., P. Le–Clech, Daugherty J.L., Caifeng Chen and Greg Leslie L., *Desalination*, **180**, 5 (2005).
57. Li L., P.Szewczykowski, Clausen L.D., Hansen K.M., Jonsson G.E., S.Ndoni, *J. Membrane. Sci.*, **384**, 126 (2011).
58. Chaoyiba, PhD thesis, University of Illinois, (2010).
59. Yen C., B.S, PhD thesis, Ohio State University, (2010).
60. Zon X., Kim K., Fang D., Ran S., Hsiao B.S., Chu B., *Polymer*, **43**, 4403 (2002).
61. Ondarçuhu T., Joachim C., Eur. Phys. Lett., **42**, 215(1998).
62. Feng L., Li S., Li Y., Li H., Zhang L., Zhai J., Song Y., Liu B., Jiang L., Zhu D., *Adv. Mater.*, **14**, 1221 (2002).
63. Ma P.X., Zhang R., *J. Biomed. Mater. Res.*, **46**, 60 (1999).
64. Liu G., Ding J., Qiao L., Guo A., Dymov B.P., Gleeson J.T., Hashimoto T.K., *Chem. Eur. J.*, **5**, 2740 (1999).
65. Doshi J., Reneker D.H., *J. Electrostat.*, **35**, 151 (1995).
66. Reneker D.H., Yarin A.L., *Polymer*, **49**, 2387 (2008).
67. Yordem O.S., Papila M., Menceloglu Y.Z., *Mater. Design*, **29**, 34 (2008).
68. Zussman E., Theron A., Yarin A., Appl. Phys. Lett., **82**, 973 (2003).
69. Gibson P.W., H.L. Schreuder–Gibson and Rivin D., *AIChE.J.*, **45**, 190 (1999).

70. Theron A., Zussman E., Yarin A.L., *Nanotechnology*, **12**, 384 (2001).

71. Teo W.E., Inai R., Ramakrishna S., *Sci. Technol. Adv. Mat.*, **12**, 1 (2011).
72. T.Subbiah, Bhat G.S., Tock R.W., Parameswaran S., Ramkumar S.S., *J. Appl. Polym. Sci.*, **96**, 557 (2005).
73. Zong X., Kim K., Fang D., Ran S., Hsiao B.S., Chu B., *Polymer*, **43**, 4403 (2002).
74. Tan S., Huang X., Wu B., *Polym. Int.*, **56**, 1330 (2007).
75. Burger C., Hsiao B.S., Chu B., Annu. *Rev. Mater. Res.*, **36**, 333 (2006).
76. Zhang C., Li Y., Wang W., Zhan N., Xiao N., Wang S., Li Y., Yang Q., *Eur. Polym. J.*, **47**, 2228 (2011).
77. CHOI J., PhD Thesis, Case Western Reserve University, (2010).
78. Reneker D.H., Yarin A.L., Zussman E., Xu H., *Advances In Applied Mechanics*, **41**, 43(2007).
79. Huang Z.M., Zhang Y.Z., Kotaki M., Ramakrishna S., Compos. Sci. Technol., **63**, 2223 (2003).

80. Zander N.E., *Polymers*, **5**, 19 (2013).
81. Bhardwaj N., Kundu S.C., *Biotechnol. Adv.*, **28**, 325 (2010).
82. Wang N., Burugapalli K., Song W., Halls J., Moussy F., Ray A., Zheng Y., *Biomaterials*, **34**, 888 (2013).
83. Jung H.R., D.H.Ju, Lee W.J., Zhang X., Kotek R., *Electrochim. Acta.*, **54**, 3630 (2009).
84. Gong Z., Ji G., Zheng M., Chang X., Dai W., Pan L., Shi Y., Zheng Y., *Nanoscale. Res. Lett.*, **4**, 1257 (2009).
85. Wang Y., Zheng M., Lu H., Feng S., Ji G., Cao J., *Nanoscale. Res. Lett.*, **5**, 913 (2010).
86. Yin G.B., *J. Fiber Bioeng. Informatics*, **3**, 137 (2010).
87. Lee J.B., Jeong S.I., Bae M.S., Yang D.H., Heo D.N., Kim C.H., Alsberg E., Kwon I.K., *Tissue. Eng. A*, **17**, 2695 (2011).
88. Taha A.A., Qiao J., Li F., Zhang B., *J. Environ. Sci–China.*, **24**, 610 (2012).
89. Kim G.H., Kim W.D., *J. Biomed. Mater. Res. B: Applied Biomaterials*, **17**, 2695 (2006).
90. Zhang Y.Z., Feng Y., Huang Z.M., Ramakrishna S., Lim C.T., *Nanotechnology*, **17**, 901 (2006).
91. Ramakrishna S., Fujihara K., Teo W.E., Lim T.C., Ma Z., World Scientific Publishing: Singapore, 2005.
92. Burger C., Hsiao B.S., Chu B., *Annu. Rev. Mater. Res*, **36**, 333 (2006).
93. Berkalp O.B., *Fibers. Text. East. Eur.*, **14**, 81 (2006).
94. Choat B., Jansen S., Zwieniecki M.A., Smets E., Holbrook N.M., *J. Exp. Bot.*, **55**, 1569 (2004).
95. Alrawi A.T., Mohammed S.J., *Int. J. Soft. Comput.*, **3**, 1 (2012).
96. Krajewska B., Olech A., *Polym. Gels Networks*, **4**, 33 (1996).
97. Esselburn J.D., MSc Thesis, Wright State University, (2009).
98. Ziabari M., Mottaghitalab V., Haghi A.K., *Korean. J. Chem. Eng.*, **25**, 923 (2008).
99. A.Shrestha, MSc Thesis, University of Colorado, (2012).
100. Borkar N., MSc Thesis, University of Cincinnati, (2010).
101. Cuperus F.P., Smolders C.A., *Adv. Colloid. Interface. Sci.*, **34**, 135 (1991).
102. L. Mart'ınez, F.J. Florido–D'ıaz, Hernández A., Prádanos P., J. Membrane. Sci., **203**, 15(2002).
103. Bloxson J.M., MSc Thesis, Kent State University, (2012).
104. Cao G.Z., Meijerink J., Brinkman H.W., Burggra A.J., *J. Membrane. Sci.*, **83**, 221 (1993).
105. Cuperus F.P., Bargeman D., Smolders C.A., *J. Membrane. Sci.*, **71**, 57 (1992).
106. Fernando J.A., Chuung D.D.L., *J. Porous. Mat.*, **9**, 211 (2002).
107. Shobana K.H., M. Suresh Kumar, Radha K.S., Mohan D., *Sch. J. Eng. Res.*, **1**, 37 (2012).
108. Cañas A., Ariza M.J., Benavente J., *J. Membrane. Sci.*, **183**, 135 (2001).
109. Frey M.W., Li L., *J. Eng. Fiber. Fabr.*, **2**, 31 (2007).
110. J. ˇSirc, Hobzov R., N.Kostina, Munzarov M., M. Jukl'ıˇckov, Lhotka M., S.Kubinov, A.Zaj'ıcov and J. Mich'alek, *J. Nanomater.*, **2012**, 1 (2012).
111. A.B.Venkatarangan, PhD Thesis, University of New York, (2000).
112. Zhou B., PhD Thesis, Massachusetts Institute of Technology, (2006).
113. Manwart C., Aaltosalmi U., Koponen A., Hilfer R., Timonen J., *Phys. Rev. E*, **66**, 016702 (2002).

114. Ekneligoda T.C., Zimmerman R.W., in proceeding of Royal Society A, March 8, 2008, pp. 759–775.
115. L. Ghasemi–Mobarakeh, Semnani D., Morshed M., *J. Appl. Polym. Sci.*, **106**, 2536 (2007).
116. He W., Ma Z., Yong T., Teo W.E., Ramakrishna S., *Biomaterials*, **26**, 7606 (2005).
117. T.Wiederkehr, Klusemann B., Gies D., Müller H., Svendsen B., *Comput. Mater. Sci.*, **47**, 881 (2010).
118. Delerue J.F., Perrie E., Yu Z.Y., Velde B., *Phys. Chem. Earth. A*, **24**, 639 (1999).
119. Diógenes A.N., L.O.E. dos Santos, Fernandes C.P., Moreira A.C., Apolloni C.R., *Therm. Eng.*, **8**, 35 (2009).
120. Faessel M., Delisee C., Bos F., Castera P., *Compos. Sci. Technol.*, **65**, 1931 (2005).

CHAPTER 3

EXPERIMENTAL TECHNIQUES FOR APPLICATION OF RECYCLED POLYMERS IN CONSTRUCTION INDUSTRIES

O. EMAMGHOLIPOUR and A. K. HAGHI

CONTENTS

3.1	Introduction	52
3.2	Experimental Procedure	52
3.3	Results and Discussion	58
3.4	Experimental Procedure	60
3.5	Results and Discussions	63
3.6	Experimental Procedure	67
3.7	Results and Discussion	72
3.8	Experimental Procedure	81
3.9	Experimental Tests	82
3.10	Results and Discussions	83
3.11	Concluding Remarks	105
	Keywords	105
	References	105

3.1 INTRODUCTION

This chapter is divided into four sections. In each section we have shown laboratory experiments in order to show the possibilities for converting the different types of polymer waste to wealth. These recycled materials have the required capabilities and potentials to be used in construction industries.

SECTION 1. PRACTICAL HINTS AND UPDATE ON PERFORMANCES OF ORDINARY PORTLAND CEMENT USING RECYCLED GLASS AND RECYCLED TEXTILE FIBERS AS A FRACTION OF AGGREGATES

3.2 EXPERIMENTAL PROCEDURE

Materials used included Ordinary Portland cement type 1, standard sand, silica fume, glass with tow particle size, rice husk ash, tap water and finally fibrillated polypropylene fibers.

The fibers included in this section were monofilament fibers obtained from industrial recycled raw materials that were cut in factory to 6 mm length. Properties of waste Polypropylene fibers are reported in Table 1 and Fig. 1.

TABLE 1 Properties of polypropylene fibers.

Property	Polypropylene
Unit weight [g/cm^3]	0.9–0.91
Reaction with water	Hydrophobic
Tensile strength [ksi]	4.5–6.0
Elongation at break [%]	100–600
Melting point [°C]	175
Thermal conductivity [W/m/K]	0.12

Also the silica fume and rice husk ash contain 91.1% and 92.1% SiO_2 with average size of 7.38 μm and 15.83 μm respectively were used. The chemical compositions of all pozzolanic materials containing the reused glass, silica fume and rice husk ash were analyzed using an X-ray microprobe analyzer and listed in Table 2.

FIGURE 1 Polypropylene fiber used in this study.

TABLE 2 Chemical composition of materials.

	Content (%)		
Oxide	**Glass C**	**Silica fume**	**Rice husk ash**
SiO_2	72.5	91.1	92.1
Al_2O_3	1.06	1.55	0.41
Fe_2O_3	0.36	2	0.21
CaO	8	2.24	0.41
MgO	4.18	0.6	0.45
Na_2O	13.1	—	0.08
K_2O	0.26	—	2.31
CL	0.05	—	—
SO_3	0.18	0.45	—
L.O.I	—	2.1	—

FIGURE 2 Ground waste glass.

To obtain this aim recycled windows clean glass was crushed and grinded in laboratory, and sieved the ground glass to the desired particle size (Fig. 2). To study the particle size effect, two different ground glasses were used, namely:
- Type I: ground glass having particles passing a #80 sieve (180μm);
- Type II: ground glass having particles passing a #200 sieve (75μm).

In addition the particle size distribution for two types of ground glass, silica fume, rice husk ash and ordinary Portland cement were analyzed by laser particle size set, have shown in Fig. 3. As it can be seen in Fig. 3 silica fume has the finest particle size. According to ASTM C618, fine ground glasses under 45μm qualify as a pozzolan due to the fine particle size. Moreover glass type I and II respectively have 42% and 70% fine particles smaller than 45 μm that causes pozzolanic behavior. SEM particle shape of tow kind of glasses is illustrated in Fig. 4.

For the present study, twenty batches were prepared. Control mixes were designed containing standard sand at a ratio of 2.25:1 to the cement in matrix. A partial replacements of cement with pozzolans include ground waste glass (GI, GII), silica fume (SF) and rice husk ash (RH) were used to examine the effects of pozzolanic materials on mechanical properties of PP reinforced mortars at high temperatures. The amount of pozzolans which replaced were 10% by weight of cement which is the rang that is most often used.

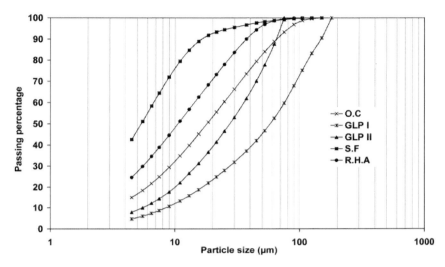

FIGURE 3 Particle size distribution of ground waste glass type I, II, silica fume, rice husk ash and ordinary cement.

FIGURE 4 Particle shape of ground waste glass type I, type II.

TABLE 3 Mixter Properties.

Batch No	Sand/c	w/c	Content (by weight)					PP fibers (by volume)
			O.C	GI	GII	SF	RH	
1	2.25	0.47	100	—	—	—	—	0
2	2.25	0.47	90	10	—	—	—	0
3	2.25	0.47	90	—	10	—	—	0
4	2.25	0.47	90	—	—	10	—	0
5	2.25	0.47	90	—	—	—	10	0
6	2.25	0.6	100	—	—	—	—	0.5
7	2.25	0.6	90	10	—	—	—	0.5
8	2.25	0.6	90	—	10	—	—	0.5
9	2.25	0.6	90	—	—	10	—	0.5
10	2.25	0.6	90	—	—	—	10	0.5
11	2.25	0.6	100	—	—	—	—	1
12	2.25	0.6	90	10	—	—	—	1
13	2.25	0.6	90	—	10	—	—	1
14	2.25	0.6	90	—	—	10	—	1
15	2.25	0.6	90	—	—	—	10	1
16	2.25	0.6	100	—	—	—	—	1.5
17	2.25	0.6	90	10	—	—	—	1.5
18	2.25	0.6	90	—	10	—	—	1.5
19	2.25	0.6	90	—	—	10	—	1.5
20	2.25	0.6	90	—	—	—	10	1.5

Meanwhile, polypropylene fibers were used as addition by volume fraction of specimens. The reinforced mixtures contained PP fiber with three designated fiber contents of 0.5%, 1% and 1.5% by total volume.

In the plain batches without any fibers, water to cementations ratio of 0.47 was used whereas in modified mixes (with different amount of PP fibers) it changed to 0.6 due to water absorption of fibers. The mix proportions of mortars are given in Table 3.

The strength criteria of mortar specimens and impacts of polypropylene fibers on characteristics of them were evaluated at the age of 60 days.

In our laboratory, the test programme mix conducted as follows:
1. The fibers were placed in the mixer.
2. Three-quarters of the water was added to the fibers while the mixer was running at 60 rpm; mixing continues for one minute.
3. The cement was gradually the cement to mix with the water.
4. The sand and remaining water were added, and the mixer was allowed to run for another two minutes.

After mixing, the samples were casted into the forms 50×50×50 mm for compressive strength and 50×50×20 mm for flexural strength tests. All the molds were coated with mineral oil to facilitate remolding. The samples were placed in two layers. Each layer was tamped 25 times using a hard rubber mallet. The sample surfaces were finished using a metal spatula. After 24 hours, the specimens were demolded and cured in water at 20°C. The suitable propagation of fibers in matrix is illustrated in Fig. 5.

FIGURE 5 Propagation of polypropylene fibers in mortar matrix (Up: 0.5% fiber, Down: 1% fiber).

The heating equipment was an electrically heated set. The specimens were positioned in heater and heated to desire temperature of 300 and 600°C at a rate of 10–12°C/min. After 3 h, heater turned off. It was allowed to cool down before the specimens were removed to prevent thermal shock to the specimens. The rate of cooling was not controlled. The testes to determine the strength were made for all specimens at the age of 60 days. At least three specimens were tested for each variable.

3.3 RESULTS AND DISCUSSION

3.3.1 DENSITY

The initial density of specimens containing polypropylene fibers was less than that of mixes without any fibers. Density of control mixes without any replacement of cement at 23,300 and 600°C are reported in Table 4. According to the results, density decrease of fiber-reinforced specimens was close to that of plain ones. The weight of the melted fibers was negligible. The weight change of mortar was mainly due to the dehydration of cement paste.

TABLE 4 Density of Control Specimens.

Heated at	23°	300°	600°	PP fibers
Density (gr/cm³)	2.57	2.45	2.45	0%
Density (gr/cm³)	2.50	2.36	2.36	0.5%
Density (gr/cm³)	2.44	2.28	2.27	1%
Density (gr/cm³)	2.41	2.23	2.21	1.5%

3.3.2 COMPRESSIVE STRENGTH

In order to asses the effect of elevated temperatures on mortar mixes under investigation, measurements of mechanical properties of test specimens were made shortly before and after heating, when specimens were cooled down to room temperature. Compressive strength of reference specimen and heated ones at the age of 60 days are illustrated in Fig. 6.

According to the results, by increasing the amount of polypropylene fibers in matrix the compressive strength of specimens reduced. Also, it's clear that the compressive strength of specimens were decreased by increasing the temperature to 300 and 600°C, respectively, as supported by previous literatures.

The rate of strength reduction in fiber-reinforced specimens is more than the plain samples and by rising the temperature it goes up.

The basic factor of strength reduction in plain specimens is related to matrix structural properties exposed to elevated temperature, but this factor for fiber-reinforced specimens is related to properties of fibers. Fibers melt at temperature higher than 190°C and generate lots of holes in the matrix. These holes are the most important reasons of strength reduction for fiber-reinforced specimens.

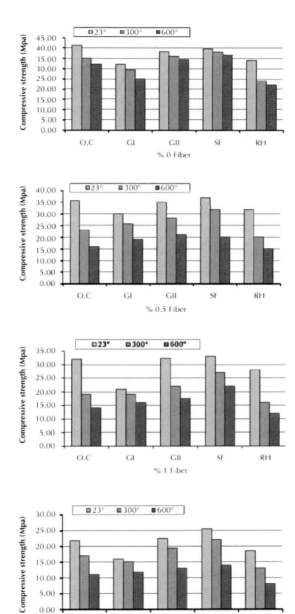

FIGURE 6 Compressive strength of samples at different temperatures.

Also, results indicate that silica fume and glass type II have an appropriate potential to apply as a partial replacement of cement due to their respective pozzolanic activity index values (according to ASTM C618 and C989, and Table2).

3.3.3 FLEXURAL STRENGTH

The specimens were used for flexural testes were 50×50×200 mm. The results of plain specimens and samples containing 1.5% fibers are shown in Table 5. The heat resistance of the flexural strength appeared to decrease when polypropylene fibers were incorporated into mortar. This is probably due to the additional porosity and small channels created in the matrix of mortar by the fibers melting like compressive strength. However the effect of the pozzolans on flexural strength is not clear but it seems that silica fume and glass types II have better impact on strength in compare with control specimens than rice husk ash and glass type I.

TABLE 5 Flexural strength of control samples with 0% and 1.5% fibers.

Batch	Flexural Strength (Mpa)			PP fibers
No	23°	300°	600°	(by volume)
1	4	3.1	2.8	0
2	3	2.4	2.1	0
3	3.7	3.2	2.7	0
4	3.9	3.4	2.7	0
5	3.3	2.8	2.6	0
16	2.2	1	0.7	1.5
17	1.6	0.8	0.6	1.5
18	2	1.1	0.8	1.5
19	2.4	1.3	0.8	1.5
20	1.8	1	0.6	1.5

SECTION 2. PRACTICAL HINTS AND UPDATE ON PERFORMANCES OF ORDINARY PORTLAND CEMENT USING RECYCLED TIRE FIBERS AND CHIPS AS A FRACTION OF AGGREGATES

3.4 EXPERIMENTAL PROCEDURE

It should be noted that concrete strength is greatly affected by the properties of its constituents and the mixture design parameters. In this section, the raw materials used included Portland cement Type 1, mixture of aggregates (coarse and medium), and sand, water and tire fibers. Also silica fume and rice husk ash as high reactive pozzolans were used in this study. Tires were cut by hand and a cuter in laboratory in form of fiber and chips. They were cut into strips of 30 mm × 5 mm × 5 mm, 60 mm × 5 mm × 5 mm as fibers and 10 mm × 10 mm × 5 mm and 20 mm × 20 mm × 5 mm as chips that illustrated in Fig. 1. Also the chemical composition of ordinary cement, silica fumes and rice husk ash are reported in Table 1.

TABLE 1 Chemical composition of materials.

Materials	Ordinary cement	Silica fume	Rice husk ash
SiO_2	21.24	91.1	92.1
Al_2O_3	5.97	1.55	0.41
Fe_2O_3	3.34	2	0.21
CaO	62.72	2.24	0.41
MgO	2.36	0.6	0.45
Na_2O	0.13	—	0.08
K_2O	0.81	—	2.31
CL	—	—	—
SO_3	1.97	0.45	—
L.O.I.	1.46	2.1	—

FIGURE 1 Rubber in form of Fibers and chips used in experiments.

The concrete mix was rated at 40 Mpa (compression strength). A control mix was designed using ACI Standard 211.1 mix design methods. The modified batches designed to be compared with. In the modified batches, 15% by volume of the coarse aggregates was replaced by tires. The mix ratio by weight for control concrete was cement: water: gravel: sand: = 1: 0.50: 3.50: 1.88. The mix ratio by weight for rubberized concrete was cement: water: gravel: waste tire: sand: = 1: 0.50: 3.40: 0.10: 1.88.

Also all rubberized batches treated with pozzolanic materials as the additional part of cement. In this study 10% and 20% of cement content by weight are selected for addition of pozzolan to the rubberized concrete mixtures.

After the concrete was mixed, it was placed in a container to set for 24 hours after that, the specimens were demolded and cured in water at 20°C.

Twenty-one batches of 15 cm radius by 30 cm height cylinders were prepared according to ACI specifications. One batch was made without waste tires and pozzolanic addition to be the control while four batches were prepared using waste tire fibers, without any pozzolanic addition.

Remain specimens prepared with waste tire and additional cementations part.

Waste tire measurement of main batches is reported in Table 2. According to this table rubberized modified batches include silica fume and rice husk ash prepared as followed.

TABLE 2 The size of tire in each batch.

Batch number	Waste tire shape	Length (mm)	Width (mm)	Height (mm)
1	—	—	—	—
2	Tires Fiber	30	5	5
3		60	5	5
4	Tire Chips	10	10	5
5		20	20	5

3.4.1 TESTING METHODS

3.4.1.1 COMPRESSIVE AND TENSILE STRENGTH

ASTM C 39 Standard was used in conducting compressive tests and ASTM C496–86 Standard was used for the split tensile strength tests. Three specimens of each mixture were tested to determine the average strength.

3.4.1.2 FLEXURAL TOUGHNESS

Flexural toughness determine according to ASTM C1018.

3.4.1.3 SLUMP

Slump tests were also conducted to measure the workability of concrete. The slump should never exceed 15 cm. Slump test was performed according to ASTM C143.

3.5 RESULTS AND DISCUSSIONS
3.5.1 COMPRESSIVE AND TENSILE STRENGTH OF CONCRETE

The results of the tests for strength performed on the samples in the experiments are shown in Fig. 2.

It can be seen from Fig. 2, that there was a significant and almost consistent decrease in the compressive strength of the rubberized concrete (*Rubcrete*) batches. Of all the batches tested, the control batch had the highest compressive strength. There was approximately 40% decrease in the compressive strength with the addition of the waste tire fibers. Also batch 4 had the highest compressive strength of all the modified samples.

Also Fig. 2 shows that the control samples had the highest split tensile strength. Batch 3 which consisted of waste tires fiber had the highest split tensile strength of all the rubberized concrete (*Rubcrete*) samples.

Totally results indicate that the size, proportions and surface texture of rubber particles noticeably affect compressive strength of Rubcrete mixtures.

Also its obvious that concrete mixtures with tire chips rubber aggregates exhibited higher compressive and lower splitting tensile strengths than other modified batches, but never has passed the regular Portland cement concrete specimens.

Generally there was approximately 40% reduction in compressive strength and 30% reduction in splitting tensile strength when 15% by volume of coarse aggregates were replaced with rubber fibers and chips.

All of modified mixtures demonstrated a ductile failure and had the ability to absorb a large amount of energy under compressive and tensile loads.

The beneficial effect of silica fume and rice husk ash as high reactive pozzolans on mechanical characteristics of rubberized concrete respect to pozzolans is exhibit in Fig. 2.

It is clear that by increasing the amount of pozzolans in Rubcrete, compressive strengths rise up. The rate of strength increase ranged from 5% to 40% for rubberized concrete, depending on the variation in size of rubber fibers and chips, amount of silica fume and rice husk ash and the fine particles in pozzolans, in general pozzolan activity index.

Results indicate that silica fume had grater impact on strength of specimens than rice husk ash. It is relevant to pozzolanic behavior and activity index.

3.5.2 MODULUS OF ELASTICITY

Static modulus of elasticity test results for reference samples as a function of rubber shape are depicted in Fig. 3.

Bar chart in Fig. 3 show that the static elastic modulus decreased by replacement of rubber whit course aggregated, similar to that observed in both compressive and splitting tensile strengths. It shows that the control samples without any rubber particle possessed the highest modulus of elasticity. Also batch 4 indicates the highest module of elasticity in compare with modified batches. In addition it's very close to batch 1 as control.

The general deviation in values for the rubberized samples was rather small. The volume and modulus of the aggregate are the factors that are mainly responsible for the modulus of elasticity of concrete. Therefore, small additions of tire fiber would not be able to significantly change the modulus of the composite, especially according to recent studies replacement up to 10% has no considerable effect on module of elasticity.

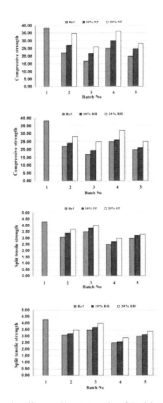

FIGURE 2 Compressive and split tensile strength of Rubbercrete (MPa).

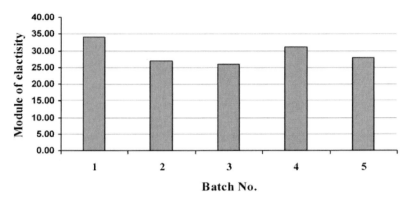

FIGURE 3 Variation of modulus of elasticity (Mpa).

3.5.3 WORKABILITY OF CONCRETE

One of the concerns when adding waste tires to the concrete was whether the workability of the concrete would be negatively affected. Workability refers to the ability of the concrete to be easily molded.

From Fig. 4 it was seen that there was a variation of approximately +/– 0.6 cm between the slum of the reference samples and the rubberized concrete (*Rubcrete*).

Results imply that the workability of the concrete was not adversely affected by the addition of waste tires.

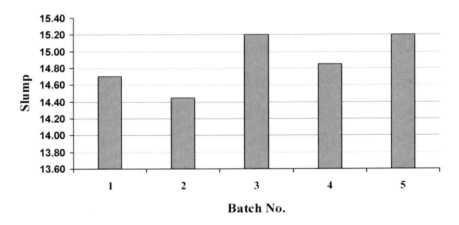

FIGURE 4 Variation of slump (mm) of concrete (MPa).

3.5.4 EFFECT OF WASTE TIRE ON TOUGHNESS OF CONCRETE

Figure 4 shows the applied load – displacement curves for split tensile testing of a control sample and a rubberized sample from batch 2. Toughness is defined as the ability of materials to sustain load after initial cracking and is measured as the total strain experienced at failure. Upon cracking, tier fibers are able to bridge the initial crack and hold the crack together and sustain the load until the fibers either pullout from the matrix (in early age) or fracture that say flexural toughness. According to Fig. 5 it can be observed that after crack initiation, fibers can still carry load and absorb the energy.

For plain concrete, the behavior was in a brittle manner. When the strain energy was high enough to cause the crack to self-propagate, fracture occurred almost while the peak load was reached (this is due to the tremendous amount of energy being released). According to Fig. 5, the tire fiber bridging effect helped to control the rate of energy release significantly. Thus, fibers still can carry load even after the peak. With the effect of fibers bridging across the crack surface, fibers were able to maintain the load carrying ability even after the concrete had been cracked. These are accordance to ASTM C1018, in which toughness or energy absorption defined as the area under the load-deflection curve from crack point to 1/150 of span.

Our laboratory results indicate that, the area beneath the curve for the concrete without rubber is very small compared to the area beneath the curve for the concrete with rubber. This implies that concrete with rubber is much tougher than concrete without rubber.

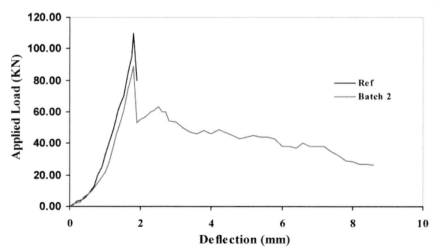

FIGURE 5 Load displacement results for split tensile testing of rubberized and plain concrete.

SECTION 3. PRACTICAL HINTS AND UPDATE ON PERFORMANCES OF ORDINARY PORTLAND CEMENT USING RECYCLED MELAMINE AS A FRACTION OF AGGREGATES

3.6 EXPERIMENTAL PROCEDURE

The materials used in present study are as follows:

Ordinary Portland Cement: Type I Portland cement conforming to ASTM C150–94.

Sand: Fine aggregate is taken from natural sand. Therefore, it was used after separating by sieve in accordance with the grading requirement for fine aggregate (ASTM C33–92). Table 1 presents the properties of the sand and its gradation is presented in Fig. 1.

TABLE 1 Properties of sand and melamine aggregates.

Properties	Sand	Melamine
Density (g/cm^3)	2.60	1.574
Bulk density (g/cm^3) (Melamine)	—	0.3–0.6
Water absorption (%)	1.64	7.2
Max size (mm)	4.75	1.77
Min size (mm)	—	0.45
Sieve 200 (%)	0.24	—
Flammability (Melamine)	Nonflammable	Nonflammable
Decomposition (Melamine)	—	At > 280°C formation of NH$_3$

Thermosetting plastic: Melamine is a widely used type of thermosetting plastic. Therefore, in the present work has been selected for application in the mixed design of composite. The mechanical and physical properties of melamine are shown in Table 1. The melamine waste was ground with a grinding machine. The ground melamine waste was separated under sieve analysis. Scanning electron micrographs (SEM) of melamine aggregates is shown in Fig. 2. Those have irregular shape and rough surface texture. The grain size distributions were then plotted as shown in Fig. 1. It was observed that the gradation curve of the combination of sand and plastic aggregates after sieve number 16 meets most of the requirements of ASTM C33–92.

The melamine aggregates were been saturated surface dry. Therefore, melamine aggregates immerse in water at approximately 21°C for 24 h and removing surface moisture by warm air bopping.

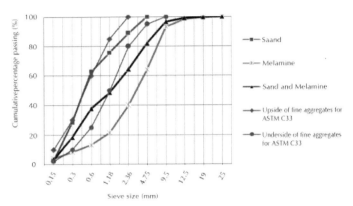

FIGURE 1 Gap-grading analysis of sand and melamine aggregates according to ASTM C33–92.

FIGURE 2 SEM photographs of materials: (up) melamine aggregates and (down) silica fume.

Experimental Techniques for Application of Recycled Polymers

Aluminum powder: In the present study, aluminum powder was selected as an agent to produce hydrogen gas (air entrainment) in the cement. This type of lightweight concrete is then called aerated concrete. The following are possible chemical reactions of aluminum with water:

$$2Al + 6H_2O \quad 2Al(OH)_3 + 3H_2 \quad (1)$$

$$2Al + 4H_2O \quad 2AlO(OH) + 3H_2 \quad (2)$$

$$2Al + 3H_2O \quad Al_2O_3 + 3H_2 \quad (3)$$

The first reaction forms the aluminum hydroxide bayerite ($Al(OH)_3$) and hydrogen, the second reaction forms the aluminum hydroxide boehmite ($AlO(OH)$) and hydrogen, and the third reaction forms aluminum oxide and hydrogen. All these reactions are thermodynamically favorable from room temperature past the melting point of aluminum (660°C). All are also highly exothermic. From room temperature to 280°C, $Al(OH)_3$ is the most stable product, while from 280–480°C, $AlO(OH)$ is most stable. Above 480°C, Al_2O_3 is the most stable product. The following equation illustrates the combined effect of hydrolysis and hydration on tricalcium silicate.

$$3CaO.SiO_2 + water \quad xCaO.ySiO_2(aq.) + Ca(OH)_2 \quad (4)$$

In considering the hydration of Portland cement it is demonstrate that the more basic calcium silicates are hydrolyzed to less basic silicates with the formation of calcium hydroxide or 'slaked lime' as a by-product. It is this lime, which reacts with the aluminum powder to form hydrogen in the making of aerated concrete from Portland cement:

$$2Al + 3Ca(OH)_2 + 6H_2O \quad 3CaO.Al_2O_3.6H_2O + 3H_2 \quad (5)$$

Hydrogen gas creates many small air (hydrogen gas) bubbles in the cement paste. The density of concrete becomes lower than the normal weight concrete due to this air entrainment.

Silica fume: In the present work, Silica fume has been used. Its chemical compositions and physical properties are being given in Tables 2 and 3, respectively. Scanning electron micrographs (SEM) of silica fume is shown in Fig. 2.

Super plasticizer: Premia 196 with a density of 1.055 ± 0.010 kg/m³ was used. It was based on modified polycarboxylate.

TABLE 2 Chemical composition of Silica fume.

Chemical composition	Silica fume
SiO_2 (%)	86–94
Al_2O_3 (%)	0.2–2
Fe_2O_3 (%)	0.2–2.5
C (%)	0.4–1.3
Na_2O (%)	0.2–1.5
K_2O (%)	0.5–3
MgO (%)	0.3–3.5
S (%)	0.1–0.3
CaO (%)	0.1–0.7
Mn (%)	0.1–0.2
SiC (%)	0.1–0.8

TABLE 3 Physical properties of Silica fume.

Items	Silica fume
Specific gravity (gr/cm^3)	2.2–2.3
Particle size (μm)	< 1
Specific surface area (m^2/gr)	15–30
Melting point (°C)	1230
Structure	Amorphous

3.6.1 MIX DESIGN

To determine the suitable composition of each material, the mixing proportions were tested in the laboratory, as shown in Table 4. In this section, the mix proportions were separated for five experimental sets. For each set, the cement and Aluminum powder contents was specified as a constant proportion. The proportion of each of the remaining materials, that is, sand, water, silica fume, aluminum powder, and melamine, was varied for each mix design.

3.6.2 EXPERIMENTAL TECHNIQUES

Mortar was mixed in a standard mixer and placed in the standard mold of 50 × 50 × 50 mm according to ASTM C109-02. In the pouring process of mortar, an

Experimental Techniques for Application of Recycled Polymers 71

expansion of volume due to the aluminum powder reaction had to be considered. The expanded portion of mortar was removed until finishing. The fresh mortar was tested for slump according to ASTM C143-03. The specimens were cured by wet curing at normal room temperature. The hardened mortar was tested for dry density, compressive strength, water absorption and voids for the curing age of 7 and 28 days. The test results for melamine, sand and water contents were reported for 7 days curing age for mix nos. 1–3, because these were very close to the results of 28 days. When silica fume was added in the latter mix nos. 4 and 5, the test results were presented for 28 days. This is because the presence of silica fume increases the duration for completion of the chemical reaction. The testing procedures of dry density, water absorption and voids were performed according to ASTM C642-97 and compressive strength was performed according to ASTM C109-02.

TABLE 4 Mix proportions of melamine lightweight composites (by weight).

Mix no.	Cement	Aluminum powder	Sand	Water	Silica fume	Melamine	*Super plasticizer*
1. Determination of melamine content (1st trial mix design)	1.0	0.004	1.0	0.35	—	1.0	—
						1.5	
						2.0	
						2.5	
						3.0	
2. Determination of sand content	1.0	0.004	1.0	0.35	—	1.0	—
			1.2				
			1.4				
			1.6				
			1.8				

TABLE 4 *(Continued)*

3. Determination of water content or water–cement ratio (w/c)	1.0	0.004	1.4	0.30	—	1.0	—
				0.35			
				0.40			
				0.45			
				0.50			
				0.55			
4. Determination of silica fume content	1.0	0.004	1.4	0.35	0.10	1.0	0.005
					0.15		0.007
					0.20		0.009
					0.25		0.012
					0.30		0.015
					0.35		0.020
5. Determination of melamine content (final mix design)	1.0	0.004	1.4	0.35	0.25	1.0	0.012
						1.2	0.011
						1.4	0.010
						1.6	0.009
						1.8	0.008
						2.0	0.007
						2.2	0.006

3.7 RESULTS AND DISCUSSION

3.7.1 MIX NUMBER 1 (DETERMINATION OF MELAMINE CONTENT FOR THE FIRST TRIAL MIX DESIGN)

Figure 3 present the variations in the compressive strength and dry density for 7 days age of mortars as a function of the value of melamine substitutes used. It can initially be seen, to increased melamine, the compressive strength and dry density of composites decreased. The reduction in the compressive strength is due to the addition of melamine aggregates or could be due to either a poor bond between the cement paste and the melamine aggregates or to the low strength that is characteristic of plastic aggregates.

TABLE 5 Specification of non-load-bearing lightweight concrete (ASTM C129).

Type	Compressive strength (MPa)	Density (kg/m³)
	Average of three unit Individual unit	
II	4.1–3.5	< 1680

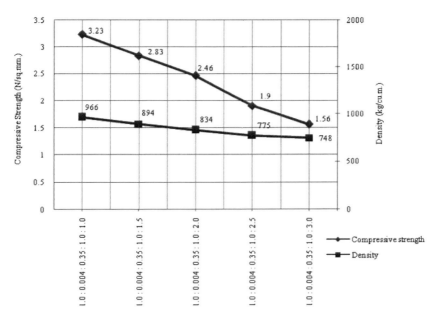

FIGURE 3 Compressive strength and density for varying melamine content (curing for 7 days).

FIGURE 4 Absorption after immersion for varying melamine content (curing for 7 days).

The absorption is an indirect parameter to examine the inside porosity of mortar. The results showed that the absorption after immersion and voids of mortar increased as the melamine content increased (see Figs. 4 and 5). Therefore, to increased melamine plastic, the inside porosity of mortar increased. This might be other reason for the reduction in the compressive strength and density.

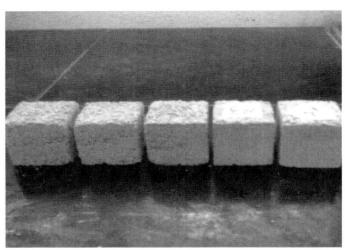

FIGURE 5 Optical photographs of samples containing varying melamine, right to left containing 1.0, 1.5, 2.0, 2.5 and 3.0 the weight percentage of melamine

3.7.2 MIX NUMBER 2 (DETERMINATION OF SAND CONTENT)

The results of compressive strength and dry density for 7 days age are shown in Fig. 6. It can be seen that a reduction of sand leads to a reduction in the strength and dry density. The compressive strength and dry density for sand content equal to or greater than 1.4 exactly satisfy the standard value.

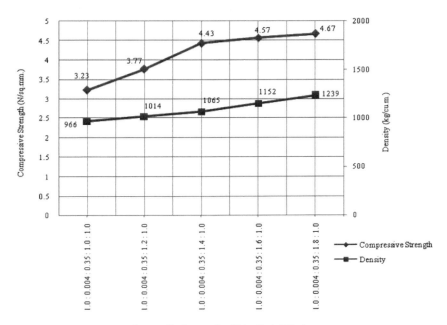

FIGURE 6 Compressive strength and density for the determination of the optimum sand content (curing for 7 days).

Figure 7 present the variations in the absorption after immersion and voids as a function of the value of sand substitutes used. The results showed that the absorption after immersion and voids of mortar decreased as the sand content increased.

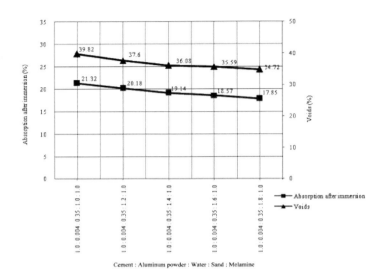

FIGURE 7 Absorption after immersion and voids for varying sand content (curing for 7 days).

3.7.3 MIX NUMBER 3 (DETERMINATION OF WATER CONTENT)

The results of compressive strength and dry density for 7 days age are shown in Fig. 8. The results showed that the compressive strength and dry density of mortar decreased as the water content increased.

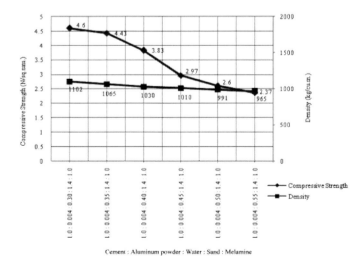

FIGURE 8 Compressive strength and density for the determination of the optimum water content (curing for 7 days).

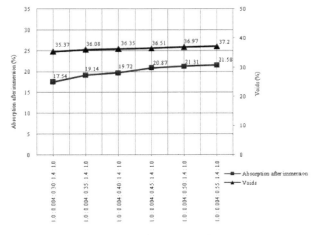

FIGURE 9 Absorption after immersion and voids for varying water content (curing for 7 days).

3.7.4 MIX NUMBER 4 (DETERMINATION OF SILICA FUME CONTENT)

Figure 10 present the variations in the compressive strength and dry density for 28 days age of mortars as a function of the value of silica fume substitutes used. It was found that the results of compressive strength for seven days age do not increase when compared with those without silica fume.

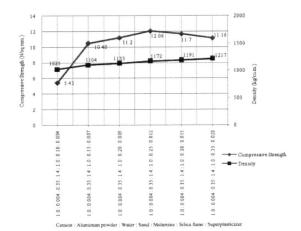

FIGURE 10 Compressive strength and density for the determination of the optimum silica fume content (curing for 28 days).

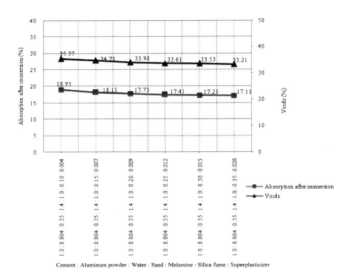

FIGURE 11 Absorption after immersion and voids for varying silica fume content (curing for 28 days).

3.7.5 MIX NUMBER 5 (DETERMINATION OF THE FINAL MELAMINE PLASTIC CONTENT)

Figure 12 presents the variations in the compressive strength and dry density for 28 days age of mortars as a function of the value of melamine substitutes used. It was found that the presence of melamine caused a reduction in the dry density and compressive strength of concretes as discussed previously. Figure 13 shows that the scanning electron microscopy analysis of composites reveals that cement paste-melamine aggregates adhesion is imperfect and weak. Therefore, the problem of bonding between plastic particles and cement paste is main reason to decrease of compressive strength. An optimum melamine content of 2.0 was selected. The results of compressive strength and dry density, which are 7.06 MPa and 887 kg /m3, are according to ASTM C129 Type II standard.

The results showed that the absorption after immersion and voids of mortar increased as the melamine content increased (see Fig. 14). Also, the structure analysis of mortars by scanning electron microscopy has revealed a low level of compactness in mortars when the value of melamine plastic increased (see Fig. 15). It was confirmed that to increased melamine plastic, the inside porosity of mortar increased.

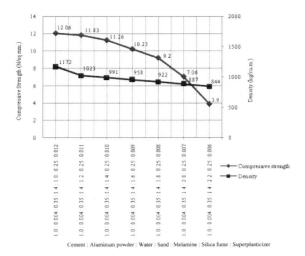

FIGURE 12 Compressive strength and density for the determination of the optimum melamine plastic content (curing for 28 days).

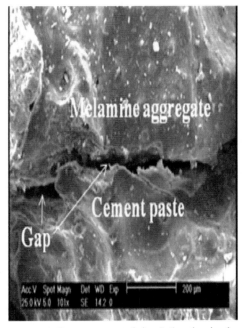

FIGURE 13 Microstructure of concrete containing 1.6 melamine by weight of cement, as obtained using SEM (enlargement: 101×).

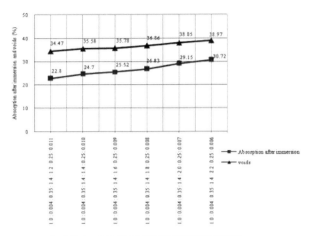

Cement : Aluminum powder : Water : Sand : Melamine : Silica fume : Superplasticizer

FIGURE 14 Absorption after immersion and voids for varying melamine plastic content (curing for 28 days).

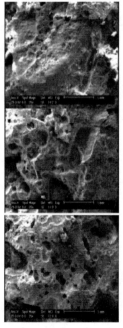

FIGURE 15 Scanning electron micrographs of various mortars containing melamine plastic aggregates. (a) 1.2 by weight of cement (enlargement: 25×). (b) 1.6 by weight of cement (enlargement: 25×). (c) 2.0 by weight of cement (enlargement: 25×).

3.7.6 COMPARISON RESEARCH FINDINGS ON THE USE OF WASTE PLASTIC IN CONCRETE

The results of this study are in a perfectly agreement with the other research findings on the use of waste plastic in concrete (see Table 6).

SECTION 4. PRACTICAL HINTS AND UPDATE ON PERFORMANCES OF CEMENT-STABILIZED SOIL USING RECYCLED TEXTILE FIBERS AS A FRACTION OF AGGREGATES

3.8 EXPERIMENTAL PROCEDURE

The soil samples used in the present study were obtained from the northern region sites of Iran. The soil is classified as clayey sand (SC) according to unified soil classification system. The general characteristics of this type of soil could be seen in Table 1. The standard penetration test was used to obtain the soil information at site. The test is carried out according to ASTM D1586 and the derived number commonly called SPT numbers of the tested samples can be seen in Table 1.

TABLE 1 Characteristics of clayey sand.

Specific Gravity (gr/cm^3)	SPT number	% Finer than 0.002 mm	Plastic Limit	Liquid Limit
1.45–1.65	15–30	15–30%	20–24	37–45

The cement used in this study is ordinary Portland cement that its physical and chemical properties are given in Table 2 and the waste fibers are Polyamide and Acrylic, which their characteristics are given in Table 3. The waste fibers are cut such that the lengths are 8 mm (0.315 inch).

TABLE 2 Physical and chemical properties of the cement.

Physical properties	Cement
Fineness	3.12
Chemical composition	
Silica (SiO$_2$)	20.44%
Alumina (Al$_2$O$_3$)	5.5%
Calcium oxide (CaO)	64.86%
Potash (K$_2$O)	22.31%
Magnesia (MgO)	1.59%
Loss on ignition	1.51%

TABLE 2 *(Continued)*

Physical properties	Cement
PH	12.06
$3CaO \cdot SiO_2$	66.48%
$2CaO \cdot SiO_2$	10.12%
$4CaO \cdot Al_2O_3 \cdot Fe_2O_3$	9.43%
Free lime	1.65%
$3CaO \cdot Al2O3$	8.06%

TABLE 3 Mechanical and physical properties of polyamide and Acrylic.

Fiber	Water Absorption	Shear Modulus	Tensile Strength	Flexural Strength	Compressive Yield Strength
Acrylic	0.3–2%	203 ksi	7,980–12,300 psi	11,700–20,000 psi	14,500–17,000 psi
Polyamide	2.1–4%	8,560 psi	2,180–12,300 psi	13,100–15,200 psi	2,470 psi

The specimens were prepared according to the standard definitions with three different cement contents (5%, 7% and 9%) and two kinds of fiber polymers (Polyamide and Acrylic). The reinforced samples were prepared with three different fiber contents of 0.1%, 0.2% and 0.3%. The specimens are prepared with maximum dry density and optimum moisture content. Testing was performed on specimens with fibers distributed uniformly in space and with as much as possible uniform distribution of fiber orientation in all directions. Such a distribution requires an elaborate technique for preparation of the specimens. This technique included a five-layer specimen construction. As the fibers tend to assume a close-to-horizontal orientation during mixing with dry soil, a specially designed tool was used for reorienting the fibers in the specimen before compaction.

For each test and period of curing, three specimens were prepared. The specimens were tested after curing in a moist chamber.

3.9 EXPERIMENTAL TESTS

In this study, compaction, unconfined compression, indirect tensile, flexure, direct shear and durability tests were carried out on reinforced and non-reinforced samples. The testing procedures used are described in table 4. Flexure tests were carried with the strain rate of 0.2mm/min on the rectangular specimens with dimension of 51×51×150 mm.

TABLE 4 Testing procedures.

Test type	Test procedure
Compaction test	ASTM D558–96
Unconfined compressive strength	ASTM D1633
Indirect tensile test	ASTM C496
Flexure tests	ASTM D1635–95
Direct shear tests	ASTM D3080–90
Durability tests	ASTM D 559–93

3.10 RESULTS AND DISCUSSIONS

3.10.1 THE EFFECT OF ADDING RECYCLED TEXTILE FIBERS ON COMPACTION

The maximum dry density (MDD) and optimum water content (OWC) vs. cement and fiber contents curves obtained in compaction tests are given in Figs. 1 and 2.

As shown in Fig. 1, the percent of optimum water content increases as the cement and fiber content increases. The increase of the cement content will result in an increment in the fine-grained content of the soil. Therefore, the specific surface of the soil would increase, and consequently much more water is needed to reach to the maximum dry density.

On the other hand, added polymer fibers to the soil-cement mixtures, will absorb a portion of the water content and correspond to the increase of fiber polymers percentage, the amount of water absorption will increase. Consequently, greater amount of water is needed to provide the OWC of the mixture.

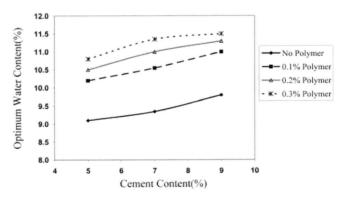

FIGURE 1 Optimum water content variation.

From Fig. 2 it can be concluded that in the specimens stabilized with 5% cement and reinforced with fibers, those specimens with 0.1% fibers, have the maximum dry density. Increasing the fibers content to 0.2% and 0.3%, the maximum dry density decreases. The maximum dry density in the reinforced specimens with 0.3% of fibers is less than those of non-reinforced ones. Moreover, in the stabilized specimens with 7% and 9% of cement, correspond to the increases of fibers up to 0.2%, maximum dry density increases. The greater amount of fibers content of 0.2%, the maximum dry density decreases. Furthermore, It is deduced from the Fig. 2 that the amount of maximum dry density in the stabilized specimens with 7% and 9% cement and reinforced with 0.3% fibers, have greater values in comparison with those of non-reinforced.

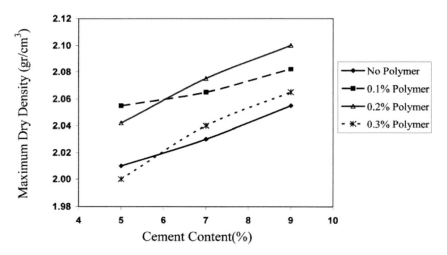

FIGURE 2 Maximum dry density variation with cement content.

3.10.2 THE EFFECT OF ADDING FIBERS ON UCS

Figure 3 shows the variations of maximum UCS (q_u) of the stabilized clayey sand-cement which were reinforced with fibers, and that of non-reinforced ones cured after 7 and 28 days. The experimental evident indicate that, in the stabilized specimens with 5% cement and reinforced with fiber polymers, the maximum value of q_u occurs for the reinforced specimens with 0.1% of fibers. In the stabilized specimens with 7% and 9% of cement, reinforced with fiber polymers, the q_u, in both of polymers, occurs in the specimens, which have 0.2% fibers. Therefore, in the compressive test, when the percent of cement changes, the percent of optimum fibers changes. By considering the following diagrams and the effect of polyamide and Acrylic on q_u, it can be seen that there is no significant difference between

q_u of the soil-cement mixtures reinforced with polyamide and acrylic. Because the non-uniform distribution of the fibers in the soil-cement mixture, the existing differences can be the result of the distribution quality of the fibers. Studies carried out by many researchers described that the variation of q_u, with increase of cement is linear. In the case of soil-cement specimens, not only this behavior can be observed but also all reinforced soil-cement specimens with polymers were showed such behavior (Fig. 3).

Based on the results of unconfined compression tests shown in Figs. 2 and 3, it can be concluded that the increment of UCS by using fiber reinforcement on specimens containing less than 5% of cement content is more considerable. The reason for this phenomenon can be explained by the fact that the UCS value of specimens containing more than 5% of cement content is more affected by cement than by fibers, while the fiber-reinforcement effect on improving UCS value of specimens containing less than 5% of cement is more significant. Comparing the results of compressive strength and maximum dry density of the reinforced and non-reinforced specimens showed that there is a direct relation between maximum UCS and maximum dry density. The stabilized and reinforced specimens with 5% of cement, the highest compressive strength and maximum dry density refers to the reinforced specimens which have 0.1% of fibers, while in the stabilized and reinforced specimens with 7 and 9% of cement, the highest compressive strength and maximum dry density refers to the specimens which have 0.2% of fibers.

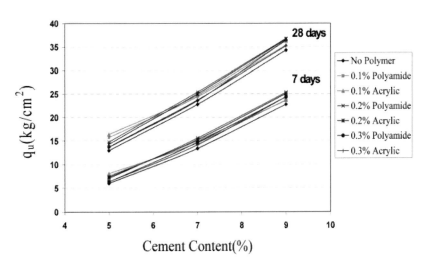

FIGURE 3 The variation of qu versus cement percentage.

3.10.3 THE EFFECT OF ADDING RECYCLED TEXTILE FIBERS ON AXIAL STRESS – STRAIN BEHAVIOR

Before performing the test, surfaces of specimens are covered with a thin layer of kaolinite powder (specimen capping) so that a smooth surface is obtained for uniform distribution of stress. In most of curves an initial reduction is evident after which curves takes the normal shape, which is due to kaolinite layer compression that lead to specimen strain. Stress-strain characteristics of cement-stabilized soil are very important to determine the behavior of these material used in pavement construction under repeated loading. Stress-strain behavior of cement-stabilized soil is nonlinear but for low stresses and also for limited loading unloading, this behavior may be idealized linear. Thus, Stress-strain curves must be corrected to determine their behavior. To do this, linear section must be extended until it intersects the strain curve. All strains are then subtracted as much as origin strain. Resulting curve is named modified stress-strain curve. Figs. 4 and 5 show the real and corrected path for a series of tests.

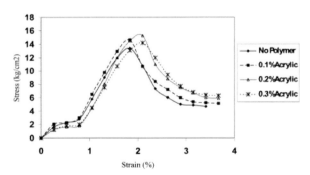

FIGURE 4 Real (non-corrected) stress-strain curve.

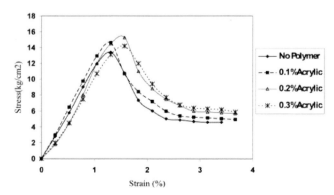

FIGURE 5 Corrected stress-strain curve.

Experimental Techniques for Application of Recycled Polymers

The corrected curve of axial stress-strain behavior of stabilized and reinforced soil with different percents of cement and fibbers after 7 and 28 days' curing are shown in Figs. 6 to 11. The results indicate that adding 0.1% of fiber to the stabilized soil increases maximum compressive strength. The strain related to the maximum stress of the fiber-reinforced specimens will grow by using greater amount of polymer content, which increases the ductility of the specimens. This subject confirms the last research in the field of stabilizing and reinforcing soil by cement and polymer, and shows the effect of fibers in making soil-cement ductile.

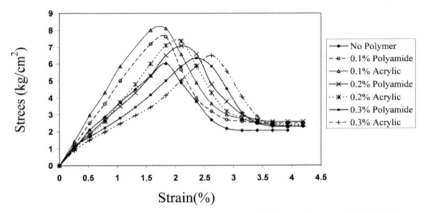

FIGURE 6 The variation of stress-strain behavior of the specimen with 5% of cement and different percents of polymer after 7 days' curing.

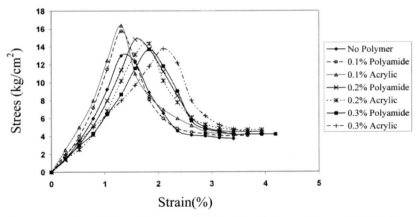

FIGURE 7 The variation of stress-strain behavior of the specimen with 5% of cement and different percents of polymer after 28 days' curing.

Evaluating the experimental evident, the residual strength in all reinforced specimens is more than that of non-reinforced specimens. The only exception is about the stabilized specimens with 5% of cement that are reinforced with 0.2% of fibers. In this case, the amount of residual strength decreases with the increase of fibers content.

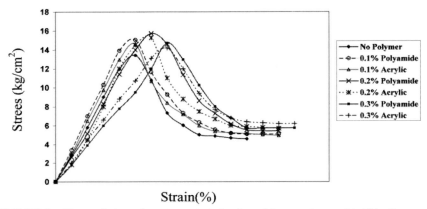

FIGURE 8 The variation of stress-strain behavior of the specimen with 7% of cement and different percents of polymer after 7 days' curing.

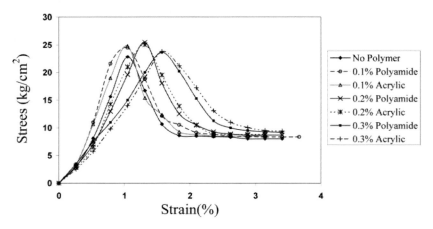

FIGURE 9 The variation of stress-strain behavior of the specimen with 7% of cement and different percents of polymer after 28 days' curing.

However, in the stabilized specimens with 7% and 9% of cement, the amount of maximum residual strength belongs to the reinforced specimens with 0.3% fibers. Generally, the reason of greater residual strength of the reinforced specimens with fibers in comparison with that of non-reinforced is that existing fibers in the soil-cement mixture prohibit the complete failure of the composite structure after reaching to maximum bearable stress of soil. In addition, adding fibers improve the load bearing capacity of the specimens. In the other word, the reinforced specimen undergoes more stress before failure in comparison with the non-reinforced specimens.

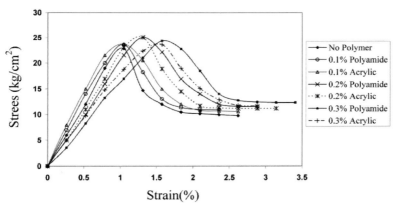

FIGURE 10 The variation of stress-strain behavior of the specimen with 9% of cement and different percents of polymer after 7 days' curing.

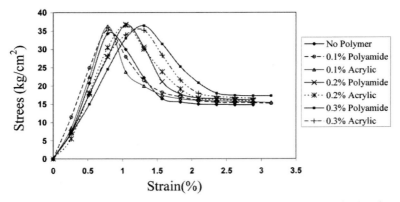

FIGURE 11 The variation of stress-strain behavior of the specimen with 9% of cement and different percents of polymer after 28 days' curing.

3.10.4 THE EFFECT OF RECYCLED TEXTILE FIBERS ON MODULUS OF ELASTICITY

Figures 6–11 show the stress-strain properties of the stabilized and reinforced specimens with different percents of cement and fibers. Secant modulus can be used as the value of elastic modulus of geomaterials and concrete. The calculating method of secant modulus at 50% of compressive strength (E_{50}) has been used to calculate modulus of elasticity. Secant modulus (E_{50}) of the stabilized specimens with different percent of cement and fibers were calculated and shown in Figs. 12 and 13. The results showed that using more cement content and curing period will increase the E_{50} value of the mixtures. The E_{50} value of reinforced soil-cement mixes will increase by using up to 0.1% of fiber content in comparison with non-reinforced specimens, while using more than 0.1% of fiber content will decrease the E_{50} to an amount less than non-reinforced specimens.

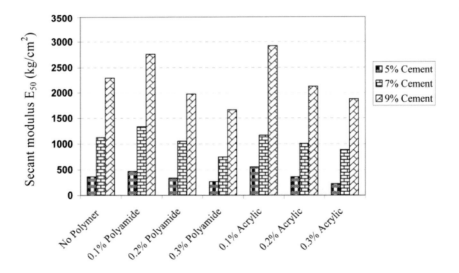

FIGURE 12 The secant modulus of the specimens after 7 days' curing.

Experimental Techniques for Application of Recycled Polymers

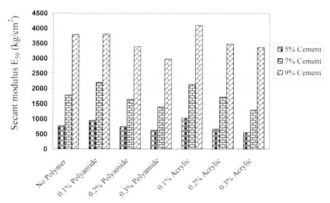

FIGURE 13 The secant modulus of the specimens after 28 days' curing.

The variations of secant modulus at 50% of compressive strength (E_{50}) based on the maximum UCS of the reinforced and non-reinforced cement-stabilized soils are shown by a power function in Fig. 14. Following numerical E–q equations were drawn for different mixtures:

E_{50} (non-reinforced) = 29.96 $q_u^{1.35}$
E_{50} (Reinforced with 0.1% polymer) = 31.467 $q_u^{1.349}$
E_{50} (Reinforced with 0.2%, 0.3% polymer) = 18.836 $q_u^{1.4463}$

Using the above equations it can be seen that adding 0.1% of fibers to the soil-cement, increase the secant modulus (E50) between 2.1% and 5.6%, while adding 0.2% and 0.3% of fibers decrease the secant modulus (E50) between 11.5% and 26.8%.

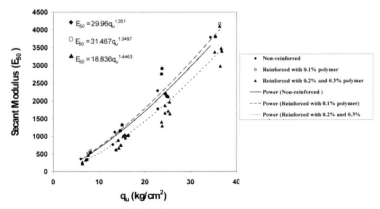

FIGURE 14 Relation between E_{50} and q_u.

3.10.5 THE EFFECT OF RECYCLED TEXTILE FIBERS ON INDIRECT TENSILE STRENGTH

The results of fiber-reinforcement on indirect tensile strength (ITS) of the specimens are shown in Figs. 15 and 16. It can be concluded from these figures that, first, tensile strength increase with the increase of cement percentage in the mixture and then the tensile strength in all of the fiber-reinforced specimens is more then that of non-reinforced specimens. In addition, with the increase of fiber polymers, the tensile strength increases by a constant percent of cement. From the Figs. 15 and 16, it is clear that the specimens with 0.3% of fibers have the greatest tensile strength.

Generally, these results show that, these two kinds of polymers are suitable to improve tensile strength. The different effects of polyamide and acrylic on the tensile strength of soil-cement mixture are, also, related to the distributing way of fibers in the mixture.

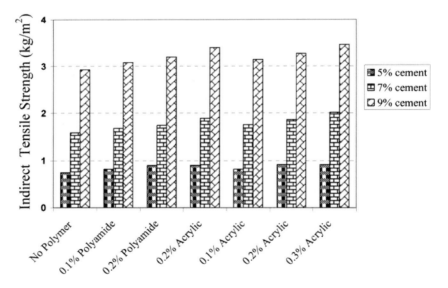

FIGURE 15 Comparison between tensile strength for reinforced and non-reinforced specimens after 7 days' curing.

Experimental Techniques for Application of Recycled Polymers 93

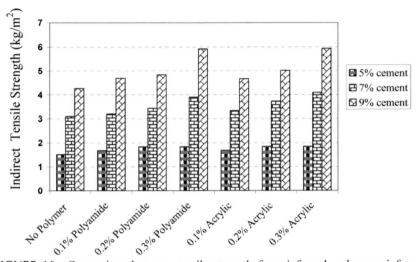

FIGURE 16 Comparison between tensile strength for reinforced and non-reinforced specimens after 28 days' curing.

Relation between compressive and tensile strengths for stabilized reinforced and non-reinforced soil specimens are shown in Figs. 17 (a and b). Many researchers proposed that tensile strength of cement-stabilized embankment is about 8% of unconfined compressive strength (Fig. 17a). As it can be seen the ratio of tensile strength to compressive strength for non-reinforced clayey sand specimens is about 12% but this ratio is approximately 15% for polymeric fiber mixed specimens. In this research it can be concluded that effect of fiber on tensile strength is greater compared to compressive strength.

The pore size distribution in a fibrous material has significant impact on the mixture transport process. These distributions are clearly shown in Fig. 18. Pore size distribution can influence the spontaneous uptake of liquids (i.e., moisture content) and therefore the strength keeps on increasing. It should be noted that the amount of pores within the fibers could have significant effects on the strength improvement as well. The greater the pores, the more moist aggregates the fibers can hold.

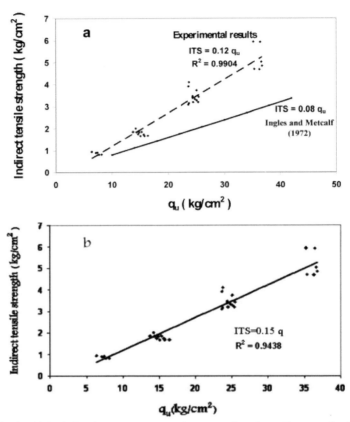

FIGURE 17 (a) Relation between compressive strength and tensile strength of cement-stabilized soil specimens, non-reinforced. (b) Relation between compressive strength and tensile strength of cement-stabilized soil specimens-reinforced.

FIGURE 18 Pore size distribution within the polyamide fibers and tensile failure mode in an indirect tensile test.

Figure 19 shows the crack distribution adjacent to the polyamide fibers. It is notable that the cracks did not spread out in the direction where the fibers are located.

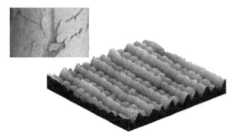

FIGURE 19 Optical micrograph of cut-polyamide fibers.

3.10.6 THE EFFECT OF RECYCLED FIBERS ON MODULUS OF RUPTURE (M_R)

The result of the flexure tests, for the specimens after 7 and 28 days' curing, are shown respectively in Figs. 20 and 21. As it is expected, the MR increases with the increase of fibers content. In addition, it is revealed that, first, in a constant percent of cement; all fiber-reinforced specimens exhibit greater MR value than non-reinforced ones. Second, by the increase of fiber's percentage MR increases. As it is illustrated in theses figures, the reinforced specimens with 0.3% of fibers possess the highest MR and tensile strength; it should be noted that adding more than a specific amount of fibers content increase the porosity of the soil. As a result, it causes MR and tensile strength to decrease. Therefore, it is possible that, increasing more percents of fibers for preparation of specimens causes MR and tensile strength to decrease in comparison with non-reinforced ones.

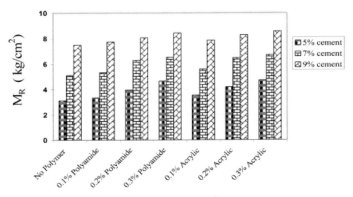

FIGURE 20 M_R for the specimens after 7 days' curing.

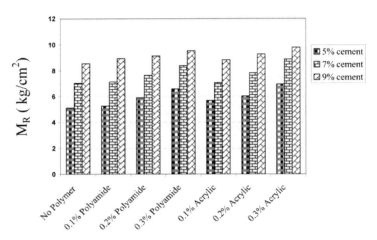

FIGURE 21 M_R for the specimens after 28 days' curing.

3.10.7 THE EFFECT OF FIBERS ON BEHAVIOR FLEXURAL LOAD-DEFORMATION

Flexural behavior of the load-displacement with different percents of cement, for the cured specimens after 7 and 28 days' curing is shown in Fig. 22 from (a) to (c) and Fig. 24 from (a) to (c). It can be seen that for all of the specimens, before the first cracking, an approximate linear relation creates within the load-displacement that coincides with the maximum load bearing of the specimens. On the basis of the experimental results it could be concluded that all of the reinforced specimens exhibit more displacement and will not exactly rupture after undergoing maximum load bearing in spite of non-reinforced specimens. In other word, deviation is seen in the direction of axial displacement. The presence of fibers in the soil-cement mixture increases the flexural strength of the soil-cement-fiber compound in comparison with the non-reinforced specimens. The main effect of fibers appears as the soil reaches to its own maximum strength. Presence of fibers will delay the rupture of the specimen. Also, the increase of displacement with the applied load is resultant of the polymer's tension, and finally the rupture of the specimen due to the maximum bearable stress is the result of the polymer rupture. This could be a result of the bridge effect of fibers used in the specimens. Generally, the presence of fibers in the soil-cement mixture causes mixture's ductility to increase and its behavior changes from fragile state to flexible behavior.

Experimental Techniques for Application of Recycled Polymers

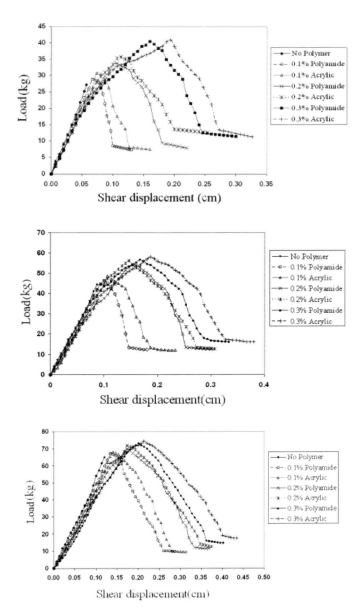

FIGURE 22 (a) Flexural load-displacement of the specimens with 5% cement after 7 days' curing. (b) Flexural load-displacement of the specimens with 7% cement after 7 days' curing. (c) Flexural load-displacement of the specimens with 9% cement after 7 days' curing.

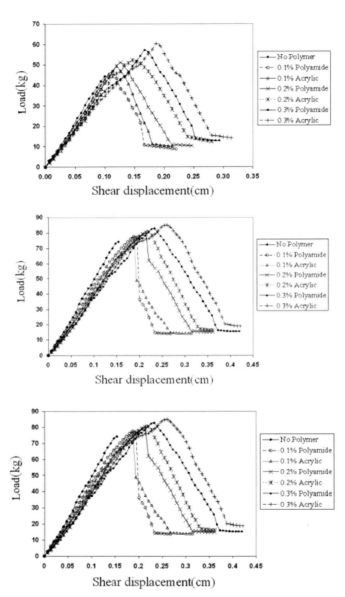

FIGURE 23 (a) Flexural load-displacement of the specimens with 5% cement after 28 days' curing. (b) Flexural load-displacement of the specimens with 7% cement after 28 days' curing. (c) Flexural load-displacement of the specimens with 9% cement after 28 days' curing.

3.10.8 THE EFFECT OF TEXTILE FIBERS ON FRACTURE ENERGY

The area under the load-deformation curves up to failure (approximately point of zero load) divided by the specimen's cross sectional area, is a measure of the energy absorption capacity or the toughness of the composite, and it is sometimes termed the fracture energy. The average fracture energy of specimens after 7 and 28 days curing are shown in Figs. 24 and 25. It can be concluded that the amount of fracture energy increases with the increase of cement percentage and, for a constant percent of cement, with the increase of fibers, it increases as well.

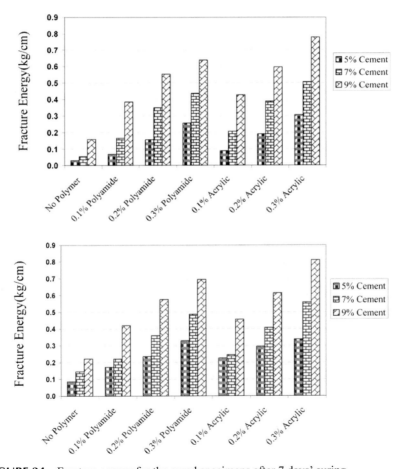

FIGURE 24 Fracture energy for the cured specimens after 7 days' curing.

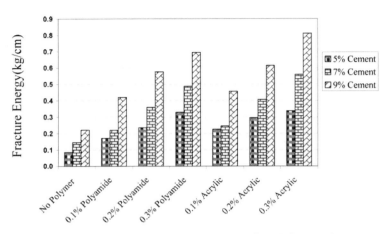

FIGURE 25 Fracture energy for the cured specimens after 28 days' curing.

On the basis of experimental results for the reinforced and non-reinforced soil-cement specimens containing different amount of cement and fibers with two different curing period, fracture energy for the specimens containing 5% of cement, by the effect of fibers, is more than that in the specimens which contain 7% of cement.

The effect of fibers in the specimens containing 7% of cement is more significant than that of the specimens containing 9% of cement. This phenomenon indicates the superiority of fiber polymers in the low percent of cement.

3.10.9 THE EFFECT OF RECYCLED TEXTILE FIBERS ON SHEAR STRENGTH

Shear strength test have been conducted on reinforced and non-reinforced soil-cement specimens containing different percents of fibers content. The applied perpendicular stress $1 kg/cm^2$ ($100 KN/m^2$) has been chosen in all tests. The shear stress-strain diagram for the specimens that contain 5% and 9% of cement are shown in Figs. 26 and 27, respectively.

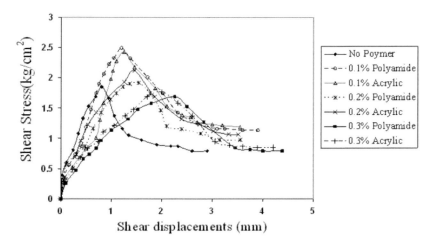

FIGURE 26 Shear stress-strain behavior of the cured specimen with 5% cement in 7 days' curing.

Figure 26 shows that, cement-stabilized specimens with 5% cement and reinforced with 0.1% of fibers have the highest shear strength, and by increasing the fibers content, shear strength decreases. This occurs in a way that the shear strength in the stabilized specimens with 5% of cements and reinforced with 0.1% and 0.2% of fibers is more than that of non-reinforced specimens. On the other hand, the shear strength of specimens containing 0.3% of fibers, is less than that of non-reinforced ones. The residual strength of reinforced specimens is more than that of non-reinforced specimens. It is readily observed from Figs. 26 and 27 that the overall soil behavior is significantly influenced by the investigated variables. Peak strength, stress-strain behavior and residual response are changes as s consequence of either the separate or the joined effects of fiber and cement inclusions. Shear strength and stress-strain behavior (stiffness) are dramatically increased by increasing the cement content for each percent of fiber contents. By using more fiber content, a moderate increase in shear strength is accompanied by a rigidity loss; the residual strength is increased and the volumetric response becomes more compressive in the early stages of loading and less expansive afterwards. For the specimens containing 0.1% of fibers the residual strength have a greater value in comparison with other specimens. Adding fibers increase the strain related to the maximum shear strength of the specimens, which results in a more ductile specimen. This ductility increases with the increase percent of fibers.

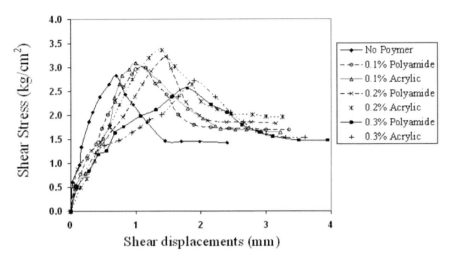

FIGURE 27 Shear stress-strain behavior of the cured specimen with 9% cement after 7 days' curing.

It can be drawn from Fig. 27 that, in the stabilized specimens with 9% of cement, adding fibers up to 0.2% of fibers content increases the shear strength to its maximum value while adding more fibers decrease the shear strength. Therefore, the shear strength of the stabilized specimens with 9% of cements and reinforced with 0.3% of fibers is less than that of non-reinforced specimens.

Also, in the stabilized specimens with 5% and 9% of cement and reinforced with fibers, the residual strength in all reinforced specimens is more than that of non-reinforced.

3.10.10 THE EFFECT OF RECYCLED TEXTILE FIBERS ON DILATION

The mechanical properties of stabilized samples in shear box (for condition of unsaturated and at optimum moisture) are similar to soft rocks and practically their dilation is a consequence of contact of teeth generated over the shear surface of samples. It is important that the numerical rates of dilation depend on amounts of the normal loads, used in direct shear tests. Dilation in direct shear test distinguishes by monitoring of the vertical displacement, among the horizontal displacement of soil shear samples. In this study, all of the vertical displacements were positive, namely volume of all samples was increased. The effect of fibers on dilation of the cured specimens after 7 days' curing is shown in Fig. 28.

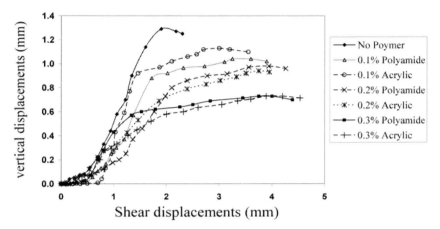

FIGURE 28 Influence of adding fibers on dilation of shear samples content of 5% cement after 7 days' curing.

It can be seen in this figure that the dilation in reinforced specimens is less than that of non-reinforced specimens and, their amount dilation decrease with the increase of fibers. This phenomenon could be the result of this matter that using fibers make a smoother shear surface in comparison with non-reinforced specimens.

3.10.11 EFFECT ON DURABILITY

The durability of stabilized soil is primarily evaluated from the degree to which the engineering properties are retained. The durability concern may arise from the exposure to wetting and drying, freezing and thawing, sulphate attack, etc. Although dominating exposure condition(s) may vary geographically, such exposures are common and, should therefore, be considered as a part of the design procedure. During construction, soil cement is compacted to a high density. As the cement hydrates, the soil cement mixture hardens in this dense state and becomes a slab-like structural material. Soil cement can bridge over small, local weak subgrade areas. If the appropriate freeze-thaw and wet-dry soil cement design criteria have been satisfied, the soil cement will not rut or shove during spring thaws and will be minimally affected by water or, freezing and thawing. Compatibility is related to the pore structure of a specimen. When a stabilized soil specimen is placed in water or dried in an oven, water penetrates into or leaves the specimen through the interconnected pore structure. It is also known that the surface tension of water is high enough to generate substantial capillary pressure to cause cracking of the surrounding matrix, especially for stabilized soils with very little tensile strength. However, to add fiber polymers, as reinforced factor, prohibits the small crack to develop and cause durability of the sol-cement specimens to increase.

Tables 5 and 6 show the rates of decreasing weight of the specimens after wet and dry cycles of durability test. The decreasing weight of specimens with 5% of cement curing after 7 days do not satisfy the criterion of the Table 4, but as it is, first, the existing fiber causes the weight loss of specimens to decrease and, second, the decreasing weight of cured specimens satisfies the criteria mentioned in Table 4. Generally, the existing fibers improve the properties of the durability of specimens and decrease the weight loss of specimens.

TABLE 5 The rate of decreasing weight after cycles of durability test for cured specimens in 7 and 28 days' curing.

	5%	7%	9%	5%	7%	9%
	7 days			28 days		
No Polymer	27.64%	4.68%	1.87%	6.90%	2.18%	0.66%
0.1% Polyamide	19.37%	4.00%	1.61%	6.50%	1.96%	0.60%
0.2% Polyamide	21.99%	3.35%	1.44%	6.65%	1.84%	0.54%
0.3% Polyamide	24.51%	4.20%	1.56%	6.83%	2.02%	0.62%
0.1% Acrylic	18.84%	4.27%	1.60%	6.56%	1.89%	0.58%
0.2% Acrylic	20.86%	3.54%	1.51%	6.78%	1.79%	0.55%
0.3% Acrylic	23.37%	4.29%	1.67%	6.87%	2.05%	0.62%

TABLE 6 The rate of decreasing weight after cycles of durability test for cured specimens in 90 and 180 days' curing.

	5%	7%	9%	5%	7%	9%
	90 Days			180 Days		
No Polymer	4.23%	1.20%	0.46%	2.62%	0.78%	0.26%
0.1% Polyamide	3.83%	1.18%	0.41%	2.53%	0.70%	0.21%
0.2% Polyamide	3.95%	0.94%	0.39%	2.35%	0.64%	0.19%
0.3% Polyamide	4.19%	1.16%	0.44%	2.59%	0.76%	0.24%
0.1% Acrylic	3.95%	1.16%	0.42%	2.55%	0.75%	0.22%
0.2% Acrylic	4.08%	1.00%	0.40%	2.28%	0.68%	0.20%
0.3% Acrylic	4.20%	1.18%	0.45%	2.40%	0.76%	0.25%

3.11 CONCLUDING REMARKS

The introduction and development of advanced composite material opened the door to new and innovative application in civil and structural engineering [1–11]. Key points of this investigation are to evaluate the application of recycled materials in concrete composites and:
- To convert waste into useful product.
- To consume wastes; this would otherwise go to landfill.
- To protect the environment from being heavily contaminated.

KEYWORDS

- **composite materials**
- **construction industries**
- **environment**
- **landfill**
- **recycled polymers**
- **waste conversion**

REFERENCES

1. Albano, C., Camacho, N., Hernandez, M., Matheus, A., Gutierrez. A., 2009. Influence of content and particle size of waste pet bottles on concrete behavior at different w/c ratios. Waste Management 29, 2707–2716.
2. Al-Salem, S.M., Lettieri, P., Baeyens, J., 2010. The valorization of plastic solid waste (PSW) by primary to quaternary routes: From re-use to energy and chemicals. Progress in Energy and Combustion Science 36, 103–129.
3. Chan, Y.N.S, Ji, X, 1999. Comparative study of the initial surface absorption and chloride diffusion of high performance zeolite, silica fume and PFA concretes. Cement and Concrete Composites 21 (1999) 293–300.
4. Choi, Y.W., Moon, D.J., Kim, Y.J., Lachemi, M., 2009. Characteristics of mortar and concrete containing fine aggregate manufactured from recycled waste polyethylene terephthalate bottles. Construction and Building Materials 23, 2829–2835.
5. Choi, Y.W., Moon, D.J., Chung, J.S., Cho, S.K., 2005. Effects of waste PET bottles aggregate on the properties of concrete. Cement and Concrete Research 35, 776–781.
6. Ismail, Z.Z., AL-Hashmi, E.A., 2008. Use of waste plastic in concrete mixture as aggregate replacement. Waste Management 28, 2041–2047.
7. Marzouk, O.Y., Dheilly, R.M., Queneudec, M., 2007. Valorization of post-consumer waste plastic in cementations concrete composites. Waste Management 27, 310–318.
8. Naik, T.R., Singh, S.S., Huber, C.O., Brodersen, B.S., 1996. Use of postconsumer waste plastics in cement-based composites. Cement and Concrete Research 26 (10), 1489–1492.

9. Panyakapo, P., Panyakapo M., 2008. Reuse of thermosetting plastic waste for lightweight concrete. Waste Management 28, 1581–1588.
10. Rao, G.A., 2003. Investigations on the performance of silica fume incorporated cement pastes and mortars. Cement and Concrete Research 33 (2003) 1765–1770.
11. Siddique, R., Khatib, J., Kaur, I., 2008. Use of recycled plastic in concrete: A review. Waste Management 28, 1835–1852.

CHAPTER 4

NANOCOMPOSITES POLYETHYLENE/ORGANOCLAY ON SUPRASEGMENTAL LEVEL

G. V. KOZLOV, K.S. DIBIROVA, G. M. MAGOMEDOV, and G. E. ZAIKOV

CONTENTS

4.1 Introduction 108
4.2 Experimental 109
4.3 Results and Discussion 109
4.4 Conclusions 113
Keywords 113
References 113

4.1 INTRODUCTION

The structural mechanism of polymer nanocomposites filled with organoclay on suprasegmental level was offered. Within the frameworks of this mechanism nanocomposites elasticity modulus is defined by local order domains (nanoclusters) sizes similarly to natural nanocomposites (polymers). Densely packed interfacial regions formation in nanocomposites at nanofiller introduction is the physical basis of nanoclusters size decreasing.

Very often a filler (nanofiller) is introduced in polymers with the purpose of the latter stiffness increase. This effect is called polymer composites (nanocomposites) reinforcement and it is characterized by reinforcement degree E_c/E_m (E_n/E_m), where E_c, E_n and E_m are elasticity moduli of composite, nanocomposite and matrix polymer, accordingly. The indicated effect significance results to a large number of quantitative models development, describing reinforcement degree: micromechanical [1], percolation [2] and fractal [3] ones. The principal distinction of the indicated models is the circumstance, that the first ones take into consideration the filler (nanofiller) elasticity modulus and the last two – don't. The percolation [2] and fractal [3] models of reinforcement assume, that the filler (nanofiller) role comes to modification and fixation of matrix polymer structure. Such approach is obvious enough, if to account for the difference of elasticity modulus of filler (nanofiller) and matrix polymer. So, for the considered in the present paper nanocomposites low density polyethylene/Na$^+$-montmorillonite the matrix polymer elasticity modulus makes up 0.2 GPa [4] and nanofiller – 400–420 GPa [5], that is, the difference makes up more than three orders. It is obvious, that at such conditions organoclay strain is equal practically to zero and nanocomposites behavior in mechanical tests is defined by polymer matrix behavior.

Lately it was offered to consider polymers amorphous state structure as a natural nanocomposite [6]. Within the frameworks of cluster model of polymers amorphous state structure it is supposed, that the indicated structure consists of local order domains (clusters), immersed in loosely-packed matrix, in which the entire polymer free volume is concentrated [7, 8]. In its turn, clusters consist of several collinear densely packed statistical segments of different macromolecules, i.e. they are an amorphous analogue of crystallites with stretched chains. It has been shown [9], that clusters are nanoworld objects (true nanoparticles-nanoclusters) and in case of polymers representation as natural nanocomposites they play nanofiller role and loosely packed matrix-nanocomposite matrix role. It is significant that the nanoclusters dimensional effect is identical to the indicated effect for particulate filler in polymer nanocomposites – sizes decrease of both nanoclusters [10] and disperse particles [11] results to sharp enhancement of nanocomposite reinforcement degree (elasticity modulus). In connection with the indicated observations the question arises: how organoclay introduction in polymer matrix influences on nanoclusters size and how the variation of the latter influences on nanocomposite elasticity modulus value. The purpose of the present paper is these two problems solution on the example of nanocomposite linear low-density polyethylene/Na⁺-montmorillonite [4].

4.2 EXPERIMENTAL

Linear low density polyethylene (LLDPE) of mark Dowlex-2032, having melt flow index 2.0 g/10 min and density 926 kg/m³, that corresponds to crystallinity degree of 0.49, used as a matrix polymer. Modified Na⁺-montmorillonite (MMT), obtained by cation exchange reaction between MMT and quaternary ammonium ions, was used as nanofiller MMT contents makes up 1–7 mass % [4].

Nanocomposites linear low-density polyethylene/Na⁺-montmorillonite (LLDPE/MMT) were prepared by components blending in melt using Haake twin-screw extruder at temperature 473 K [4].

Tensile specimens were prepared by injection molding on Arburg Allounder 305-210-700 molding machine at temperature 463 K and pressure 35 MPa. Tensile tests were performed by using tester Instron of the model 1137 with direct digital data acquisition at temperature 293 K and strain rate ~3.35 × 10⁻³ s⁻¹. The average error of elasticity modulus determination makes up 7%, yield stress – 2% [4].

4.3 RESULTS AND DISCUSSION

For the solution of the first from the indicated problems the statistical segments number in one nanocluster n_{cl} and its variation at nanofiller contents change should be estimated. The parameter n_{cl} calculation consistency includes the following stages. At first the nanocomposite structure fractal dimension d_f is calculated according to the equation [12]:

$$d_f = (d-1)(1+\nu), \qquad (1)$$

where d is dimension of Euclidean space, in which a fractal is considered (it is obvious, that in our case $d=3$), ν is Poisson's ratio, which is estimated according to mechanical tests results with the aid of the relationship [13]:

$$\frac{\sigma_Y}{E_n} = \frac{1-2\nu}{6(1+\nu)}, \qquad (2)$$

where σ_Y and E_n are yield stress and elasticity modulus of nanocomposite, accordingly.

Then nanoclusters relative fraction φ_{cl} can be calculated by using the following equation [8]:

$$d_f = 3 - 6\left(\frac{\phi_{cl}}{C_\infty S}\right)^{1/2}, \tag{3}$$

where C_∞ is characteristic ratio, which is a polymer chain statistical flexibility indicator [14], S is macromolecule cross-sectional area.

The value C_∞ is a function of d_f according to the relationship [8]:

$$\tilde{N}_\infty = \frac{2d_f}{d(d-1)(d-d_f)} + \frac{4}{3} \tag{4}$$

The value S for low-density polyethylenes is accepted equal to 14.9 Å² [15]. Macromolecular entanglements cluster network density v_{cl} can be estimated as follows [8]:

$$v_{cl} = \frac{\phi_{cl}}{C_\infty l_0 S}, \tag{5}$$

where l_0 is the main chain skeletal bond length, which for polyethylenes is equal to 0.154 nm [16].

Then the molecular weight of the chain part between nanoclusters M_{cl} was determined according to the equation [8]:

$$M_{cl} = \frac{\rho_p N_A}{v_{cl}}, \tag{6}$$

where ρ_p is polymer density, which for the studied polyethylenes is equal to ~ 930 kg/m³, N_A is Avogadro number.

And at last, the value n_{cl} is determined as follows [8]:

$$n_{cl} = \frac{2M_e}{M_{cl}}, \tag{7}$$

where M_e is molecular weight of a chain part between entanglements traditional nodes ("binary hookings"), which is equal to 1390 g/mole for low-density polyethylenes [17].

In Fig. 1, the dependence of nanocomposite elasticity modulus E_n on value n_{cl} is adduced, from which E_n enhancement at n_{cl} decreasing follows. Such behavior of nanocomposites LLDPE/MMT is completely identical to the behavior of both particulate-filled [11] and natural [10] nanocomposites.

In Ref. [18], the theoretical dependences of E_n as a function of cluster model parameters for natural nanocomposites was obtained:

$$E_n = c\left(\frac{\phi_{cl} V_{cl}}{n_{cl}}\right), \tag{8}$$

where c is constant, accepted equal to 5.9×10^{-26} m^3 for LLDPE.

In Fig. 1, the theoretical dependence $E_n(n_{cl})$, calculated according to the equation (8), for the studied nanocomposites is adduced, which shows a good enough correspondence with the experiment (the average discrepancy of theory and experiment makes up 11.6%, that is comparable with mechanical tests experimental error). Therefore, at organoclay mass contents W_n increasing within the range of 0–7 mass % n_{cl} value reduces from 8.40 up to 3.17 that are accompanied by nanocomposites LLDPE/MMT elasticity modulus growth from 206 up to 569 MPa.

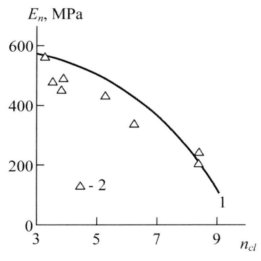

FIGURE 1 The dependences of elasticity modulus E_n on Statistical segments number per one nanocluster n_{cl} for nanocomposites LLDPE/MMT. 1 – calculation according to the Eq. (8), 2 – the experimental data.

Let us consider the physical foundations of n_{cl} reduction at W_n growth. The main equation of the reinforcement percolation model is the following one [9]:

$$\frac{E_n}{E_m} = 1 + 11(\phi_n + \phi_{if})^{1.7}, \qquad (9)$$

where ϕ_n and ϕ_{if} are relative volume fractions of nanofiller and interfacial regions, accordingly.

The value ϕ_n can be determined according to the equation [5]:

$$\phi_n = \frac{W_n}{\rho_n}, \qquad (10)$$

where ρ_n is nanofiller density, which is equal to ~1700 kg/m³ for Na^+-montmorillonite [5].

Further the Eq. (9) allows to estimate the value ϕ_{if}. In Fig. 2, the dependence $n_{cl}(\phi_{if})$ for nanocomposites LLDPE/MMT is adduced. As one can see, n_{cl} reduction at ϕ_{if} increasing is observed, i.e. formed on organoclay surface densely packed (and, possibly, subjecting to epitaxial crystallization [9]) interfacial regions as if pull apart nanoclusters, reducing statistical segments number in them. As it follows from the Eqs. (8) and (9), these processes have the same direction, namely, nanocomposite elasticity modulus increase.

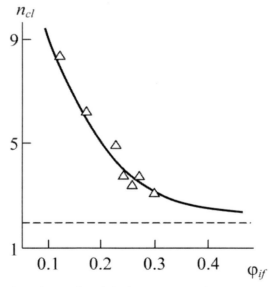

FIGURE 2 The dependence of statistical segments number per one nanocluster n_{cl} on interfacial regions relative fraction ϕ_{if} for nanocomposites LLDPE/MMT. Horizontal shaded line indicates the minimum value $n_{cl}=2$.

4.4 CONCLUSIONS

Hence, the obtained in the present paper results demonstrated common reinforcement mechanism of natural and artificial (filled with inorganic nanofiller) polymer nanocomposites. The statistical segments number per one nanocluster reduction at nanofiller contents growth is such a mechanism on suprasegmental level. The indicated effect physical foundation is the densely packed interfacial regions formation in artificial nanocomposites.

KEYWORDS

- elasticity modulus
- interfacial regions
- nanoclusters
- nanocomposite
- organoclay
- polyethylene

REFERENCES

1. Ahmed S., Jones F.R. J. Mater. Sci., 1990, v. 25, № 12, 4933–4942.
2. Bobryshev A.N., Kozomazov V.N., Babin L.O., Solomatov V.I. Synergetics of Composite Materials. Lipetsk, NPO ORIUS, 1994, 154 p.
3. Kozlov G.V., Yanovskii Yu.G., Zaikov G.E. Structure and Properties of Particulate–Filled Polymer Composites: the Fractal Analysis. New York, Nova Science Publishers, Inc., 2010, 282 p.
4. Hotta S., Paul D.R. Polymer, 2004, v. 45, № 21, 7639–7654.
5. Sheng N., Boyce M.C., Parks D.M., Rutledge G.C., Abes J.I., Cohen R.E. Polymer, 2004, v. 45, № 2, 487–506.
6. Bashorov M.T., Kozlov G.V., Mikitaev A.K. Materialovedenie, 2009, № 9, 39–51.
7. Kozlov G.V., Novikov V.U. Uspekhi Fizicheskikh Nauk, 2001, v. 171, № 7, 717–764.
8. Kozlov G.V., Zaikov G.E. Structure of the Polymer Amorphous State. Utretch, Boston, Brill Academic Publishers, 2004, 465 p.
9. Mikitaev A.K., Kozlov G.V., Zaikov G.E. Polymer Nanocomposites: Variety of Structural Forms and Applications. New York, Nova Science Publishers, Inc., 2008, 319 p.
10. Kozlov G.V., Mikitaev A.K. Polymers as Natural Nanocomposites: Unrealized Potential. Saarbrücken, Lambert Academic Publishing, 2010, 323 p.
11. Edwards D.C. J. Mater. Sci., 1990, v. 25, № 12, 4175–4185.
12. Balankin A.S. Synergetics of Deformable Body. Moscow, Publishers of Ministry Defence SSSR, 1991, 404 p.
13. Kozlov G.V., Sanditov D.S. Anharmonic Effects and Physical–Mechanical Properties of Polymers. Novosibirsk, Nauka, 1994, 261 p.

14. Budtov V.P. Physical Chemistry of Polymer Solutions. Sankt–Peterburg, Khimiya, 1992, 384 p.
15. Aharoni S.M. Macromolecules, 1985, v. 18, № 12, 2624–2630.
16. Aharoni S.M. Macromolecules, 1983, v. 16, № 9, 1722–1728.
17. Wu S. J. Polymer Sci.: Part B: Polymer Phys., 1989, v. 27, № 4, 723–741.
18. Kozlov G.V. Recent Patents on Chemical Engineering, 2011, v. 4, № 1, p. 53–77.

CHAPTER 5

QUANTUM AND WAVE CHARACTERISTICS OF SPATIAL-ENERGY INTERACTIONS

G. A. KORABLEV and G. E. ZAIKOV

CONTENTS

5.1	Introduction	116
5.2	On Two Principles of Adding Energy Characteristics of Interactions	116
5.3	Initial criteria	120
5.4	Wave Equation of P-Parameter	121
5.5	Quantum Properties of P-Parameter	123
5.6	Conclusions	126
Keywords		127
References		127

5.1 INTRODUCTION

It is demonstrated that for two-particle interactions the principle of adding reciprocals of energy characteristics of subsystems is performed for processes flowing by the potential gradient, and the principle of their algebraic addition – for the processes against the potential gradient.

The equation of the dependence of spatial-energy parameter of free atoms on their wave, spectral and frequency characteristics has been obtained.

Quantum conceptualizations on the composition of atoms and molecules make the foundation of modern natural science theories. Thus, the electronic angular momentum in stationary condition equals the integral multiple from Planck's constant. This main quantum number and three other combined explicitly characterize the state of any atom. The repetition factors of atomic quantum characteristics are also expressed in spectral data for simple and complex structures.

It is known that any periodic processes of complex shape can be shown as separate simple harmonic waves. "By Fourier theory, oscillations of any shape with period T can be shown as the total of harmonic oscillations with periods T_1, T_2, T_3, T_4, etc. Knowing the periodic function shape, we can calculate the amplitude and phases of sinusoids, with this function as their total" [1].

Therefore, many regularities in intermolecular interactions, complex formation and nanothermodynamics are explained with the application of functional divisible quantum or wave energy characteristics of structural interactions.

In this research we tried to apply the conceptualizations on spatial-energy parameter (P-parameter) for this.

5.2 ON TWO PRINCIPLES OF ADDING ENERGY CHARACTERISTICS OF INTERACTIONS

The analysis of the kinetics of various physic-chemical processes demonstrates that in many cases the reciprocals of velocities, kinetic or energy characteristics of the corresponding interactions are added.

Here are some examples: ambipolar diffusion, total rate of topochemical reaction, change in the light velocity when transiting from vacuum into the given medium, effective permeability of biomembranes.

In particular, such assumption is confirmed by the formula of electron transport probability (W_∞) due to the overlapping of wave functions 1 and 2 (in stationary state) during electron-conformation interactions:

$$W_\infty = \frac{1}{2} \frac{W_1 W_2}{W_1 + W_2} \tag{1}$$

Equation (1) is applied when evaluating the characteristics of diffusion processes accompanied with non-radiating electron transport in proteins [2].

Also: "It is known from the traditional mechanics that the relative motion of two particles with the interaction energy U(r) is the same as the motion of a material point with the reduced mass μ:

$$\frac{1}{\mu} = \frac{1}{m_1} + \frac{1}{m_2} \tag{2}$$

in the field of central force U(r), and total translational motion – as the free motion of the material point with the mass:

$$m = m_1 + m_2 \tag{3}$$

Such situation can be also found in quantum mechanics" [3].

The problem of two-particle interactions flowing by the bond line was solved in the time of Newton and Lagrange:

$$\mathring{A} = \frac{m_1 v_1^2}{2} + \frac{m_2 v_2^2}{2} + U(\vec{r}_2 - \vec{r}_1), \tag{4}$$

where E – system total energy, first and second components – kinetic energies of the particles, third – potential energy between particles 1 and 2, vectors $\vec{r}_2$ and $\vec{r}_1$ characterize the distance between the particles in final and initial states.

For moving thermodynamic systems the first law of thermodynamics can be shown as follows [4]:

$$\delta \mathring{A} = d\left(U + \frac{mv^2}{2}\right) \pm \delta A, \tag{5}$$

where: δE – amount of energy transferred to the system; component $d\left(U + \frac{mv^2}{2}\right)$ characterizes changes in internal and kinetic energies of the system; $+\delta A$ – work performed by the system; $-\delta A$ – work performed on the system.

Since the work numerically equals the change in the potential energy, then:

$$+\delta A = -\Delta U \text{ и } -\delta A = +\Delta U \tag{6 and 7}$$

Probably not only the value of potential energy but its changes are important in thermodynamic and also in many other processes in the dynamics of interactions of moving particles. Therefore, by the analogy with Eq. (4) the following should be fulfilled for two-particle interactions:

$$\delta E = d\left(\frac{m_1 v_1^2}{2} + \frac{m_2 v_2^2}{2}\right) \pm \Delta U \tag{8}$$

Here $$\Delta U = U_2 - U_1, \tag{9}$$

where U_2 and U_1 – potential energies of the system in final and initial states.

At the same time, the total energy (E) and kinetic energy $\left(\dfrac{mv^2}{2}\right)$ can be calculated from their zero value. In this case only the last component is modified in the equation (4).

The character of the changes in the potential energy value (ΔU) was analyzed by its index for different potential fields as given in Table 1.

From the table it is seen that the values of $-\Delta U$ and consequently $+\delta A$ (positive work) correspond to the interactions taking place by the potential gradient, and ΔU and $-\delta A$ (negative work) take place during the interactions against the potential gradient.

The solution of two-particle problem of the interaction of two material points with masses m_1 and m_2 obtained under the condition of no external forces available corresponds to the interactions taking place by the gradient, the positive work is performed by the system (similar to attraction process in the gravitation field).

The solution for this equation through the reduced mass (μ) [5] is Lagrangian equation for the relative motion of the isolated system of two interacting material points with masses m_1 and m_2, in coordinate x it looks as follows:

$$\mu \cdot x'' = -\frac{\partial U}{\partial x}; \quad \frac{1}{\mu} = \frac{1}{m_1} + \frac{1}{m_2}.$$

Here: U – mutual potential energy of material points; μ – reduced mass. At the same time $x'' = a$ (characteristic of system acceleration). For elementary regions of interactions Δx can be taken as follows:

$$\frac{\partial U}{\partial x} \approx \frac{\Delta U}{\Delta x} \quad \text{That is: } \mu a \Delta x = -\Delta U. \text{ Then:}$$

$$\frac{1}{1/(a\Delta x)} \frac{1}{1/m_1 + 1/m_2} \approx -\Delta U; \quad \frac{1}{a/(m_1 a \Delta x)} \approx -\Delta U$$

or:
$$\frac{1}{\Delta U} \approx \frac{1}{\Delta U_1} + \frac{1}{\Delta U_2} \tag{10}$$

Quantum and Wave Characteristics of Spatial-Energy Interactions

where ΔU_1 and ΔU_2 – potential energies of material points on the elementary region of interactions, ΔU – resulting (mutual) potential energy of these interactions.

TABLE 1 Directedness of interaction processes.

No	Systems	Potential field type	Process	U	r_2/r_1 (x_2/x_1)	U_2/U_1	Index ΔU	Index δA	Process directedness in the potential field
1	Opposite electric charges	Electrostatic	Attraction	$-k\frac{q_1q_2}{r}$	$r_2 < r_1$	$U_2 > U_1$	—	+	By gradient
			Repulsion	$-k\frac{q_1q_2}{r}$	$r_2 > r_1$	$U_2 < U_1$	+	—	Against gradient
2	Same electric charges	Electrostatic	Attraction	$k\frac{q_1q_2}{r}$	$r_2 < r_1$	$U_2 > U_1$	+	—	Against gradient
			Repulsion	$k\frac{q_1q_2}{r}$	$r_2 > r_1$	$U_2 < U_1$	—	+	By gradient
3	Elementary masses m_1 and m_2	Gravitational	Attraction	$-\gamma\frac{m_1m_2}{r}$	$r_2 < r_1$	$U_2 > U_1$	—	+	By gradient
			Repulsion	$-\gamma\frac{m_1m_2}{r}$	$r_2 > r_1$	$U_2 < U_1$	+	—	Against gradient
4	Spring deformation	Field of spring forces	Compression	$k\frac{\Delta x^2}{2}$	$x_2 < x_1$	$U_2 > U_1$	+	—	Against gradient
			Stretching	$k\frac{\Delta x^2}{2}$	$x_2 > x_1$	$U_2 > U_1$	+	—	Against gradient
5	Photoeffect	Electrostatic	Repulsion	$k\frac{q_1q_2}{r}$	$r_2 > r_1$	$U_2 < U_1$	—	+	By gradient

Thus:
1. In systems in which the interaction takes place by the potential gradient (positive work), the resultant potential energy is found by the principle of adding the reciprocals of the corresponding energies of subsystems [6]. The reduced mass for the relative motion of isolated system of two particles is calculated in the same way.

2. In systems in which the interaction takes place against the potential gradient (negative work), their masses and corresponding energies of subsystems (similar to Hamiltonian) are added algebraically.

5.3 INITIAL CRITERIA

From the Eq. (10) it is seen that the resultant energy characteristic of the system of interaction of two material points is found by the principle of adding the reciprocals of initial energies of interacting subsystems.

"Electron with the mass m moving near the proton with the mass M is equivalent to the particle with the mass $m_r = \dfrac{mM}{m+M}$ " [7].

Therefore modifying the Eq. (10), we can assume that the energy of atom valence orbitals (responsible for interatomic interactions) can be calculated [6] by the principle of adding the reciprocals of some initial energy components based on the equations:

$$\frac{1}{q^2/r_i} + \frac{1}{W_i n_i} = \frac{1}{P_E} \qquad (11)$$

or

$$\frac{1}{P_0} = \frac{1}{q^2} + \frac{1}{(Wrn)_i}; \qquad (12)$$

$$P_E = P_0/r_i \qquad (13)$$

where: W_i – orbital energy of electrons [8]; r_i – orbital radius of i orbital [9]; $q = Z^*/n^*$ – by [10,11], n_i – number of electrons of the given orbital, Z^* and n^* – nucleus effective charge and effective main quantum number, r – bond dimensional characteristics.

P_0 is called a spatial-energy parameter (SEP), and P_E – effective P-parameter (effective SEP). Effective SEP has a physical sense of some averaged energy of valence orbitals in the atom and is measured in energy units, for example, in electron-volts (eV).

The values of P_0 parameter are tabulated constants for electrons of the given atom orbital.

For SEP dimensionality:

$$[P_0] = [q] = [E] \cdot [r] = [h] \cdot [v] = \frac{kgm^3}{s^2} = Jm,$$

where [E], [h] and [υ] – dimensionalities of energy, Plank's constant and velocity.

The introduction of P-parameter should be considered as further development of quasi-classical concepts with quantum-mechanical data on atom structure to obtain the criteria of phase-formation energy conditions. For the systems of similarly charged (e.g., orbitals in the given atom) homogeneous systems the principle of algebraic addition of such parameters is preserved:

$$\sum P_E = \sum (P_0/r_i); \quad (14)$$

$$\sum P_E = \frac{\sum P_0}{r} \quad (15)$$

or:

$$\sum P_0 = P_0' + P_0'' + P_0''' + \ldots; \quad (16)$$

$$r \sum P_E = \sum P_0 \quad (17)$$

Here P-parameters are summed up by all atom valence orbitals.

To calculate the values of P_E-parameter at the given distance from the nucleus either the atomic radius (R) or ionic radius (r_I) can be used instead of r depending on the bond type.

Let us briefly explain the reliability of such an approach. As the calculations demonstrated the values of P_E-parameters equal numerically (in the range of 2%) the total energy of valence electrons (U) by the atom statistic model. Using the known correlation between the electron density (β) and intra-atomic potential by the atom statistic model [12], we can obtain the direct dependence of P_E-parameter on the electron density at the distance r_i from the nucleus.

The rationality of such technique was proved by the calculation of electron density using wave functions by Clementi [13] and comparing it with the value of electron density calculated through the value of P_E-parameter.

5.4 WAVE EQUATION OF P-PARAMETER

To characterize atom spatial-energy properties two types of P-parameters are introduced. The bond between them is a simple one:

$$P_E = \frac{P_0}{R}$$

where R – atom dimensional characteristic. Taking into account additional quantum characteristics of sublevels in the atom, this equation can be written down in coordinate x as follows:

$$\Delta P_E \approx \frac{\Delta P_0}{\Delta x} \text{ or } \partial P_E = \frac{\partial P_0}{\partial x}$$

where the value ΔP equals the difference between P_0-parameter of i orbital and P_{CD} – countdown parameter (parameter of main state at the given set of quantum numbers).

According to the established [6] rule of adding P-parameters of similarly charged or homogeneous systems for two orbitals in the given atom with different quantum characteristics and according to the energy conservation rule we have:

$$\Delta P_E'' - \Delta P_E' = P_{E,\lambda}$$

where $P_{E,\lambda}$ – spatial-energy parameter of quantum transition.

Taking for the dimensional characteristic of the interaction $\Delta\lambda = \Delta x$, we have:

$$\frac{\Delta P_0''}{\Delta\lambda} - \frac{\Delta P_0'}{\Delta\lambda} = \frac{P_0}{\Delta\lambda} \text{ or: } \frac{\Delta P_0'}{\Delta\lambda} - \frac{\Delta P_0''}{\Delta\lambda} = -\frac{P_0\lambda}{\Delta\lambda}$$

Let us again divide by $\Delta\lambda$ term by term:

where:
$$\left(\frac{\Delta P_0'}{\Delta\lambda} - \frac{\Delta P_0''}{\Delta\lambda}\right) \Big/ \Delta\lambda = -\frac{P_0}{\Delta\lambda^2},$$

$$\left(\frac{\Delta P_0'}{\Delta\lambda} - \frac{\Delta P_0''}{\Delta\lambda}\right) \Big/ \Delta\lambda \sim \frac{d^2 P_0}{d\lambda^2},$$

That is,
$$\frac{d^2 P_0}{d\lambda^2} + \frac{P_0}{\Delta\lambda^2} \approx 0$$

Taking into account only those interactions when $2\pi\Delta x = \Delta\lambda$ (closed oscillator), we have the following equation:

Quantum and Wave Characteristics of Spatial-Energy Interactions

$$\frac{d^2 P_0}{dx^2} + 4\pi^2 \frac{P_0}{\Delta\lambda^2} \approx 0$$

Since $\Delta\lambda = \frac{h}{mv}$,

then: $\frac{d^2 P_0}{dx^2} + 4\pi^2 \frac{P_0}{h^2} m^2 v^2 \approx 0$

or
$$\frac{d^2 P_0}{dx^2} + \frac{8\pi^2 m}{h^2} P_0 E_k = 0 \tag{18}$$

where $E_k = \frac{mV^2}{2}$ – electron kinetic energy.

Schrodinger equation for the stationery state in coordinate x:

$$\frac{d^2 \psi}{dx^2} + \frac{8\pi^2 m}{h^2} \psi E_k = 0$$

When comparing these two equations we see that P_0-parameter numerically correlates with the value of Ψ-function:

$$P_0 \approx \Psi,$$

and is generally proportional to it: $P_0 \sim \Psi$. Taking into account the broad practical opportunities of applying the P-parameter methodology, we can consider this criterion as the materialized analogue of Ψ-function [14, 15].

Since P_0-parameters like Ψ-function have wave properties, the superposition principles should be fulfilled for them, defining the linear character of the equations of adding and changing P-parameter.

5.5 QUANTUM PROPERTIES OF P-PARAMETER

According to Planck, the oscillator energy (E) can have only discrete values equaled to the whole number of energy elementary portions-quants:

$$nE = h\nu = hc/\lambda \tag{19}$$

where h – Planck's constant, v – electromagnetic wave frequency, c – its velocity, λ – wavelength, $n = 0, 1, 2, 3\ldots$

Planck's equation also produces a strictly definite bond between the two ways of describing the nature phenomena – corpuscular and wave.

P_0-parameter as an initial energy characteristic of structural interactions, similarly to the Eq. (19), can have a simple dependence from the frequency of quantum transitions:

$$P_0 \sim \hbar(\lambda v_0) \quad (20)$$

where: λ – quantum transition wavelength [16]; $\hbar = h/(2\pi)$; v_0 – kayser, the unit of wave number equaled to $2.9979\cdot 10^{10}$ Hz.

In accordance with Rydberg equation, the product of the right part of this equation by the value $(1/n^2 - 1/m^2)$, where n and m – main quantum numbers – should result in the constant.

Therefore the following equation should be fulfilled:

$$P_0(1/n_1^2 - 1/m_1^2) = N\hbar(\lambda v_0)(1/n^2 - 1/m^2) \quad (21)$$

where the constant N has a physical sense of wave number and for hydrogen atom equals $2\times 10^2 \text{Å}^{-1}$.

The corresponding calculations are demonstrated in Table 2. There: $r_i' = 0.5292$ Å – orbital radius of 1S-orbital and $r_i'' = 2^2 \times 0.5292 = 2.118$ Å – the value approximately equaled to the orbital radius of 2S-orbital.

The value of P_0-parameter is obtained from the equation (12), for example, for 1S-2P transition:

$$1/P_0 = 1/(13.595 \times 0.5292) + 1/14.394 \rightarrow P_0 = 4.7985 \text{ eVÅ}$$

The value q^2 is taken from Refs. [10, 11], for the electron in hydrogen atom it numerically equals the product of rest energy by the classical radius.

The accuracy of the correlations obtained is in the range of percentage error 0.06 (%), that is, the Eq. (21) is in the accuracy range of the initial data.

In the Eq. (21) there is the link between the quantum characteristics of structural interactions of particles and frequencies of the corresponding electromagnetic waves.

But in this case there is the dependence between the spatial parameters distributed along the coordinate. Thus in P_0-parameter the effective energy is multiplied by the dimensional characteristic of interactions, and in the right part of the Eq. (21) the kayser value is multiplied by the wavelength of quantum transition.

In Table 2 you can see the possibility of applying the Eq. (21) and for electron Compton wavelength ($\lambda_\kappa = 2.4261 \times 10^{-12}$ m), which in this case is as follows:

$$P_0 = 10^7 \hbar(\lambda_\kappa v_0) \qquad (22)$$

(with the relative error of 0.25%).

Integral-valued decimal values are found when analyzing the correlations in the system "proton-electron" given in Table 3:

1. Proton in the nucleus, energies of three quarks $5 + 5 + 7 \approx 17$ (MeV) → $P_p \approx 17$ MeV × 0.856×10^{-15} m $\approx 14.552 \times 10^{-9}$ eVm. Similarly for the electron $P_e = 0.511$ (MeV) × 2.8179×10^{-15} m (electron classic radius) → $P_e = 1.440 \times 10^{-9}$ eVm.

Therefore:

$$P_p \approx 10\, P_E \qquad (23)$$

2. Free proton $P_n = 938.3$ (MeV) × 0.856×10^{-15} (m) = 8.0318×10^{-7} eVm. For electron in the atom $P_a = 0.511$ (MeV) × 0.5292×10^{-5}(m) = 2.7057×10^{-5} eVm.

Then:

$$3P_a \approx 10^2 P_n \qquad (24)$$

The relative error of the calculations by these equations is found in the range of the accuracy of initial data for the proton ($\delta \approx 1\%$).

From Tables 2 and 3 we can see that the wave number N is quantized by the decimal principle:

$$N = n10^Z,$$

where n and Z – whole numbers.

Other examples of electrodynamics equations should be pointed out in which there are integral-valued decimal functions, for example, in the formula:

$$4\pi\varepsilon_0 c^2 = 10^7,$$

where ε_0 – electric constant.

In [17] the expression of the dependence of constants of electromagnetic interactions from the values of electron P_e-parameter was obtained:

$$k\mu_0 c = k/(\varepsilon_0 c) = P_e^{1/2} c^2 \approx 10/\alpha \qquad (25)$$

where: $k = 2\pi/\sqrt{3}$; μ_0 – magnetic constant; c – electromagnetic constant; α – fine structure constant.

All the above conclusions are based on the application of rather accurate formulas in the accuracy range of initial data.

TABLE 2 Quantum properties of hydrogen atom parameters.

Orbitals	W_i (eV)	r_i (Å)	q_i^2 (eVÅ)	P_0 (eVÅ)	$P_0(1/n_1^2 - 1/m_1^2)$ (eVÅ)	N (Å^{-1})	λ (Å)	Quantum transition	$Nh\lambda v_0$ (eVÅ)	$Nh\lambda v_0 \times (1/n^2 - 1/m^2)$ (eVÅ)
1S	13.595	0.5292	14.394	4.7985	3.5989	2×10^2	1215	1S-2P	4.7951	3.5963
1S						2×10^2	1025	1S-3P	4.0452	3.5954
1S						2×10^2	912	1S-nP	3.5990	3.5990
2S	3.3988	2.118	14.394	4.7985	3.5990	2×10^2	6562	2S-3P		3.5967
2S						2×10^2	4861	2S-4P		3.5971
2S						2×10^2	3646	2S-nP		3.5973
1S	13.595	0.5292	14.394	4.7985		10^7	2.4263×10^{-2}	–	4.7878	

TABLE 3 Quantum ratios of proton and electron parameters.

Particle	E (eV)	r (Å)	P = Er (eVÅ)	Ratio
Free proton	938.3×10^6	0.856×10^{-5}	$8.038 \times 10^3 = P_n$	$3P_a/P_n \approx 10^2$
Electron in an atom	0.511×10^6	0.5292	$2.7042 \times 10^5 = P_a$	
Proton in atom nuclei	$(5 + 5 + 7) \times 10^6 = 17 \times 10^6$	0.856×10^{-5}	$145.52 = P_p$	$P_p/P_e \approx 10$
Electron	0.511×10^6	2.8179×10^{-5}	$14.399 = P_e$	

5.6 CONCLUSIONS

1. Two principles of adding interaction energy characteristics are functionally defined by the direction of interaction by potential gradient (positive work) or against potential gradient (negative work).
2. Equation of the dependence of spatial-energy parameter on spectral and frequency characteristics in hydrogen atom has been obtained.

KEYWORDS

- frequency characteristics
- interaction energy
- negative work
- quantum
- spatial-energy
- spectral characteristics

REFERENCES

1. Gribov L.A., Prokofyeva N.I. Basics of physics. M.: Vysshaya shkola, 1992, 430 p.
2. Rubin A.B. Biophysics. Book 1. Theoretical biophysics. M.: Vysshaya shkola, 1997, 319 p.
3. Blokhintsev D.I. Basics of quantum mechanics. M.: Vysshaya shkola, 1991, 512 p.
4. Yavorsky B.M., Detlaf A.A. Reference-book in physics. M.: Nauka, 1998, 939 p.
5. Christy R.W., Pytte A. The structure of matter: an introduction to odern physics. Translated from English. M.: Nauka, 1999, 596 p.
6. Korablev G.A. Spatial–Energy Principles of Complex Structures Formation, Netherlands, Brill Academic Publishers and VSP, 2005, 426p. (Monograph).
7. Eyring G., Walter J., Kimball G. Quantum chemistry. M., F. L., 1998, 528 p.
8. Fischer C.F. Atomic Data, 1992, № 4, 301–399.
9. Waber J.T., Cromer D.T. J. Chem. Phys, 1965, vol 42, № 12, 4116–4123.
10. Clementi E., Raimondi D.L. Atomic Screening constants from S.C.F. Functions, J. Chem. Phys., 1993, v.38, №11, 2686–2689.
11. Clementi E., Raimondi D.L. J. Chem. Phys., 1997, v. 47, № 4, 1300–1307.
12. Gombash P. Statistic theory of an atom and its applications. M.: I.L., 1991, 398 p.
13. Clementi E. J.B.M. S. Res. Develop. Suppl., 1995, v. 9, № 2, 76.
14. Korablev G.A., Zaikov G.E. J. Appl. Polym. Sci., USA, 2006, v. 101, № 3, 2101–2107.
15. Korablev G.A., Zaikov G.E. Progress on Chemistry and Biochemistry, Nova Science Publishers, Inc. New York, 2009, 355–376.
16. Allen K.W. Astrophysical values. M.: Mir, 1997, 446 p.
17. Korablev G.A. Exchange spatial-energy interactions. Izhevsk. Publishing house "Udmurt University", 2010, 530 p. (Monograph).

CHAPTER 6

KEY ELEMENTS ON NANOPOLYMERS—FROM NANOTUBES TO NANOFIBERS

A. K. HAGHI and G. E. ZAIKOV

CONTENTS

6.1	Introduction	130
6.2	Graphene	132
6.3	Carbon Nanotubes	133
6.4	Fullerenes	135
6.5	Classification of Nanotubes	136
6.6	Chirality	137
6.7	Effective Parameters	140
6.8	Slippage of the Fluid Particles Near the Wall	152
6.9	The Density of the Liquid Layer Near a Wall of Carbon Nanotube	155
6.10	The Effective Viscosity of the Liquid in a Nanotube	156
6.11	Energy Release Due to the Collapse of the Nanotube	161
6.12	Fluid Flow in Nanotubes	165
6.13	Some of the Ideas and Approaches for Modeling in Nanohydromecanics	173
6.14	Modeling and Simulation Techniques for Nanofibers	183
6.15	Mechanism of Nanostructure Formation	218
6.16	Wet Spinning Technique (A Case Study)	228
6.17	Electrospinning of Blend Nanofibers	234
6.18	Results and Discussion	236
6.19	Effect of Pani Content	236
6.20	Effect of Electrospinning Temperature	239
6.21	Effect of Applied Voltage	243
6.22	Electrical Conductivity	243
Keywords		244
References		244

6.1 INTRODUCTION

The appearance of "nanoscience" and "nanotechnology" stimulated the burst of terms with "nano" prefix. Historically the term "nanotechnology" appeared before and it was connected with the appearance of possibilities to determine measurable values up to 10^{-9} of known parameters: 10^{-9}m-nm (nanometer), 10^{-9}s-ns (nanosecond), 10^{-9} degree (nanodegree, shift condition). Nanotechnology and molecular nanotechnology comprise the set of technologies connected with transport of atoms and other chemical particles (ions, molecules) at distances contributing the interactions between them with the formation of nanostructures with different nature. Although Nobel laureate Richard Feyman (1959) showed the possibility to develop technologies on nanometer level, Eric Drexler is considered to be the founder and ideologist of nanotechnology. When scanning tunnel microscope was invented by Nobel laureates Rorer and Binig (1981) there was an opportunity to influence atoms of a substance thus stimulating the work in the field of probe technology, which resulted in substantiation and practical application of nanotechnological methods in 1994. With the help of this technique it is possible to handle single atoms and collect molecules or aggregates of molecules, construct various structures from atoms on a certain substrate (base). Naturally, such a possibility cannot be implemented without preliminary computer designing of so-called "nanostructures architecture." Nanostructures architecture assumes a certain given location of atoms and molecules in space that can be designed on computer and afterwards transferred into technological programme of nanotechnological facility.

The notion "science of nanomaterials" assumes scientific knowledge for obtaining, composition, properties and possibilities to apply nanostructures, nanosystems and nanomaterials. A simplified definition of this term can be as follows: material science dealing with materials comprising particles and phases with nanometer dimensions. To determine the existence area for nanostructures and nanosystems it is advisable to find out the difference of these formations from analogous material objects.

From the analysis of literature the following can be summarized: the existence area of nanosystems and nanoparticles with any structure is between the particles of molecular and atomic level determined in picometers and aggregates of molecules or per molecular formations over micron units. Here, it should be mentioned that in polymer chemistry particles with nanometer dimensions belong to the class of per molecular structures, such as globules and fibrils by one of parameters, for example, by diameter or thickness. In chemistry of complex compounds clusters with nanometer dimensions are also known.

In 1991, Japanese researcher Idzhima was studying the sediments formed at the cathode during the spray of graphite in an electric arc. His attention was attracted by the unusual structure of the sediment consisting of microscopic fibers and filaments. Measurements made with an electron microscope showed that the diameter of these filaments does not exceed a few nanometers and a length of one to several microns.

Having managed to cut a thin tube along the longitudinal axis, the researchers found that it consists of one or more layers, each representing a hexagonal grid of graphite, which is based on hexagon with vertices located at the corners of the carbon atoms. In all cases, the distance between the layers is equal to 0.34 nm that is the same as that between the layers in crystalline graphite.

Typically, the upper ends of tubes are closed by multilayer hemispherical caps, each layer is composed of hexagons and pentagons, reminiscent of the structure of half a fullerene molecule.

The extended structure consisting of rolled hexagonal grids with carbon atoms at the nodes are called nanotubes.

Lattice structure of diamond and graphite are shown in Fig. 1. Graphite crystals are built of planes parallel to each other, in which carbon atoms are arranged at the corners of regular hexagons. Each intermediate plane is shifted somewhat toward the neighboring planes, as shown in Fig. 1.

The elementary cell of the diamond crystal represents a tetrahedron, with carbon atoms in its center and four vertices. Atoms located at the vertices of a tetrahedron form a center of the new tetrahedron, and thus, are also surrounded by four atoms each, etc. All the carbon atoms in the crystal lattice are located at equal distance (0.154 nm) from each other.

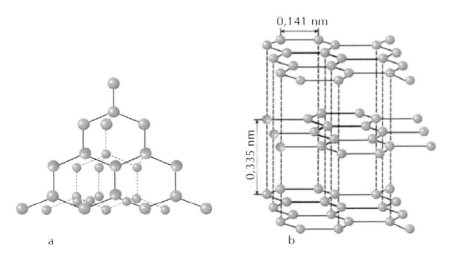

FIGURE 1 The structure of (a) the diamond lattice and (b) graphite.

Nanotubes are rolled into a cylinder (hollow tube) graphite plane, which is lined with regular hexagons with carbon atoms at the vertices of a diameter of several nanometers. Nanotubes can consist of one layer of atoms – single-wall nanotubes SWNT and represent a number of "nested" one into another layer pipes – multi-walled nanotubes – MWNT.

Nanostructures can be built not only from individual atoms or single molecules, but the molecular blocks. Such blocks or elements to create nanostructures are graphene, carbon nanotubes and fullerenes.

6.2 GRAPHENE

Graphene is a single flat sheet, consisting of carbon atoms linked together and forming a grid, each cell is like a bee's honeycombs (Fig. 2). The distance between adjacent carbon atoms in graphene is about 0.14 nm.

Graphite, from which slates of usual pencils are made, is a pile of graphene sheets (Fig. 3). Graphenes in graphite is very poorly connected and can slide relative to each other. So, if you conduct the graphite on paper, then after separating graphene from sheet the graphite remains on paper. This explains why graphite can write.

FIGURE 2 Schematic illustration of the graphene. Light balls – the carbon atoms, and the rods between them – the connections that hold the atoms in the graphene sheet.

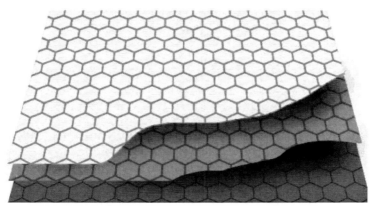

FIGURE 3 Schematic illustration of the three sheets of graphene, which are one above the other in graphite.

6.3 CARBON NANOTUBES

Many perspective directions in nanotechnology are associated with carbon nanotubes.

Carbon nanotubes are a carcass structure or a giant molecule consisting only of carbon atoms.

Carbon nanotube is easy to imagine, if we imagine that we fold up one of the molecular layers of graphite – graphene (Fig. 4).

FIGURE 4 Carbon nanotubes.

The way of folding nanotubes – the angle between the direction of nanotube axis relative to the axis of symmetry of graphene (the folding angle) – largely determines its properties.

Of course, no one produces nanotubes, folding it from a graphite sheet. Nanotubes formed themselves, for example, on the surface of carbon electrodes during arc discharge between them. At discharge, the carbon atoms evaporate from the surface, and connect with each other to form nanotubes of all kinds – single, multi-layered and with different angles of twist (Fig. 5).

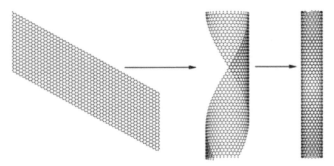

FIGURE 5 One way of imaginary making nanotube (right) from the molecular layer of graphite (left).

The diameter of nanotubes is usually about 1 nm and their length is a thousand times more, amounting to about 40 microns. They grow on the cathode in perpendicular direction to surface of the butt. The so-called self-assembly of carbon nanotubes from carbon atoms occurs (Fig. 6). Depending on the angle of folding of the nanotube they can have conductivity as high as that of metals, and they can have properties of semiconductors.

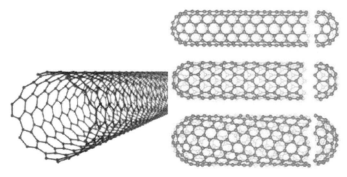

FIGURE 6 Left – schematic representation of a single-layer carbon nanotubes, on the right (top to bottom) – two-ply, straight and spiral nanotubes.

Carbon nanotubes are stronger than graphite, although made of the same carbon atoms, because carbon atoms in graphite are located in the sheets. And everyone knows that sheet of paper folded into a tube is much more difficult to bend and break than a regular sheet. That's why carbon nanotubes are strong. Nanotubes can be used as a very strong microscopic rods and filaments, as Young's modulus of single-walled nanotube reaches values of the order of 1–5 TPa, which is much more than steel! Therefore, the thread made of nanotubes, the thickness of a human hair is capable to hold down hundreds of kilos of cargo.

It is true that at present the maximum length of nanotubes is usually about a hundred microns – which is certainly too small for everyday use. However, the length of the nanotubes obtained in the laboratory is gradually increasing – now scientists have come close to the millimeter border. So there is every reason to hope that in the near future, scientists will learn how to grow a nanotube length in centimeters and even meters.

6.4 FULLERENES

The carbon atoms, evaporated from a heated graphite surface, connecting with each other, can form not only nanotube, but also other molecules, which are closed convex polyhedral, for example, in the form of a sphere or ellipsoid. In these molecules, the carbon atoms are located at the vertices of regular hexagons and pentagons, which make up the surface of a sphere or ellipsoid.

The molecules of the symmetrical and the most studied fullerene consisting of 60 carbon atoms (C_{60}), form a polyhedron consisting of 20 hexagons and 12 pentagons and resembles a soccer ball (Fig. 7). The diameter of the fullerene C_{60} is about 1 nm.

FIGURE 7 Schematic representation of the fullerene C_{60}.

6.5 CLASSIFICATION OF NANOTUBES

The main classification of nanotubes is conducted by the number of constituent layers.

Single-walled nanotubes – the simplest form of nanotubes. Most of them have a diameter of about 1 nm in length, which can be many thousands of times more. The structure of the nanotubes can be represented as a "wrap" a hexagonal network of graphite (graphene), which is based on hexagon with vertices located at the corners of the carbon atoms in a seamless cylinder. The upper ends of the tubes are closed by hemispherical caps, each layer is composed of hexa- and pentagons, reminiscent of the structure of half of a fullerene molecule. The distance d between adjacent carbon atoms in the nanotube is approximately equal to $d = 0.15$ nm.

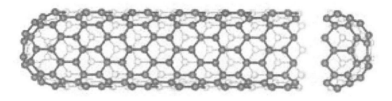

FIGURE 8 Graphical representation of single-walled nanotube.

Multi-walled nanotubes consist of several layers of graphene stacked in the shape of the tube. The distance between the layers is equal to 0.34 nm, which is the same as that between the layers in crystalline graphite.

Carbon nanotubes could find applications in numerous areas:
- additives in polymers;
- catalysts (autoelectronic emission for cathode ray lighting elements, planar panel of displays, gas discharge tubes in telecom networks);
- absorption and screening of electromagnetic waves;
- transformation of energy;
- anodes in lithium batteries;
- keeping of hydrogen;
- composites (filler or coating);
- nanosondes;
- sensors;
- strengthening of composites;
- supercapacitors.

FIGURE 9 Graphic representation of a multiwalled nanotube.

More than a decade, carbon nanotubes, despite their impressive performance characteristics have been used, in most cases, for scientific research.

To date, the most developed production of nanotubes has Asia, the production capacity, which is 2–3 times higher than in North America and Europe combined. Is dominated by Japan, which is a leader in the production of MWNT. Manufacturing North America, mainly focused on the SWNT. Growing at an accelerated rate production in China and South Korea. In the coming years, China will surpass the level of production of the U.S. and Japan, and by 2013, a major supplier of all types of nanotubes, according to experts, could be South Korea.

6.6 CHIRALITY

Chirality – a set of two integer positive indices (n, m), which determines how the graphite plane folds and how many elementary cells of graphite at the same time fold to obtain the nanotube.

From the value of parameters (n, m) are distinguished direct (achiral) high-symmetry carbon nanotubes
- armchair $n = m$
- zigzag $m = 0$ or $n = 0$
- helical (chiral) nanotube

In Fig. 10a is shown a schematic image of the atomic structure of graphite plane – graphene, and shown how a nanotube can be obtained the from it. The nanotube is fold up with the vector connecting two atoms on a graphite sheet. The cylinder is obtained by folding this sheet so that were combined the beginning and end of the vector. That is, to obtain a carbon nanotube from a graphene sheet, it should turn so that the lattice vector $\overline{R}$ has a circumference of the nanotube in Fig. 10b. This vector can be expressed in terms of the basis vectors of the elementary cell graphene sheet $\vec{R} = n\vec{r_1} + m\vec{r_2}$. Vector $\overline{R}$, which is often referred to simply by a pair of indices (n, m), called the chiral vector. It is assumed that $n > m$. Each pair of numbers (n, m) represents the possible structure of the nanotube.

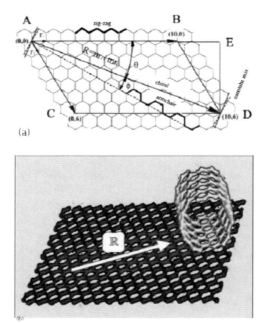

FIGURE 10 Schematic representation of the atomic structure of graphite plane.

In other words the chirality of the nanotubes (n, m) indicates the coordinates of the hexagon, which as a result of folding the plane has to be coincide with a hexagon, located at the beginning of coordinates (Fig. 11).

Many of the properties of nanotubes (for example, zonal structure or space group of symmetry) strongly depend on the value of the chiral vector. Chirality indicates what property has a nanotube – a semiconductor or metallicheskm. For example, a nanotube (10,10) in the elementary cell contains 40 atoms and is the type of metal, whereas the nanotube (10,9) has already in 1084 and is a semiconductor (Fig. 12).

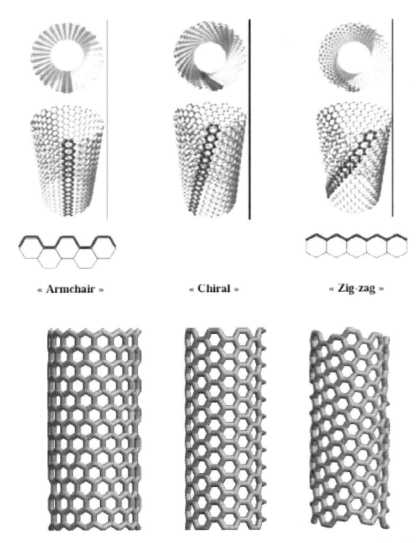

FIGURE 11 Single-walled carbon nanotubes of different chirality (in the direction of convolution). Left to right: the zigzag (16,0), armchair (8,8) and chiral (10,6) carbon nanotubes.

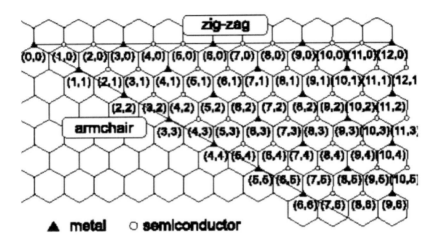

FIGURE 12 The scheme of indices (n,m) of lattice vector $\overline{R}$ tubes having semiconductor and metallic properties.

If the difference $n - m$ is divisible by 3, then these CNTs have metallic properties. Semimetals are all achiral tubes such as "chair". In other cases, the CNTs show semiconducting properties. Just type chair CNTs $(n = m)$ are strictly metal.

6.7 EFFECTIVE PARAMETERS

In addition to size effects that occur in micro and nano, we should note the following factors that determine the processes in low-dimensional systems:
- surface roughness (resistance to flow effects, interactions with the particles, etc.);
- dissolved gases (formation of bubbles, sticking to the surface, etc.);
- chemical surface properties (chemical reactions, etc.);
- hydrophobic – hydrophilic of surface;
- contaminants;
- heating due to uncontrollable processes;
- electrical properties of the surface (double layer, the change of surface and volume charge, charge transfer, etc.).

Viscous forces in the fluid can lead to large dispersion flow along the axis of motion. They have a significant impact, both on the scale of individual molecules, and the scale of microflows – near the borders of the liquid-solid (beyond a few molecular layers), during the motion on a complex and heterogeneous borders.

Influence of the effect of boundary regions on the particles and fluxes have been observed experimentally in the range of molecular thicknesses up to hundreds of nanometers. If the surface has a superhydrophobic properties, this range can extend to the micron thickness. *Molecular theory can predict the effect of hydrophobic surfaces in the system only up to tens of nanometers.*

Fluids, the flow of liquid or gas, have properties that vary continuously under the action of external forces. In the presence of fluid shear forces are small in magnitude, leads large changes in the relative position of the element of fluid. In contrast, changes in the relative positions of atoms in solids remain small under the action of any small external force. Termination of action of the external forces on the fluid does not necessarily lead to the restoration of its initial form.

6.7.1 CAPILLARY EFFECTS

To observe the capillary effects, you must open the nanotube, that is, remove the upper part – lids. Fortunately, this operation is quite simple.

The first study of capillary phenomena have shown that there is a relationship between the magnitude of surface tension and the possibility of its being drawn into the channel of the nanotube. It was found that the liquid penetrates into the channel of the nanotube, if its surface tension is not higher than 200 mN/m. Therefore, for the entry of any substance into the nanotube using solvents having a low surface tension. For example concentrated nitric acid with surface tension of 43 mN/m is used to inject certain metals into the channel of a nanotube. Then annealing is conducted at 4000°C for 4 h in an atmosphere of hydrogen, which leads to the recovery of the metal. Thus, the obtained nanotubes containing nickel, cobalt and iron.

Along with the metals carbon nanotubes can be filled with gaseous substances, such as hydrogen in molecular form. This ability is of great practical importance, since opening the ability to safely store hydrogen, which can be used as a clean fuel in internal combustion engines.

6.7.2 SPECIFIC ELECTRICAL RESISTANCE OF CARBON NANOTUBES

Due to small size of carbon nanotubes only in 1996 they succeeded to directly measure their electrical resistivity ρ. The results of direct measurements showed that the resistivity of the nanotubes can be varied within wide limits to 0.8 ohm/cm. The minimum value is lower than that of graphite. Most of the nanotubes have metallic conductivity, and the smaller shows properties of a semiconductor with a band gap of 0.1 to 0.3 eV.

The resistance of single-walled nanotube is independent of its length, because of this it is convenient to use for the connection of logic elements in microelectronic devices. The permissible current density in carbon nanotubes is much greater than in metallic wires of the same cross section, and one hundred times better achievement for superconductors.

The results of the study of emission properties of the material, where the nanotubes were oriented perpendicular to the substrate, have been very interesting for practical use. Attained values of the emission current density of the order of 0.5 mA/mm^2. The value obtained is in good agreement with the Fowler–Nordheim expression.

The most effective and common way to control microflow substances are *electrokinetic* and *hydraulic*. At the same time the most technologically advanced and automated considered electrokinetic.

Charges transfer in mixtures occurs as a result of the directed motion of charge carriers – ions. There are different mechanisms of such transfer, but usually are *convection, migration and diffusion*.

Convection is called mass transfer the macroscopic flow. *Migration* – the movement of charged particles by electrostatic fields. The velocity of the ions depends on field strength. In microfluidics a special role is played *electrokinetic processes* that can be divided into four types: *electro-osmosis, electrophoresis, streaming potential and sedimentation potential*. These processes can be qualitatively described as follows:

1. *Electro-osmosis* – the movement of the fluid volume in response to the applied electric field in the channel of the electrical double layers on its wetted surfaces.
2. *Electrophoresis* – the forced motion of charged particles or molecules, in mixture with the acting electric field.
3. *Streamy potential* – the electric potential, which is distributed through a channel with charged walls, in the case when the fluid moves under the action of pressure forces. Joule electric current associated with the effect of charge transfer is flowing stream.
4. *The potential of sedimentation* – an electric potential is created when charged particles are in motion relative to a constant fluid. The driving force for this effect – usually gravity.

In general, for the microchannel cross-section S amount of introduced probe (when entering electrokinetic method) depends on the applied voltage U, time t during which the received power, and mobility of the sample components μ:

$$Q = \frac{\mu S U t}{L} \cdot c$$

where: c – probe concentration in the mixture, L – the channel length.

Amount of injected substance is determined by the electrophoretic and total electro-osmotic mobilities μ.

In the hydrodynamic mode of entry by the pressure difference in the channel or capillary of circular cross section, the volume of injected probe V_c:

$$V_c = \frac{4}{128} \cdot \frac{\Delta p \pi dt}{\eta L}$$

where: Δp – pressure differential, d – diameter of the channel, η – viscosity.

In the simulation of processes in micron-sized systems the following basic principles are fundamental:
1. hypothesis of *laminar* flow (sometimes is taken for granted when it comes to microfluidics);
2. continuum hypothesis (detection limits of applicability);
3. laws of formation of the velocity profile, mass transfer, the distribution of electric and thermal fields;
4. boundary conditions associated with the geometry of structural elements (walls of channels, mixers zone flows, etc.).

Since we consider the physical and chemical transport processes of matter and energy, mathematical models, most of them have the form of systems of differential equations of second order partial derivatives. Methods for solving such equations are analytical (Fourier and its modifications, such as the method of Greenberg, Galerkin, in some cases, the method of d'Alembert and the Green's functions, the Laplace operator method, etc.) or numerical (explicit or, more effectively, implicit finite difference schemes) – traditional. The development involves, basically, numerical methods and follows the path of saving computing resources, and increasing the speed of modern computers.

Laminar flow – a condition in which the particle velocity in the liquid flow is not a random function of time. The small size of the microchannels (typical dimensions of 5 to 300 microns) and low surface roughness create good conditions for the establishment of laminar flow. Traditionally, the image of the nature of the flow gives the dimensionless characteristic numbers: the Reynolds number and Darcy's friction factor.

In the motion of fluids in channels the turbulent regime is rarely achieved. At the same time, the movement of gases is usually turbulent.

Although the liquid – are quantized in the length scale of intermolecular distances (about 0.3 nm to 3 nm in liquids and for gases), they are assumed to be continuous in most cases, microfluidics. Continuum hypothesis (continuity, continuum) suggests that the macroscopic properties of fluids consisting of molecules, the same as if the fluid was completely continuous (structurally homogeneous). Physical characteristics: mass, momentum and energy associated with the volume of fluid containing a sufficiently large number of molecules must be taken as the sum of all the relevant characteristics of the molecules.

Continuum hypothesis leads to the concept of fluid particles. In contrast to the ideal of a point particle in ordinary mechanics, in fluid mechanics, particle in the fluid has a finite size.

At the atomic scale we would see large fluctuations due to the molecular structure of fluids, but if we the increase the sample size, we reach a level where it is possible to obtain stable measurements. This volume of probe must contain a sufficiently large number of molecules to obtain reliable reproducible signal with small statistical fluctuations. For example, if we determine the required volume as a cube with sides of 10 nm, this volume contains some of the molecules and determines the level of fluctuations of the order of 0.5%.

The most important position in need of verification is to analyze the admissibility of mass transfer on the basis of the continuum model that can be used instead of the concentration dependence of the statistical analysis of the ensemble of individual particles. The position of the continuum model is considered as a necessary condition for microfluidics.

The applicability of the hypothesis is based on comparison of free path length of a particle λ in a liquid with a characteristic geometric size d. The ratio of these lengths – the Knudsen number: $Kn = \lambda / d$. Based on estimates of the Knudsen number defined two important statements:
1. $Kn < 10^{-3}$ – justified hypothesis of a continuous medium; and
2. $Kn < 10^{-1}$ – allowed the use of adhesion of particles to the solid walls of the channel.

Wording of the last condition can also be varied: both in form $U = 0$ and in a more complex form, associated with shear stresses. The calculation of λ can be carried out as

$$\lambda \approx \sqrt[3]{\overline{V} / Na},$$

where: $\overline{V}$ – molar volume, Na – Avogadro's number.

Under certain geometrical approximations of the particles of substance free path length can be calculated as $\lambda \approx 1/\left(\sqrt{2}\pi r_s^2 Na\right)$, if used instead r_S Stokes radius, as a consequence of the spherical approximation of the particle. On the other hand, for a rigid model of the molecule r_S should be replaced by the characteristic size of the particles R_g – the radius of inertia, calculated as $R_g = n_i \cdot \delta_i / \sqrt{6}$. Here δ_i – the length of a fragment of the chain (link), n_i – the number of links.

Of course, the continuum hypothesis is not acceptable when the system under consideration is close to the molecular scale. This happens in nanoliquid, such as liquid transport through nanopores in cell membranes or artificially made nanochannels.

In contrast to the continuum hypothesis, the essence of modeling the molecular dynamics method is as follows. We consider a large ensemble of particles, which simulate atoms or molecules, i.e., all atoms are material points. It is believed that the particles interact with each other and, moreover, may be subject to external influence. Interatomic forces are represented in the form of the classical potential force (the gradient of the potential energy of the system).

The interaction between atoms is described by means of van der Waals forces (intermolecular forces), mathematically expressed by the Lennard–Jones potential:

$$V(r) = \frac{Ae^{-\sigma r}}{r} - \frac{C_6}{r^6}$$

where: A and C_6 – some coefficients depending on the structure of the atom or molecule, σ – the smallest possible distance between the molecules.

In the case of two isolated molecules at a distance of r_0 the interaction force is zero, that is, the repulsive forces balance attractive forces. When $r > r_0$ the resultant force is the force of gravity, which increases in magnitude, reaching a maximum at $r = r_m$ and then decreases. When $r < r_0$ – a repulsive force. Molecule in the field of these forces has potential energy $V(r)$, which is connected with the force of $f(r)$ by the differential equation

$$dV = -f(r)dr$$

At the point $r = r_0$, $f(r) = 0$, $V(r)$ reaches an extremum (minimum).

The chart of such a potential is shown below in Fig. 13. The upper (positive) half-axis r corresponds to the repulsion of the molecules, the lower (negative) half-plane – their attraction. We can say simply: at short distances the molecules mainly repel each, on the long – draw each other. Based on this hypothesis, and now an obvious fact, the van der Waals received his equation of state for real gases.

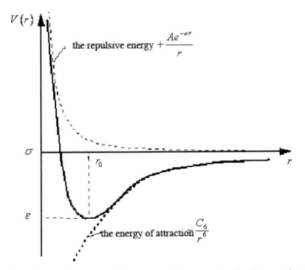

FIGURE 13 The chart of the potential energy of intermolecular interaction.

The exponential summand in the expression for the potential describing the repulsion of the molecules at small distances, often approximated as

$$\frac{Ae^{-\sigma r}}{r} \approx \frac{C_{12}}{r^{12}}$$

In this case we obtain the Lennard–Jones potential:

$$V(r) = \frac{C_{12}}{r^{12}} - \frac{C_6}{r^6} \tag{1}$$

The interaction between carbon atoms is described by the potential

$$V_{CC}(r) = K(r-b)^2,$$

where: K – constant tension (compression) connection, $b = 1,4A$ – the equilibrium length of connection, r – current length of the connection.

The interaction between the carbon atom and hydrogen molecule is described by the Lennard–Jones

$$V(r) = 4\varepsilon\left[\left(\frac{\sigma}{r}\right)^{12} - \left(\frac{\sigma}{r}\right)^{6}\right]$$

For all particles (Fig. 14) the equations of motion are written:

$$m\frac{d^2\overline{r_i}}{dt^2} = \overline{F}_{T-H_2}(\overline{r_i}) + \sum_{j \neq i}\overline{F}_{H_2-H_2}(\overline{r_i}-\overline{r_j}),$$

where: $\overline{F}_{T-H_2}(\overline{r})$ – force, acting by the CNT, $\overline{F}_{H_2-H_2}(\overline{r_i}-\overline{r_j})$ – force acting on the i-th molecule from the j-th molecule.

The coordinates of the molecules are distributed regularly in the space, the velocity of the molecules are distributed according to the Maxwell equilibrium distribution function according to the temperature of the system:

$$f(u,v,w) = \frac{\beta^3}{\pi^{3/2}}\exp\left(-\beta^2\left(u^2+v^2+w^2\right)\right) \quad \beta = \frac{1}{\sqrt{2RT}}$$

The macroscopic flow parameters are calculated from the distribution of positions and velocities of the molecules:

$$\overline{V} = \langle \overline{v}_i \rangle = \frac{1}{n}\sum_i \overline{v}_i, \quad \frac{3}{2}RT = \frac{1}{2}\langle |\overline{v}_i'|^2 \rangle, \quad \overline{v}_i' = \overline{v}_i - \overline{V},$$

$$\rho = \frac{m}{V_0},$$

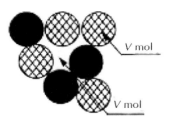

FIGURE 15 Position of particles.

The resulting system of equations is solved numerically. However, the molecular dynamics method has limitations of applicability:

1. the de Broglie wavelength h/mv (where h – Planck's constant, m – the mass of the particle, v – velocity) of the atom must be much smaller than the interatomic distance;
2. Classical molecular dynamics cannot be applied for modeling systems consisting of light atoms such as helium or hydrogen;
3. at low temperatures, quantum effects become decisive for the consideration of such systems must use quantum chemical methods;
4. necessary that the time at which we consider the behavior of the system were more than the relaxation time of the physical quantities.

In 1873, Van der Waals proposed an equation of state is qualitatively good description of liquid and gaseous systems. It is for one mole (one mole) is:

$$\left(p + \frac{a}{v^2}\right)(v - b) = RT \tag{2}$$

Note that at $p > \frac{a}{v^2}$ and $v \gg b$ this equation becomes the equation of state of ideal gas

$$pv = RT \tag{3}$$

Van der Waals equation can be obtained from the Clapeyron equation of Mendeleev by an amendment to the magnitude of the pressure a/v^2 and the amendment

b to the volume, both constant a and b independent of T and v but dependent on the nature of the gas.

The amendment b takes into account:
1. the volume occupied by the molecules of real gas (in an ideal gas molecules are taken as material points, not occupying any volume);
2. so-called "dead space", in which can not penetrate the molecules of real gas during motion, i.e. volume of gaps between the molecules in their dense packing.

Thus, $b = v_{мол.} + v_{заз.}$ (Fig. 16). The amendment a/v^2 takes into account the interaction force between the molecules of real gases. It is the internal pressure, which is determined from the following simple considerations. Two adjacent elements of the gas will react with a force proportional to the product of the quantities of substances enclosed in these elementary volumes.

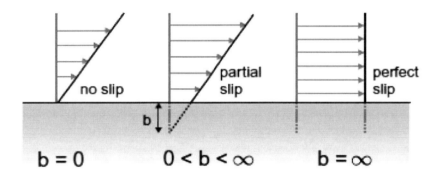

FIGURE 16 Location scheme of molecules in a real gas.

Therefore, the internal pressure P_{BH} is proportional to the square of the concentration n:

$$P_{BH} \sim n^2 \sim \rho^2 \sim \frac{1}{v^2},$$

where: ρ – the gas density.

Thus, the total pressure consists of internal and external pressures:

$$p + P_{BH} = p + \frac{a}{v^2}$$

Equation (3) is the most common for an ideal gas. Under normal physical conditions and from Eq. (3) we obtain:

$$R = \frac{R\mu}{\mu} = \frac{8314}{\mu}$$

Knowing $R\mu$ we can find the gas constant for any gas with the help of the value of its molecular mass μ (Table 1):

TABLE 1 The molecular weight of some gases.

Gas	N	Ar	H_2	O_2	CO	CO_2	Ammonia	Air
μ	28	40	2	32	28	44	17	29

For gas mixture with mass M state equation has the form:

$$pv = MR_{CM}T = \frac{8314MT}{\mu_{CM}} \qquad (4)$$

where: R_{CM} – gas constant of the mixture.

The gas mixture can be given by the mass proportions g_i, voluminous r_i or mole fractions n_i respectively, which are defined as the ratio of mass m_i, volume v_i or number of moles N_i of i gas to total mass M, volume v or number of moles N of gas mixture. Mass fraction of component is $g_i = \frac{m_i}{M}$, where $i = 1, n$. It is obvious that $M = \sum_{i=1}^{n} m_i$ and $\sum_{i=1}^{n} g_i = 1$. The volume fraction is $r_i = \frac{v_i}{v_{CM}}$, where v_i – partial volume of component mixtures.

Similarly, we have $\sum_{i=1}^{n} v_i = v_{CM}, \sum_{i=1}^{n} r_i = 1$.

Depending on specificity of tasks the gas constant of the mixture may be determined as follows:

$$R_{CM} = \sum_{i=1}^{n} g_i R_i; \qquad R_{CM} = \frac{1}{\sum_{i=1}^{n} r_i R_i^{-1}}$$

If we know the gas constant R_{CM}, the seeming molecular weight of the mixture is equal to

$$\mu_{CM} = \frac{8314}{R_{CM}} = \frac{8314}{\sum_{i=1}^{n} g_i R_i} = 8314 \sum_{i=1}^{n} r_i R_i^{-1}$$

The pressure of the gas mixture p is equal to the sum of the partial pressures of individual components in the mixture p_i:

$$p = \sum_{i=1}^{n} p_i \qquad (5)$$

Partial pressure p_i – pressure that has gas, if it is one at the same temperature fills the whole volume of the mixture ($p_i v_{CM} = RT$).

With various methods of setting the gas mixture partial pressures

$$p_i = pr_i; \; p_i = \frac{p g_i \mu_{CM}}{\mu_i} \qquad (6)$$

From the Eq. (6) we see that for the calculation of the partial pressures p_i necessary to know the pressure of the gas mixture, the volume or mass fraction i of the gas component, as well as the molecular weight of the gas mixture μ and the molecular weight of i of gas μ_i.

The relationship between mass and volume fractions are written as follows:

$$g_i = \frac{m_i}{m_{CM}} = \frac{\rho_i v_i}{\rho_{CM} v_{CM}} = \frac{R_{CM}}{R_i} r_i = \frac{\mu_i}{\mu_{CM}} r_i$$

We rewrite Eq. (2) as

$$v^3 - \left(b + \frac{RT}{p}\right) v^2 + \frac{a}{p} v - \frac{ab}{p} = 0 \qquad (7)$$

When $p = p_k$ and $T = T_k$, where p_k and T_k – critical pressure and temperature, all three roots of Eq. (7) are equal to the critical volume v_k

$$v^3 - \left(b + \frac{RT_k}{p_k}\right) v^2 + \frac{a}{p_k} v - \frac{ab}{p_k} = 0 \qquad (8)$$

Because the $v_1 = v_2 = v_3 = v_k$, then Eq. (8) must be identical to the equation

$$(v-v_1)(v-v_2)(v-v_3) = (v-v_k)^3 = v^3 - 3v^2 v_k + 3vv_k^2 - v_k^3 = 0 \qquad (9)$$

Comparing the coefficients at the equal powers of v in both equations leads to the equalities

$$b + \frac{RT_k}{p_k} = 3v_k; \quad \frac{a}{p_k} = 3v_k^2; \quad \frac{ab}{p_k} = v_k^3 \qquad (10)$$

Hence

$$a = 3v_k^2 p_k; \quad b = \frac{v_k}{3} \qquad (11)$$

Considering Eq. (10) as equations for the unknowns p_k, v_k, T_k, we obtain

$$p_k = \frac{a}{27b^2}; \quad v_k = 3b; \quad T_k = \frac{8a}{27bR} \qquad (12)$$

From Eqs. (10) and (11) or (12) we can find the relation

$$\frac{RT_k}{p_k v_k} = \frac{8}{3} \qquad (13)$$

Instead of the variables p, v, T let's introduce the relationship of these variables to their critical values (leaden dimensionless parameters)

$$\pi = \frac{p}{p_k}; \quad \omega = \frac{v}{V_k}; \quad \tau = \frac{T}{T_k} \qquad (14)$$

Substituting Eqs. (12) and (14) in Eq. (7) and using Eq. (13), we obtain

$$\left(\pi p_k + \frac{3v_k^2 p_k}{\omega^2 v_k^2} \right)\left(\omega v_k - \frac{v_k}{3} \right) = RT_k \tau,$$

$$\left(\pi + \frac{3}{\omega^2} \right)(3\omega - 1) = 3 \frac{RT_k}{p_k v_k} \tau,$$

$$\left(\pi + \frac{3}{\omega^2}\right)(3\omega - 1) = 8\tau \quad (15)$$

6.8 SLIPPAGE OF THE FLUID PARTICLES NEAR THE WALL

Features of the simulation results of Poiseuille flow in the microtubules, when the molecules at the solid wall and the wall atoms at finite temperature of the wall make chaotic motion lies in the fact that in the intermediate range of Knudsen numbers there is slippage of the fluid particles near the wall.

Researchers describe three possible cases:
1. the liquid can be stable (no slippage),
2. slides relative to the wall (with slippage flow),
3. the flow profile is realized; this is when the friction of the wall is completely absent (complete slippage).

In the framework of classical continuum fluid dynamics, according to the Navier boundary condition the velocity slip is proportional to fluid velocity gradient at the wall:

$$v\big|_{y=0} = L_S \, dv/dy\big|_{y=0} \quad (16)$$

Here and in Fig. 17, L_S represents the "slip length" and has a dimension of length.

Because of the slippage, the average velocity in the channel $\langle v_{pdf} \rangle$ increases.

In a rectangular channel (of width >> height h and viscosity of the fluid η) due to an applied pressure gradient – dp/dx the authors of that article obtained:

$$\langle v_{pdf} \rangle = \frac{h^2}{12\eta}\left(-\frac{dp}{dx}\right)\left(1 + \frac{6L_S}{h}\right) \quad (17)$$

The results of molecular dynamics simulation for nanosystems with liquid, with characteristic dimensions of the order of the size of the fluid particles, show that a large slippage lengths (of the order of microns) should occur in the carbon nanotubes of nanometer diameter and, consequently, can increase the flow rate by three orders of ($6L_S/h > 1000$). Thus, the flow with slippage is becoming more and more important for hydrodynamic systems of small size.

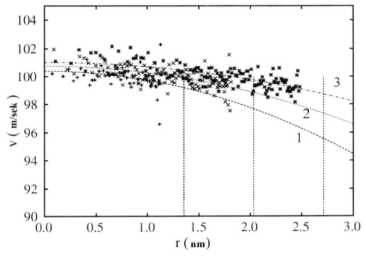

FIGURE 17 Three cases of slip flow past a stationary surface. The slip length b is indicated.

The results of molecular dynamics simulation of unsteady flow of mixtures of water – water vapor, water and nitrogen in a carbon nanotube are reported by many researchers.

Carbon nanotubes have been considered "zigzag" with chiral vectors (20, 20), (30, 30) and (40, 40), corresponding to pipe diameters of 2,712, 4,068 and 5,424 nm, respectively.

Knowing the value of the flow rate and the system pressure, which varies in the range of 600–800 bars, are high enough to ensure complete filling of the tubes. This pressure can be achieved by the total number of water molecules 736, 904, and 1694, respectively.

The effects of slippage of various liquids on the surface of the nanotube were studied in detail.

The length of slip, can be calculated using the current flow velocity profiles of liquid, shown in Fig. 18, were 11, 13, and 15 nm for the pipes of 2,712, 4,068 and 5,424 nm respectively. The dotted line marked by theoretical modeling data. The vertical lines indicate the position of the surface of carbon nanotubes.

It was found out that as the diameter decreases, the speed of slippage of particles on the wall of nanotube also decreases. The authors attribute this to the increase of the surface friction.

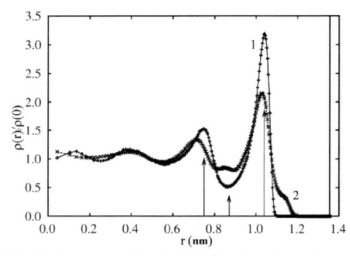

FIGURE 18 Time average streaming velocity profiles of water in a carbon nanotubes of different diameter: 2.712 nm, curve 1; 4.068 nm, curve 2; 5.424 nm, curve 3.

Experiments with various pressure drops in nanotubes demonstrated slippage of fluid in micro- and nanosystems. The most remarkable were the two recent experiments, which were conducted to improve the flow characteristics of carbon nanotubes with the diameters of 2 and 7 nm respectively. In the membranes in which the carbon nanotubes were arranged in parallel, there was a slip of the liquid in the micrometer range. This led to a significant increase in flow rate – up to three – four orders of magnitude.

In the experiments for the water moving in microchannels on smooth hydrophobic surfaces, there are slidings at about 20 nm. If the wall of the channel is not smooth but twisty or rough, and at the same time, hydrophobic, such a structure would lead to an accumulation of air in the cavities and become superhydrophobic (with contact angle greater than 160°). It is believed that this leads to creation of contiguous areas with high and low slippage, which can be described as "effective slip length". This effective length of the slip occurring on the rough surface can be several tens of microns, which was indeed experimentally confirmed by many researchers.

Another possible problem is filling of the hydrophobic systems with liquid. Filling of micron size hydrophobic capillaries is not a big problem, because pressure of less than 1 atm is sufficient. Capillary pressure, however, is inversely proportional to the diameter of the channel, and filling for nanochannels can be very difficult.

6.9 THE DENSITY OF THE LIQUID LAYER NEAR A WALL OF CARBON NANOTUBE

Researchers showed radial density profiles of oxygen averaged in time and hydrogen atoms in the "zigzag" carbon nanotube with chiral vector (20, 20) and a radius $R = 1,356$ nm (Fig. 19). The distribution of molecules in the area near the wall of the carbon nanotube indicated a high density layer near the wall of the carbon nanotube. Such a pattern indicates the presence of structural heterogeneity of the liquid in the flow of the nanotube. In Fig. (19) $\rho^*(r) = \rho(r)/\rho(0)$, 2,712 nm diameter pipe is completely filled with water molecules at $300^0 K$. The overall density $r(0) = 1000 \kappa s/M^3$. The arrows denote the location of distinguishable layers of the water molecules and the vertical line the position of the CNT wall.

The distribution of molecules in the region of $0.95 \leq r \leq R$ nm indicates a high density layer near the wall carbon nanotube. Such a pattern indicates the presence of structural heterogeneity of the liquid in the flow of the nanotube.

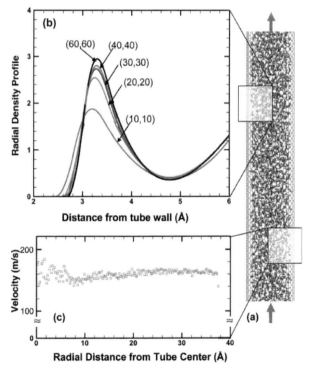

FIGURE 19 Radial density profiles of oxygen (curve 1) and hydrogen (curve 2) atoms for averaged in the time interval.

Many researchers obtained similar results for the flow of the water in the carbon nanotubes using molecular dynamic simulations.

Figure 20 shows the scheme of the initial structure and movement of water molecules in the model carbon nanotube (a), the radial density profile of water molecules inside nanotubes with different radii (b) and the velocity profile of water molecules in a nanotube chirality (60,60) (c).

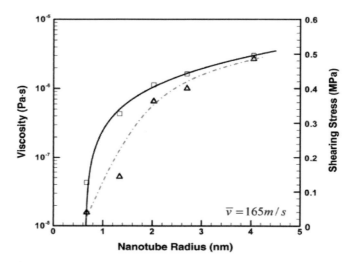

FIGURE 20 Dependences of the density and velocity profiles of water molecules inside the nanotubes with different radii (Xi Chen et al., 2008).
Schematic of the initial structure and transport of water molecules in a model CNT.
The radial density profile (RDP) of water molecules inside CNTs with different radii.
The representative radial velocity profile (RVP) of water molecules inside a (60,60) nanotube.

It is mentioned that the area available for molecules of the liquid is less than the area bounded by a solid wall, primarily due to the van der Waals interactions.

6.10 THE EFFECTIVE VISCOSITY OF THE LIQUID IN A NANOTUBE

Scientists showed a significant increase in the effective viscosity of the fluid in the nanovolumes compared to its macroscopic value. It was shown that the effective viscosity of the liquid in a nanotube depends on the diameter of the nanotube.

The effective viscosity of the liquid in a nanotube is defined as follows.

Let's establish a conformity nanotubes filled with liquid, possibly containing crystallites with the same size tube filled with liquid, considered as a homogeneous medium (i.e., without considering the crystallite structure), in which Poiseuille flow is realized at the same pressure drop and consumption rate. The viscosity of a homogeneous fluid, which ensures the coincidence of these parameters, will be called the effective viscosity of the flow in the nanotube.

Researchers showed that while flowing in the narrow channels of width less than 2 nm, water behaves like a viscous liquid. In the vertical direction water behaves as a rigid body, and in a horizontal direction it maintains its fluidity.

It is known that at large distances the van der Waals interaction has a magnetting tendency and occurs between any molecules like polar as well as nonpolar. At small distances it is compensated by repulsion of electron shells. Van der Waals interaction decreases rapidly with distance. Mutual convergence of the particles under the influence magnetting forces continues until these forces are balanced with the increasing forces of repulsion.

Knowing the deceleration of the flow (Fig. 20) of water a, the authors of the cited article calculates the effective shear stress between the wall of the pipe length l and water molecules by the formula

$$\tau = Nma/(2\pi Rl) \tag{18}$$

Here, the shear stress is a function of tube radius and flow velocity $\bar{v}$, m – mass of water molecules, the average speed is related to volumetric flow $\bar{v} = Q/(\pi R^2)$.

Denoting n_0 the density of water molecules number, we can calculate the shear stress in the form of:

$$\tau\big|_{r=R} = n_0 mRa/2 \tag{19}$$

Figure 21 shows the results of calculations of the authors of the cited article – influence of the size of the tube R_0 on the effective viscosity (squares) and shear stress τ (triangles), when the flow rate is approximately 165 m/sec.

According to classical mechanics of liquid flow at different pressure drops Δp along the tube length l is given by Poiseuille formula

$$Q_P = \frac{\pi R^4 \Delta p}{8\eta l} \qquad Q_P = \pi R^2 \bar{v} \tag{20}$$

Therefore,

$$\tau = \frac{\Delta p R}{2l} \tag{21}$$

and the effective viscosity of the fluid can be estimated as $\eta = \tau \cdot R/(4\bar{v})$.

The change in the value of shear stress directly causes the dependence of the effective viscosity of the fluid from the pipe size and flow rate. In this case the effective viscosity of the transported fluid can be determined from Eqs. (20) and (21) as

$$\eta = \frac{\tau \cdot R}{4\bar{v}} \qquad (22)$$

With increasing radius τ according to Eq. (21) increases.

Calculations showed that the magnitude of the shear stress τ is relatively small in the range of pipe sizes considered. This indicates that the surface of carbon nanotubes is very smooth and the water molecules can easily slide through it.

In fact, shear stress is primarily due to van der Waals interaction between the solid wall and the water molecules. It is noted that the characteristic distance between the near-wall layer of fluid and pipe wall depends on the equilibrium distance between atoms O and C, the distribution of the atoms of the solid wall and bend of the pipe.

From Fig. 21 we can see, that the effective viscosity η increases by two orders of magnitude when R_0 changes from 0.67 to 5.4 nm. The value of the calculated viscosity of water in the tube (10,10) is 8.5 × 10^{-8} Pa/s, roughly four orders of magnitude lower than the viscosity of a large mass of water.

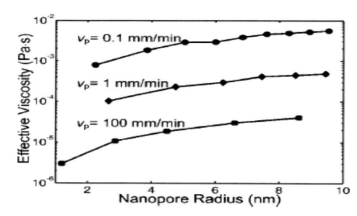

FIGURE 21 Size effect of shearing stress (triangle) and viscosity (square), with $\bar{v}$=165 м/с.

According to Eqs. (20)–(22), the effective viscosity can be calculated as $\eta = \frac{\pi R^4 \Delta p}{8QL}$. The results of calculations (Xi Chen et al., 2008) are shown in Fig. 22.

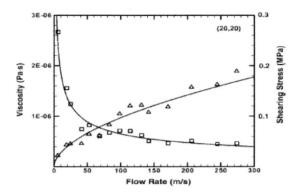

FIGURE 22 Effective viscosity as a function of the nanopore radius and the loading rate.

The dependence of the shear stress on the flow rate is illustrated in Fig. 23. From the example of the tube (20,20) it's clear that τ increases with v. The growth rate slowed down at higher values $\bar{v}$.

At high speeds $\bar{v}$, while water molecules are moving along the surface of the pipe, the liquid molecules do not have enough time to fully adjust their positions to minimize the free energy of the system. Therefore, the distance between adjacent carbon atoms and water molecules may be less than the equilibrium van der Waals distances. This leads to an increase in van der Waals forces of repulsion and leads to higher shear stress.

Scientists showed that, even though the equation for viscosity is based on the theory of the continuum, it can be extended to a complex flow to determine the effective viscosity of the nanotube.

Figure 23 also shows a dependence η on $\bar{v}$ on the inside of the nanotube (20,20). It is seen that η decreases sharply with increasing flow rate and begins to asymptotically approaches a definite value when $\bar{v}$>150 m/sec. For the current pipe size and flow rate ranges $\eta \sim 1/\sqrt{\bar{v}}$, this trend is the result of addiction $\tau - \bar{v}$, contained the same Fig. 23. According to Fig. 22 high-speed effects are negligible.

One can easily see that the dependence of viscosity on the size and speed is consistent qualitatively with the results of molecular dynamic simulations. In all studied cases, the viscosity is much smaller than its macroscopic analogy. When the radius of the pores varies from about 1 nm to 10 nm, the value of the effective viscosity increases by an order of magnitude respectively. A more significant change occurs when the speed increases from 0.1 mm/min up to 100 mm/min. This results in a change in the value of viscosity η, respectively, by 3–4 orders. The discrepancy between simulation and test data can be associated with differences in the structure of the nanopores and liquid phase.

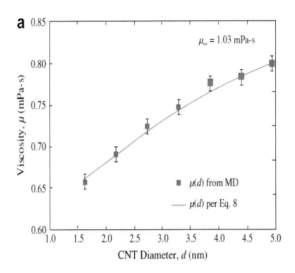

FIGURE 23 Flow rate effect of shearing stress (triangle) and viscosity (square), with R_0 =1.336 нм.

Figure 24 shows the viscosity dependence of water, calculated by the method of DM, the diameter of the CNT. The viscosity of water, as shown in the figure, increases monotonically with increasing diameter of the CNT.

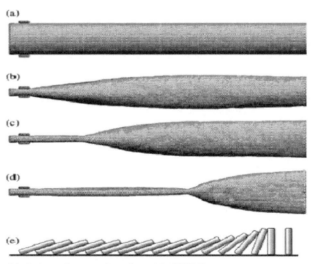

FIGURE 24 Variation of water viscosity with CNT diameter.

6.11 ENERGY RELEASE DUE TO THE COLLAPSE OF THE NANOTUBE

Scientists theoretically predicted the existence of a "domino effect" in single-walled carbon nanotube.

Squashing it at one end by two rigid moving to each other by narrow graphene planes (about 0.8 nm in width and 8.5 nm in length), one can observe it rapidly (at a rate exceeding 1 km/s) release its stored energy by collapsing along its length like a row of dominoes. The effect resembles a tube of toothpaste squeezing itself (Fig. 25).

The structure of a single-walled carbon nanotube has two possible stable states: circular or collapsed. Chang realized that for nanotubes wider than 3.5 nanometers, the circular state stores more potential energy than the collapsed state as a result of van der Waal's forces. He performed molecular dynamics simulations to find out what would happen if one end of a nanotube was rapidly collapsed by clamping it between two graphene bars.

This phenomenon occurs with the release of energy, and thus allows for the first time to talk about carbon nanotubes as energy sources. This effect can also be used as a accelerator of molecules.

The tube does not collapse over its entire length at the same time, but sequentially, one after the other carbon ring, starting from the end, which is tightened (Fig. 25). It happens just like a domino collapses, arranged in a row (this is known as the "domino effect"). Only here the role of bone dominoes is performed by a ring of carbon atoms forming the nanotube, and the nature of this phenomenon is quite different.

Recent studies have shown that nanotubes with diameters ranging from 2 to 6 nm, there are two stable equilibrium states – cylindrical (tube no collapses) and compressed (imploded tube) – with difference values of potential energy, the difference between which and can be used as an energy source.

Researchers found that switching between these two states with the subsequent release of energy occurs in the form of arising domino effect wave. The scientists have shown that such switching is not carried out in carbon nanotubes with diameters of 2 nm and not more, as evident from previous studies, but with a little more, starting from 3.5 nm.

A theoretical study of the "domino effect" was conducted using a special method of classical molecular dynamics, in which the interaction between carbon atoms was described by van der Waals forces (intermolecular forces), mathematically expressed by the Lennard–Jones potential.

The main reason for the observed effect, in author's opinion, is the competition of the potential energy of the van der Waals interactions, which "collapses" the nanotube with the energy of elastic deformation, which seeks to preserve the geometry of carbon atoms, which eventually leads to a bistable (collapsed and no collapsed) configuration of the carbon nanotube.

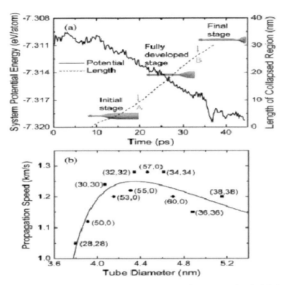

FIGURE 25 "Domino effect" in a carbon nanotube. (a) The initial form of carbon nanotubes – cylindrical. (b) One end of the tube is squeezed. (c), (d) Propagation of domino waves – the configuration of the nanotube 15 and 25 picoseconds after the compression of its end. (e) Schematic illustration of the "domino effect" under the influence of gravity.

For small diameter tubes the dominant is the energy of elastic deformation, the cylindrical shape of such a nanotube is stable. For nanotubes with sufficiently large diameter the van der Waals interaction energy is dominant. This means more stability and less compressed nanotube stability, or, as physicists say, metastability (i.e., apparent stability) of its cylindrical shape.

Thus, "domino effect" wave can be produced in a carbon nanotube with a relatively large diameter (more than 3.5 nm, as the author's calculations), because only in such a system the potential energy of collapsing structures may be less than the potential energy of the "normal" nanotube. In other words, the cylindrical and collapsing structure of a nanotubes with large diameters are, respectively, its metastable and stable states.

Change of the potential energy of a carbon nanotube with a propagating "domino effect" wave in it with time is represented as a graph in Fig. 26a.

This chart shows three sections of features in the change of potential energy. The first (from 0 ps to 10 ps) are composed of elastic strain energy, which appears due to changes in the curvature of the walls of the nanotubes in the process of collapsing, the energy change of van der Waals interactions occurring between the opposite walls of the nanotubes, as well as the interaction between the tube walls and graphene planes, compressing her end.

The second region (from 10 ps to 35 ps) corresponds to the "domino effect" – "domino effect" wave spreads along the surface of the carbon nanotube. Energetically, it looks like this: at every moment, when carbon ring collapses, some of the potential energy of the van der Waals is converted to kinetic energy (the rest to the energy of elastic deformation), which is kind of stimulant to support and "falling domino" – following the collapse of the rings, which form the nanotube, with each coagulated ring reducing the total potential energy of the system.

Finally, the third segment (from 35 ps to 45 ps) corresponds to the ended "domino" process – carbon tube collapsed completely. We emphasize that the nanotube, which collapsed (as seen from the Fig. 26a), has less potential energy than it was before beginning of the "domino effect".

In other words, the spread "domino effect" waves – a process that goes with the release of energy: about 0.01 eV per atom of carbon. This is certainly not comparable in any way with the degree of energy yield in nuclear reactions, but the fact of power generation carbon nanotube is obvious.

Later scientists analyzed the kinematic characteristics of the process – what the rate of propagation of the wave of destruction or collapse of carbon nanotubes and their characteristics is it determined?

Calculations show that the wave of dominoes in a tube diameter of 4–5 nm is about 1 km/s (as seen from the Fig. 26b) and depends on its geometry – the diameter and chirality in a non-linear manner. The maximum effect should be observed in the tube with a diameter slightly less than 4.5 nm – it will be carbon rings to collapse at a speed of 1.28 km/sec. The theoretical dependence obtained by the author, shows the blue solid line. And now an example of how energy is released in such a system with a "domino effect" can be used in nanodevices. The author offers an original way to use – "nanogun" (Fig. 27a). Imagine that at our disposal there is a carbon nanotube with chirality of (55.0) and corresponding to observation of the dominoes diameter.

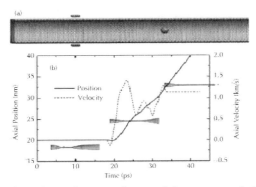

FIGURE 26 (a) Time dependences of potential energy and the length of the collapsed nanotube, (b) The velocity of propagation of the domino wave for carbon nanotubes with different diameters and chirality.

Put a C_{60} fullerene inside a nanotube. A little imagination, and a carbon nanotube can be considered as the gun trunk, and the C_{60} molecule – as its shell.

The molecule located inside will extruded from it into the other, open end under the influence of squeezing nanotube (see Fig. 5.6).

The question is, what is the speed of the "core"? Chang estimated that, depending on the initial position of the fullerene molecule when leaving a nanotube, it can reach speeds close to the velocity of "domino effect" waves – about 1 km/s (Fig. 27b). Interestingly, this speed is reached by the "core" for just 2 picoseconds and at a distance of 1 nm. It is easy to calculate that the observed acceleration is of great value $0,5 \cdot 10^{15}{}_M / c^2$. For comparison, the speed of bullets in an AK-47 is 1.5 times lower than the rate of fullerene emitted from a gun.

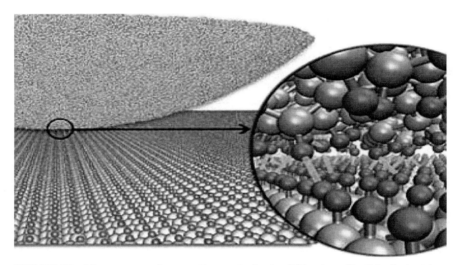

FIGURE 27 Nanocannon scheme acting on the basis of "domino effect" in the incision. (a) Inside a carbon nanotube (55.0) is the fullerene molecule C_{60}. (b) The initial position and velocity of the departure of the "core" (a fullerene molecule), depending on the time. The highest rate of emission of C_{60} (1.13 km/s) comparable to the velocity of the domino wave.

Necessary be noted that the simulation took place in nanogun at assumption of zero temperature in Kelvin. However, this example is not so abstract and may be used in the injecting device.

Thus, for the first time been demonstrated, albeit only in theory, the use of single-walled carbon nanotubes as energy sources.

6.12 FLUID FLOW IN NANOTUBES

The friction of surface against the surface in the absence of the interlayer between the liquid material (so-called dry friction) is created by irregularities in the given surfaces that rub one another, as well as the interaction forces between the particles that make up the surface.

As part of their study, the researchers built a computer model that calculates the friction force between nanosurfaces (Fig. 28). In the model, these surfaces were presented simply as a set of molecules for which forces of intermolecular interactions were calculated.

As a result, scientists were able to establish that the friction force is directly proportional to the number of interacting particles. The researchers propose to consider this quantity by analogue of so-called true macroscopic contact area. It is known that the friction force is directly proportional to this area (it should not be confused with common area of the contact surfaces of the bodies).

In addition, the researchers were able to show that the friction surface of the nanosurfaces can be considered within the framework of the classical theories of friction of non-smooth surfaces.

A literature review shows that nowadays molecular dynamics and mechanics of the continuum in are the main methods of research of fluid flow in nanotubes.

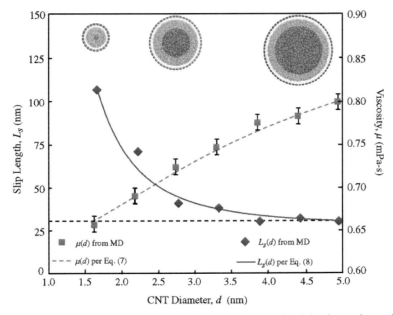

FIGURE 28 Computer model of friction at the nanoscale (the right shows the surfaces of interacting particles).

Although the method of molecular dynamics simulations is effective, it at the same time requires enormous computing time especially for large systems. Therefore, simulation of large systems is more reasonable to carry out nowadays by the method of continuum mechanics.

The fluid flow in the channel can also be considered in the framework of the continuum hypothesis. The Navier–Stokes equation can be used and the velocity profile can be determined for Poiseuille flow.

However the water flow by means of pressure differential through the carbon nanotubes with diameters ranging from 1.66 to 4.99 nm is studies using molecular dynamics simulation study. For each nanotube the value enhancement predicted by the theory of liquid flow in the carbon nanotubes is calculated. This formula is defined as a ratio of the observed flow in the experiments to the theoretical values without considering slippage on the model of Hagen–Poiseuille. The calculations showed that the enhancement decreases with increasing diameter of the nanotube.

Important conclusion is that by constructing a functional dependence of the viscosity of the water and length of the slippage on the diameter of carbon nanotubes, the experimental results in the context of continuum fluid mechanics can easily be described. The aforementioned is true even for carbon nanotubes with diameters of less than 1.66 nm.

The theoretical calculations use the following formula for the steady velocity profile of the viscosity η of the fluid particles in the CNT under pressure gradient $\partial p / \partial z$:

$$v(r) = \frac{R^2}{4\eta}\left[1 - \frac{r^2}{R^2} + \frac{2L_S}{R}\right]\frac{\partial p}{\partial z} \qquad (23)$$

The length of the slip, which expresses the speed heterogeneity at the boundary of the solid wall and fluid is defined as:

$$L_S = \frac{v(r)}{dv/dr}\bigg|_{r=R} \qquad (24)$$

Then the volumetric flow rate, taking into account the slip Q_S is defined as:

$$Q_S = \int_0^R 2\pi r \cdot v(r) dr = \frac{\pi\left[(d/2)^4 + 4(d/2)^3 \cdot L_S\right]}{8\eta} \cdot \frac{\partial p}{\partial z} \qquad (25)$$

Equation (25) is a modified Hagen–Poiseuille equation, taking into account slippage. In the absence of slip $L_S = 0$ (Eq. (25)) coincides with the Hagen–

Poiseuille flow (Eq. (20)) for the volumetric flow rate without slip Q_P. In some works the parameter enhancement flow ε is also introduced. It is defined as the ratio of the calculated volumetric flow rate of slippage to Q_p (calculated using the effective viscosity and the diameter of the CNT). If the measured flux is modeled using Eq. (25), the degree of enhancement takes the form:

$$\varepsilon = \frac{Q_S}{Q_P} = \left[1 + 8\frac{L_S(d)}{d}\right]\frac{\eta_\infty}{\eta(d)} \qquad (26)$$

where $d = 2R$ – diameter of CNT, η_∞ – viscosity of water, $L_S(d)$ – CNT slip length depending on the diameter, $\eta(d)$ – the viscosity of water inside CNTs depending on the diameter.

If $\eta(d)$ finds to be equal to η_∞, then the influence of the effect of slip on ε is significant, if $L_S(d) \geq d$. If $L_S(d) < d$ and $\eta(d) = \eta_\infty$, then there will be no significant difference compared to the Hagen–Poiseuille flow with no slip.

Table 2 shows the experimentally measured values of the enhancement water flow. Enhancement flow factor and the length of the slip were calculated using the equations given above.

TABLE 2 Experimentally measured values of the enhancement water flow.

Nanosystems	Diameter (nm)	Enhancement, ε	slip length, L_S, (nm)
carbon nanotubes	300–500	1	0
	44	22–34	113–177
carbon nanotubes	7	10^4–10^5	3900–6800
	1,6	560–9600	140–1400

Figure 29 depicts the change in viscosity of the water and the length of the slip in diameter. As can be seen from the figure, the dependence of slip length to the diameter of the nanotube is well described by the empirical relation

$$L_S(d) = L_{S,\infty} + \frac{C}{d^3} \qquad (27)$$

where $L_{S,\infty} = 30$ nm – slip length on a plane sheet of graphene, a C – const.

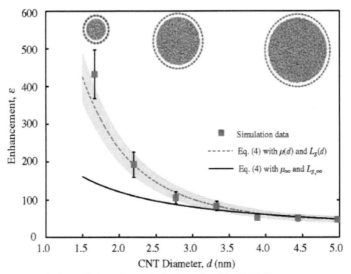

FIGURE 29 Variation of viscosity and slip length with CNT diameter.

Figure 30 shows dependence of the enhancement of the flow rate ε on the diameter for all seven CNTs.

There are three important features in the results. First, the enhancement of the flow decreases with increasing diameter of the CNT. Second, with increasing diameter, the value tends to the theoretical value of Eqs. (26) and (27) with a slip $L_{s\infty} = 30$ nm and the effective viscosity $\mu(d) = \mu_\infty$. The dotted line shown the curve of 15% in the second error in the theoretical data of viscosity and slip length. Third, the change ε in diameter of CNTs cannot be explained only by the slip length.

To determine the dependence the volumetric flow of water from the pressure gradient along the axis of single-walled nanotube with the radii of 1.66, 2.22, 2.77, 3.33, 3.88, 4.44 and 4.99 nm in the method of molecular modeling was used. Snapshot of the water–CNT is shown in Fig. 31.

Figure 31 shows the results of calculations to determine the pressure gradient along the axis of the nanotube with the diameter of 2.77 nm and a length of 20 nm. Change of the density of the liquid in the cross sections was less than 1%.

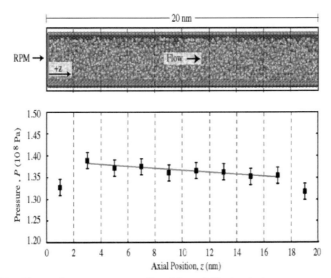

FIGURE 30 Flow enhancement as predicted from MD simulations.

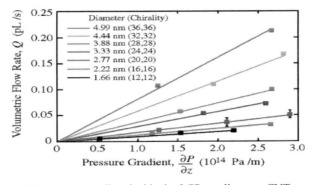

FIGURE 31 Axial pressure gradient inside the 2.77 nm diameter CNT.

Figure 32 shows the dependence of the volumetric flow rate from the pressure gradient for all seven CNTs. The flow rate ranged from 3–14 m/sec. In the range considered here the pressure gradient $(0-3) \times 10^9 \, atm/m$ Q ($pl/sek = 10^{-15} \, m^3/sek$) is directly proportional $\partial p / \partial z$. Coordinates of chirality for each CNT are indicated in the figure legend. The linearity of the relations between flow and pressure gradient confirms the validity of calculations of the Eq. (25).

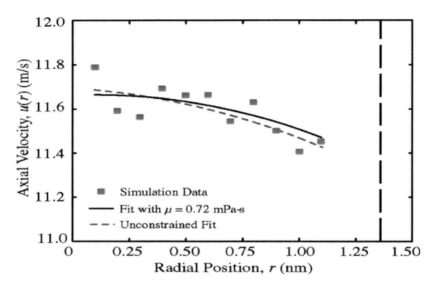

FIGURE 32 Relationship between volumetric liquid flow rate in carbon nanotubes with different diameters and applied pressure gradient.

Figure 33 shows the profile of the radial velocity of water particles in the CNT with diameter 2.77 nm. The vertical dotted line at 1.38 nm marked surface of the CNT. It is seen that the velocity profile is close to a parabolic shape.

Researchers consider the flow of water under a pressure gradient in the single-walled nanotubes of "chair" type of smaller radii: 0.83, 0.97, 1.10, 1.25, 1.39 and 1.66 nm.

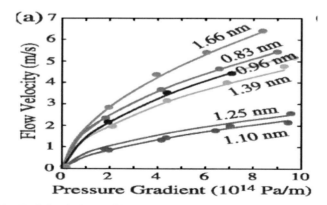

FIGURE 33 Radial velocity profile inside 2.77 nm diameter CNT.

Figure 34 shows the dependence of the mean flow velocity $\bar{v}$ on the applied pressure gradient $\Delta P/L$ in the long nanotubes – 75 nm at 298 K. A similar picture pattern occurs in the tube with the length of 150 nm.

As we can see, there is conformance with the Darcy law, the average flow rate for each CNT increases with increasing pressure gradient. For a fixed value of $\Delta p/L$, however, the average flow rate does not increase monotonically with increasing diameter of the CNTs, as follows from Poiseuille equation. Instead, when at the same pressure gradient, decrease of the average speed in a CNT with the radius of 0.83 nm to a CNT with the radius of 1.10 nm, similar to the CNTs 1.10 and 1.25 nm, then increases from a CNT with the radius of 1.25 nm to a CNT of 1.66 nm.

The nonlinearity of the relationship between $\bar{v}$ and $\Delta P/L$ are the result of inertia losses (i.e., insignificant losses) in the two boundaries of the CNT. Inertial losses depend on the speed and are caused by a sudden expansion, abbreviations, and other obstructions in the flow.

We note the important conclusion of the work of scientists in which the method of molecular modeling shows that the Eq. (23) (Poiseuille parabola) correctly describes the velocity profile of liquid in a nanotube when the diameter of a flow is 5–10 times more than the diameter of the molecule (≈ 0.17 nm for water).

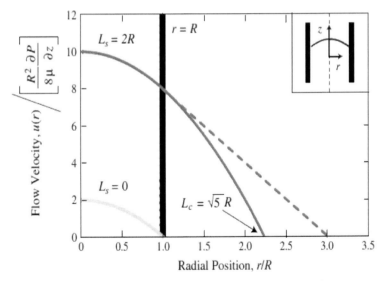

FIGURE 34 Relationship between average flow velocity and applied pressure gradient for the 75 nm long CNTs.

In Fig. 35, we can see the effect of slip on the velocity profile at the boundary of radius R of the pipe and fluid. When $L_S = 0$ the fluid velocity at the wall vanishes, and the maximum speed (on the tube axis) exceeds flow speed twice.

The figure shows the velocity profiles for Poiseuille flow without slip ($L_S = 0$) and with slippage $L_S = 2R$. The flow rate is normalized to the speed corresponding to the flow without slip. Thick vertical lines indicate the location of the pipe wall. The thick vertical lines indicate the location of the tube wall. As the length of the slip, the flow rate increases, decreases the difference between maximum and minimum values of the velocity and the velocity profile becomes more like a plug.

Velocity of the liquid on a solid surface can also be quantified by the coefficient of slip L_c. The coefficient of slippage – there is a difference between the radial position in which the velocity profile would be zero and the radial position of the solid surface. Slip coefficient is equal to $L_C = \sqrt{R^2 + 2RL_S} = \sqrt{5}R$.

For linear velocity profiles (e.g., Couette flow), the length of the slip and slip rate are equal. These values are different for the Poiseuille flow.

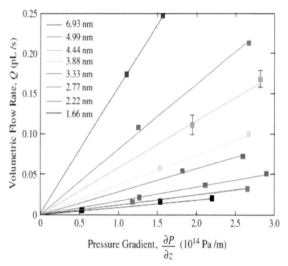

FIGURE 35 No-slip Poiseuille flow and slip Poiseuille flow through a tube.

Figure 36 shows dependence of the volumetric flow rate Q from the pressure gradient $\partial p / \partial z$ in long nanotubes with diameters between 1.66 nm and 6.93 nm. As can be seen in the studied range of the pressure gradient Q is proportional to $\partial p / \partial z$. As in the Poiseuille flow, volumetric flow rate increases monotonically with the diameter of CNT at a fixed pressure gradient. Magnitude of calculations error for all the dependencies are similar to the error for the CNT diameter 4.44 nm, marked in the figure.

Many researchers considered steady flow of incompressible fluids in a channel width $2h$ under action of the force of gravity ρg or pressure gradient $\partial p / \partial y$, which is described by the Navier–Stokes equations. The velocity profile has a parabolic form:

$$U_y(z) = \frac{\rho g}{2\eta} \cdot \left[(\delta + h)^2 - z^2\right]$$

where δ – length of the slip, which is equal to the distance from the wall to the point at which the velocity extrapolates to zero.

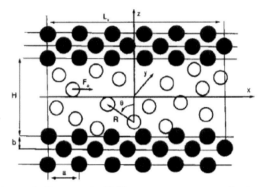

FIGURE 36 Volumetric flow rate in CNTs versus pressure gradient.

6.13 SOME OF THE IDEAS AND APPROACHES FOR MODELING IN NANOHYDROMECANICS

Let's consider the fluid flow through the nanotube. Molecules of a substance in a liquid state are very close to each other (Fig. 37).

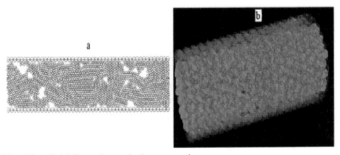

FIGURE 37 The fluid flow through the nanotube.

Most liquid's molecules have a diameter of about 0,1 nm. Each molecule of the fluid is "squeezed" on all sides by neighboring molecules and for a period of time $(10^{-10} - 10^{-13} s)$ fluctuates around certain equilibrium position, which itself from time to time is shifted in distance commensurating with the size of molecules or the average distance between molecules l_{cp}:

$$l_{cp} \approx \sqrt[3]{\frac{1}{n_0}} = \sqrt[3]{\frac{\mu}{N_A \rho}},$$

where: n_0 – number of molecules per unit volume of fluid, N_A – Avogadro's number, ρ – fluid density, μ – molar mass.

Estimates show that one cubic of nanowater contains about 50 molecules. This gives a basis to describe the mass transfer of liquid in a nanotube-based continuum model. However, the specifics of the complexes, consisting of a finite number of molecules, should be kept in mind. These complexes, called clusters in literatures, are intermediately located between the bulk matter and individual particles – atoms or molecules. The fact of heterogeneity of water is now experimentally established.

There are groups of molecules in liquid – "microcrystals" containing tens or hundreds of molecules. Each microcrystal maintains solid form. These groups of molecules, or "clusters" exist for a short period of time, then break up and are re-created again. Besides, they are constantly moving so that each molecule does not belong at all times to the same group of molecules, or "cluster".

Modeling predicts that gas molecules bounce off the perfectly smooth inner walls of the nanotubes as billiard balls, and water molecules slide over them without stopping. Possible cause of unusually rapid flow of water is maybe due to the small-diameter nanotube molecules move on them orderly, rarely colliding with each other. This "organized" move is much faster than usual chaotic flow. However, while the mechanism of flow of water and gas through the nanotubes is not very clear, and only further experiments and calculations can help understand it.

The model of mass transfer of liquid in a nanotube proposed in this paper is based on the availability of nanoscale crystalline clusters in it.

A similar concept was developed in which the model of structured flow of fluid through the nanotube is considered. It is shown that the flow character in the nanotube depends on the relation between the equilibrium crystallite size and the diameter of the nanotube.

Figure 38 shows the results of calculations by the molecular dynamics of fluid flow in the nanotube in a plane (a) and three-dimensional (b) statement. The figure shows the ordered regions of the liquid.

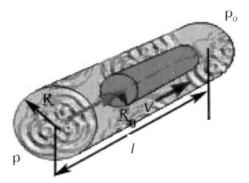

FIGURE 38 The results of calculations of fluid flow in the nanotube.

The typical size of crystallite is 1–2 nm, i.e. compared, for example, with a diameter of silica nanotubes of different composition and structure. The flow model proposed in the present work is based on the presence of "quasi-solid" phase in the central part of the nanotube and liquid layer, non-autonomous phases.

Consideration of such a structure that is formed when fluid flows through the nanotube, is also justified by the aforementioned results of the experimental studies and molecular modeling.

When considering the fluid flow with such structure through the nanotube, we will take into account the aspect ratio of "quasi-solid" phase and the diameter of the nanotube so that a character of the flow is stable and the liquid phase can be regarded as a continuous medium with viscosity η.

Let's establish relationship between the volumetric flow rate of liquid Q flowing from a liquid layer of the nanotube length l, the radius R and the pressure drop $\Delta p / l$, $\Delta p = p - p_0$, where p_0 is the initial pressure in the tube (Fig. 39).

Let R_0 be a radius of the tube from the "quasi-solid" phase, v – velocity of fluid flow through the nanotube.

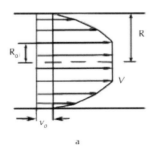
a

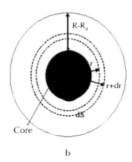
b

FIGURE 39 Flow through liquid layer of the nanotube.

Structural regime of fluid flow (Fig. 40) implies the existence of the continuous laminar layer of liquid (the liquid layer in the nanotube) along the walls of a pipe. In the central part of a pipe a core of the flow is observed, where the fluid moves, keeping his former structure, i.e. as a solid ("quasi-solid" phase in the nanotube). The velocity slip is indicated in Fig. 40a through v_0.

Let's find the velocity profile $v(r)$ in a liquid interlayer $R_0 \leq r \leq R$ of the nanotube. We select a cylinder with radius r and length l in the interlayer, located symmetrically to the center line of the pipe (see Fig. 40b).

At the steady flow, the sum of all forces acting on all the volumes of fluid with effective viscosity η, is zero.

The following forces are applied on the chosen cylinder: the pressure force and viscous friction force affects the side of the cylinder with radius r, calculated by the Newton formula.

Thus,

$$(p-p_0)\pi r^2 = -\eta \frac{dv}{dr} 2\pi r l \qquad (28)$$

Integrating expression (28) between r to R with the boundary conditions $r = R : v = v_0$, we obtained a formula to calculate the velocity of the liquid layers located at a distance r from the axis of the tube:

$$v(r) = (p-p_0)\frac{R^2 - r^2}{4\eta l} + v_0 \qquad (29)$$

Maximum speed $v_я$ has the core of the nanotube $0 \leq r \leq R_0$ and is equal to:

$$v_я = (p-p_0)\frac{R^2 - R_0^2}{4\eta l} + v_0 \qquad (30)$$

Such structure of the liquid flow through nanotubes considering the slip is similar to a behavior of viscoplastic liquids in the tubes. Indeed, as we know, for viscoplastic fluids a characteristic feature is that they are to achieve a certain critical internal shear stresses τ_0 behave like solids, and only when internal stress exceeds a critical value above begin to move as normal fluid. Scientists shown that liquid behaves in the nanotube the similar way. A critical pressure drop is also needed to start the flow of liquid in a nanotube.

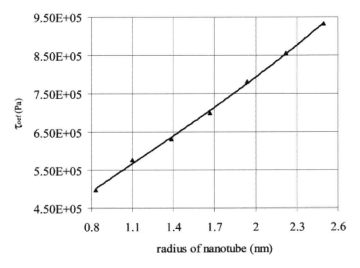

FIGURE 40 Structure of the flow in the nanotube.

Structural regime of fluid flow requires existence of continuous laminar layer of liquid along the walls of pipe. In the central part of the pipe is observed flow with core radius R_{00}, where the fluid moves, keeping his former structure, i.e. as a solid.

The velocity distribution over the pipe section with radius R of laminar layer of viscoplastic fluid is expressed as follows:

$$v(r) = \frac{\Delta p}{4\eta l}\left(R^2 - r^2\right) - \frac{\tau_0}{\eta}(R - r) \qquad (31)$$

The speed of flow core in $0 \leq r \leq R_{00}$ is equal

$$v_я = \frac{\Delta p}{4\eta l}\left(R^2 - R_{00}^2\right) - \frac{\tau_0}{\eta}(R - R_{00}) \qquad (32)$$

Let's calculate the flow or quantity of fluid flowing through the nanotube cross-section S at a time unit. The liquid flow dQ for the inhomogeneous velocity field flowing from the cylindrical layer of thickness dr, which is located at a distance r from the tube axis is determined from the relation

$$dQ = v(r)dS = v(r)2\pi r dr \qquad (33)$$

where ds – the area of the cross-section of cylindrical layer (between the dotted lines in Fig. 40).

Let's place Eq. (29) in Eq. (33), integrate over the radius of all sections from R_0 to R and take into account that the fluid flow through the core flow is determined from the relationship $Q_{\dot y} = \pi R_0^2 v_y$. Then we get the formula for the flow of liquid from the nanotube:

$$Q = Q_P \left[v_* \frac{8l\eta R_*^2}{\Delta p R^4} + \left(\frac{R_*}{R}\right)^4 \right] \quad (34)$$

If $(R_0/R)^4 \ll 1$ (no nucleus) and $v_0 \Delta p R^2 / 8l\eta \ll 1$ (no slip), then Eq. (34) coincides with Poiseuille formula (Eq. 20). When $R_0 \approx R$ (no of a viscous liquid interlayer in the nanotube), the flow rate Q is equal to volumetric flow $Q \approx \pi R^2 v_0$ of fluid for a uniform field of velocity (full slip).

Accordingly, flow rate of the viscoplastic fluid flowing with a velocity (4.7), is equal to:

$$Q = -\frac{\pi R^3 \tau_0}{3\eta}\left[1 - \left(\frac{R_{00}}{R}\right)^3\right] + Q_P\left[1 - \left(\frac{R_{00}}{R}\right)^4\right] \quad (35)$$

Comparing Eqs. (29)–(32) and (34), (35), we can see that the structure of the flow of the liquid through the nanotubes considering the slippage, is similar to that of the flow of viscoplastic fluid in a pipe of the same radius R.

Given that the size of the central core flow of viscoplastic fluid (radius R_{00}) is defined by

$$R_{00} = \frac{2\tau_0 l}{\Delta p} \quad (35)$$

for viscoplastic fluid flow we obtain Buckingham formula:

$$Q = Q_P\left[1 + \frac{1}{3}\left(\frac{2l\tau_0}{R\Delta p}\right)^4 - \frac{4}{3}\left(\frac{2l\tau_0}{R\Delta p}\right)\right] \quad (36)$$

We'll establish a conformity of the pipe that implements the flow of a viscoplastic fluid with a fluid-filled nanotube, the same size and with the same pressure drop. We say that an effective internal critical shear stress τ_{0ef} of viscoplastic fluid flow, which ensures the coincidence rate with the flow of fluid in the nanotube. Then from Eq. (36) we obtain equation of fourth order to determine τ_{0ef}:

$$\left(\frac{2l\tau_{0ef}}{R\Delta p}\right)^4 - 4\left(\frac{2l\tau_{0ef}}{R\Delta p}\right) = A, \ A = 3(\varepsilon-1), \varepsilon = Q/Q_P \quad (37)$$

The solution of Eq. (37) can be found, for example, the iteration method of Newton:

$$\overline{\tau}_{0ef\,n} = \overline{\tau}_{0ef\,n-1} - \frac{\overline{\tau}_{0ef\,n-1}^{-4} - 4\overline{\tau}_{0ef\,n-1} - A}{4\overline{\tau}_{0ef\,n-1}^{-3} - 4}, \ \overline{\tau}_{0ef} = \frac{2l\tau_{0ef}}{R\Delta p} \quad (38)$$

The first component in Eq. (34) represents the contribution to the fluid flow due to the slippage, and it becomes clear that the slippage significantly enhances the flow rate in the nanotube, when $l\eta v_0 >\approx \Delta pR^2$.

This result is consistent with experimental and theoretical results (which show that water flow in nanochannels can be much higher than under the same conditions, but for the liquid continuum.

In the absence of slippage $\varepsilon = 1$ the equation (4.16) has a trivial solution $\overline{\tau}_{0ef} = 0$.

THE RESULTS OF THE CALCULATION:

Let's determine the dependence of the effective critical inner shear stress τ_{0ef} on the radius of the nanotubes, by taking necessary values for calculations $\varepsilon = Q/Q_P$. The results of calculations at $\Delta p/l = 2,1 \cdot 10^{14} \ Pa/m$ are in the table below:

R, м	τ_{0ef} (Па)	$\varepsilon = Q/Q_P$
$0,83 \cdot 10^{-9}$	498498	350
$0,83 \cdot 10^{-9}$	577500	200
$0,83 \cdot 10^{-9}$	632599	114
$1,665 \cdot 10^{-9}$	699300	84
$1,94 \cdot 10^{-9}$	782208	68
$2,22 \cdot 10^{-9}$	855477	57
$2,495 \cdot 10^{-9}$	932631	50

Calculations show that the value of effective internal shear stress depends on the size of the nanotube.

Figure 41 shows the dependence τ_{0ef} on the nanotube radius.

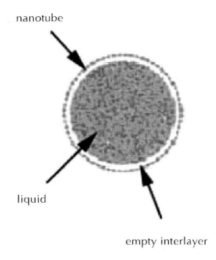

FIGURE 41 Dependence of the effective inner shear stress from the radius of the nanotube.

As you can see, this dependence $\tau_{0ef}(R)$ is almost linear. Within the range of the considered nanotube sizes τ_{0ef} has relatively low values, which indicates the smoothness of the surface of carbon nanotubes.

THE FLOW OF FLUID WITH AN EMPTY INTERLAYER:

The works of EM Kotsalis et al. (2004) and Xi Chen et al. (2008) were analyzed in the aforementioned analysis of the structure of liquid flow in carbon nanotubes. The results of the calculations of the cited works (Figs. 42 and 43) showed that during the flow of the liquid particles, an empty layer between the fluid and the nanotube is formed. The area near the walls of the carbon nanotube $R_* \leq r \leq R$ becomes inaccessible for the molecules of the liquid due to van der Waals repulsion forces of the heterogeneous particles of the carbon and water (Fig. 2). Moreover, according to the results of EM Kotsalis et al. (2004) and Xi Chen et al. (2008) thicknesses of the layers $R_* \leq r \leq R$ regardless of radiuses of the nanotubes are practically identical: $R_*/R \approx 0{,}8$.

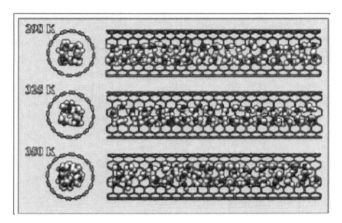

FIGURE 42 The structure of the flow.

A similar result was obtained in Hongfei Ye et al. (2011), which is an image (Fig. 43) of the configuration of water molecules inside (8, 8) single-walled carbon nanotubes at different temperatures: 298, 325, and 350 ^0K.

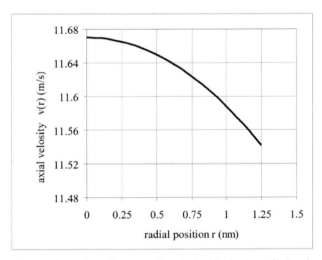

FIGURE 43 The configuration of water molecules inside single-walled carbon nanotubes.

Integrating expression (28) between r to R_* at the boundary conditions $r = R_*$: $v = v_*$, we obtain a formula to calculate the velocity of the liquid layers located at a distance r from the axis of the tube:

$$v(r) = v_* + \frac{2Q_p}{\pi R^2}\left[\left(\frac{R_*}{R}\right)^2 - \left(\frac{r}{R}\right)^2\right] \qquad (39)$$

Let's insert Eq. (39) in Eq. (33), integrate over the radius of all sections from 0 to R_*. Then we get a formula for the flow of liquid from the nanotube:

$$Q = Q_P\left[v_* \frac{8l\eta R_*^2}{\Delta p R^4} + \left(\frac{R_*}{R}\right)^4\right] \qquad (40)$$

or

$$\varepsilon = \frac{8l\eta v_*}{\Delta p R^2}\left(\frac{R_*}{R}\right)^2 + \left(\frac{R_*}{R}\right)^4$$

from which we can determine the unknown v_*:

$$v_* = \frac{Q_P}{\pi R^2}\left[\varepsilon - \left(\frac{R_*}{R}\right)^4\right]\left(\frac{R}{R_*}\right)^2 \qquad (41)$$

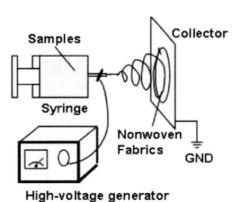

FIGURE 44 Profile of the radial velocity of water in carbon nanotubes.

Figure 44 shows the profile of the radial velocity of water particles in a carbon nanotube with a diameter of 2.77 nm, calculated using the formula (39) at $Q_P = 4,75 \cdot 10^{-19} n/m^2$, $\varepsilon = 114$. The velocity at the border v_* is equal to 11.55 m/s.

It is seen that the velocity profile is similar to a parabolic shape, and at the same time agrees with the numerical values.

The calculations suggest the following conclusions. Flow of liquid in a nanotube was investigated using synthesis of the methods of the continuum theory and molecular dynamics. Two models are considered. The first is based on the fact that fluid in the nanotube behaves like a viscoplastic. A method of calculating the value of limiting shear stress is proposed, which was dependent on the nanotube radius. A simplified model agrees quite well with the results of the molecular simulations of fluid flow in carbon nanotubes. The second model assumes the existence of an empty interlayer between the liquid molecules and wall of the nanotube. This formulation of the task is based on the results of experimental works known from the literature. The velocity profile of fluid flowing in the nanotube is practically identical to the profile determined by molecular modeling.

As seen from the results of the calculations, the velocity value varies slightly along the radius of the nanotube. Such a velocity distribution of the fluid particles can be explained by the lack of friction between the molecules of the liquid and the wall due to the presence of an empty layer. This leads to an easy slippage of the liquid and, consequently, anomalous increase in flow compared to the Poiseuille flow.

6.14 MODELING AND SIMULATION TECHNIQUES FOR NANOFIBERS

Symbols	Definition	Units
R	Radius of jet	m
n	Jet velocity	m/s
Q	Flow rate	m^3/s
I	Jet current	A
J	Current density	A/m^2
$A(s), S$	Cross-sectional area	m^2
K	The conductivity of the liquid	S/m
E	Electric field	V/m
L	Spinning distance	M
s	Electric density	C/m^2
P	Linear momentum	kg m/s, (N.s)
F	Force	N
t	Time	S
m	Mass	Kg

Symbol	Description	Units
r	Density	kg/m³
z	Jet axial position	m
p	Pressure	N/m²
t_{zz}	Axial viscous normal stress	N/m²
g	Surface tension	N/m
R'	Slope of the jet surface	—
t_t^e	Tangential electric force	N/m²
t_n^e	Normal electric force	N/m²
ε	Dielectric constant of the jet	—
$\bar{\varepsilon}$	Dielectric constant of the ambient air	—
W_e	Electric work	J
l	Distance	M
U_e	Electric potential energy	J
t_{rr}	Radial normal stress	N/m²
r_p	Polarized charge density	C/m²
P'	Polarization	C/m²
$q, q_0 \& Q_b$	Charge	C
r	Distance between two charges	m
V	Electric potential	kV
t	Shear stress	N/m²
K'	Flow consistency index	—
m	Flow behavior index	—
m	Constant	—
$\dot{\gamma}$	Rate of strain tensor	s⁻¹
S_k	Excess stress	N/m²
C_k	Configurational tensor	N/m²
h	Viscosity	P
E_k	Young's modulus	N/m²
m_k	Shear modulus	N/m²

b_k	Constitutive mobility	m²/(V·s)
N	Number of beads per unit volume	m⁻³
T	Temperature	K
G	Elastic modulus	N/m²
l	Filament	m
ε'	Lagrangian axial strain	—
b	Power exponent	—

6.14.1 INTRODUCTION

Electrospinning is a procedure in which an electrical charge to draw very fine (typically on the micro or nanoscale) fibers from polymer solution or molten. Electrospinning shares characteristics of both electrospraying and conventional solution dry spinning of fibers. The process does not require the use of coagulation chemistry or high temperatures to produce solid threads from solution. This makes the process more efficient to produce the fibers using large and complex molecules. Recently, various polymers have been successfully electrospun into ultrafine fibers mostly in solvent solution and some in melt form [1–2]. Optimization of the alignment and morphology of the fibers is produced by fitting the composition of the solution and the configuration of the electrospinning apparatus such as voltage, flow rate, and etc. As a result, the efficiency of this method can be improved [3]. Mathematical and theoretical modeling and simulating procedure will assist to offer an in-depth insight into the physical understanding of complex phenomena during electrospinningand might be very useful to manage contributing factors toward increasing production rate [4–5].

Despite the simplicity of the electrospinning technology, industrial applications of it are still relatively rare, mainly due to the notable problems of very low fiber production rate and difficulties in controlling the process [6].

Modeling and simulation (M&S) give information about how something will act without actually testing it in real. The model is a representation of a real object or system of objects for purposes of visualizing its appearance or analyzing its behavior. Simulation is transition from a mathematical or computational model to description of the system behavior based on sets of input parameters [7–8]. Simulation is often the only means for accurately predicting performance of the modeled system [9]. Using simulation is generally cheaper and safer than conducting experiments with a prototype of the final product. Also simulation can often be even more realistic than traditional experiments, as they allow the free configuration of environmental and operational parameters and can often be run faster than in real time. In a situation with different alternatives analysis, simulation can improve the efficiency, in particular when the necessary data to initialize can easily be obtained from operational data. Applying simulation adds decision support systems to the toolbox of traditional decision support systems [10].

Simulation permits set up a coherent synthetic environment that allows for integration of systems in the early analysis phase for a virtual test environment in the final system. If managed correctly, the environment can be migrated from the development and test domain to the training and education domain in real system under realistic constraints [11].

A collection of experimental data and their confrontation with simple physical models appears as an effective approach towards the development of practical tools for controlling and optimizing the electrospinning process. On the other hand, it is necessary to develop theoretical and numerical models of electrospinning because of demanding a different optimization procedure for each material [12]. Utilizing a model to express the effect of electrospinning parameters will assist researchers to make an easy and systematic way of presenting the influence of variables and by means of that, the process can be controlled. Additionally, it causes to predict the results under a new combination of parameters. Therefore, without conducting any experiments, one can easily estimate features of the product under unknown conditions [13].

6.14.2 ELECTROSPINNING

Spinning is the processes applied for drawing a fiber from polymer into filaments by passing through a spinneret, which is classified into solution spinning (wet or dry) and melt spinning. Conventional fiber-forming techniques have limitations of controlling the fiber diameter due to their dependency on the devices such as spinneret diameter. The major technique that can be used to make fibers thinner than 100 μm is electrospinning which is capable of giving very long continuous fibers by electrostatically drawing a polymer jet through a virtual spatio-temporal orifice [14]. An overview on invention history of this technique have reported for obtaining more clear vision and then the principle of electrospinning methodology is discussed.

6.14.2.1 THE HISTORY OF THE SCIENCE AND TECHNOLOGY OF ELECTROSPINNING

William Gilbert discovered the first record of the electrostatic attraction of a liquid in 1600 [15]. The first electrospinning patent was submitted by John Francis Cooley in 1900 [16]. After that in 1914 John Zeleny studied on the behavior of fluid droplets at the end of metal capillaries, which caused the beginning of the mathematically model the behavior of fluids under electrostatic forces [17]. Between 1931 and 1944 Anton Formhals took out at least 22 patents on electrospinning [16]. In 1938, N.D. Rozenblum and I.V. Petryanov–Sokolov generated electrospun fibers, which they developed into filter materials [18]. Between 1964 and 1969 Sir Geoffrey Ingram Taylor produced the beginnings of a theoretical foundation of electrospinning by mathematically modeling the shape of the (Taylor) cone formed by the fluid droplet under the effect of an electric field [19–20]. In the early 1990s several research groups (such as Reneker) demonstrated electrospun nanofibers. Since 1995, the number of publications about electrospinning has been increasing exponentially every year [16].

6.14.2.2 THE BASIC PRINCIPLES OF ELECTROSPINNING

As mentioned before, electrospinning gives the impression of being a very simple and easily controlled technique for the production of fibers with dimensions down to the nanometer range. Electrospinning fiber precursors are classified in two polymers as fiber forming substantial and materials such as metals, ceramics, and glasses. In a typical electrospinning experiment in a laboratory, a polymer solution or melt is pumped through a thin nozzle with an inner diameter on the order of 100 mm. In most of laboratory systems, the nozzle simultaneously serves as an electrode, to which a high electric field of 100–500 kVm^{-1} is applied, and the distance to the counter electrode is 10–25 cm [21]. The currents that flow during electrospinning range from a few hundred nanoamperes to microamperes. A high voltage is applied to the solution such that at a critical voltage, typically more than 5 kV, the repulsive force within the charged solution is larger than its surface tension and a jet would erupt from the tip of the spinneret. Although the jet is stable near to the tip of the spinneret, it soon enters a bending instability stage with further stretching of the solution jet under the electrostatic forces in the solution as the solvent evaporates [22–23]. The substrate on which the electrospun fibers are collected is typically brought into contact with the counter electrodes that are rotating and flat types [22]. The vertical alignment of the electrodes "from top to bottom" is not insignificant with respect to the process, but in principal, electrospinning can also be carried out "from bottom to top" or horizontally (Fig. 45) [24].

The formation of nanofibers is determined by many operating parameters, which are included in Table 1.

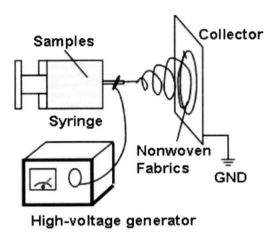

FIGURE 45 Electrospinning set up.

TABLE 1 Classification of affecting parameters and nozzle configuration types on electrospinning.

Affecting parameters		
Process parameters	System parameters	Ambient parameters
✓ electric potential	✓ molecular weight	✓ temperature
✓ flow rate	✓ molecular weight distribution	✓ humidity
✓ concentration	✓ architecture of the polymer	✓ air velocity
✓ spinning distance		
Type of electrospinning nozzle configuration		
✓ single		
✓ side by side		
✓ co axial		

6.14.3 MODELING AND SIMULATION CONCEPTS

Numerous processing operations of complex fluids involve free surface deformations: examples include spraying and atomization of fertilizers and pesticides, fiber-spinning operations, paint application, roll coating of adhesives and food processing operations. Systematically understanding such flows in various processes such as electrospinning can be extremely difficult because of the large number of different forces that may be involved: including capillarity, viscosity, inertia, gravity as well as the additional stresses resulting from the extensional deformation of the microstructure within the fluid [25].

Theoretical and numerical understanding of flows in these processes can help to dominate their drawbacks. For example, in our study, inspite of individual applications of electrospinning process, its mass production is still presented a challenge [6]. For achieving higher mass production and orientation of nanofibers for special application like tissue engineering and microelectronics, it is necessary to comprehend and control dynamic and mechanic behavior of electrospinning jet [26–27]. Modeling and simulations will give a better understanding of electrospinning jet mechanics. For example, the effect of secondary external field can be surveyed using simulation studies. As well, poor deposition control may be in part owing to the lack of understanding of mechanisms of dynamic interactions of the fast moving jets with the electric field and collectors [27].

Electrospinning modeling and simulation are discussed in detail in following sections.

6.14.3.1 ELECTROSPINNING MODELING

The electrospinning process is a fluid dynamics related problem. Controlling the property, geometry, and mass production of the nanofibers, is essential to comprehend quantitatively how the electrospinning process transforms the fluid solution through a millimeter diameter capillary tube into solid fibers which are four to five orders smaller in diameter [28]. Although information on the effect of various processing parameters and constituent material properties can be obtained experimentally, theoretical models offer in-depth scientific understanding which can be useful to clarify the affecting factors that cannot be exactly measured experimentally. Results from modeling also explained how processing parameters and fluid behavior lead to the nanofiber of appropriate properties. The term "properties" refers to basic properties (such as fiber diameter, surface roughness, fiber connectivity, etc.), physical properties (such as stiffness, toughness, thermal conductivity, electrical resistivity, thermal expansion coefficient, density, etc.) and specialized properties (such as biocompatibility, degradation curve, etc. for biomedical applications) [23, 29].

For example, the developed models can be used for the analysis of mechanisms of jet deposition and alignment on various collecting devices in arbitrary electric fields [27].

The various method formulated by researchers are prompted by several applications of nanofibers. It would be sufficient to briefly describe some of these methods to observed similarities and disadvantages of these approaches. An abbreviated literature review of these models will be discussed in Section 6.14.5.

6.14.3.2 ELECTROSPINNING SIMULATION

Electrospun polymer nanofibers demonstrate outstanding mechanical and thermodynamic properties as compared to macroscopic-scale structures. These features are attributed to nanofiber microstructure [30–31]. Theoretical modeling predicts the nanostructure formations during electrospinning. This prediction could be verified by various experimental condition and analysis methods, which called simulation. Numerical simulations can be compared with experimental observations as the last evidence [27, 32].

Parametric analysis and accounting complex geometries in simulation of electrospinning are extremely difficult due to the non-linearity nature in the problem. Therefore, a lot of researches have done to develop an existing electrospinning simulation for viscoelastic liquids [33].

6.14.4 THE BASIC OF ELECTROSPINNING MODELING

Balance of the producing accumulation is, particularly, a basic source of quantitative models of phenomena or processes. Differential balance equations are formulated for momentum, mass and energy through the contribution of local rates of transport expressed by principle of Newton's, Fick's and Fourier laws. For description of more complex systems like electrospinning that involved strong turbulence of the fluid flow, characterization of product property is necessary and various balances are required [34].

The basic principle used in modeling of chemical engineering process is a concept of balance of momentum, mass and energy, which can be expressed in a general form as:

$$A = I + G - O - C \qquad (42)$$

where: A = accumulation built up within system, I = input entering through system surface, G = generation produced in system volume, O = output leaving through system boundary, and C = consumption used in system volume.

The form of expression depends on the level of the process phenomenon description [34–35].

According to the electrospining models, the jet dynamics is governed by a set of three equations representing mass, energy and momentum conservation for the electrically charge jet [36].

6.14.4.1 MASS CONSERVATION

The concept of mass conservation is widely used in many fields such as chemistry, mechanics, and fluid dynamics. Historically, mass conservation was discovered in chemical reactions by Antoine Lavoisier in the late 18th century, and was of decisive importance in the progress from alchemy to the modern natural science of chemistry. The concept of matter conservation is useful and sufficiently accurate for most chemical calculations, even in modern practice [37].

The equations for the jet follow from Newton's Law and the conservation laws obey, namely, conservation of mass and conservation of charge [38].

According to the conservation of mass equation

$$\pi R^2 v = Q \tag{43}$$

For incompressible jets, by increasing the velocity the radius of the jet decreases. At the maximum level of the velocity, the radius of the jet reduces. The macromolecules of the polymers are compacted together closer while the jet becomes thinner as it shown in Fig. 46. When the radius of the jet reaches the minimum value and its speed becomes maximum to keep the conservation of mass equation, the jet dilates by decreasing its density, which called electrospinning dilation [39–40].

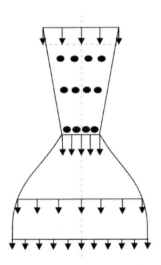

FIGURE 46 Macromolecular chains are compacted during the electrospinning [40].

6.14.4.2 ELECTRIC CHARGE CONSERVATION

An electric current is a flow of electric charge. Electric charge flows when there is voltage present across a conductor. In physics, charge conservation is the principle that electric charge can neither be created nor destroyed. The net quantity of electric charge, the amount of positive charge minus the amount of negative charge in the universe, is always conserved. The first written statement of the principle was by American scientist and statesman Benjamin Franklin in 1747 [41]. Charge conservation is a physical law, which states that the change in the amount of electric charge in any volume of space is exactly equal to the amount of charge in a region and the flow of charge into and out of that region [42].

During the electrospinning process, the electrostatic repulsion between excess charges in the solution stretches the jet. This stretching also decreases the jet diameter that this leads to the law of charge conservation as the second governing equation [43].

In electrospinning process, the electric current, which induced by electric field included two parts, conduction and convection.

The conventional symbol for current is I:

$$I = I_{conduction} + I_{convection} \tag{44}$$

Electrical conduction is the movement of electrically charged particles through a transmission medium. The movement can form an electric current in response to an electric field. The underlying mechanism for this movement depends on the material.

$$I_{conduction} = J_{cond} \times S = KE \times \pi R^2 \tag{45}$$

$$J = \frac{I}{A(s)} \tag{46}$$

$$I = J \times S \tag{47}$$

Convection current is the flow of current with the absence of an electric field.

$$I_{convection} = J_{conv} \times S = 2\pi R(L) \times \sigma v \tag{48}$$

$$J_{conv} = \sigma v \tag{49}$$

So, the total current can be calculated as:

$$\pi R^2 KE + 2\pi R v \sigma = I \tag{50}$$

6.14.4.3 MOMENTUM BALANCE

In classical mechanics, linear momentum or translational momentum is the product of the mass and velocity of an object. Like velocity, linear momentum is a vector quantity, possessing a direction as well as a magnitude:

$$P = mv \tag{51}$$

Linear momentum is also a conserved quantity, meaning that if a closed system (one that does not exchange any matter with the outside and is not acted on by outside forces) is not affected by external forces, its total linear momentum cannot change. In classical mechanics, conservation of linear momentum is implied by Newton's laws of motion; but it also holds in special relativity (with a modified formula) and, with appropriate definitions, a (generalized) linear momentum conservation law holds in electrodynamics, quantum mechanics, quantum field theory, and general relativity [35, 44]. For example, according to the third law, the forces between two particles are equal and opposite. If the particles are numbered 1 and 2, the second law states:

$$F_1 = \frac{dP_1}{dt} \tag{52}$$

$$F_2 = \frac{dP_2}{dt} \tag{53}$$

Therefore:

$$\frac{dP_1}{dt} = -\frac{dP_2}{dt} \tag{54}$$

$$\frac{d}{dt}(P_1 + P_2) = 0 \tag{55}$$

If the velocities of the particles are v_{11} and v_{12} before the interaction, and afterwards they are v_{21} and v_{22}, then

$$m_1 v_{11} + m_2 v_{12} = m_1 v_{21} + m_2 v_{22} \tag{56}$$

This law holds no matter how complicated the force is between particles. Similarly, if there are several particles, the momentum exchanged between each pair of particles adds up to zero, so the total change in momentum is zero. This conservation law applies to all interactions, including collisions and separations caused by explosive forces. It can also be generalized to situations where Newton's laws do not hold, for example in the theory of relativity and in electrodynamics [44–45].

The momentum equation for electrospinning modeling is formulated by considering the forces on a short segment of the jet [46–47].

$$\frac{d}{dz}(\pi R^2 \rho v^2) = \pi R^2 \rho g + \frac{d}{dz}\left[\pi R^2(-p + \tau_{zz})\right] + \frac{\gamma}{R}.2\pi RR' + 2\pi R(t_t^e - t_n^e R') \quad (57)$$

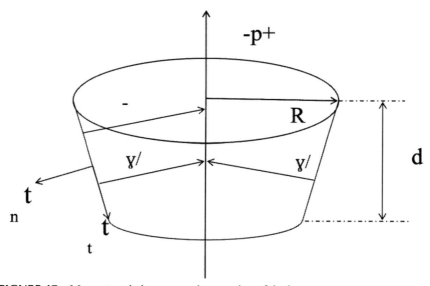

FIGURE 47 Momentum balance on a short section of the jet.

As it is shown in the Fig. 47, the element's angels could be defined as α and β. According to the mathematical relationships, it is obvious that:

$$\alpha + \beta = \pi \quad (58)$$

$$\sin\alpha = \tan\alpha \qquad (59)$$
$$\cos\alpha = 1$$

The relationships (Fig. 47) between these electric forces are given below:

$$t_n^e \sin\alpha \cong t_n^e \tan\alpha \cong -t_n^e \tan\beta \cong -\frac{dR}{dz}t_n^e = -R't_n^e \qquad (60)$$

$$t_t^e \cos\alpha \cong t_t^e \qquad (61)$$

So the effect of the electric forces in the momentum balance equation can be presented as:

$$2\pi RL(t_t^e - R't_n^e)dz \qquad (62)$$

(Notation: In the main momentum equation, final formula is obtained by dividing into dz)

Generally, the normal electric force is defined as:

$$t_n^e \cong \frac{1}{2}\bar{\varepsilon}E_n^2 = \frac{1}{2}\bar{\varepsilon}(\frac{\sigma}{\bar{\varepsilon}})^2 = \frac{\sigma^2}{2\bar{\varepsilon}} \qquad (63)$$

A few amount of electric forces is perished in vicinity of the air.

$$E_n = \frac{\sigma}{\bar{\varepsilon}} \qquad (64)$$

The electric force can be presented by:

$$F = \frac{\Delta We}{\Delta l} = \frac{1}{2}(\varepsilon - \bar{\varepsilon})E^2 \times \Delta S \qquad (65)$$

The force per surface unit is:

$$\frac{F}{\Delta S} = \frac{1}{2}(\varepsilon - \bar{\varepsilon})E^2 \qquad (66)$$

Generally the electric potential energy is obtained by:

$$Ue = -We = -\int F.ds \qquad (67)$$

$$\Delta We = \frac{1}{2}(\varepsilon - \bar{\varepsilon})E^2 \times \Delta V = \frac{1}{2}(\varepsilon - \bar{\varepsilon})E^2 \times \Delta S.\Delta l \qquad (68)$$

So, finally it could be resulted:

$$t_n^e = \frac{\sigma^2}{2\bar{\varepsilon}} - \frac{1}{2}(\varepsilon - \bar{\varepsilon})E^2 \qquad (69)$$

$$t_t^e = \sigma E \qquad (70)$$

6.14.4.4 COULOMB'S LAW

Coulomb's law is a mathematical description of the electric force between charged objects which is formulated by the 18th-century French physicist Charles–Augustin de Coulomb. It is analogous to Isaac Newton's law of gravity. Both gravitational and electric forces decrease with the square of the distance between the objects, and both forces act along a line between them [48]. In Coulomb's law, the magnitude and sign of the electric force are determined by the electric charge, more than the mass of an object. Thus, charge which is a basic property matter determines how electromagnetism affects the motion of charged targets [49].

Coulomb force is thought to be the main cause for the instability of the jet in the electrospinning process [50]. This statement is based on the Earnshaw's theorem, named after Samuel Earnshaw [51] which claims that "A charged body placed in an electric field of force cannot rest in stable equilibrium under the influence of the electric forces alone". This theorem can be notably adapted to the electrospinning process [50]. The instability of charged jet influences on jet deposition and as a consequence on nanofiber formation. Therefore, some researchers applied developed models to the analysis of mechanisms of jet deposition and alignment on various collecting devices in arbitrary electric fields [52].

The equation for the potential along the centerline of the jet can be derived from Coulomb's law. Polarized charge density is obtained:

$$\rho_{p'} = -\vec{\nabla}.\vec{P}' \qquad (71)$$

Where P' is polarization:

$$\vec{P}' = (\varepsilon - \bar{\varepsilon})\vec{E} \qquad (72)$$

By substituting P' in equation:

$$\rho_{p'} = -(\bar{\varepsilon} - \varepsilon)\frac{dE}{dz'} \tag{73}$$

Beneficial charge per surface unit can be calculated as below:

$$\rho_{p'} = \frac{Q_b}{\pi R^2} \tag{74}$$

$$Q_b = \rho_b . \pi R^2 = -(\bar{\varepsilon} - \varepsilon)\pi R^2 \frac{dE}{dz'} \tag{75}$$

$$Q_b = -(\bar{\varepsilon} - \varepsilon)\pi \frac{d(ER^2)}{dz'} \tag{76}$$

$$\rho_{sb} = Q_b . dz' = -(\bar{\varepsilon} - \varepsilon)\pi \frac{d}{dz'}(ER^2)dz' \tag{77}$$

The main equation of Coulomb's law:

$$F = \frac{1}{4\pi\varepsilon_0}\frac{qq_0}{r^2} \tag{78}$$

The electric field is:

$$E = \frac{1}{4\pi\varepsilon_0}\frac{q}{r^2} \tag{79}$$

The electric potential can be measured:

$$\Delta V = -\int E.dL \tag{80}$$

$$V = \frac{1}{4\pi\varepsilon_0}\frac{Q_b}{r} \tag{81}$$

According to the beneficial charge equation, the electric potential could be rewritten as:

$$\Delta V = Q(z) - Q_\infty(z) = \frac{1}{4\pi\bar{\varepsilon}} \int \frac{(q - Q_b)}{r} dz' \qquad (82)$$

$$Q(z) = Q_\infty(z) + \frac{1}{4\pi\bar{\varepsilon}} \int \frac{q}{r} dz' - \frac{1}{4\pi\bar{\varepsilon}} \int \frac{Q_b}{r} dz' \qquad (83)$$

$$Q_b = -(\bar{\varepsilon} - \varepsilon)\pi \frac{d(ER^2)}{dz'} \qquad (84)$$

The surface charge density's equation is:

$$q = \sigma \cdot 2\pi RL \qquad (85)$$

$$r^2 = R^2 + (z - z')^2 \qquad (86)$$

$$r = \sqrt{R^2 + (z - z')^2} \qquad (87)$$

The final equation, which obtained by substituting the mentioned equations is:

$$Q(z) = Q_\infty(z) + \frac{1}{4\pi\bar{\varepsilon}} \int \frac{\sigma \cdot 2\pi R}{\sqrt{(z-z')^2 + R^2}} dz' - \frac{1}{4\pi\bar{\varepsilon}} \int \frac{(\bar{\varepsilon} - \varepsilon)\pi}{\sqrt{(z-z')^2 + R^2}} \frac{d(ER^2)}{dz'} \qquad (88)$$

It is assumed that β is defined:

$$\beta = \frac{\varepsilon}{\bar{\varepsilon}} - 1 = -\frac{(\bar{\varepsilon} - \varepsilon)}{\bar{\varepsilon}} \qquad (89)$$

So, the potential equation becomes:

$$Q(z) = Q_\infty(z) + \frac{1}{2\bar{\varepsilon}} \int \frac{\sigma \cdot R}{\sqrt{(z-z')^2 + R^2}} dz' - \frac{\beta}{4} \int \frac{1}{\sqrt{(z-z')^2 + R^2}} \frac{d(ER^2)}{dz'} \qquad (90)$$

The asymptotic approximation of χ is used to evaluate the integrals mentioned above:

$$\chi = \left(-z + \xi + \sqrt{z^2 - 2z\xi + \xi^2 + R^2} \right) \qquad (91)$$

Where χ is "aspect ratio" of the jet (L = length, R_0 = Initial radius).

This leads to the final relation for the axial electric field:

$$E(z) = E_{\infty}(z) - \ln \chi \left(\frac{1}{\bar{\varepsilon}} \frac{d(\sigma R)}{dz} - \frac{\beta}{2} \frac{d^2(ER^2)}{dz^2} \right) \quad (92)$$

6.14.4.5 CONSTITUTIVE EQUATIONS

In modern condensed matter physics, the constitutive equation plays a major role. In physics and engineering, a constitutive equation or relation is a relation between two physical quantities that is specific to a material or substance, and approximates the response of that material to external stimulus, usually as applied fields or forces [53]. There are a sort of mechanical equation of state, and describe how the material is constituted mechanically. With these constitutive relations, the vital role of the material is reasserted [54]. There are two groups of constitutive equations: Linear and nonlinear constitutive equations [55]. These equations are combined with other governing physical laws to solve problems; for example in fluid mechanics the flow of a fluid in a pipe, in solid state physics the response of a crystal to an electric field, or in structural analysis, the connection between applied stresses or forces to strains or deformations [53].

The first constitutive equation (constitutive law) was developed by Robert Hooke and is known as Hooke's law. It deals with the case of linear elastic materials. Following this discovery, this type of equation, often called a "stress-strain relation" in this example, but also called a "constitutive assumption" or an "equation of state" was commonly used [56]. Walter Noll advanced the use of constitutive equations, clarifying their classification and the role of invariance requirements, constraints, and definitions of terms like "material", "isotropic", "aeolotropic", etc. The class of "constitutive relations" of the form stress rate = f (velocity gradient, stress, density) was the subject of Walter Noll's dissertation in 1954 under Clifford Truesdell [53]. There are several kinds of constitutive equations, which are applied commonly in electrospinning. Some of these applicable equations are discussed as following:

6.14.4.5.1 OSTWALD-DE WAELE POWER LAW

Rheological behavior of many polymer fluids can be described by power law constitutive equations [55]. The equations that describe the dynamics in electrospinning constitute, at a minimum, those describing the conservation of mass, momentum and charge, and the electric field equation. Additionally a constitutive equation for the fluid behavior is also required [57]. A Power-law fluid, or the

Ostwald-de Waele relationship, is a type of generalized Newtonian fluid for which the shear stress, τ, is given by:

$$\tau = K'\left(\frac{\partial v}{\partial y}\right)^m \quad (93)$$

Which $\partial v/\partial y$ is the shear rate or the velocity gradient perpendicular to the plane of shear. The power law is only a good description of fluid behavior across the range of shear rates to which the coefficients are fitted. There are a number of other models that better describe the entire flow behavior of shear-dependent fluids, but they do so at the expense of simplicity, so the power law is still used to describe fluid behavior, permit mathematical predictions, and correlate experimental data [47, 58].

Nonlinear rheological constitutive equations applicable for polymer fluids (Ostwald-de Waele power law) were applied to the electrospinning process by Spivak and Dzenis [59–61].

$$\hat{\tau}^c = \mu \left[tr\left(\dot{\hat{\gamma}}^2\right)\right]^{(m-1)/2} \dot{\hat{\gamma}} \quad (94)$$

$$\mu = K\left(\frac{\partial v}{\partial y}\right)^{m-1} \quad (95)$$

Viscous Newtonian fluids are described by a special case of equation above with the flow index $m=1$. Pseudoplastic (shear thinning) fluids are described by flow indices $0 \leq m \leq 1$. Dilatant (shear thickening) fluids are described by the flow indices $m>1$ [60].

6.14.4.5.2 GIESEKUS EQUATION

In 1966, Giesekus established the concept of anisotropic forces and motions into polymer kinetic theory. With particular choices for the tensors describing the anisotropy, one can obtained Giesekus constitutive equation from elastic dumbbell kinetic theory [62–63]. The Giesekus equation is known to predict, both qualitatively and quantitatively, material functions for steady and non-steady shear and elongational flows. However, the equation sustains two drawbacks: it predicts that the viscosity is inversely proportional to the shear rate in the limit of infinite shear rate and it is unable to predict any decrease in the elongational viscosity with increasing elongation rate in uniaxial elongational flow. The first one is not serious because of retardation time, which is included in the constitutive equation but the second one is more critical because the elongational viscosity of some polymers decrease with increasing of elongation rate [64–65].

In the main Giesekus equation, the tensor of excess stresses depending on the motion of polymer units relative to their surroundings was connected to a sequence of tensors characterizing the configurational state of the different kinds of network structures present in the concentrated solution or melt. The respective set of constitutive equations indicates [66–67]:

$$S_k + \eta \frac{\partial C_k}{\partial t} = 0 \tag{96}$$

The equation below indicates the upper convected time derivative (Oldroyd derivative):

$$\frac{\partial C_k}{\partial t} = \frac{DC_k}{Dt} - \left[C_k \nabla v + (\nabla v)^T C_k \right] \tag{97}$$

(Note: The upper convective derivative is the rate of change of some tensor property of a small parcel of fluid that is written in the coordinate system rotating and stretching with the fluid.)

C_k also can be measured as following:

$$C_k = 1 + 2E_k \tag{98}$$

According to the concept of "recoverable strain" S_k may be understood as a function of E_k and vice versa. If linear relations corresponding to Hooke's law are adopted.

$$S_k = 2\mu_k E_k \tag{99}$$

So:

$$S_k = \mu_k (C_k - 1) \tag{100}$$

The Eq. (96) becomes:

$$S_k + \lambda_k \frac{\partial S_k}{\partial t} = 2\eta D \tag{101}$$

$$\lambda_k = \frac{\eta}{\mu_k} \tag{102}$$

As a second step in order to rid the model of the shortcomings is the scalar mobility constants B_k, which are contained in the constants η. This mobility constant can be represented as:

$$\frac{1}{2}(\beta_k S_k + S_k \beta_k) + \tilde{\eta}\frac{\partial C_k}{\partial t} = 0 \qquad (103)$$

The two parts of Eq. (103) reduces to the single constitutive equation:

$$\beta_k + \tilde{\eta}\frac{\partial C_k}{\partial t} = 0 \qquad (104)$$

The excess tension tensor in the deformed network structure where the well-known constitutive equation of a so-called Neo-Hookean material is proposed [66, 68]:

Neo-Hookean equation: $S_k = 2\mu_k E_k = \mu_k(C_k - 1)$ (105)

$$\mu_k = NKT$$

$$\beta_k = 1 + \alpha(C_k - 1) = (1 - \alpha) + \alpha C_k \qquad (106)$$

where K is Boltzmann's constant.

By substitution Eqs. (105) and (106) in the equation (105), it can obtained where the condition $0 \leq \alpha \leq 1$ must be fulfilled, the limiting case α=0 corresponds to an isotropic mobility [69].

$$0 \leq \alpha \leq 1 \quad [1 + \alpha(C_k - 1)](C_k - 1) + \lambda_k \frac{\partial C_k}{\partial t} = 0 \qquad (107)$$

$$\alpha = 1 \quad C_k(C_k - 1) + \lambda_k \frac{\partial C_k}{\partial t} = 0 \qquad (108)$$

$$0 \leq \alpha \leq 1 \quad C_k = \frac{S_k}{\mu_k} + 1 \qquad (109)$$

By substituting equations above in Eq. (102), it becomes:

$$\left[1+\frac{\alpha S_k}{\mu_k}\right]\frac{S_k}{\mu_k}+\lambda_k\frac{\partial C_k}{\partial t}=0 \qquad (110)$$

$$\frac{S_k}{\mu_k}+\frac{\alpha S_k^2}{\mu_k^2}+\lambda_k\frac{\partial(S_k/\mu_k+1)}{\partial t}=0 \qquad (111)$$

$$\frac{S_k}{\mu_k}+\frac{\alpha S_k^2}{\mu_k^2}+\frac{\lambda_k}{\mu_k}\frac{\partial S_k}{\partial t}=0 \qquad (112)$$

$$S_k+\frac{\alpha S_k^2}{\mu_k}+\lambda_k\frac{\partial S_k}{\partial t}=0 \qquad (113)$$

D means the rate of strain tensor of the material continuum [66].

$$D=\frac{1}{2}\left[\nabla v+(\nabla v)^T\right] \qquad (114)$$

The equation of the upper convected time derivative for all fluid properties can be calculated as:

$$\frac{\partial \otimes}{\partial t}=\frac{D\otimes}{Dt}-\left[\otimes.\nabla v+(\nabla v)^T.\otimes\right] \qquad (115)$$

$$\frac{D\otimes}{Dt}=\frac{\partial \otimes}{\partial t}+\left[(v.\nabla).\otimes\right] \qquad (116)$$

By replacing S_k instead of the symbol:

$$\lambda_k\frac{\partial S_k}{\partial t}=\lambda_k\frac{DS_k}{Dt}-\lambda_k\left[S_k\nabla v+(\nabla v)^T S_k\right]=\lambda_k\frac{DS_k}{Dt}-\lambda_k(v.\nabla)S_k \qquad (117)$$

By simplification the equation above:

$$S_k+\frac{\alpha S_k^2}{\mu_k}+\lambda_k\frac{DS_k}{Dt}=\lambda_k(v.\nabla)S_k \qquad (118)$$

$$S_k=2\mu_k E_k \qquad (119)$$

The assumption of $E_k=1$ would lead to the next equation:

$$S_k + \frac{\alpha \lambda_k S_k^2}{\eta} + \lambda_k \frac{DS_k}{Dt} = \frac{\eta}{\mu_k}(2\mu_k)D = 2\eta D = \eta\left[\nabla v + (\nabla v)^T\right] \quad (120)$$

In electrospinning modeling articles τ is used commonly instead of S_k [36, 39, 70].

$$S_k \leftrightarrow \tau$$

$$\tau + \frac{\alpha \lambda_k \tau^2}{\eta} + \lambda_k \tau_{(1)} = \eta\left[\nabla v + (\nabla v)^T\right] \quad (121)$$

6.14.4.5.3 MAXWELL EQUATION

Maxwell's equations are a set of partial differential equations that, together with the Lorentz force law, form the foundation of classical electrodynamics, classical optics, and electric circuits. These fields are the bases of modern electrical and communications technologies. Maxwell's equations describe how electric and magnetic fields are generated and altered by each other and by charges and currents. They are named after the Scottish physicist and mathematician James Clerk Maxwell who published an early form of those equations between 1861 and 1862 [71–72].

The simplest model of flexible macromolecules in a dilute solution is the elastic dumbbell (or bead-spring) model. This has been widely used for purely mechanical theories of the stress in electrospinning modeling [73].

Maxwell constitutive equation was first applied by Reneker et al. (2000). Consider an electrified liquid jet in an electric field parallel to its axis. They modeled a segment of the jet by a viscoelastic dumbbell. They used Gaussian electrostatic system of units. According to this model each particle in the electric field exerts repulsive force on another particle. Therefore the stress between these particles can be measured by [52]:

$$\dot{\tau} = G\left(\varepsilon' - \frac{\tau}{\eta}\right) \quad (122)$$

The stress can be calculated by a Maxwell viscoelastic constitutive equation [74]:

$$\dot{\tau} = G\left(\varepsilon' - \frac{\tau}{\eta}\right) \quad (123)$$

Where ε' is the Lagrangian axial strain:

$$\varepsilon' \equiv \frac{\partial \dot{x}}{\partial \xi}.\hat{i}. \tag{124}$$

6.14.4.6 SCALING

Physical aspect of a phenomenon can use the language of differential equation, which represents the structure of the system by selecting the variables that characterize the state of it, and certain mathematical constraint on the values of those variables can take on. These equations can predict the behavior of the system over a quantity like time. For an instance, a set of continuous functions of time that describe the way the variables of the system developed over time starting from a given initial state [75]. In general, the renormalization group theory, scaling and fractal geometry, are applied to the understanding of the complex phenomena in physics, economics and medicine [76].

In more recent times, in statistical mechanics, the expression "scaling laws" has referred to the homogeneity of form of the thermodynamic and correlation functions near critical points, and to the resulting relations among the exponents that occur in those functions. From the viewpoint of scaling, electrospinning modeling can be studied in two ways, allometric and dimensionless analysis. Scaling and dimensional analysis actually started with Newton, and allometry exists everywhere in our day life and scientific activity [76–77].

6.14.4.6.1 ALLOMETRIC SCALING

Electrospinning applies electrically generated motion to spin fibers. So, it is difficult to predict the size of the produced fibers, which depends on the applied voltage in principal. Therefore, the relationship between radius of jet and the axial distance from nozzle is always the subject of investigation [78–79]. It can be described as an allometric equation by using the values of the scaling exponent for the initial steady, instability and terminal stages [80].

The relationship between r and z can be expressed as an allometric equation of the form:

$$r \approx z^b \tag{125}$$

When the power exponent, $b = 1$ the relationship is isometric and when $b \neq 1$ the relationship is allometric [78, 81]. In another view, $b = -1/2$ is considered for the straight jet, $b = -1/4$ for instability jet and $b = 0$ for finally stage [55, 79].

Due to high electrical force acting on the jet, it can be illustrated [78]:

$$\frac{d}{dz}\left(\frac{v^2}{2}\right) = \frac{2\sigma E}{\rho r} \qquad (126)$$

Equations of mass and charge conservations applied here as mentioned before [78, 81–82]

From the above equations it can be seen that [39, 78]

$$v \approx r^{-2},\ \sigma \approx r,\ E \approx r^{-2},\ \frac{dv^2}{dz} \approx r^{-2},\ r \approx z^{-1/2} \qquad (127)$$

The charged jet can be considered as a one-dimensional flow as mentioned. If the conservation equations modified, they would change as [78]:

$$2\pi r \sigma^\alpha v + K\pi r^2 E = I \qquad (128)$$

$$r \approx z^{-\alpha/(\alpha+1)} \qquad (129)$$

Where α is a surface charge parameter, the value of α depends on the surface charge in the jet. When $\alpha = 0$ no charge in jet surface, and in $\alpha = 1$ use for full surface charge.

Allometric scaling equations are more widely investigated by different researchers. Some of the most important allometric relationships for electrospinning are presented in Table 2.

TABLE 2 Investigated scaling laws applied in electrospinning model.

Parameters	Equation	Ref.
The conductance and polymer concentration	$g \approx c^\beta$	[39]
The fiber diameters and the solution viscosity	$d \approx \eta^\alpha$	[79]
The mechanical strength and threshold voltage	$\bar{\sigma} \approx E_{threshold}^{-\alpha}$	[83]
The threshold voltage and the solution viscosity	$E_{threshold} \approx \eta^{1/4}$	[83]
The viscosity and the oscillating frequency	$\eta \approx \omega^{-0.4}$	[83]
the volume flow rate and the current	$I \approx Q^b$	[82]
The current and the fiber radius	$I \approx r^2$	[84]

TABLE 2 *(Continued)*

Parameters	Equation	Ref.
The surface charge density and the fiber radius	$\sigma \approx r^3$	[84]
The induction surface current and the fiber radius	$\phi \approx r^2$	[84]
The fiber radius and AC frequency	$r \approx \Omega^{1/4}$	[55]

β, α and b= scaling exponent

6.14.4.6.2 DIMENSIONLESS ANALYSIS

One of the simplest, yet most powerful, tools in the physics is dimensional analysis in which there are two kinds of quantities: dimensionless and dimensional. Dimensionless quantities, which are without associated physical dimensions, are widely used in mathematics, physics, engineering, economics, and in everyday life (such as in counting). Numerous well-known quantities, such as π, e, and φ, are dimensionless. They are "pure" numbers, and as such always have a dimension of 1 [85–86].

Dimensionless quantities are often defined as products or ratios of quantities that are not dimensionless, but whose dimensions cancel out when their powers are multiplied [87].

In non-dimensional scaling, there are two key steps:
(a) Identify a set of physically-relevant dimensionless groups, and
(b) Determine the scaling exponent for each one.

Dimensional analysis will help you with step (a), but it cannot be applicable possibly for step (b).

A good approach to systematically getting to grips with such problems is through the tools of dimensional analysis (Bridgman, 1963). The dominant balance of forces controlling the dynamics of any process depends on the relative magnitudes of each underlying physical effect entering the set of governing equations [88]. Now, the most general characteristics parameters, which used in dimensionless analysis in electrospinning are introduced in Table 3.

TABLE 3 Characteristics parameters employed and their definitions.

Parameter	Definition
Length	R_0
Velocity	$v_0 = \dfrac{Q}{\pi R_0^2 K}$

TABLE 3 *(Continued)*

Parameter	Definition
Electric field	$E_0 = \dfrac{I}{\pi R_0^2 K}$
Surface charge density	$\sigma_0 = \bar{\varepsilon} E_0$
Viscose stress	$\tau_0 = \dfrac{\eta_0 \upsilon_0}{R_0}$

For achievement of simplified form of equations and reduction a number of unknown variables, the parameters should be subdivided into characteristics scales in order to become dimensionless. Electrospinning dimensionless groups are shown in Table 4 [89].

TABLE 4 Dimensionless groups employed and their definitions.

Name	Definition	Field of application
Froude number	$Fr = \dfrac{\upsilon_0^2}{g R_0}$	The ratio of inertial to gravitational forces
Reynolds number	$\mathrm{Re} = \dfrac{\rho \upsilon_0 R_0}{\eta_0}$	The ratio of the inertia forces to the viscos forces
Weber number	$We = \dfrac{\rho \upsilon_0^2 R_0}{\gamma}$	The ratio of the surface tension forces to the inertia forces
Deborah number	$De = \dfrac{\lambda \upsilon_0}{R_0}$	The ratio of the fluid relaxation time to the instability growth time
Electric Peclet number	$Pe = \dfrac{2 \bar{\varepsilon} \upsilon_0}{K R_0}$	The ratio of the characteristic time for flow to that for electrical conduction
Euler number	$Eu = \dfrac{\varepsilon_0 E^2}{\rho \upsilon_0^2}$	The ratio of electrostatic forces to inertia forces

TABLE 4 *(Continued)*

Name	Definition	Field of application
Capillary number	$Ca = \dfrac{\eta v_0}{\gamma}$	The ratio of inertia forces to viscose forces
Ohnesorge number	$oh = \dfrac{\eta}{(\rho \gamma R_0)^{1/2}}$	The ratio of viscose force to surface force
Viscosity ratio	$r_\eta = \dfrac{\eta_p}{\eta_0}$	The ratio of the polymer viscosity to total viscosity
Aspect ratio	$\chi = \dfrac{L}{R_0}$	The ratio of the length to the primary radius of jet
Electrostatic force parameter	$\varepsilon = \dfrac{\overline{\varepsilon} E_0^2}{\rho v_0^2}$	The relative importance of the electrostatic and hydrodynamic forces
Dielectric constant ratio	$\beta = \dfrac{\varepsilon}{\overline{\varepsilon}} - 1$	The ratio of the field without the dielectric to the net field with the dielectric

The governing and constitutive equations can be transformed into dimensionless form using the dimensionless parameters and groups.

6.14.5 SOME OF ELECTROSPINNING MODELS

The most important mathematical models for electrospinning process are classified in the Table 5. According to the year, advantages and disadvantages of the models:

TABLE 5 The most important mathematical models for electrospinning.

Researchers	Model	Year	Ref.
Taylor, G. I.	Leaky dielectric model	1969	[90]
Melcher, J. R.	• Dielectric fluid		
	• Bulk charge in the fluid jet considered to be zero		
	• Only axial motion		
	• Steady state part of jet		

TABLE 5 *(Continued)*

Researchers	Model	Year	Ref.
Ramos	Slender body	1996	[91]
	• Incompressible and axi-symertric and viscose jet under gravity force		
	• No electrical force		
	• Jet radius decreases near zero		
	• Velocity and pressure of jet only change during axial direction		
	• Mass and volume control equations and Taylor expansion were applied to predict jet radius		
Saville, D. A.	Electrohydrodynamic model	1997	[90]
	• The hydrodynamic equations of dielectric model was modified		
	• Using dielectric assumption		
	• This model can predict drop formation		
	• Considering jet as a cylinder (ignoring diameter reduction)		
	• Only for steady state part of the jet		
Spivak, A.	Spivak and Dzenis model	1998	[92]
Dzenis, Y.	• The motion of viscose fluid jet with the low conductivity were surveyed in a external electric field		
	• Single Newtonian Fluid jet		
	• the electric field assumed to be uniform and constant, unaffected by the charges carried by the jet		
	• Use asymptotic approximation were applied in a long distance from nozzle		
	• Tangential electric force assumed to be zero		
	• Using non-linear rheological constitutive equation (Ostwald_dewaele law), non-linear behavior of fluid jet were investigated		

TABLE 5 *(Continued)*

Researchers	Model	Year	Ref.
Jong Wook	Droplet formation model • Droplet formation of charged fluid jet was studied in this model • The ratio of mass, energy and electric charge transition are the most important parameters on droplet formation • Deformation and break-up of droplets were investigated too • Newtonian and Non–Newtonian fluids • Only for high conductive and viscose fluids	2000	[93]
Reneker, D. H. Yarin, A. L.	Reneker model • For description of instabilities in visco-elastic jets • Using molecular chains theory, behavior of polymer chain by spring-bead model in electric field was studied • Electric force based on electric field cause instability of fluid jet while repulsion force between surface charges make perturbation and bending instability • The motion path of these two cases were studied • Governing equations: momentum balance, motion equations for each bead, Maxwell tension and columbic eqs.	2000	[52]

TABLE 5 *(Continued)*

Researchers	Model	Year	Ref.
Hohman, M. Shin, M.	Stability theory • This model is based on dielectric model with some modification for Newtonian fluids. • This model can describe whipping, bending and Rayleigh instabilities and introduced new ballooning instability. • 4 motion regions were introduced: dipping mode, spindle mode, oscillating mode, precession mode. • Surface charge density introduced as the most effective parameter on instability formation. • Effect of fluid conductivity and viscosity on nanofibers diameter were discussed. • Steady solutions may be obtained only if the surface charge density at the nozzle is set to zero or a very low value	2001	[94]
Feng, J. J	Modifying Hohman model • For both Newtonian and non–Newtonian fluids • Unlike Hohman model, the initial surface charge density was not zero, so the "ballooning instability" did not accrue. • Only for steady state part of the jet • Simplifying the electric field equation which Hohman used in order to eliminating Ballooning instability.	2002	[46]
Wan-Guo-Pan	Wan–Guo–Pan model • They introduced thermo-electro-hydro dynamics model in electrospinning process • This model is modification on Spivak model which mentioned before • The governing equations in this model: Modified Maxwell equation, Navier–Stocks equations, and several rheological constitutive equation.	2004	[61]

TABLE 5 (Continued)

Researchers	Model	Year	Ref.
Ji-Haun	AC-electrospinning model	2005	[55]
	• Whipping instability in this model was distinguished as the most effective parameter on uncontrollable deposition of nanofibers • Applying AC current can reduce this instability so make oriented nanofibers • This model found a relationship between axial distance from nozzle and jet diameter • This model also connected AC frequency and jet diameter		
Roozemond (Eindhoven University and Technology)	Combination of slender body and dielectric model • In this model, a new model for viscoelastic jets in electrospinning were presented by combining these two models • All variables were assumed uniform in cross section of the jet but they changed in during z direction • Nanofiber diameter can be predicted	2007	[95]
Wan	Electromagnetic model • Results indicated that the electromagnetic field which made because of electrical field in charged polymeric jet is the most important reason of helix motion of jet during the process	2012	[26]
Dasri	Dasri model • This model was presented for description of unstable behavior of fluid jet during electrospinning • This instability causes random deposition of nanofiber on surface of the collector • This model described dynamic behavior of fluid by combining assumption of Reneker and Spivak models	2012	[96]

The most frequent numeric mathematical methods, which were used in different models are listed in Table 6.

TABLE 6 Applied numerical methods for electrospinning.

Method	Ref.
Relaxation method	[33, 36, 46]
Boundary integral method (boundary element method)	[74, 93]
Semi-inverse method	[36, 55]
(Integral) control-volume formulation	[91]
Finite element method	[90]
Kutta-Merson method	[97]
Lattice Boltzmann method with finite difference method	[43]

6.14.6 ELECTROSPINNING SIMULATION EXAMPLE

In order to survey of electrospinning modeling application, its main equations were applied for simulating the process according to the constants, which summarized in Table 7.

Mass and charge conservations allow v and σ to be expressed in terms of R and E, and the momentum and E-field equations can be recast into two second-order ordinary differential equations for R and E. The slender-body theory (the straight part of the jet) was assumed to investigate jet behavior during the spinning distance. The slope of the jet surface (R') is maximum at the origin of the nozzle. The same assumption has been used in most previous models concerning jets or drops. The initial and boundary conditions, which govern the process are introduced as:

Initial values ($z=0$):

$$R(0) = 1$$
$$E(0) = E_0$$
$$\tau_{prr} = 2r_\eta \frac{R_0'}{R_0^3} \tag{130}$$
$$\tau_{pzz} = -2\tau_{prr}$$

Feng [46] indicated that E(0) effect is limited to a tiny layer below the nozzle which its thickness is a few percent of R_0. It was assumed that the shear inside the nozzle is effective in stretching of polymer molecules as compered with the following elongation.

Boundary values ($z=\chi$):

$$R(\chi) + 4\chi R'(\chi) = 0$$
$$E(\chi) = E_\chi$$
$$\tau_{prr} = 2r_\eta \frac{R'_\chi}{R^3_\chi} \qquad (90)$$
$$\tau_{pzz} = -2\tau_{prr}$$

The asymptotic scaling can be stated as [46]:

$$R(z) \propto z^{-1/4} \qquad (91)$$

Just above the deposit point ($z=\chi$), asymptotic thinning conditions applied. R drops towards zero and E approaches E. The electric field is not equal to E, so we assumed a slightly larger value, E_χ.

TABLE 7 Constants used in electrospinning simulation.

Constant	Quantity
Re	$2.5 \cdot 10^{-3}$
We	0.1
Fr	0.1
Pe	0.1
De	10
ε	1
β	40
χ	20
E_0	0.7
E_χ	0.5
r_η	0.9

The momentum, electric field and stress equations could be rewritten into a set of four coupled first order ordinary differential equations (ODE's) with the above-mentioned boundary conditions. Numerical relaxation method has been chosen to solve the generated boundary value problem.

The results of these systems of equations are presented in Figs. 48 and 49 that matched quite well with the other studies that have been published [36, 38, 46, 70, 98].

The variation prediction of R, R', ER², ER²' and E versus axial position (z) are shown in Fig. 48. Physically, the amount of counductable charges reduces with decreasing jet radius. Therefore, to maintain the same jet current, more surface charges should be carried by the convection. Moreover, in the considered simulation region, the density of surface charge gradually increased. As the jet gets thinner and faster, electric conduction gradually transfers to convection. The electric field is mainly induced by the axial gradient of surface charge, thus, it is insensitive to the thinning of the electrospun jet:

$$\frac{d(\sigma R)}{dz} \approx -\left(2R\frac{dR}{dz}\right)/Pe \qquad (131)$$

Therefore, the variation of E versus z can be written as:

$$\frac{d(E)}{dz} \approx \ln \chi \left(\frac{d^2 R^2}{dz^2}\right)/Pe \qquad (132)$$

Downstream of the origin, E shoots up to a peak and then relax due to the decrease of electrostatic pulling force in consequence of the reduction of surface charge density, if the current was held at a constant value. However, in reality, the increase of the strength of the electric field also increases the jet current, which is relatively linearly [46, 98]. As the jet becomes thinner downstream, the increase of jet speed reduces the surface charge density and thus E, so the electric force exerted on the jet and thus R' become smaller. The rates of R and R' are maximum at z=0, and then relaxes smoothly downstream toward zero [36, 46]. According to the relation between R, E and z, ER² and ER²' vary in accord with parts (c) and (d) in Fig. 48.

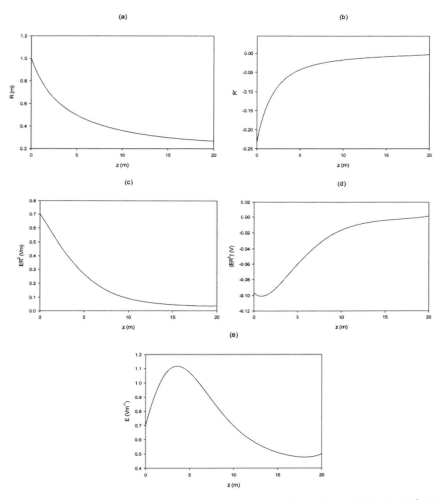

FIGURE 48 Solutions given by the electrospinning model for (a) R; (b) R'; (c) ER^2; (d) $ER^{2'}$ and (e) E.

Figure 49 shows the changes of axial, radial shear stress and the difference between them, the tensile force (T) versus z. The polymer tensile force is much larger in viscoelastic polymers because of the strain hardening. T also has an initial rise, because the effect of strain hardening is so strong that it overcomes the shrinking radius of the jet. After the maximum value of T, it reduces during the jet thinning. As expected, the axial polymer stress rises, because the fiber is stretched in axial direction, and the radial polymer stress declines. The variation of T along the jet can be nonmonotonic, however, meaning the viscous normal stress may promote or resist stretching in a different part of the jet and under different conditions [36, 46].

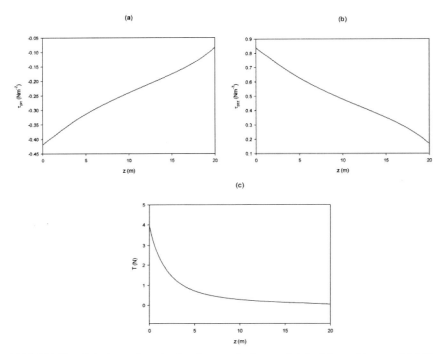

FIGURE 49 Solutions given by the electrospinning model for (a) t_{prr}; (b) t_{pzz} and (c) T.

6.15 MECHANISM OF NANOSTRUCTURE FORMATION

The notion "cluster" assumes energy-wise compensated nucleus with a shell, the surface energy of which is rather small, as a result under given conditions the cluster represents a stable formation. Nanocomposites of various shapes can be obtained either by the dispersion of a substance in a definite medium, or condensation or synthesis from low-molecular chemical particles (Fig. 50). Thus, the techniques for obtaining nanostructures can be classified by the mechanism of their formation. Other features, by which the methods for nanostructure production can be classified, comprise the variants of nanoproduct formation process by the change of energy consumption. A temperature or an energy factor is usually evident here. Besides, a so-called apparatus factor plays an important role together with the aforesaid features. At present, nanostructural "formations" of various shapes and compositions are obtained in a rather wide region of actions upon the chemical particles and substances.

In chemical literature a cluster is equated with a complex compound containing a nucleus and a shell. Usually a nucleus consists of metal atoms combined with metallic bond, and a shell of ligands. Manganese carbonyls [$(Co)_5MnMn(Co)_5$] and cobalt carbonyls [$Co_6(Co)_{18}$], nickel pentadienyls [$Ni_6(C_5H_6)_6$] belong to elementary clusters.

In recent years, the notion "cluster" has got an extended meaning. At the same time the nucleus can contain not only metals or not even contain metals. In some clusters, for instance, carbon ones there is no nucleus at all. In this case their shape can be characterized as a sphere (icosahedron, to be precise)—fullerenes, or as a cylinder—fullerene tubules. Surely a certain force field is formed by atoms on internal walls inside such particles. It can be assumed that electrostatic, electromagnetic, and gravitation fields conditioned by corresponding properties of atoms contained in particle shells can be formed inside tubules and fullerenes. If analyze papers recently published, it should be noted that a considerable exceeding of surface size over the volume and, consequently, a relative growth of the surface energy in comparison with the growth of volume and potential energy is the main feature of clusters. If particle dimensions (diameters of "tubes" and "spheres") change from 1 up to several hundred nanometers, they would be called nanoparticles. In some papers, the area of nanoclusters existence is within 1–10 nm.

Based on classical definitions, in given paper, metal nanoparticles and nanocrystals are referred to as nanoclusters. Apparently, the difference of nanoparticles from other particles (smaller or larger) is determined by their specific characteristics. The search of nanoworld distinctions from atomic-molecular, micro- and macroworld can lead to finding analogies and coincidences in colloid chemistry, chemistry of polymers, and coordination compounds. Firstly, it should be noted that nanoparticles usually represent a small collective aggregation of atoms being within the action of adjacent atoms, thus conditioning the shape of nanoparticles. A nanoparticle shape can vary depending upon the nature of adjacent atoms and character of formation medium. Obviously, the properties of separate atoms and molecules (of small size) are determined by their energy and geometry characteristics, the determinative role being played by electron properties. In particular, electron interactions determine the geometry of molecules and atomic structures of small size and mobility of these chemical particles in media, as well as their activity or reactivity.

When the number of atoms in chemical particle exceeds 30, a certain stabilization of its shape being also conditioned by collective influence of atoms constituting the particle is observed. Simultaneously, the activity of such a particle remains high but the processes with its participation have a directional character. The character of interactions with the surroundings of such structures is determined by their formation mechanism.

During polymerization or co-polymerization the influence of macromolecule growth parameters changes with the increase of the number of elementary acts of its growth. According to scientists, after 7–10 acts the shape or geometry of nanoparticles formed becomes the main determinative factor providing the further growth of macromolecule (chain development). A nanoparticle shape is usually determined not only by its structural elements but also by its interactions with surrounding chemical particles.

Due to the overlapping of different classification features, it is appropriate to present a set of diagrams by main features. For instance, the methods for nanoparticle formation by substance dispersion and chemical particle condensation can be identified with physical and chemical methods, though such decision is incorrect, since substance destruction methods can contain both chemical and physical impacts. In turn, when complex nanostructures are formed from simple ones, both purely physical and chemical factors are possible. However, in the process of substance dispersion high-energy sources, such as electric arc, laser pyrolysis, plasma sources, mechanical crushing or grinding should be applied.

From the aforesaid, it can be concluded that the possibility of self-organization of nanoparticles with the formation of corresponding nanosystems and nanomaterials is the main distinction of nanoworld from pico-, micro-, and macroworld. Recently, much attention has been paid to synergetics or the branch of science dealing with self-organization processes since these processes, in many cases, proceeds with small energy consumption and, consequently, is more ecologically clear in comparison with existing technological processes.

In turn, nanoparticle dimensions are determined by its formation conditions. When the energy consumed for macroparticle destruction or dispersion over the surface increases, the dimensions of nanomaterials are more likely to decrease. The notion "nanomaterial" is not strictly defined. Several researchers consider nanomaterials to be aggregations of nanocrystals, nanotubes, or fullerenes. Simultaneously, there is a lot of information available that nanomaterials can represent materials containing various nanostructures. The most attention researchers pay to metallic nanocrystals. Special attention is paid to metallic nanowires and nanofibers with different compositions.

Here are some names of nanostructures:

1) fullerenes, 2) gigantic fullerenes, 3) fullerenes filled with metal ions, 4) fullerenes containing metallic nucleus and carbon (or mineral) shell, 5) one-layer nanotubes, 6) multi-layer nanotubes, 7) fullerene tubules, 8) "scrolls", 9) conic nanotubes, 10) metal-containing tubules, 11) "onions", 12) "Russian dolls", 13) bamboo-like tubules, 14) "beads", 15) welded nanotubes, 16) bunches of nanotubes, 17) nanowires, 18) nanofibers, 19) nanoropes, 20) nanosemi-spheres (nanocups), 21) nanobands and similar nanostructures, as well as various derivatives from enlisted structures. It is quite possible that a set of such structures and notions will be enriched.

Key Elements on Nanopolymers—From Nanotubes to Nanofibers 221

In most cases nanoparticles obtained are bodies of rotation or contain parts of bodies of rotation. In natural environment there are minerals containing fullerenes or representing thread-like formations comprising nanometer pores or structures. In the first case, it is talked about schungite that is available in quartz rock in unique deposit in Prionezhje. Similar mineral can also be found in the river Lena basin, but it consists of micro- and macro-dimensional cones, spheroids, and complex fibers. In the second case, it is talked about kerite from pegmatite on Volyn (Ukraine) that consists of polycrystalline fibers, spheres, and spirals mostly of micron dimensions, or fibrous vetcillite from the state of Utah (USA); globular anthraxolite and asphaltite.

Diameters of some internal channels are up to 20–50 nm. Such channels can be of interest as nanoreactors for the synthesis of organic, carbon, and polymeric substances with relatively low energy consumption. In case of directed location of internal channels in such matrixes and their inner-combinations the spatial structures of certain purpose can be created. Terminology in the field of nanosystems existence is still being developed, but it is already clear that nanoscience obtains qualitatively new knowledge that can find wide application in various areas of human practice thus, significantly decreasing the danger of people's activities for themselves and environment.

The system classification by dimensional factor is known, based on which we consider the following:

- microobjects and microparticles 10^{-6}–10^{-3} m in size;
- nanoobjects and nanoparticles 10^{-9}–10^{-6} m in size;
- picoobjects and picoparticles 10^{-12}–10^{-9} m in size.

Assuming that nanoparticle vibration energies correlate with their dimensions and comparing this energy with the corresponding region of electromagnetic waves, we can assert that energy action of nanostructures is within the energy region of chemical reactions. System self-organization refers to synergetics. Quite often, especially recently, the papers are published, for example, it is considered that nanotechnology is based on self-organization of metastable systems. As assumed, self-organization can proceed by dissipative (synergetic) and continual (conservative) mechanisms. Simultaneously, the system can be arranged due to the formation of new stable ("strengthening") phases or due to the growth provision of the existing basic phase. This phenomenon underlies the arising nanochemistry. Below is one of the possible definitions of nanochemistry.

Nanochemistry is a science investigating nanostructures and nanosystems in metastable ("transition") states and processes flowing with them in near-"transition" state or in "transition" state with low activation energies.

To carry out the processes based on the notions of nanochemistry, the directed energy action on the system is required, with the help of chemical particle field as well, for the transition from the prepared near-"transition" state into the process product state (in our case-into nanostructures or nanocomposites). The perspective area of nanochemistry is the chemistry in nanoreactors. Nanoreactors can be compared with specific nanostructures representing limited space regions in which chemical particles orientate creating "transition state" prior to the formation of the desired nanoproduct. Nanoreactors have a definite activity, which predetermines the creation of the corresponding product. When nanosized particles are formed in nanoreactors, their shape and dimensions can be the reflection of shape and dimensions of the nanoreactor.

In the last years a lot of scientific information in the field of nanotechnology and science of nanomaterials appeared. Scientists defined the interval from 1 to 1000 nm as the area of nanostructure existence, the main feature of which is to regulate the system self-organization processes. However, later some scientists limited the upper threshold at 100 nm. At the same time, it was not well substantiated. Now many nanostructures varying in shapes and sizes are known. These nanostructures have sizes that fit into the interval, determined by Smally, and are active in the processes of self-organization, and also demonstrate specific properties.

Problems of nanostructure activity and the influence of nanostructure super small quantities on the active media structural changes are explained.

The molecular nanotechnology ideology is analyzed. In accordance with the development tendencies in self-organizing systems under the influence of nanosized excitations the reasons for the generation of self-organization in the range 10^{-6}–10^{-9}m should be determined.

Based on the law of energy conservation the energy of nanoparticle field and electromagnetic waves in the range 1–1,000 nm can transfer, thus corresponding to the range of energy change from soft X-ray to near IR radiation. This is the range of energies of chemical reactions and self-organization (structuring) of systems connected with them.

Apparently the wavelengths of nanoparticle oscillations near the equilibrium state are close or correspond to their sizes. Then based on the concepts of ideologists of nanotechnology in material science the definition of nanotechnology can be as follows:

Nanotechnology is a combination of knowledge in the ways and means of conducting processes based on the phenomenon of nanosized system self-organization and utilization of internal capabilities of the systems that results in decreasing the energy consumption required for obtaining the targeted product while preserving the ecological cleanness of the process.

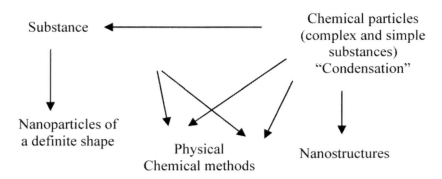

FIGURE 50 Classification diagram of nanostructure formation techniques by the features "dispersion" and "condensation."

At the same time, the conceptions "dispersion" and "condensation" are conditional and can be explained in various ways (Figs. 51 and 52). Among the physical methods of "dispersion" high-temperature methods and methods with relatively low temperatures or high-energy and low-energy ones are distinguished.

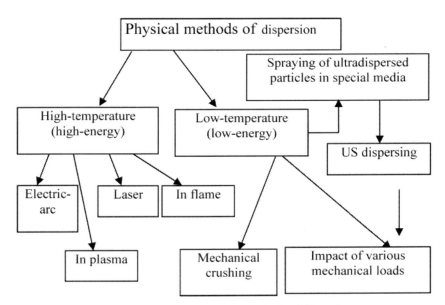

FIGURE 51 Classification diagram of physical methods of dispersion.

Considerably fewer chemical methods of dispersion are known, though, in the physical methods listed chemical processes are surely present, since it is difficult to imagine spraying and grinding of a substance without chemical reactions of destruction. Therefore, it is more appropriate to speak about physical methods of impact upon the substance that lead to their dispersion (decomposition, destruction at high temperatures, radiations or mechanical loads). Then, chemical methods are identified with methods of impact upon the substances of chemical particles and media.

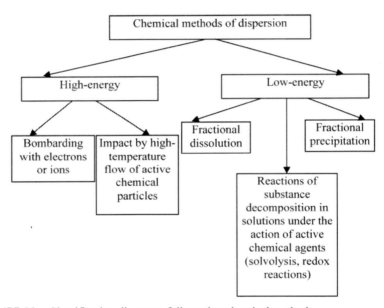

FIGURE 52 Classification diagram of dispersion chemical methods.

Basically nanostructures are obtained from pico-sized chemical particles by means of chemical or physical-chemical techniques, in which the activity of a chemical particle but not its activation during "condensation" is determinant (Fig. 52). Therefore, the classification diagram of chemical techniques for nanostructure formation under the action of chemical media can be given as follows:

Key Elements on Nanopolymers—From Nanotubes to Nanofibers 225

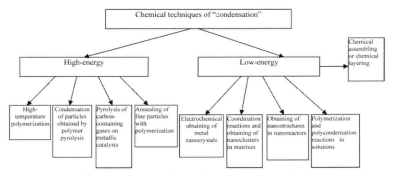

FIGURE 53 Classification diagram of chemical techniques of "condensation."

Actually, such diagrams are approximate in the same way as conditional are separate points of separation and difference between them. Several techniques can be referred to combined or physical methods of impact upon active chemical particles. For instance, CVD method comprises high-energy technique that leads to the formation of gaseous phase that can be attributed to chemical methods of dispersion, and then active chemical particles formed during the pyrolysis "transform" into nanostructures. The polymerization processes of gaseous phase particles are carried out by means of probe technological stations, and this can be referred to the physical techniques of "condensation." So, under the physical methods of formation of various nanostructures, including nanocrystals, nanoclusters, fullerenes, nanotubes, and so on, it means the techniques in which physical impact results in the formation of nanostructures from active pico-sized chemical particles (Fig. 54).

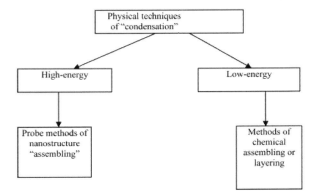

FIGURE 54 Physical impact results in the formation of nanostructures from active pico-sized chemical particles.

The proposed classification does not reflect multi-vicissitude of methods for nanostructure formation. Usually nanostructure production comprises the following:
1) preparation of "embryos" or precursors of nanostructures;
2) production of nanostructures; and
3) isolation and refining of nanostructures of a definite shape.

The production of nanostructures by various methods, including mechanical ones, for instance, extrusion, grinding and similar operations can proceed in several stages.

Mechanical crushing, combined methods for grinding in media and influence of action power and medium, where the substance is crushed and sprayed, upon the size and shape of nanostructures formed will be discussed in this chapter. At the same time, multi-stepped combined methods for obtaining nanostructures with different shapes and sizes have been applied in recent years. Ways of substance dispersion by mechanical methods till nanoproducts are obtained (fine powders or ultradispersed particles) can be given in the following diagram (Fig. 55):

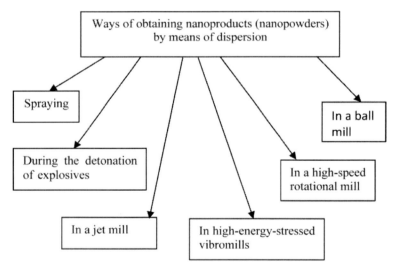

FIGURE 55 Diagram of classification of ways for obtaining powder-like nanoproducts by means of dispersion.

One of the most widely applied methods used for grinding different substances is the crushing and grinding in mills: ball, rod, colloid combined with the action of shock force, abrasion force, centrifugal force and vibration upon the materials. Since, initially, carbon nanostructures, fullerenes and nanotubes were extracted from the carbon dust obtained in electrical charge, the possibility of the formation of corresponding nanostructures when using conventional mechanical methods of action upon substances seems doubtful. However, the investigation of the products obtained after graphite crushing in ball mills allows making the conclusion that the method proposed can be quite competitive high-temperature way for forming nanostructures when the mechanical power provides practically the same conditions as the heat. In this case, it is difficult to imagine a directed action of the combined forces upon the substance. However, at the set speed of the mill drum the directed action of the "stream" of steel balls upon the material can be predicted. As an example, let us give the description of one of such processes during graphite grinding. Carbon nanostructures resembling "onions" by their shape are obtained when grinding graphite in ball mills. The product is ground in planetary ball mill in the atmosphere of pure argon. Mass ratio of steel balls to the powder of pure graphite equals 40. The rotation speed of the drum is 270 rpm. The grinding time changed from 150 to 250 hrs. It was observed that after 150 hrs of grinding the nanoproduct obtained resembles by characteristics and appearance the nanoparticles obtained in electric-arc method.

Iron-containing nanostructures were also obtained in planetary ball mills after milling the iron powders in heptane adding oleic acid.

The distinctive feature of the considered technique for producing metal/carbon nanocomposites is a wide application of independent, modern, experimental and theoretical analysis methods to substantiate the proposed technique and investigation of the composites obtained (quantum-chemical calculations, methods of transmission electron microscopy and electron diffraction, method of X-ray photoelectron spectroscopy, X-ray phase analysis, etc.). The technique developed allows synthesizing a wide range of metal/carbon nanocomposites by composition, size and morphology depending on the process conditions. In its application it is possible to use secondary metallurgical and polymer raw materials. Thus, the nanocomposite structure can be adjusted to extend the function of its application without pre-functionalization. Controlling the sizes and shapes of nanostructures by changing the metal-containing phase, to some extent, apply completely new, practicable properties to the materials which sufficiently differ from conventional materials.

The essence of the method consists in coordination interaction of functional groups of polymer and compounds of 3d-metals as a result of grinding of metal-containing and polymer phases Further, the composition obtained undergoes thermolysis following the temperature mode set with the help of thermogravimetric and differential thermal analyses. At the same time, one observes the polymer carbonization, partial or complete reduction of metal compounds and structuring of carbon material in the form of nanostructures with different shapes and sizes.

Metal/carbon nanocomposite (Me/C) represents metal nanoparticles stabilized in carbon nanofilm structures. In turn, nanofilm structures are formed with carbon amorphous nanofibers associated with metal containing phase. As a result of stabilization and association of metal nanoparticles with carbon phase, the metal chemically active particles are stable in the air and during heating as the strong complex of metal nanoparticles with carbon material matrix is formed.

6.16 WET SPINNING TECHNIQUE (A CASE STUDY)

Low voltage actuating materials ("artificial muscles") are required for many applications in robotics, medical devices and machines. One of the biggest limitations to date for the application of conducting polymers, such as polyaniline, is their low breaking strength- typically less than 10 MPa during an electrochemical actuation cycle under external load. We report here the significant improvement in actuator strength, stress generation and work-per-cycle through the incorporation of small amounts (up to 0.6% w/w) of carbon nanotubes as reinforcement in the polyaniline matrix. A wet spinning and drawing process has been developed to produce continuous lengths of these carbon nanotube reinforced fibers. As actuators, these composites fibers continue to operate at applied stresses in excess of 100 MPa producing a maximum work per cycle of over 300 kJ/m^3. This performance is 3 times higher than previously produced conducting polymer actuators and exceeds skeletal muscle in terms of stress generation by 300 times.

Actuating materials capable of producing useful movement and forces are recognized as the "missing link" in the development of a wide range of frontier technologies including haptic devices, microelectromechanical systems (MEMS) and even molecular machines. Immediate uses for these materials include an electronic Braille screen, a rehabilitation glove, tremor suppression and a variable camber propeller. Most of these applications could be realized with actuators that have equivalent performance to natural skeletal muscle. Although many actuator materials are available, none have the same mix of speed, movement and force as skeletal muscle. Indeed the actuator community was challenged to produce a material capable of beating a human in an arm wrestle. This challenge remains to be met.

One class of materials that has received considerable attention as actuators is the low voltage electrochemical systems utilizing conducting polymers and carbon nanotubes. Low voltage sources are convenient and safe and power inputs are potentially low. One deficiency of conducting polymers and nanotubes compared with skeletal muscle is their low actuation strains: less than 15% for conducting polymers and less than 1% for nanotubes. It has been argued that the low strains can be mechanically amplified (levers, bellows, hinges, etc.) to produce useful movements, but higher forces are needed to operate these amplifiers.

In recent studies of the forces and displacements generated from conducting polymer actuators, it has become obvious that force generation is limited by the breaking strength of the actuator material. Researchers predicted that the maximum stress generated by an actuator can be estimated as 50% of the breaking stress so that for highly drawn polyaniline (PANi) fibers stresses of the order of 190 MPa should be achievable. However, in practice the breaking stresses of conducting polymer fibers when immersed in electrolyte and operated electromechanically are significantly lower than their dry-state strengths. The reasons for the loss of strength are not well known, but the limitations on actuator performance are severe. The highest reported stress that can be sustained by conducting polymers during actuator work cycles is in the range 20–34 MPa for polypyrrole (PPy) films. However, the maximum stress that can be sustained by PPy during actuation appears very sensitive to the dopant ion used and the preparation conditions, with many studies showing maximum stress values of less than 10 MPa.

The low stress generation from conducting polymers, limited by the low breaking strengths, mean that the application of mechanical amplifiers is also very limited. To improve the mechanical performance, we have investigated the use of carbon nanotubes as reinforcing fibers in a polyaniline (PANi) matrix. Previous work has shown that the addition of single wall nanotubes (SWNTs) and multiwall nanotubes (MWNTs) to various polymer matrices have produced significant improvements in strength and stiffness. It has been shown that the modulus of PAni can be increased by up to 4 times with the addition of small (<2%) amounts of nanotubes. Similar improvements in the modulus of actuating polymers may lead to significant increases in the stress generated and work per cycle. Other previous studies have shown that PANi can be wet spun into continuous fibers and that these may be used as actuators. Isotonic strains of 0.3% and isometric stresses of 2 MPa were obtained from these fibers when operated in ionic liquid electrolytes. The aim of the present study was to develop methods for incorporating carbon nanotubes into PAni fibers and to determine the effects on actuator performance.

A wet spinning technique was used to prepare the composite fibers. Firstly, the NTs (HIPCO single wall carbon nanotubes from Carbon Nanotechnology Inc.) were dispersed by sonication for 30 minutes in a mixture of 2-acrylamido-2-methyl-1-propane sulphonic acid (AMPSA, Aldrich 99%) and dichloroacetic acid (DCAA). Polyaniline (PAni, Santa Fe Science and Technology Inc.) and additional AMPSA were then dissolved in the dispersion by high speed mixing. After degassing, the spinning solution was injected using N2 pressure into an acetone coagulation bath. The spun fibers were hand drawn to approximately 5 times their original length across a soldering iron wrapped in Teflon® tape at 100°C.

The isotonic actuation strains were measured using a Dual Mode lever system (Aurora Scientific) and were determined at different (constant) stresses as shown in Fig. 56 for the neat PANi and NT/PANi composite fibers. During isotonic actuation testing, each sample was tested at increasing loads until rupture occurred. It is shown that the addition of nanotubes greatly increases the breaking strength of the composite fibers under actuation conditions. The NT-reinforced fibers could sustain stresses up to 120 MPa without failure during electrochemical cycling. Even the neat PANi could sustain stresses to 90 MPa during work cycles.

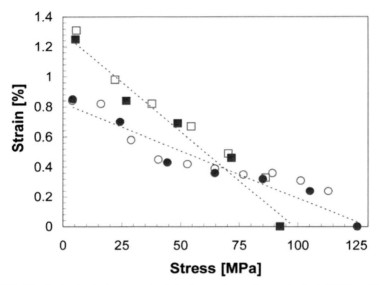

FIGURE 56 Isotonic strains measured during cyclic voltammetry in M HCl at 50 mV/s between 0.0 and 0.6V (vs. Ag/AgCl reference). Open and filled symbols show the results from two separate samples. Square symbols are for neat Pani fibers and circles are for NT-reinforced fibers.

The reinforcing effect afforded by small additions of NTs to the PANi was demonstrated by dry-state tensile testing, as shown in Fig. 57 and summarized in Table 8. A very large increase in the breaking strength from 170 MPa for neat PAni to 255 MPa was produced with the addition of 0.6 wt% NTs. Similarly, the modulus of the composite fibers was doubled from 3.4 GPa to 7.3 GPa. The change in modulus can be analyzed by the rule-of-mixtures approach used in fiber composites. As a starting point, the NTs are treated as fully dispersed and aligned in the fiber direction. Taking densities of 1.4 g/cm^3 and 1.5 g/cm^3 for the PAni and NTs, respectively [28, 29], the volume fraction of NTs in the composite fibers was 0.56%. The simple rule of mixtures approach predicts that the moduli of the composites fibers should be 7 GPa using a modulus of 640 GPa for the isolated NTs. The measured modulus is in good agreement with the calculated value, suggesting a high degree of separation, alignment and bonding of the NTs to the PAni matrix.

The results in Fig. 56 also show that the addition of nanotubes to the PAni matrix affects the actuation strain at low stress levels. The "free stroke" (strain at zero external load) for the NT/PAni sample was approximately 60% of the neat PAni free stroke. In this case the higher modulus nanotubes restrict the volume changes occurring within the PAni matrix. The actuator strains of the composite fibers may be calculated using the same iso-strain assumption inherent in the rule-of-mixtures approach. The expected actuator strain in the composite materials is estimated from:

$$\varepsilon_c^* = \frac{V_m \varepsilon_m^* E_m + V_f \varepsilon_f^* E_f}{E_c} \tag{133}$$

where subscripts c, m and f refer to the composite, PAni matrix and NT fibers and E and V are the moduli and volume fraction for each phase. Actuation in both the matrix and the NT fibers are considered in the analysis, since it is known that the NTs also produce appreciable strains when electrochemically charged. In previous work on NT/PAni composites, where unaligned mats of NTs were dip coated with PANi to give composites of 75% (w/w) NTs, it has been shown that both the nanotubes and the polymer contribute to the actuation strain.

The calculated composite actuation strain (at zero applied stress) was found to be 0.57% where the matrix actuation strain was taken as 1.3% and a NT actuation strain was assumed to be –0.06% for an anodic current pulse. The measured actuation strain (0.85%) was appreciably larger than the calculated strain. Clearly, the stiffening effect caused by the nanotubes cannot on its own account for the observed actuation behavior.

The larger than expected actuation in the composite fibers suggest that the NTs increase the efficiency of the actuation mechanism, perhaps through improved charge transfer. The magnitude of the increase in efficiency is demonstrated by calculating the matrix strain needed to account for the measured composite actuation strain. Thus, a matrix strain of 1.8% would be required to generate a composite strain of 0.85%. The higher matrix strains that occur in the NT composites compared with the neat PAni fibers, suggest a higher charge transfer efficiency possibly due to the higher conductivity of the NT–PAni composites (Table 8). For comparison, neat PAni fibers were also sputter coated with platinum to improve their conductivity and these samples showed actuation strains of 2% compared with 1.3% for uncoated PAni fibers.

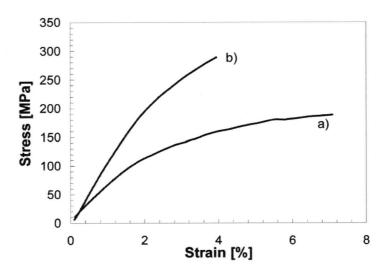

FIGURE 57 Stress-strain curves (dry state) for (a) neat Pani and (b) NT-reinforced Pani fibers.

TABLE 8 Measured and calculated properties of NT/PANi fiber composites.

	0% NT	0.62% NT
Elastic Modulus (GPa)	3.4 ± 0.4	7.3 ± 0.4
Tensile Strength (MPa)	170 ± 22	255 ± 32
Elongation at Break (%)	9 ± 3	4 ± 0.6
Max. Work Per Cycle (kJ/m3)	365	320
Adjusted PANi matrix actuation (%)	NA	1.8%
4-probe conductivity (S/cm)	497 ± 55	716 ± 36

Improved electrical conductivity has previously been shown to increase actuation performance by reducing iR losses along the fiber length and so increasing the active portion of the fiber.

Furthermore, strong interactions between the PAni and NTs has been previously suggested as enhancing the electroactivity of PAni/NT composites. As shown in Fig. 58, the electroactivity of the composites is improved by the addition of only 0.62% NTs. The more pronounced redox processes occurring in the composite fibers indicate an improved electrochemical efficiency that leads to an increase in the PAni matrix actuation.

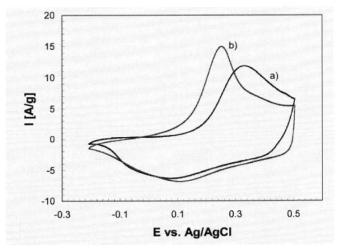

FIGURE 58 Cyclic voltammograms showing current density (I) for potential (E) scans between –0.2V and +0.5V (vs. Ag/AgCl) in M HCl at 50 MV/s: (a) neat Pani and (b) NT-reinforced Pani fibers.

Enhancement of electroactivity of PAni/NT composite films has been attributed to a possible increase in the degree of protonation of the PAni due to strong-stacking interactions between the NTs and the PAni. Clearly the improved electroactivity of the composite fibers will induce larger actuation strains in the PAni matrix, as described above.

The higher load capacity of the PAni/NT fibers also translates to higher work-per-cycles (Table 8). The maximum work-per-cycle of the neat PAni was 365 kJ/m^3 compared with other reported values of up to 83 kJ/m^3 for polypyrrole. With the addition of 0.62% NTs the work-per-cycle decreases slightly to 325 kJ/m^3, which is still more than 10 times higher than the work capacity of skeletal muscle. The dramatic increase in work per cycle compared with previous results from conducting polymers is a consequence of the higher strains at much higher stresses that are produced by the PAni and composite fibers.

In summary, it has been shown that the addition of small amounts of carbon nanotubes to PAni fibers produces significant improvements in their electroactivity, which translates to enhanced actuation performance. While the neat PAni produced an actuation strain of 1.3% at near zero loads, the presence of the NTs decreased this actuation to 0.85%. The resultant actuation strain of the composite fibers was determined by a balance between the increased electroactivity leading to higher strains in the PAni matrix (up to 1.8%) and the increased modulus, restricting matrix deformation and producing lower strains. The most significant effect of NT additions was, however, the much improved breaking strength and much higher operating stress levels. A 5-fold increase in work-per-cycle of compared with other conducting polymer actuators was achieved with the composite fibers. Useful actuation strains could be obtained at up to 100 MPa applied stress, which is 3 times higher than other conducting polymer actuators and 300 times higher than skeletal muscle. The improved strength and stiffness of the composite fibers can be utilized in various applications where high force operation is required, such as in strain amplification systems or biomimetic musculoskeletal systems.

6.17 ELECTROSPINNING OF BLEND NANOFIBERS

Polyaniline was synthesized by the oxidative polymerization of aniline in acidic media. 3 ml of distilled aniline was dissolved in 150 mL of 1N HCl and kept at 0–5°C. 7.325g of $(NH_4)_2S_2O_8$ was dissolved in 35 mL of 1N HCl and added drop wise under constant stirring to the aniline/HCl solution over a period of 20 minutes. The resulting dark green solution was maintained under constant stirring for 4 hours. The prepared suspension was dialyzed in a cellulose tubular membrane (Dialysis Tubing D9527, molecular cutoff = 12,400, Sigma) against distilled water for 48 hours. Then it was filtered and washed with water and methanol. The synthesized Polyaniline was added to 150 mL of 1N (NH4) OH solution. After an additional 4 hours the solution was filtered and a deep blue emeraldine base form of Polyaniline was obtained (PANIEB). The synthesized Polyaniline was dried and crushed into fine powder and then passed trough a 100 mesh. Intrinsic viscosity of the synthesized Polyaniline dissolved at Sulfuric acid (98%) was 1.18 dl/g at 25°C.

The PANI solution with concentration of 5% (W/W) was prepared by dissolving exact amount of PANI in NMP. The PANI was slowly added to the NMP with constant stirring at room temperature. This solution was then allowed to stir for 1 hour in a sealed container. 20% (W/W) solution of PAN in NMP was prepared separately and was added drop wise to the well-stirred PANI solution. The blend solution was allowed to stir with a mechanical stirrer for an additional 1 hour.

Various polymer blends with PANI content ranging from 10 wt% to 30 wt% were prepared by mixing different amount of 5% PANI solution and 20% PAN solution. Total concentrations of the blend solutions were kept as 12.5%.

Polymeric nanofibers can be made using the electrospinning process, which has been described in the literature and patent. Electrospinning uses a high electric field to draw a polymer solution from tip of a capillary toward a collector. A voltage is applied to the polymer solution, which causes a jet of the solution to be drawn toward a grounded collector. The fine jets dry to form polymeric fibers, which can be collected as a web.

Our electrospinning equipment used a variable high voltage power supply from Gamma High Voltage Research (USA). The applied voltage can be varied from 1–30 kV. A 5-mL syringe was used and positive potential was applied to the polymer blend solution by attaching the electrode directly to the outside of the hypodermic needle with internal diameter of 0.3 mm. The collector screen was a 20×20 cm aluminum foil, which was placed 10 cm horizontally from the tip of the needle. The electrode of opposite polarity was attached to the collector. A metering syringe pump from New Era pump systems Inc. (USA) was used. It was responsible for supplying polymer solution with a constant rate of 20 µL/min.

Electrospinning was done in a temperature-controlled chamber and temperature of electrospinning environment was adjusted on 25, 50 and 75°C. Schematic diagram of the electrospinning apparatus was shown in Fig. 59. Factorial experiment was designed to investigate and identify the effects of parameters on fiber diameter and morphology (Table 9).

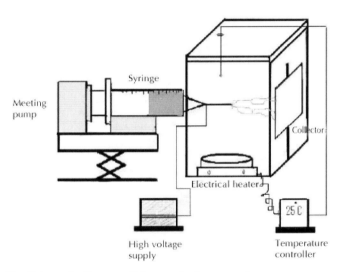

FIGURE 59 Schematic diagram of electrospinning apparatus.

6.17.1 CHARACTERIZATION

Shear viscosities of the fluids were measured at shear rate of 500 sec^{-1} and 22°C using a Brookfield viscometer (DVII+, USA). Fiber formation and morphology of the electrospun PANI/PAN fibers were determined using a scanning electron microscope (SEM) Philips XL-30A (Holland). Small section of the prepared samples was placed on SEM sample holder and then coated with gold by a BAL–TEC

SCD 005 sputter coater. The diameter of electrospun fibers was measured with image analyzer software (manual microstructure distance measurement). For each experiment, average fiber diameter and distribution were determined from about 100 measurements of the random fibers. Electrical conductivity of the electrospun mats was measured by the standard four- probe method after doping with HCl vapor.

TABLE 9 Factorial design of experiment.

Factor	Factor Level
PANI Content (wt%)	10,20,30
Electrospinning temperature (°C)	25,50,75
Applied voltage (kV)	20,25,30

6.18 RESULTS AND DISCUSSION

Published literature have shown that in the electrospinning process, the system configuration and operation conditions differ vastly from one material to another, depending on the material and the choice of solvent. Physical and chemical parameters of polymer solution such as viscosity, electrical conductivity, surface tension and air temperature can determinedly affect the formability and morphology of electrospun fibers. In the following sections effects of some electrospinning parameters on the fiber formation and morphology of PANI/PAN blend solutions were discussed and the best condition for obtaining PANI/PAN fibers was examined.

6.19 EFFECT OF PANI CONTENT

We were not able to obtain the fibers from the pure PANI solution because a stable drop at the end of the needle was not maintained. Figure 60 shows SEM micrographs of PANI nanoparticles electrospun from pure PANI solution. As seen in Fig. 60, most of PANI particles have a round shape, while the fibrous structure is not observed. The major complication in electrospinning of PANI is the poor solubility of PANI. At low polymer concentration, the solution does not contain sufficient material to produce stable solid fibers. With increasing polymer concentration, insoluble PANI particles in the solution increase rapidly, result the unspinnable solution. Therefore, we prepared PANI/PAN blend solutions with different PANI content using NMP as solvent. At PANI content above 30% regardless of electrospinning conditions drops were formed instead of fibers. A series of experiments were carried out when the PANI weight percent was varied from 10% to 30%. The applied voltage was 20 to 30 kV and the chamber temperature was held at 25, 50 and 75°C. Figure 60 shows the SEM micrographs and the surface morphology of obtained fibers at 25°C and 25 kV. At a solution containing 30% PANI, the fibrous structure was not completely stabilized and a bead-on-string structure with non-uniform morphology was obtained. The fibers between the beads had a circular cross section, with a diameter typically between 60 nm and 460 nm and mean fiber diameter of 164 nm. As the PANI content decreases to lower than 20%, a fibrous structure was stabilized. At 20% PANI content, fibers mean diameter increased to 425 nm with some beads on the fibers. At 10% PANI content, continuous fibers without beads were resulted regardless of electric field with the mean fiber diameter of 602 nm at 25 kV. Smooth and uniform fibers with average diameter of 652 nm were electrospun from PAN solution at the same electrospinning condition. These results reveal that as the PANI contents in the blends increase up to 30% the average diameter of blend nanofiber gradually decreases from 602 to 164 nm and its distribution becomes significantly broader with higher standard deviation as shown in Fig 60. It is also observed that fibers with not uniform morphology are electrospun at 25°C. Figure 61 shows SEM photomicrographs of electrospun PANI/PAN blend fibers at 50°C at various blend ratios. This figure shows that fibers with uniform morphology without remarkable beads are formed regardless of PANI content. It is also observed that at 50°C average diameter of electrospun fibers decreases from 194 nm at 10% PANI content to 124 nm at 30% PANI content at 50°C. Similar to the results obtained at 25°C fiber formation from pure PANI solution and blends containing more than 30% PANI was not possible. In electrospinning, the coiled polymer chains in the solution are transformed by the elongational flow of the jet into oriented entangled networks. Experimental observations in electrospinning confirm that for fiber formation to occur, a minimum chain entanglement is required. Below this critical chain entanglement, application of voltage results beads and droplets due to jet instability. The gradual increase in fiber diameter with content of PAN in the blends may be explained by the increase of solution viscosity due to higher viscosity of PAN solution. The ranges of shear viscosity of the PANI/PAN blends are shown at Table 10. It is obvious that shear viscosity of the solutions decrease with PANI content in the blends. Therefore, as the concentration of PAN in the blend is increased; the solution viscosity and resulted polymer chain entanglements increase significantly. During electrospinning, the stable jet ejected from Taylor's cone is subjected to tensile stresses and may undergo significant elengational flow. The nature of this elongational flow may determine the degree of stretching of the jet. The characteristics of this elengational flow can be determined by elasticity and viscosity of the solution. The results show that viscosity of the PAN solution is higher than PANI solution. Hence viscosity of the blend solution decreases with an increase in PANI content. Therefore, jet stretching during the electrospinning is more effective at higher PANI content. As a result, the fibers diameters decrease with increasing PANI content in the blends. On the other hand, at the high PANI content, an insufficiently deformable entangled network of polymer chain exists and the ejected jet reaches the collector before the solvent fully evaporates. Therefore, at low solution viscosity ejected jet breaks into droplets and a mixture of beads and fibers is obtained. This explains the formation of droplets and beads at high PANI content. Effect of electrospinning temperature is discussed in the following section. Researchers showed that the diameters of electrospun nanofibers are greatly affected by solution viscosity, and solution viscosity has an allometric relationship with its concentration. Our results shows that the electrospun nanofibers diameters (d) of PANI/PAN blends has a relationship with PANI content in the form of

$$d \propto (PANI\%)^2 \qquad (134)$$

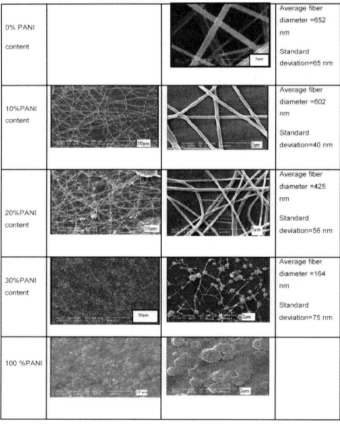

FIGURE 60 SEM micrographs of electrospun fibers at applied voltage of 25 kV and temperature of 25°C with a constant spinning distance of 10 cm.

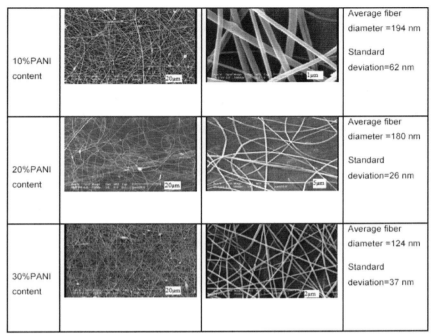

FIGURE 61 SEM micrographs of electrospun nanofibers at applied voltage of 25 kV and temperature of 50°C with a constant spinning distance of 10 cm.

6.20 EFFECT OF ELECTROSPINNING TEMPERATURE

Studies on the electrospinning show that many parameters may influence the transformation of polymer solution into nanofibers. Some of these parameters include (1) the solution related properties such as viscosity and surface tension, (2) process variables such as electric potential at the capillary tip and (3) ambient parameters such as air temperature in the electrospinning chamber. In order to study the effect of electrospinning temperature on the morphology and texture of electrospun PANI/PAN nanofibers, solution containing 20% PANI was electrospun at temperatures 25, 50 and 75°C. SEM micrographs of electrospun fibers at 20 kV are shown in Fig. 62. Interestingly, the electrospinning of the solution shows bead free fiber morphology at 50°C and 75°C, whereas fibers with large beads are observed at 25°C especially at high PANI contents (Fig. 62). The electrospun sample at 25°C shows fibers with several beads and not uniform surface morphology. With an increase in electrospinning temperature fibers morphology changes gradually from mixture of beads and fibers through uniform fibers. As shown in Fig. 62 at 50°C continuous fibers with uniform morphology were obtained while increasing the electrospinning temperature to 75°C caused bead free but fragile and cracked fibers. Diameter measurement of electrospun fibers at 25°C showed a size range of approximately 400 to 700 nm with 480 nm being the most frequently occurring. They were within the same range of reported size for electrospun PANI/PEO nanofibers [18]. With increasing the electrospinning temperature to 50°C, fiber diameter was decreased to a range of approximately 110 to 290 nm with 170 nm the most occurring frequency. At 75°C, fibers dimensions were 70 to 170 nm with 110 nm the most occurring frequency. It was obvious that diameter of electrospun fibers were decreased with increasing of electrospinning temperature. The distributions of fibers diameters electrospun at 25, 50 and 75°C are shown in Fig. 63. At 25°C broad distribution of fibers diameters was obtained, while a narrow distribution in fibers diameters was observed at 50 and 75°C.

TABLE 10 Shear viscosity of the PANI/PAN blend solutions at 22°C and shear rate of 500 sec^{-1} and average diameter of electrospun nanofibers.

PANI/PAN blend ratio % (w/w)	Shear viscosity (Pa.s)	Average nanofiber diameter (nm)
100/0 (5% solution)	0.159	No fiber
30/70	0.413	164
20/80	0.569	425
10/90	0.782	602
0/100	1.416	652

Several factors with PANI/PAN blends may explain the effects of electrospinning temperature and PANI content on morphology of the electrospun fibers. Since nanofibers are resulted from evaporation of solvent from polymer solution jets, the fiber diameters will depend on the jet sizes, elongation of the jet and evaporation rate of the solvent [24]. At a constant PANI content, as the electrospinning temperature is increased, the rate of solvent evaporation from the ejected jet increases significantly. In the case of electrospinning at 25°C due to the high boiling point of NMP (approximately 202°C), the fibers with relatively high solvent content travels during electrospinning process and reach the collector. Therefore, the collected fibers have irregular morphology due to contraction of the fibers during the electrospinning and on the collector. At higher electrospinning temperature rate of solvent evaporation from the ejected jet increases significantly and a skin is formed on the surface of the jet, which results collection of dry fiber with smooth surface. Presence of a thin, mechanically distinct polymer skin on the liquid jet during electrospinning has been discussed by researchers. On the other hand higher electrospinning temperature results higher degree of stretching and more uniform elongation of the ejected jet due to higher mobility and lower Viscosity of the solution. Therefore fibers with smaller diameters and narrower diameters distribution will be electrospun at higher electrospinning temperature.

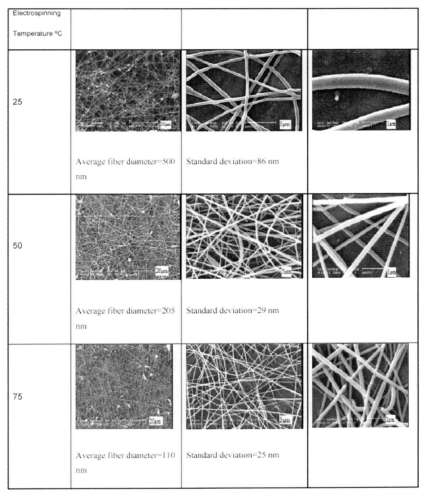

FIGURE 62 SEM micrographs of electrospun nanofibers at applied voltage of 20 kV and PANI content of 20% with a constant spinning distance of 10 cm.

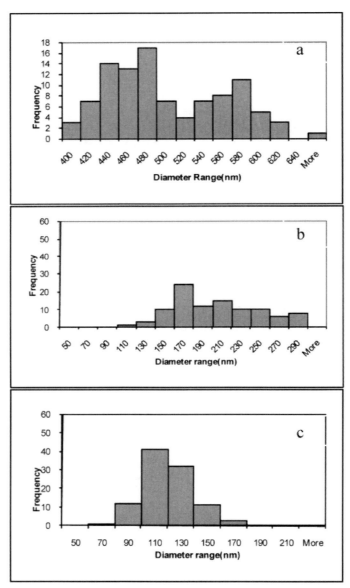

FIGURE 63 Distribution of fiber diameter electrospun at PANI content of 20%, applied voltage of 20 kV, spinning distance of 10 cm and electrospinning temperature of (a) 25°C, (b) 50°C and (c) 75°C.

6.21 EFFECT OF APPLIED VOLTAGE

In order to study the effects of applied voltage, the blend solutions were electrospun at various applied voltages and temperatures. From the results shown in Fig 64, it is obvious that the diameter of electrospun PANI/PAN fibers at 50°C decreased as the applied voltage increased. Similar results were observed for electrospun fibers at 25 and 70°C (results were not shown).

6.22 ELECTRICAL CONDUCTIVITY

Figure 64 shows electrical conductivity of the electrospun mats at various PANI/PAN blend ratios. As expected, electrical conductivity of the mats was found to increase with an increase in PANI content in the blends. Figure 65 shows that the electrical conductivity of the mats increases sharply when the PANI content in the blends is less than 5%, after which it will gradually reach to 10^{-1} S/cm at higher PANI content. This result is in agreement with the observations of Yang and co workers [35], which reported the electrical conductivity of PANI/PAN blend composites. Yang et al. [35] proposed the classical law of percolation theory, $\sigma(f)=c(f-f_p)^t$, where c is a constant, t is critical exponent of the equation, f is the volume fraction of the filler particle and f_p is the volume fraction at percolation threshold. The results of Fig. 65 indicate that the conductivity of the mats follows the scaling law of percolation theory mentioned above as shown in Eq. (135), which results a value of 0.5 wt% of PANI for f_p. This value for the percolation threshold is much lower than that reported by some scientists which may be due to the difference in the studied sample form. Their measurements were performed on the prepared films whereas our measurements were performed on the nanofiber mats. It is worth noting that the classical percolation theory predicts a percolation threshold of f_p =0.16 for conducting particles dispersed in an insulating matrix in three dimensions which is in agreement of our finding.

$$\sigma = 9\times10^{-7}(f-0.5)^{3.91} \quad R^2= 0.99 \tag{135}$$

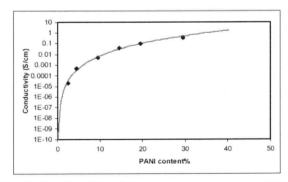

FIGURE 64 Electrical conductivity of electrospun mats at various PANI contents.

The electrospinning of PANI/PAN blend in NMP was processed and fibers with diameter ranging from 60 to 600 nm were obtained based on electrospinning conditions. Morphology of fibers was investigated at various blends ratios and electrospinning temperature. At 30% PANI content and 25°C fibers with average diameter of 164 nm were formed with beads (droplets of polymer over the woven mat) and not uniform morphology. At this condition solution viscosity and chain entanglements may not be enough, resulting in spraying of large droplets connected with very thin fibers. Averages of fibers diameters were decreased with PANI content in the solutions but PANI/PAN solution containing more than 30% PANI did not form a stable jet regardless of applied voltage and electrospinning temperature. For pure PANI solution, since the viscosity is too low to get stable drops and jets, we could not get the fibers. It was found that at 25°C fiber morphology was changed to beaded fibers when PANI content was higher than 20%. With increasing the electrospinning temperature, the morphology was changed from beaded fibers to uniform fibrous structure and the fiber diameter was also decreased from 500 nm to 100 nm when the electrospinning temperature changes from 25°C to 75°C. The mean of fiber diameter is the smallest and the fiber diameter distribution is the narrowest for the electrospun fibers at 75°C. However some cracks are observed on the surface of the electrospun fibers. There was a slightly decrease in average fiber diameter with increasing applied voltage. It is concluded that the optimum condition for nanoscale and uniform PANI/PAN fiber formation is 20% PANI content and 50°C electrospinning temperature regardless of the applied voltage. The conductivity of the mats follows the scaling law of percolation theory, which predicts a value of 0.5 wt% of PANI as percolation threshold for the blend of PANI/PAN.

KEYWORDS

- electrospinning
- mathematical modeling
- nanoelements
- nanofibers
- nanotubes
- polymers

REFRENCES

1. Reneker, D.H. and A.L. Yarin, *Electrospinning Jets and Polymer Nanofibers.* Polymer, 2008, **49**(10): p. 2387–2425.
2. Reneker, D.H., et al., *Electrospinning of Nanofibers from Polymer Solutions and Melts.* Advances in Applied Mechanics, 2007, **41**: p. 343–346.

3. Haghi, A.K. and G. Zaikov, *Advances in Nanofiber Research*. 2012: Smithers Rapra Technology. 194.
4. Haghi, A.K., *Electrospinning of nanofibers in textiles*. 2011, North Calorina: Apple Academic PressInc. 132.
5. Maghsoodloo, S., et al., *A Detailed Review on Mathematical Modeling of Electrospun Nanofibers*. Polymers Research Journal **6**: p. 361–379.
6. Frenot, A. and I.S. Chronakis, *Polymer Nanofibers Assembled by Electrospinning*. Current Opinion in Colloid and Interface Science, 2003, **8**(1): p. 64–75.
7. Fritzson, P., *Principles of object-oriented modeling and simulation with Modelica 2.1*. 2010: Wiley–IEEE Press.
8. Collins, A.J., et al., *The Value of Modeling and Simulation Standards*. 2011, Virginia Modeling, Analysis and Simulation Center, Old Dominion University: Virginia. p. 1–8.
9. Robinson, S., *Simulation: the practice of model development and use*. 2004: Wiley. 722.
10. Carson, I.I. and S. John, *Introduction to modeling and simulation*, in *Proceedings of the 36th conference on Winter simulation*. 2004, Winter Simulation Conference: Washington, DC. p. 9–16.
11. Banks, J., *Handbook of simulation*. 1998: Wiley Online Library. 342.
12. Pritsker, A.B. and B. Alan, *Principles of Simulation Modeling*. 1998, New York: Wiley. 426.
13. Carroll, C.P., *The development of a comprehensive simulation model for electrospinning*. Vol. 70. 2009, Cornell University 300.
14. Menon, A. and K. Somasekharan, *Velocity, acceleration & jerk in electrospinning*, in *The Internet Journal of Bioengineering*. 2009.
15. Gilbert, W., *De Magnete* Transl. PF Mottelay, Dover, UK. 1958, New York: Dover Publications, Inc. 366.
16. Tucker, N., et al., *The History of the Science and Technology of Electrospinning from 1600 to 1995*. Journal of Engineered Fibers and Fabrics, 2012, **7**: p. 63–73.
17. Zeleny, J., *The electrical discharge from liquid points, and a hydrostatic method of measuring the electric intensity at their surfaces*. Physical Review, 1914, **3**(2): p. 69–91.
18. Hassounah, I., *Melt electrospinning of thermoplastic polymers*. 2012: Aachen: Hochschulbibliothek Rheinisch–Westfälische Technischen Hochschule Aachen. 650.
19. Taylor, G.I., *The Scientific Papers of Sir Geoffrey Ingram Taylor*. Mechanics of Fluids, 1971, **4**.
20. Yeo, L.Y. and J.R. Friend, *Electrospinning Carbon Nanotube Polymer Composite Nanofibers*. Journal of Experimental Nanoscience, 2006, **1**(2): p. 177–209.
21. Miao, J., et al., *Electrospinning of nanomaterials and applications in electronic components and devices*. Journal of Nanoscience and Nanotechnology, 2010, **10**(9): p. 5507–5519.
22. Teo, W.E. and S. Ramakrishna, *A review on electrospinning design and nanofiber assemblies*. Nanotechnology, 2006, **17**(14): p. R89–R106.
23. Bhardwaj, N. and S.C. Kundu, *Electrospinning: a fascinating fiber fabrication technique*. Biotechnology Advances, 2010, **28**(3): p. 325–347.

24. Greiner, A. and J.H. Wendorff, *Electrospinning: a fascinating method for the preparation of ultrathin fibers.* Angewandte Chemie International Edition, 2007, **46**(30): p. 5670–5703.
25. Thoppey, N.M., et al., *Effect of Solution Parameters on Spontaneous Jet Formation and Throughput in Edge Electrospinning from a Fluid–Filled Bowl.* Macromolecules, 2012, **45**: p. 6527–6537.
26. Wan, Y., et al., *Modeling and Simulation of the Electrospinning Jet with Archimedean Spiral.* Advanced Science Letters, 2012, **10**(1): p. 590–592.
27. Liu, L. and Y.A. Dzenis, *Simulation of Electrospun Nanofiber Deposition on Stationary and Moving Substrates.* Micro & nano–Letters, 2011, **6**(6): p. 408–411.
28. Huang, Z.M., et al., *A review on polymer nanofibers by electrospinning and their applications in nanocomposites.* Composites Science and Technology, 2003, **63**(15): p. 2223–2253.
29. Ramakrishna, S., *An Introduction to Electrospinning and Nanofibers.* 2005: World Scientific Publishing Company. 396.
30. Arinstein, A., et al., *Effect of supramolecular structure on polymer nanofiber elasticity.* Nature Nanotechnology, 2007, **2**(1): p. 59–62.
31. Lu, C., et al., *Computer Simulation of Electrospinning. Part I. Effect of Solvent in Electrospinning.* Polymer, 2006, **47**(3): p. 915–921.
32. Greenfeld, I., et al., *Polymer dynamics in semidilute solution during electrospinning: A simple model and experimental observations.* Physical Review 2011, **84**(4): p. 41806–41815.
33. Solberg, R.H.M., *Position-controlled deposition for electrospinning.* 2007, Eindhoven University of Technology: Eindhoven. p. 75.
34. Gradoń, L., *Principles of Momentum, Mass and Energy Balances.* Chemical Engineering and Chemical Process Technology. **1**: p. 1–6.
35. Bird, R.B., W.E. Stewart, and E.N. Lightfoot, *Transport Phenomena.* Vol. 2. 1960, New York: Wiley & Sons, Incorporated, John 808.
36. Peters, G.W.M., M.A. Hulsen, and R.H.M. Solberg, *A Model for Electrospinning Viscoelastic Fluids*, in *Department of Mechanical Engineering.* 2007, Eindhoven University of Technology: Eindhoven p. 26.
37. Whitaker, R.D., *An historical note on the conservation of mass.* Journal of Chemical Education, 1975, **52**(10): p. 658.
38. Hohman, M.M., et al., *Electrospinning and electrically forced jets. I. Stability theory.* Physics of Fluids, 2001, **13**: p. 2201–2221.
39. He, J.H., et al., *Mathematical models for continuous electrospun nanofibers and electrospun nanoporous microspheres.* Polymer International, 2007, **56**(11): p. 1323–1329.
40. Xu, L., F. Liu, and N. Faraz, *Theoretical model for the electrospinning nanoporous materials process.* Computers and Mathematics with Applications, 2012, **64**(5): p. 1017–1021.
41. Heilbron, J.L., *Electricity in the 17th and 18th Century: A Study of Early Modern Physics.* 1979: Univ of California Press.
42. Orito, S. and M. Yoshimura, *Can the universe be charged?* Physical review letters, 1985, **54**(22): p. 2457–2460.
43. Karra, S., *Modeling electrospinning process and a numerical scheme using Lattice Boltzmann method to simulate viscoelastic fluid flows.* 2012.

44. Feynman, R.P., et al., *The Feynman Lectures on Physics; Vol. I.* American Journal of Physics, 1965, **33**: p. 750.
45. Bennett, C.O. and J.E. Myers, *Momentum, Heat, and Mass Transfer.* Vol. 370. 1982, New York: McGraw–Hill 848.
46. Feng, J.J., *The stretching of an electrified non–Newtonian jet: A model for electrospinning.* Physics of Fluids, 2002, **14**(11): p. 3912–3927.
47. Hou, S.H. and C.K. Chan, *Momentum Equation for Straight Electrically Charged Jet.* Applied Mathematics and Mechanics, 2011, **32**(12): p. 1515–1524.
48. Maxwell, J.C., *Electrical Research of the Honorable Henry Cavendish, 426*, in *Cambridge University Press*, Cambridge, Editor. 1878, Cambridge University Press, Cambridge, UK: UK.
49. Heilbron, J.L., *Electricity in the 17th and 18th Century: A Study of Early Modern Physics.* 1979: University of California Press. 437.
50. Vught, R.V., *Simulating the dynamical behavior of electrospinning processes*, in *Department of Mechanical Engineering*. 2010, Eindhoven University of Technology: Eindhoven. p. 68.
51. Jeans, J.H., *The Mathematical Theory of Electricity and Magnetism.* 1927, London: Cambridge University Press. 536.
52. Reneker, D.H., et al., *Bending Instability of Electrically Charged Liquid Jets of Polymer Solutions in Electrospinning.* Journal of Applied physics, 2000, **87**: p. 4531.
53. Truesdell, C. and W. Noll, *The non-linear field theories of mechanics.* 2004: Springer. 579.
54. Roylance, D., *Constitutive equations*, in *Lecture Notes. Department of Materials Science and Engineering*. 2000, Massachusetts Institute of Technology: Cambridge. p. 10.
55. He, J.H., Y. Wu, and N. Pang, *A mathematical model for preparation by AC-electrospinning process.* International Journal of Nonlinear Sciences and Numerical Simulation, 2005, **6**(3): p. 243–248.
56. Little, R.W., *Elasticity.* 1999: Courier Dover Publications. 431.
57. Bhattacharjee, P., V. Clayton, and A.G. Rutledge, *Electrospinning and Polymer Nanofibers: Process Fundamentals*, in *Comprehensive Biomaterials*. 2011, Elsevier. p. 497–512.
58. Clauset, A., C.R. Shalizi, and M.E.J. Newman, *Power-law Distributions in Empirical Data.* SIAM Review, 2009, **51**(4): p. 661–703.
59. Garg, K. and G.L. Bowlin, *Electrospinning jets and nanofibrous structures.* Biomicrofluidics, 2011, **5**: p. 13403–13421.
60. Spivak, A.F. and Y.A. Dzenis, *Asymptotic decay of radius of a weakly conductive viscous jet in an external electric field.* Applied Physics Letters, 1998, **73**(21): p. 3067–3069.
61. Wan, Y., Q. Guo, and N. Pan, *Thermo-electro-hydrodynamic model for electrospinning process.* International Journal of Nonlinear Sciences and Numerical Simulation, 2004, **5**(1): p. 5–8.
62. Giesekus, H., *Die elastizität von flüssigkeiten.* Rheologica Acta, 1966, **5**(1): p. 29–35.
63. Giesekus, H., *The physical meaning of Weissenberg's hypothesis with regard to the second normal-stress difference*, in *The Karl Weissenberg 80th Birthday Celebration*

Essays, J. Harris and K. Weissenberg, Editors. 1973, East African Literature Bureau p. 103–112.
64. Wiest, J.M., *A differential constitutive equation for polymer melts.* Rheologica Acta, 1989, **28**(1): p. 4–12.
65. Bird, R.B. and J.M. Wiest, *Constitutive Equations for Polymeric Liquids.* Annual Review of Fluid Mechanics, 1995, **27**(1): p. 169–193.
66. Giesekus, H., *A simple constitutive equation for polymer fluids based on the concept of deformation-dependent tensorial mobility.* Journal of Non–Newtonian Fluid Mechanics, 1982, **11**(1): p. 69–109.
67. Oliveira, P.J., *On the Numerical Implementation of Nonlinear Viscoelastic Models in a Finite–Volume Method.* Numerical Heat Transfer: Part B: Fundamentals, 2001, **40**(4): p. 283–301.
68. Simhambhatla, M. and A.I. Leonov, *On the Rheological Modeling of Viscoelastic Polymer Liquids with Stable Constitutive Equations.* Rheologica Acta, 1995, **34**(3): p. 259–273.
69. Giesekus, H., *A unified approach to a variety of constitutive models for polymer fluids based on the concept of configuration-dependent molecular mobility.* Rheologica Acta, 1982, **21**(4–5): p. 366–375.
70. Feng, J.J., *Stretching of a straight electrically charged viscoelastic jet.* Journal of Non–Newtonian Fluid Mechanics, 2003, **116**(1): p. 55–70.
71. Eringen, A.C. and G.A. Maugin, *Electrohydrodynamics*, in *Electrodynamics of Continua II.* 1990, Springer. p. 551–573.
72. Hutter, K., *Electrodynamics of Continua (A. Cemal Eringen and Gerard A. Maugin).* SIAM Review, 1991, **33**(2): p. 315–320.
73. Marrucci, G., *The free energy constitutive equation for polymer solutions from the dumbbell model.* Journal of Rheology, 1972, **16**: p. 321–331.
74. Kowalewski, T.A., S. Barral, and T. Kowalczyk, *Modeling Electrospinning of Nanofibers*, in *IUTAM Symposium on Modeling Nanomaterials and Nanosystems.* 2009, Springer: Aalborg, Denmark. p. 279–292.
75. Kuipers, B., *Qualitative reasoning: modeling and simulation with incomplete knowledge.* 1994: the MIT press. 554.
76. West, B.J., *Comments on the renormalization group, scaling and measures of complexity.* Chaos, Solitons and Fractals, 2004, **20**(1): p. 33–44.
77. De Gennes, P.G. and T.A. Witten, *Scaling Concepts in Polymer Physics.* Vol. Cornell University Press. 1980, 324.
78. He, J.H. and H.M. Liu, *Variational approach to nonlinear problems and a review on mathematical model of electrospinning.* Nonlinear Analysis, 2005, **63**: p. e919-e929.
79. He, J.H., Y.Q. Wan, and J.Y. Yu, *Allometric scaling and instability in electrospinning.* International Journal of Nonlinear Sciences and Numerical Simulation 2004, **5**(3): p. 243–252.
80. He, J.H., Y.Q. Wan, and J.Y. Yu, *Allometric Scaling and Instability in Electrospinning.* International Journal of Nonlinear Sciences and Numerical Simulation, 2004, **5**: p. 243–252.
81. He, J.H. and Y.Q. Wan, *Allometric scaling for voltage and current in electrospinning.* Polymer, 2004, **45**: p. 6731–6734.

82. He, J.H., Y.Q. Wan, and J.Y. Yu, *Scaling law in electrospinning: relationship between electric current and solution flow rate.* Polymer, 2005, **46**: p. 2799–2801.
83. He, J.H., Y.Q. Wanc, and J.Y. Yuc, *Application of vibration technology to polymer electrospinning.* International Journal of Nonlinear Sciences and Numerical Simulation, 2004, **5**(3): p. 253–262.
84. Kessick, R., J. Fenn, and G. Tepper, *The use of AC potentials in electrospraying and electrospinning processes.* Polymer, 2004, **45**(9): p. 2981–2984.
85. Boucher, D.F. and G.E. Alves, *Dimensionless numbers, part 1 and 2.* 1959.
86. Ipsen, D.C., *Units Dimensions And Dimensionless Numbers.* 1960, New York: McGraw Hill Book Company Inc. 466.
87. Langhaar, H.L., *Dimensional analysis and theory of models.* Vol. 2. 1951, New York: Wiley. 166
88. McKinley, G.H., *Dimensionless groups for understanding free surface flows of complex fluids.* Bulletin of the Society of Rheology, 2005, **2005**: p. 6–9.
89. Carroll, C.P., et al., *Nanofibers from Electrically Driven Viscoelastic Jets: Modeling and Experiments.* Korea–Australia Rheology Journal, 2008, **20**(3): p. 153–164.
90. Saville, D.A., *Electrohydrodynamics: the Taylor–Melcher leaky dielectric model.* Annual Review of Fluid Mechanics, 1997, **29**(1): p. 27–64.
91. Ramos, J.I., *Force Fields on Inviscid, Slender, Annular Liquid.* International Journal for Numerical Methods in Fluids, 1996, **23**: p. 221–239.
92. Spivak, A. and Y. Dzenis, *Asymptotic decay of radius of a weakly conductive viscous jet in an external electric field.* Applied Physics Letters, 1998, **73**(21): p. 3067–3069.
93. Ha, J.W. and S.M. Yang, *Deformation and breakup of Newtonian and non–Newtonian conducting drops in an electric field.* Journal of Fluid Mechanics, 2000, **405**: p. 131–156.
94. Hohman, M.M., et al., *Electrospinning and electrically forced jets. I. Stability theory.* Physics of Fluids, 2001, **13**: p. 2201.
95. Peters, G., M. Hulsen, and R. Solberg, *A Model for Electrospinning Viscoelastic Fluids.*
96. Dasri, T., *Mathematical Models of Bead–Spring Jets during Electrospinning for Fabrication of Nanofibers.* Walailak Journal of Science and Technology, 2012, **9**.
97. Holzmeister, A., A.L. Yarin, and J.H. Wendorff, *Barb formation in electrospinning: Experimental and theoretical investigations.* Polymer, 2010, **51**(12): p. 2769–2778.
98. Angammana, C.J. and S.H. Jayaram, *A Theoretical Understanding of the Physical Mechanisms of Electrospinning,* in *Proc. ESA Annual Meeting on Electrostatics.* 2011: Case Western Reserve University, Cleveland OH, p. 1–9.

CHAPTER 7

POLYACETYLENE

V. A. BABKIN, G. E. ZAIKOV, M. HASANZADEH, and A. K. HAGHI

CONTENTS

7.1 Synthesis, Structure, Physicochemical Properties and Application of Polyacetylene.. 252
7.2 The Nature of Kinetic Properties of Quasi-One-Dimensional Polymers (The Review of Properties and Methods of their Studying).. 268
7.3 The Nature of Kinetic Properties of Polyacetylene. 275
7.4 Continual Model of Polyacetylene. 280
7.5 Two-Dimensional Models of Relativistic Quantum Theory of Field and of Polyacetylene 288
7.6 Phase Transitions in Bidimentional Models of the Theory of a Field and Polyacetylene 293
7.7 Solitons with Fractional Charges in Polyacetylene 302
7.8 Change of a Fractional Charge of Soliton................................... 306
7.9 Transition Dielectric-Metal in Polyacetylene 307
7.9.4 Conclusion 309
Keywords 309
References... 310

7.1 SYNTHESIS, STRUCTURE, PHYSICOCHEMICAL PROPERTIES AND APPLICATION OF POLYACETYLENE

7.1.1 INTRODUCTION

Features of synthesis, structure, properties and use of polyacetylenes are considered in this monography. The catalytic polymerization of acetylene using different catalysts is shown. Plasmachemical synthesis of carbines is considered. The results of studying the structure of polyacetylenes by electron spectroscopy are presented. The results of the research of the surface morphology of polyacetylene are presented. Agency of receiving methods of polyacetylene on its properties is shown. It should be noted that the first chapter of present monography "Synthesis, structure, physicochemical properties and application of polyacetylene" prepared for publication by Professor Rakhimov A.I. and Associate Professor Titova E.S. (Volgograd State Technical University). The remaining chapters prepared for publication by Professor Ponomarev O.A. (Bashkirskii State University), Professor Babkin V.A. (Volgograd State Architect-build University, Sebrykov Departament), Associate Professor Titova E.S. and Professor Zaikov G.E. (Moscow, Institute of Biochemical Physics, Russian Academy of Sciences).

7.1.2 SYNTHESIS AND STRUCTURE OF POLYACETYLENE

The chemical element in periodic system of D.I. Mendeleev – carbon possesses a variety of unique properties. It is the reason of that, as carbon and its compounds, and materials on its basis serve as objects of basic researches and are applied in the most various areas.

Scientists' thought that two forms exists of crystal carbon only – diamond and graphite (opinion beginning of 60-th years of 20 century). These forms are widespread in the nature and they are known to mankind with the most ancient times.

The question on an opportunity of existence of forms of carbon with sp-hybridization of atoms was repeatedly considered theoretically. In 1885 German chemist A. Bayer tried to synthesize chained carbon from derivatives of acetylene by a step method. However Bayer's attempt to receive polyin has appeared unsuccessful. He received the hydrocarbon consisting from four molecules of acetylene, associated in a chain, and appeared extremely unstable.

A. M. Sladkov, V. V. Korshak, V. I. Kasatochkin and Yu. P. Kudryavcev [1] observed loss of a black sediment of polyin compound of carbon having the linear form at transmission acetylene in a water-ammonia solution of salt Cu (II) (oxidizing dehydropolycondensation of acetylene led obviously to polyacetylenides of copper). This powder blew up at heating in a dry condition, and in damp – at a detonation. Process of oxidative dehydropolycondensation of acetylene can be written down in a following kind schematically [1] at $x + y + z = n$:

$$n\ H-C\equiv C-H$$
$$\downarrow Cu^{2+}$$
$$H(-C\equiv C-)_x Cu\ +\ H(-C\equiv C-)_y H\ +\ Cu(-C\equiv C-)_z H$$
$$\downarrow FeCl_3$$
$$H(-C\equiv C-)-H$$
$$n$$

At surplus of ions Cu^{2+} the mix of various polyins and polyacetylenides of copper various molecular weights are formed. Additional oxidation of products received at this stage (with help $FeCl_3$ or $K_3[Fe(CN)_6]$) leads to formation polyins with the double molecular weight. The last do not blow up any more at heating and impact, but contain a plenty of copper. Possibly, trailer atoms of copper stabilize polyins by dint of to complexation.

The content by carbon was 90% of clean polyin (cleaning cleared from copper and impurity of other components of the reactionary medium). Only multi-hours heating of samples of polyin at 1000 °C in vacuum has allowed to receive analytically pure samples of α-carbyne. Similar processing results not only in purification, but also to partial crystallization of polyacetylene.

Under A.M. Sladkov's offer such polyacetylene have named "carbyne" (from Latin *carboneum* (carbon) with the termination "in", accepted in organic chemistry for a designation of acetylene bond).

By acknowledgment of polyin structures in chains is formation of oxalic acid after ozonation hydrolysis of carbine [2, 3]:

$$(-C\equiv C-)_n \xrightarrow{O_3} \left(\begin{array}{c} -C\equiv C- \\ | | \\ O O \\ \diagdown \diagup \\ O \end{array} \right)_n \xrightarrow{H_2O} nHOOC-COOH$$

New linear polymer with cumulene bonds was received [2, 3]. It has named polycumulene. The proof of such structure became that fact that at ozonation of polycumulene is received only carbon dioxide:

$$(=C=C=)_n \xrightarrow{O_3} 2nCO_2$$

Cumulene modification of carbyne (β-carbyne) has been received on specially developed by Sladkov two-stage method [3]. At the first stage spent polycondensation of suboxide of carbon (C_3O_2) with dimagnesium dibromine acetylene as Grignard reaction with formation polymeric glycol:

$$nO=C=C=C=O + nBrMgC\equiv CMgBr \rightarrow \left(-C\equiv C-\underset{OH}{C}=C=\underset{OH}{C}-\right)_n.$$

At the second stage this polymeric glycol reduced by stannous chloride hydrochloric acid:

$$\left(-C\equiv C-\underset{OH}{C}=C=\underset{OH}{C}-\right)_n \xrightarrow[-(HCl + SnO_2)]{+ SnCl_2} (=C=C=C=C=C=)_n.$$

High-molecular cumulene represents an insoluble dark-brown powder with the developed specific surface (200–300 m^2/g) and density 2.25 g/cm^3. At multi-hours heating at 1000°C and the depressed pressure polycumulene partially crystallizes. Two types of monocrystals have been found out in received after such annealing a product by means of transmission electronic microscopy. Crystals corresponded to α- and β-modifications of carbyne.

One of the most convenient and accessible methods of reception carbyne or its fragments – reaction of dehydrohalogenation of the some polymers content of halogens (GP). Feature of this method is formation of the carbon chain at polymerization corresponding monomers. The problem at synthesis carbyne consists only in that at full eliminating of halogen hydride with formation of linear carbon chain. Exhaustive dehydrohalogenation is possible, if the next atoms of carbon have equal quantities of atoms of halogen and hydrogen. Therefore convenient GP for reception of carbyne were various polyvinyliden halogenides (bromides, chlorides and fluorides), poly(1,2-dibromoethylene), poly (1,1,2 or 1,2,3-trichlorobutadiene), for example:

$$(-CH_2-CHal_2)_n \xrightarrow[-nHHal]{+B^-} (-CH-CHal-)_n \xrightarrow[-nHHal]{+B^-} (=CH=CH=)_n$$

The reaction of dehydrohalogenation typically carry out at presence of solutions of alkalis (B^-) in ethanol with addition of polar solvents. At use of tetrahydrofuran synthesis goes at a room temperature. This method allows to avoid course of collateral reactions. The amorphous phase only cumulene modification of carbyne is received as a result. Then, crystal of β-carbyne is synthesized from amorphous carbine by solid-phase crystallization.

Next method is dehydrogenation of polyacetylene. At interaction of polyacetylene with metallic potassium at 800°C and pressure 4 GPa led to dehydrogenation and formation of potassium hydride, the carbon matrix containing potassium. After removal potassium from products (acid processing) precipitate out brown plate crystals of β-carbyne in hexagonal forms by diameter ~1 mm and thickness up to 1 micrometer.

Carbyne also can be received by various methods of chemical sedimentation from a gas phase.

Plasmochemical synthesis of Carbyne. At thermal decomposition of hydrocarbons (acetylene, propane, heptane, benzol), carbone tetrachloride, carbon bisulphide, acetone in a stream nitrogen plasma is received the disperse carbon powders containing carbyne. Monocrystals of white color and (white or brown) polycrystals remain after selective oxidation of aromatic hydrocarbon. It is positioned, that formation of carbyne does not depend by nature initial organic compound. The moderate temperature of plasma (~3200K) and small concentration of reagents promote process.

Laser sublimations of carbon. Carbyne has been received at sedimentation on a substrate of steams of negative ions of the carbon after laser evaporation of graphite in 1971. The silvery-white layer was received on a substrate. This layer, according to data X-ray and diffraction researches, consists from amorphous and crystal particles of carbyne with the average size of crystallites $> 10^{-5}$ cm.

Arc cracking of carbon. Evaporation in electric arc spectrally pure coals with enough slow polymerization and crystallization of a carbon steam on a surface of a cold substrate yields to product in which prevail carbyne forms of carbon.

Ion-stimulated precipitation of carbyne. At ion-stimulated condensation of carbon on a lining simultaneously or alternately the stream of carbon and a stream of ions of an inert gas moves. The stream of carbon is received by thermal or ionic evaporation of graphite. This method allows receiving carbyne films with a different degree of orderliness (from amorphous up to monocrystalline layers), carbynes of the set updating, and also a film of other forms of carbon. Annealing of films of amorphous carbon with various near order leads to crystallization of various allotropic forms carbon, including carbyne.

Sladkov has drawn following conclusions on the basis of results of experiments on synthesis carbyne by methods of chemical sedimentation from gas phase:

- White sediments of carbyne are received, possibly, in the softest conditions of condensation of carbon: high enough vacuum, small intensity of a stream and low energy of flying atoms or groups of atoms, small speed of sedimentation;

- Chains, apparently, grow perpendicularly to a lining, not being cross-linked among themselves;
- Probably, being an environment monovalent heteroatoms stabilize chains, do not allow them to be cross-linked.

Reception of carbyne from carbon graphite materials leads by heating of cores from pyrolytic graphite at temperature 2700–3200K in argon medium. This leads to occurrence on the ends of cores a silvery-white strike (already through 15–20 s). This strike consists of crystals carbyne that is confirmed by data of method electron diffraction.

In 1958 Natta with employees are polymerized acetylene on catalyst system $Al(C_2H_5))_3 — Ti(OC_3H_7)_4$ [4, 5].

The subsequent researches [6–8] led to reception of films stereo regular polyacetylene. The catalyst system $Al(Et)_3 – Ti(OBu)_4$ provides reception of films of polyacetylene predominantly (up to 98%).

Films of polyacetylene are formed on a surface of the catalyst or practically to any lining moistened by a solution of the catalyst (it is preferable in toluene), in an atmosphere of the cleared acetylene [8]. The temperature and pressure of acetylene control growth of films [9, 10]. Homogeneous catalyst system before use typically maintain at a room temperature. Thus reactions of maturing of the catalyst occur [11]:

$$Ti(OBu)_4 + AlEt_3 \rightarrow EtTi(OBu)_3 \rightarrow AlEt_2(OBu)$$
$$2EtTi(OBu)_3 \rightarrow 2Ti(OBu)_3 + CH_4 + C_2H_6$$
$$Ti(OBu)_3 + AlEt_3 \rightarrow EtTi(OBu)_2 + Al(Et)_2OBu$$
$$EtTi(OBu)_2 + AlEt_3 \rightarrow EtTi(OBu)_2 \cdot Al(Et)_3$$

Ageing of the catalyst in the beginning raises its activity. However eventually the yield of polyacetylene falls because of the further reduction of the titan:

$$EtTi(OBu)_3 + Al(Et)_3 \rightarrow Ti(OBu)_4 + Al(Et)_2(OBu) + C_2H_4 + C_2H_6$$

The jelly-like product of red color is formed if synthesis led at low concentration of the catalyst. This product consists from confused fibrils in the size up to 800 Å. Foam material with density from 0.04 to 0.4 g/cm^3 is possible to receive from the dilute gels by sublimation of solvent at temperature below temperature of its freezing [12].

Research of speed dependence for formation of films of polyacetylene from catalyst concentration and pressure of acetylene has allowed to find an optimum parity of components for catalyst Al/Ti = 4. Increase of this parity up to 10 leads to increase in the sizes of fibrils [13]. Falling of speed of reaction in the end of process speaks deterioration of diffusion a monomer through a layer of the film formed on a surface of the catalyst [14]. During synthesis the film is formed simultaneously on walls of a flask and on a surface a catalytic solution. Gel collects in a cortex. The powder of polyacetylene settles at the bottom of a reactor. The molecular weight (M_n) a powder below, than gel, also is depressed with growth of concentration of the catalyst up to 400–500 [15]. The molecular weight of jealous polyacetylene slightly decreases with growth of concentration of the catalyst and grows from 2×10^4 up to 3.6×10^4 at rise in temperature from $-78°C$ up to $-10°C$. The molecular weight of polyacetylene in a film is twice less, than in gel [16].

Greater sensitivity of the catalyst to impurity does not allow to estimate unequivocally influence of various factors on M_p of polyacetylene. Low-molecular products with $M_p \sim -1200$ are formed at carrying out of synthesis in the medium of hydrogen [17]. Concentration of acetylene renders significant influence on M_p: at increase of its pressure up to 760 mm Hg increases M_p up to 120,000 [11, 18]. Apparently, the heterogeneity of a substratum arising because of imperfection of technics for synthesis is the reason for some irreproducibility properties of the received polymers. It is supposed, that synthesis is carried out on "surface" of catalytic clusters. Research by EPR method has allowed distinguishing in the catalyst up to four types of complexes [19]. A polyacetylene *cis*-transoid structures is formed as a result.

Formation of *trans*-structure at heats speaks thermal isomerization. The alternative opportunity – *trans*-disclosing of triple bond in a catalytic complex for a transitive condition is forbidden spatially [19, 20]. The structure of a complex and a kinetics of polymerization are considered in works in more detail [11, 21].

Parshakov A. S. with co-authors [22] have offered a new method of synthesis organo-inorganic composites – nanoclusters transitive metals in an organic matrix – by reactions of compounds of transitive metals of the maximum degrees of oxidation with monomers which at the first stage represent itself as a reducer. Formed thus clusters metals of the lowest degrees of oxidation are used for catalysis of polymerization a monomer with formation of an organic matrix.

Thus it is positioned, that at interaction $MoCl_5$ with acetylene in not polar mediums there is allocation HCl, downturn of a degree of oxidation of molybdenum and formation metalloorganic nanoclusters. Two distances Mo–Mo are found out in these nanoclusters by method EXAFS spectroscopy. In coordination sphere Mo there are two nonequivalent atom of chlorine and atom of carbon. On the basis of results MALDI–TOF mass-spectrometry the conclusion is made, that cluster of molybdenum has 12 or the 13-nuclear metal skeleton and its structure can be expressed by formulas $[Mo_{12}Cl_{24}(C_{20}H_{21})]^-$ or $[Mo_{13}Cl_{24}(C_{13}H_8)]^-$.

Set of results of Infrared-, Raman-, MASS-, NMR-^{13}C- and RFES has led to a conclusion, that the organic part of a composite represents polyacetylene a *trans*-structure. Polymeric chains lace, and alongside with the interfaced double bonds, are present linear fragments of twinned double –HC=C=CH- and triple –C≡C- bonds.

Reactions $NbCl_5$ with acetylene also is applied to synthesis of the organo-inorganic composites containing in an organic matrix clusters of transitive elements not only VI, but other groups of periodic system.

Solutions $MoCl_5$ sated at a room temperature in benzene or toluene used for reception of organo-inorganic composites with enough high concentration of metal. Acetylene passed through these solutions during 4–6 hours. Acetylene preliminary refined and drained from water and possible impurity. Color of a solution is varied from dark-yellow-green up to black in process of transmission acetylene. The solution heated up, turned to gel of black color and after a while the temperature dropped up to room. Reaction was accompanied by formation HCl. Completeness of interaction pentachloride with acetylene judged on the termination of its allocation.

The sediment, similar to gel, settled upon termination of transmission acetylene. It filtered off in an atmosphere of argon, washed out dry solvent and dried up under vacuum.

The received substances are fine-dispersed powders of black color, insoluble in water and in usual organic solvents.

Solutions after branch of a deposit represented pure solvent according to NMR. Formed compounds of molybdenum and products of oligomerization acetylene precipitated completely. The structure of products differed under the maintenance of carbon depending on speed and time of transmission acetylene a little. Thus relation C:H was conserved close to unit, and Cl:Mo – close to two. The structure of products of reaction differed slightly in benzene and toluene and was close to $MoCl_{1,9\pm0,1}(C_{30\pm1}H_{30\pm1})$.

TABLE 1 Data of the element analysis of products for reaction $MoCl_5$ with acetylene in benzene and toluene.

Solution	Percentage, weight %							
	C		H		Cl		Mo	
	Findings	Calculated	Findings	Calculated	Findings	Calculated	Findings	Calculated
Benzene	65.94	64.67	5.38	5.38	11.70	12.73	16.98	17.22
Toluene	66.23		5.24		11.87		16.66	

Presence on diffraction patterns the evolved products of a wide maximum at small corners allowed to assume X-ray amorphous or nanocrystalline a structure of the received substances. By a method of scanning electronic microscopy (SEM) it was revealed, that substances have low crystallinity and nonfibrillary morphology (Fig. 1a).

 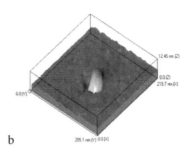

FIGURE 1 Photomicrographes $MoCl_{1,9\pm0,1}(C_{30\pm1}H_{30\pm1})$ according to SEM (a) and ASM (b).

7.1.3 PHYSICOCHEMICAL PROPERTIES AND APPLICATION OF POLYACETYLENE

By results of atomic-power microscopy (ASM) particle size can be estimated within the limits of 10÷15 nm (Fig. 1b). By means of translucent electronic microscopy has been positioned, that the minimal size of morphological element $MoCl_{1,9\pm0,1}(C_{30\pm1}H_{30\pm1})$ makes 1÷2 nm.

Substances are steady and do not fly in high vacuum and an inert atmosphere up to 300°C. Formation of structures $[Mo_{12}Cl_{24}(C_{20}H_{21})]^-$ and $[Mo_{13}Cl_{24}(C_{13}H_8)]$ is supposed also on the basis of mass-spectral of researches.

Spectrum EPR of composite $MoCl_{1,9\pm0,1}(C_{30\pm1}H_{30\pm1})$ at 300 K (Fig. 1.2.a) consists of two isotropic lines. The intensive line g=1.935 is carried to unpaired electrons of atoms of molybdenum. The observable size of the g-factor is approximately equal to values for some compounds of trivalent molybdenum. For example, in $K_3[InCl_6]\cdot 2H_2O$, where the ion of molybdenum Mo (+3) isomorphically substitutes In (+3), and value of the g-factor makes 1.93±0.06.

The line of insignificant intensity with g=2.003, close to the g-factor free electron –2.0023, has been carried to unpaired electrons atoms of carbon of a polyacetylene matrix. Intensity of electrons signals for atoms molybdenum essentially above, than for electrons of carbon atoms of a matrix. It is possible to conclude signal strength, that the basic contribution to paramagnetic properties of a composite bring unpaired electrons of atoms of molybdenum.

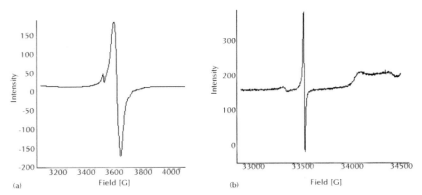

FIGURE 2 Spectrum EPR of composite $MoCl_{1.9\pm0.1}(C_{30\pm1}H_{30\pm1})$ at 300 K, removed in a continuous mode in X-(a) and W-range (b).

Spectrum EPR which has been removed in a continuous mode at 30 K, (Fig. 2b) has a little changed at transition from X to a high-frequency W-range. Observable three wide lines unpaired electrons atoms of molybdenum have been carried to three axial components with $g_1=1.9528$, $g_2=1.9696$ and $g_3=2.0156$, accordingly. Unpaired electrons atoms of carbon of a polyacetylene matrix the narrow signal $g=2.0033$ answers. In a pulse mode of shooting of spectra EPR at 30 K (Fig. 3) this line decomposes on two signals with $g_1=2.033$ and $g_2=2.035$. Presence of two signals EPR testifies to existence in a polyacetylene matrix of two types of the paramagnetic centers of the various natures which can be carried to distinction in their geometrical environment or to localized and delocalized unpaired electrons atoms of carbon polyacetylene chain.

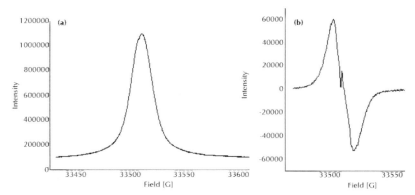

FIGURE 3 Spectrum absorption EPR of composite $MoCl_{1.9\pm0.1}(C_{30\pm1}H_{30\pm1})$, removed in a pulse mode in a W-range at 30 K (a) and its first derivative (b).

Measurement of temperature dependence of a magnetic susceptibility X_g in the field of temperatures 77÷300 K has shown, that at decrease in temperature from room up to 108K the size of a magnetic susceptibility of samples is within the limits of sensitivity of the device or practically is absent. The sample started to display a magnetic susceptibility below this temperature. The susceptibility sharply increased at the further decrease in temperature.

The magnetic susceptibility a trance-polyacetylene submits to Curie law and is very small on absolute size. Comparison of a temperature course a composite and pure allows to conclude a trance-polyacetylene, that the basic contribution to a magnetic susceptibility of the investigated samples bring unpaired electrons atoms of molybdenum in cluster. Sharp increase of a magnetic susceptibility below 108K can be connected with reduction of exchange interactions between atoms of metal.

The size of electroconductivity compressed samples $MoCl_{1,9\pm0,1}(C_{30\pm1}H_{30\pm1})$, measured at a direct current at a room temperature – $(1.3 \div 3.3) \times 10^{-7}\ \Omega^{-1}\cdot cm^{-1}$ is in a range of values for a trance-polyacetylene and characterizes a composite as weak dielectric or the semiconductor. The positioned size of conductivity of samples at an alternating current $\sigma = (3.1 \div 4.7) \times 10^{-3}\ \Omega^{-1}\cdot cm^{-1}$ can answer presence of ionic (proton) conductivity that can be connected with presence of mobile atoms of hydrogen at structure of polymer.

Research of composition, structure and properties of products of interaction $NbCl_5$ with acetylene in a benzene solution also are first-hand close and differ a little with the maintenance of carbon (Table 2).

TABLE 2 Data of the element analysis of products of interaction $NbCl_5$ with acetylene in a solution and at direct interaction.

The weights content, %							
C		H		Cl		Nb	
Findings	Calculated	Findings	Calculated	Findings	Calculated	Findings	Calculated
In a solution							
45.00	45.20	3.60	3.76	20.50	22.20	26.50	28.80
At direct interaction							
41.60	45.20	4.02	3.76	22.81	22.20	27.78	28.80

To substances formula $NbCl_{2\pm0,1}(C_{12\pm1}H_{12\pm1})$ can be attributed on the basis of the received data. Interaction can be described by the equation:

$$NbCl_{5(solv/solid)} + nC_2H_2 \rightarrow NbCl_{2\pm0,1}(C_{12\pm1}H_{12\pm1})\downarrow + (n-12)C_6H_6 + 3HCl\uparrow + Q$$

The wide line was observed on diffraction pattern $NbCl_{2\pm0,1}(C_{12\pm1}H_{12\pm1})$ at $2\theta = 23-24°C$. It allowed assuming a nanocrystalline structure of the received products.

Studying of morphology of surface $NbCl_{2\pm0,1}(C_{12\pm1}H_{12\pm1})$, received by direct interaction, method SEM has shown, that particles have predominantly the spherical form, and their sizes make less than 100 nm (Fig. 4).

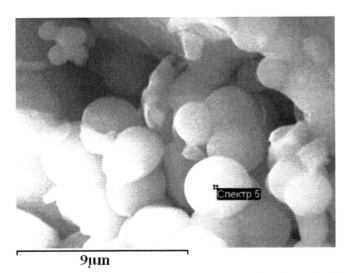

FIGURE 4 Microphoto of particles $NbCl_{2\pm0,1}(C_{12\pm1}H_{12\pm1})$, received by method SEM.

The Globular form of particles and their small size testify to the big size of their specific surface. It will be coordinated with high catalytic activity $NbCl_{2\pm0,1}(C_{12\pm1}H_{12\pm1})$.

Fibrils are formed in many cases as a result of synthesis [21]. The morphology of polyacetylene films practically does not depend on conditions of synthesis. Diameter of fibrils can change depending on these conditions and typically makes 200–800 Å [11, 21]. At cultivation of films on substrates the size fibrils decreases. The same effect is observed at reception of polyacetylene in the medium of other polymers. Time of endurance (ageing) of the catalyst especially strongly influences the size of fibrils. The size of fibrils increases with increase in time of ageing. Detailed research of growth fibrils on thin films a method of translucent electronic microscopy has allowed to find out microfibrillar branching's on the basic fibril (the size 30–50 Å) and thickenings in places of its gearing, and also presence of rings on the ends of fibrils.

Essential changes in morphology of a film at isomerization of polyacetylene are not observed. The film consists from any way located fibrils. Fibrils sometimes are going in larger formations [20]. Formation of a film is consequence of interaction fibrils among themselves due to adhesive forces.

The big practical interest is represented catalytic system $AlR_3-Ti(oBu)_4$ [23]. Polyacetylene is received on it at $-60°C$, possesses fibrous structure and can be manufactured usual, accepted in technology of polymers by methods. Particle size increases from 100 up to 500 Å at use of the mixed catalyst and the increase in density of films is observed. A filtration of suspension it is possible to receive films of any sizes. On the various substrates possessing good adhesive properties, it is possible to receive films dispersion of gel. Polymer easily doping AsF_5, $FeCl_3$, I_2 and others electron acceptors. Preliminary tests have shown some advantages of such materials at their use in accumulators [23]. Gels polyacetylene with the similar properties, received on others catalytic systems, represent the big practical interest [24]. Data on technological receptions of manufacturing of polyacetylene films from gels with diameter of particles 0.01–1.00 mm are in the patent literature [25]. Films are received at presence of the mixed catalyst at an interval of temperatures from -100 up to $-48°C$. These results testify to an opportunity of transition to enough simple and cheap technology of continuous process of reception of polyacetylene films. Rather accessible catalysts are WCl_6 and $MoCl_6$. Acetylene polymerizes at $20°C$ and pressure up to 14 atm at their presence. However the received polymers contain carbonyl groups because of presence of oxygen and possess low molecular weight [26]. The complex systems including in addition to WCl_6 or $MoCl_6$ tetraphenyltin are more perspective for reception of film materials. The Film of doped predominantly (90%) a trance-structure with fibrous morphology is formed on a surface of a concentrated solution of the vanadic catalyst. Diameter of fibrils of the polymer received on catalyst $MoCl_5 - Ph_4Sn$, can vary within the limits of 300–10000 Å; in case of catalyst $WCl_6 - Ph_4Sn$ it reaches 1.2×10^5 Å [27].

Research of a kinetics of polymerization of acetylene on catalysts $Ti(OBu)_5-Al(Et)_3$, $WCl_6-PhtSn$, $MoCl_5-Ph_4Sn$, $Ti(CH_2CeH_5)$, has shown, that speed of process falls in the specified number [31]. Films, doped by various acceptors [CH $(SbF_5)_{0,7}]_n$, $[CH (CF_3SO_3H)_{0,8}]_m$ – had conductivity 10–20 $\Omega^{-1}\cdot cm^{-1}$ at $20°C$.

New original method of reception of films doped in a trance-form is polymerization of a 7,8-bis(trifluoromethyl)tricyclo[4,2,2,0]deca-3,7,9-trien (BTFM) with disclosing a cycle.

Polymerization occurs on catalytic system $WCl_6 - Sn(CH_3)_4$, precipitated on surfaces of a reactor. The film prepolymer as a result of heating in vacuum at 100–150° detaches trifluoromethylbenzene. The silvery film polyacetylene is formed. The density of polymer reaches 1.1 g/cm^3 and comes nearer to flotation density polyacetylene, received in other ways. Received this method of amorphous polyacetylene has completely a trance-configuration and does not possess fibrous structure. The rests of 1,2-bis(trifluoromethylbenzene) are present at polymer according to Infrared-spectroscopy. Crystal films of polyacetylene with monoclinic system and $\beta = 91.5°$ are received at long heating of prepolymer on networks of an electronic microscope at 100° in vacuum. Improvement of a method has allowed to receive films and completely oriented crystal polyacetylene. In the further ways of synthesis of polymers from others monomers have been developed.

Naphthalene evolves at heat treatment of prepolymer in the first case. Anthracene evolves in the second case. Purification polymer from residual impurities occurs when the temperature of heat treatment rise. According to spectral researches, the absorption caused by presence of sp-hybrid carbon is not observed in films. Absorption in the field of 1480 cm^{-1} caused by presence of C = C bonds in Spectra KP of considered polymers, is shifted compared to the absorption observed in polyacetylene, received by other methods (1460 cm^{-1}).

It is believed that it is connected with decrease of size of interface blocks. The obtained films are difficult doping in a gas phase due to its high density. Conductivity of initial films reaches 10–200 (Ω^{-1}.cm^{-1}) when doped with bromine or iodine in a solution [26].

Solutions of complex compound cyclopentadienyl-dititana in hexane and sodium cyclopentadienyl complex have high catalytic activity. Films with a metallic luster can be obtained by slowly removing the solvent from the formed gel polyacetylene in vacuum. It is assumed that the active complex has a tetrahedral structure. Polymerization mechanism is similar to the mechanism of olefin polymerization on catalyst Ziegler – Natta. Obtained at –80°C *cis*-polyacetylene films after doping had a conductivity of 240 (Ω^{-1}.cm^{-1}).

Classical methods of ionic and radical polymerization do not allow to receive high-molecular polymers with system of the conjugated bonds because of isomerization the active centers [2] connected with a polyene chain. Affinity to electron and potential of ionization considerably varies with increase in effective conjugated. One of the methods, allowing to lead a cation process of polymerization, formation of a complex with the growing polyconjugated chains during synthesis. Practical realization of such process probably at presence of the big surplus of a strong acceptor of electrons in the reaction medium. In this case the electronic density of polyene chain falls. The probability of electron transfer from a chain on the active center decreases accordingly.

Polyacetylene films were able to synthesize on an internal surface of a reactor in an interval of temperatures from –78°C up to –198°C at addition of acetylene to arsenic pentafluoride. Strips of absorption *cis*-polyacetylene are identified in the field of 740 cm^{-1} and doped complexes in the field of 900 and 1370 cm^{-1}.

Similar in composition films were prepared by polymerization vinylacetylene in the gas phase in the presence SbF_5. However, to achieve a metallic state has failed. Soluble polyenes were obtained in solution AsF_3 in the by cationic polymerization, including soluble agents. Polymerization was carried out at the freezing temperature of acetylene. From the resulting solution were cast films with low conductivity, characteristic of weakly doped polyacetylene with 103 molecular weight. Spectral studies confirmed the presence of the polymer obtained in the *cis*-structure. Practical interest are insoluble polymers with a conductivity of 10^{-3}

($\Omega^{-1}.cm^{-1}$), obtained at $-78°C$ polymerization of acetylene in the presence AsF_5, NaAsF, SiF, AsF_3, BF_3, SbF_5, PF_6.

Effective co-catalyst of cationic polymerization of acetylene and its derivatives are compounds of bivalent mercury and its organic derivatives. As a result, the reactions of oxide or mercury salts with proton and aprotic acids into saturated hydrocarbon formed heterogeneous complexes. They are effectively polymerized acetylene vinylacetylene, phenylacetylene, propargyl alcohol. Polyacetylene predominantly *trans*-structure with a crystallinity of 70% was obtained in the form of films during the polymerization of acetylene on the catalyst surface. According to X-ray studies, the main reflection corresponds to the interplanar distance $d = 3.22$ Å. Homogeneous catalyst obtained in the presence of aromatic ligands. Active complex of the catalyst with acetylene is stable at low temperatures. The alkylation of solvent and its interpolymerization with acetylene is in the presence of aromatic solvents (toluene, benzene). In the infrared spectrum of the films revealed absorption strias corresponding to the aromatic cycle. Patterns of polymerization of acetylene monomers, the effect of temperature and composition of the catalyst on the structure and properties of the resulting polymers were studied. Morphology of the films showed the absence of fibrils.

Almost all of the above methods of synthesis polyacetylene with high molecular weight lead to the formation of insoluble polymers. Their insolubility due to the high intermolecular interaction and form a network structures. For the polymers obtained by Ziegler catalyst systems, the crosslink density of the NMR data of 3–5%. This is confirmed by ozonolysis. Quantum chemical calculations confirm that the isomerization process intermolecular bonds are formed.

The most accessible and promising method for synthesis of the polyacetylene of linear structure is the polymerization of acetylene in the presence of metals VIII group, in the presence with reducing agents (catalyst Luttlnger). Polyacetylene with high crystallinity was obtained by polymerization of acetylene on the catalyst system $Co(NO_3)_2$ $NaBH_4$ with component ratio 1:2. Both components are injected into the substrate containing monomer, to prevent the death of a catalyst. Raising the temperature and the concentration of sodium borane leads to partial reduction of the polymer. The activity of nickel complex can be significantly improved if the polymerization leads in the presence of $NaBH_2$. Crystalline polymers, obtained at low temperatures, do not contain fibrils. Crystallites have dimensions of 70 Å.

A typical reflex observed at $23.75°C$ ($d=3.74$ Å) confirms the *trans*-structure of the polymer. Palladium complexes are ineffective in obtaining high molecular weight polymers. Catalyst Luttlnger enables one to obtain linear polymers of *cis*-structure, characterized by high crystallinity. The yield of polyacetylene is 25–30 g/g catalyst.

Chlorination of the freshly prepared polymers at low temperatures allows to obtain soluble chlorpolymers about 104 molecular weight [29]. Although the authors argue that the low-temperature chlorination, in contrast to hydrogenation, there is no polymer degradation, data suggest that an appropriate choice of temperature and solvent derived chlorinated polymers have a molecular weight up to 2.5×10^5. Destruction more visible at chlorination on light and on elevated temperatures. The proof of the linear structure of polyacetylene is the fact that soluble iodinated polymers are obtained by iodination polyacetylene suspension in ethanol.

Systematic studies of methods for the synthesis of polyacetylene allowed to develop a simple and convenient method of obtaining the films on various substrates wetted by an ethereal solution of the catalyst. The disadvantages of these films, as well as films produced by Shirakava [21, 22] are difficult to clean them of residual catalyst and the dependence of properties on the film thickness. Much more manufacturable methods for obtaining films of polyacetylene spray pre-cleaned from residues of the catalyst in a stream of polyacetylene gels inert gas or a splash of homogenized gels [24]. The properties of such films depend on the conditions of their formation. Free film thickness of 2–3 mm is filtered under pressure in an inert atmosphere containing a homogenized suspension of polyacetylene 5–10 g/L. The suspension formed in organic media at low temperatures in the presence of the catalyst $Co(NO_3)_2$ $NaBH_4$. The films obtained by spraying a stream of inert gas, homogeneous, have good adhesion to substrates made of metal, polyurethane, polyester polyethylenetereftalate, polyimide, etc.

Suspension of polyacetylene changes its properties with time significantly. Cross-linking and aggregation of fibrils observed in an inert atmosphere at temperatures above –20 °C. This leads to a decrease in the rate of oxidation and chlorination. The morphology of the films changes particularly striking during the ageing of the suspension in the presence of moisture and oxygen: increasing the diameter of the fibrils, decreasing their length, breaks and knots are formed. The suspension does not change its properties in two weeks.

Preparation of soluble polymers with a system of conjugated double bonds, and high molecular weight is practically difficult because of strong intermolecular interactions. Sufficiently high molecular weight polyacetylene were obtained in the form of fine-dispersed particles during the synthesis on a Luttinger catalyst in the presence surfactants: copolymer of styrene and polyethylene with polyethylene oxide. Acetylene was added to the catalyst and the copolymer solution in a mixture of cyclohexane – tetrahydrofuran at –60°C and heated to –30°C. Stable colloidal solutions with spherical particles in size from 40 to 2000 Å, formed. The density of selected films is 1.15 g/cm^3. Colloidal solutions of polyacetylene can be obtained in the presence of other polymers that prevent aggregation of the forming molecules of polyacetylene.

Polyacetylene obtained in the presence of Group VIII metals, in combination with $NaBH_4$, has almost the same morphology, as a polymer synthesized by Shirakawa [22]. The dimensions of fibrils lay in the range 300–800 Å, and depend on the concentration of the catalyst, the synthesis temperature and medium [29].

Thermogravimetric curves for the polyacetylene, there are two exothermic peaks at 145 and 325°C [13]. The first of these corresponds to an irreversible *cis*-trans isomerization. Migration of hydrogen occurs at 325°C, open chain and cross-linking without the formation of polyacetylene volatile products. The color of the polymer becomes brown. A large number of defects appear. In the infrared spectrum there are absorption bands characteristic of the CH_2, CH_3, –C = C– and –C_2H_5–groups [13].

Structuring polymer occurs in the temperature range 280–380°C. But 72% of initial weight of polyacetylene losses at 720°C. The main products of the decomposition of polyacetylene are benzene, hydrogen and lower hydrocarbons [12]. The crystallinity of polyacetylene reduced when heated in air to 90°C after several hours. The brown amorphous substance, similar cuprene, obtained after 70 hours.

Catalytic hydrogenation of polyacetylene leads to the formation of cross-linked product [30]. Non-cross-linked and soluble products are obtained in the case of hydrogenation of polyacetylene doped with alkali metals [27, 28]. Polyacetylenes are involved in redox reactions that occur in processing strong oxidizing and reducing agents (iodine, bromine, AsF_5, Na-naphthalene) in order to significantly increase the electrical conductivity [31].

Practical use of polyacetylene is complicated by its easy oxidation by air oxygen [32]. Oxidation is easily exposed to the polymer obtained by polymerization of acetylene [33, 34]. The *cis*- or *trans*-$(CH)_x$ in air or oxygen for about an hour exposed to the irreversible oxidative degradation [35, 36]. The limiting value of weight gain due to absorption of polyacetylene (absorption) of oxygen from air oxidation at room temperature is 35% [32]. The resulting product is characterized by the formula $[(C_2H_2)O_{0.9}]_n$. The ease of oxidation depends on the morphology of the polymer and changes in the series of crystal < amorphous component < the surface of the fibrils [34]. The absorption of oxygen begins at the surface of fibrils, and then penetrates. Polyacetylene globular morphology is more stable to the effects of O_2 than polymer fibrillar structures [37].

Polyacetylene obtained by polymerization of acetylene in Ziegler–Natta catalysts, after doping Cl_2, Br_2, I_2, AsF_5 becomes a semiconductor in the form of flexible, silvery films "organic metals" [38]. Doping with iodine increases the amorphous samples d 6×10^{-5}, and crystal – to 7×10^2 $\Omega^{-1}.cm^{-1}$ [39]. The highest electrical conductivity of the polyacetylene compared with those obtained by other methods, the authors [40] explain the presence of catalyst residues. In their view, the concentration of the structure of sp^3 – hybridized carbon atoms is relatively little effect on the conductivity as compared with the influence of catalyst residues. Doping with iodine films of polyacetylene obtained by metathesis polymerizing cyclooctatetraene leads to an increase in their electrical conductivity

10^{-8} to 50–350 $\Omega^{-1}.cm^{-1}$ [41], and have received polymerization of benzvalene with ring opening from 10^{-8}–10^{-5} to 10^{-4}–10^{-1} $\Omega^{-1}.cm^{-1}$ [42].

Conductivity increases when pressure is applied to polyacetylene, obtained by polymerization of acetylene [43] and interphase dehydrochlorination of PVC [44]. Anomalously large (up to ten orders of magnitude) an abrupt increase in conductivity when the load is found for iodine-doped crystalline polivinilena – conversion product of PVC [45].

Magnetic properties of polyacetylene significantly depend on the configuration of chains [46]. In the EPR spectrum of the polymerization of polyacetylene singlet line with g-factor of 2.003 [23] and a line width (ΔH) of 7 to 9.5 Oe for the *cis*-isomer [24] and from 0.28 to 5 Oe for the *trans*-isomer [25] observed. According to other reports [47], the *cis*-isomer, syn-synthesized by polymerization of acetylene at 195 K, the EPR signal with g-factor=2.0025 is not observed. This signal appears when the temperature of polymerization increases, when the *trans*-isomer in the form of short chains mainly at the ends of the molecules is 5–10 wt% [26]. The morphology of polyacetylene also has an effect on the paramagnetic properties. The concentration of PMC in the amorphous polyacetylene is ~1018 spin/g, and in the crystal – 1019 spin/g [26].

7.2 THE NATURE OF KINETIC PROPERTIES OF QUASI-ONE-DIMENSIONAL POLYMERS (THE REVIEW OF PROPERTIES AND METHODS OF THEIR STUDYING)

7.2.1 INTRODUCTION

Development of a science about kinetic properties of firm bodies and liquids always led to revealing of the big role of a stable particles in kinetic processes. In chemistry this opening of radicals and ion- radicals, development of chain processes an establishment, that catalytic properties of substances are defined by defects of structure by which energy of activation is essentially reduced. Durability, conductivity, painting and so on at crystals is defined basically by defects of structure. Properties enough greater molecules, especially kinetic properties, are connected substantially with defects and multifocal uptake. As examples polyacetylene and a number of other polymers can serve.

In the given review we shall stop only on three aspects of a problem: we shall give a general characteristic of quasi-one-dimensional systems, we shall describe the basic properties of polyacetylene and we shall formulate problems in the given area.

7.2.2 A GENERAL CHARACTERISTIC OF ONE-DIMENSIONAL SYSTEMS

A number of exact results are known for quasi-one-dimensional systems. These results are received on the basis of exact calculation of modeling systems, or are consequence of the general theorems of interaction. Results are little. The most important results are: 1) the proof of absence of phase transition at $T_c \neq 0$, that is the proof of destruction of the first order; 2) an establishment, that the account of correlations leads to collectivization of conditions; 3) the proof, that localization of conditions occurs in as much as weak casual floor; 4) an establishment of the fact, that results 1) −3) lose force already at infinitesimal interaction in three measurements.

Quasi-one-dimensional systems have high density of the raised conditions and fluctuations. It leads to a number of features, such as formation of superstructures, infringement of the concept of quasi-particles, inapplicability of a method of the self-coordinated field. We shall consider these features in more detail.

7.2.2.1 THE FORMATION OF SUPERSTRUCTURES

The formation of superstructures is universal property for one-dimensional materials: metal passes in semimetal, then in dielectric. Structural instability of an one-dimensional metal condition is proved by Peierls (1937). It leads high permittivity's $\varepsilon=10^4$ (Frelih, 1954). Peierls instability this phenomenon when at downturn of temperature in a lattice there are stationary distortions with a wave vector k, equal to the double vector of electron at the top borrowed level of metal, that is $k = 2CF$. This phenomenon is closely connected with huge anomaly—reduction of density of oscillatory conditions at a wave vector $q = 2CF$. Peierls instability is shown that frequency of the some infrared-fluctuations decreases and addresses in zero. Frequencies of fluctuations in molecules are defined by the formula:

$$\omega_q^2 = \omega_{0q}^2 \left(1 - 2\Pi(q,\omega_q)/\omega_{0q}\right)$$

$\omega_{0q}^2 -$ Not indignant frequency, $\Pi(q,\omega_q)$ − own energy of fluctuations appearing due to kernels with electrons. The equation $\left(1-2\Pi(q,\omega_q)/\omega_{0q}\right)=0$ defines those frequencies and wave a vector at which before all there is an instability. It appears at $q = 2CF$. Reorganization of structure of a molecule occurs at this wave vector.

Two tendencies are distinctly shown in one-dimensional gas of electrons, kernels being a floor: superconducting pairing of electrons with opposite backs and impulses and Peierls pairing electron and holes with opposite impulses (doubling a handrail) (Brazovskiy, Dzyaloginskiy, 1974; Brazovskiy, 1981). Impurities in system suppress dielectric and superconducting transitions.

7.2.2.2 INAPPLICABILITY OF THE CONCEPT OF ONE-PARTIAL CONDITIONS

Strong collectivization of electrons occurs in one-dimensional system. It follows that speed of a sound c comparable whit speed of electrons v_F on top level (for two and three-dimensional systems they are accordingly equal $v_F/\sqrt{2}$ and $v_F/\sqrt{3}$). In model with linear spectrum $H_0 = v_F \sum p(a_{ps}^+ a_{ps} - b_{ps}^+ b_{ps})$ in one-dimensional case all degrees of freedom appear collective. While interaction is not present, it is possible to use H_0 and H_{10} – other representation H_0, written down in representation of operators of density

$$H_{10} = \frac{2\pi V_F}{L} \sum [\rho_s^{(1)}(p)\rho_s^{(1)}(-p) + \rho_s^{(2)}(p)\rho_s^{(2)}(-p)]$$

$$p_s^{(1)}(p) = \sum a_{p+k,s}^+ a_{ks} \quad p_s^{(2)}(p) = \sum b_{p+k,s}^+ b_{ks}$$

At presence of interaction $H_{int} \approx a^+b + ab, a^+a^+bb, b^+b^+aa, a^+a^+aa, b^+b^+bb$ for one-dimensional systems the picture of the description is meaningful only H_{10}. Quasi-particle is done inapplicable at transition to one-dimensional systems. We shall consider one-partial Green's function $G(q, \omega)$. If quasi-particle the description was fair, $-\pi^{-1} Jm G(0, \omega + i0) = \sum Z_i \delta(\omega - \omega_i) + b(\omega)$. This kind of function of Green allowed to enter the concept of quasi-particles because of presence $\delta(\omega - \omega_i)$. For one-dimensional systems it is received for precisely solved model of Luttinger–Tomonug (Luther, 1974)

$$-\pi^{-1} jm G(0, \omega + i0) = \frac{r}{c}(\frac{wr}{2c})^{\gamma-1}\Gamma(1-\gamma)\sin(\pi\gamma/2) + (\frac{2r}{\gamma c})(\frac{\omega r}{2c})^{2r-1}[\Gamma(1-\gamma)\sin \pi\gamma]^2$$

The member of type $\delta(\omega - \omega_i)$ in this expression is not present. In this case (q, ω) has no poles. Concept of quasi-particle to enter it is impossible.

Though the quasi-particle description is impossible in case of one-dimensional systems because of strong interaction of electrons with deformations of a lattice. Thus new collective conditions arise – solitones, which are decisions of the nonlinear equations and replace one-partial, conditions at the description of properties one-dimensional and quasi-one-dimensional systems.

7.2.2.3 INAPPLICABILITY OF A METHOD OF THE SELF-COORDINATED FIELD

Used of Hartree-Fok approximation is too rough approach a one-dimensional case. It is established from comparison with exact decisions for modeling systems and some exact results received in the general approach. For example, Hartree-Fok approximation leads to temperature of phase transition distinct from zero though according to the general q^{-2}-Bogoliubov thereof it not can be.

The method of the self-coordinated field in "ladder" approach and representation one-partial excitement in which it results does not describe behavior of polyacetylene as insufficiently full considers collective phenomena of fluctuation. Its

Polyacetylene

noncritical application has led to a number of mistakes at calculation Peierls transition and one-dimensional superconductivity. Even the account of "parquet" diagrams and a method of multiplicate renormalizations lead to wrong results (1973).

From stated follows, that it is necessary to use the methods of calculation considering collective pheromones of fluctuation in a much greater degree, than in standard methods. These methods were stated by us in the beginning of a rate.

7.2.3 A GENERAL CHARACTERISTIC OF ONE-DIMENSIONAL SYSTEMS

Polyacetylene is the elementary linear interfaced polymer with stable trans–conformation and not leveled communications. Schematically its structure can be represented in the form of:

A, B – conditions of polyacetylene, C^* – division of a circuit due to instability. At big enough distance between divisions of a circuit (domain walls) conditions A and B do not differ on energies. It provides an opportunity of existence topological solitones – domain sides from a radical which extend on odd number of atoms of carbon (that the A – structure has been connected with B – structure). The domain wall can grasp electron, forming carbanion, or lose electron, forming carbocation. All these of heterogeneity are active particles and can give chemical transformations. However chemical kinetics though is of interest here it will not be considered.

Except for the structure specified above, there is still a structure which consists of two parts A and A, disconnected by a wall from even number of atoms of carbon.

Research of solitones in polyacetylene leads to the decision of equations of Bogolubov-de Jen.

$$\varepsilon_n u_n = -ic\frac{\partial}{\partial x}u_n + \Delta(x)v_n,$$

$$\varepsilon_n v_n = ic\frac{\partial}{\partial x}v_n + \Delta(x)u_n,$$

$$\Delta(x) = -\frac{g^2}{\omega^2}\sum v_n^* u_n$$

They are fair for any size Δ, Ginzburg–Landau equations follow from them at small Δ (1980).

7.2.3.1 THE BAND STRUCTURE

The Band structure without taking into account an alternation of communications has a usual appearance cosine. It is a start structure. The account of Peierls instability leads to structure of a zone, which differs a crack about an impulse equal $\pi/2a$.

The structure of a zone depends on number of electrons, falling unit, which can change at addition zones by electrons or holes. The Solitones condition lays inside of the forbidden zone, a little below its center. In polyacetylene takes place strong electron-vibrational interaction. In this case it is shown, that Peierls transition at length of circuit $N \rangle N_c = T_c^{-1}$ (Bulaevskiy, 1974).

7.2.3.2 OPTICAL PROPERTIES OF POLYACETYLENE

Anisotropy in factors of reflection is observed in many works: 80% of radiation is reflected at falling light on a plane of a molecule and 10% at falling in a perpendicular direction in the field of from 0.001eV up to 6eV. For energy of quantums of light, it is more than reflection does not occur, as in this area there is a strong absorption by plasma fluctuations. We have for dielectric polarization:

$$\varepsilon_{11}(\omega) = \varepsilon_\infty(1 - \frac{\omega_p^2}{\omega(\omega + i/\tau)}) \quad \varepsilon(\omega) = 2.2$$

where: ω_p – plasma frequency, $\omega_\infty = 2.0$, $\omega_k^2 = 2.6$, $\tau = 3 \times 10^{-15}$ second. The reason of the big dielectric susceptibility is the wave of charging density (VCD), cooperating with impurity (pinning), which destroys the distant order. Otherwise the susceptibility would be infinite, and due to interaction with impurity it is done final, though also big. Along the allocated axis period VCD is equal CF/2. If this

Polyacetylene

period and the period of the basic lattice are commensurable, VCD can cooperate and with fluctuations of a lattice. This effect is small, if CF/2 and the period of the basic lattice are incommensurable (Larkin, 1977).

Stationary raised conditions of polyacetylene are (because of incommensurabilities) peak solitons (e=0, s=½). Solitons is conditions when wave function of electron is localized, the lattice is deformed. However, in these conditions the full density of a charge and energy is delocalize also constant on length of system (Brazovskiy, 1980). At a double commensurability (polyacetylene without impurity) can arise solitons with charges e = 0, +1, −1 and spin s = 0, ½. Batching of impurity it is possible to receive even a fractional charge.

At a premise of system in a magnetic field in it there is a magnetic moment as backs are divided on subzones and on a miscellaneous they are filled.

Solitons it is possible to consider as the connected conditions of exciting and local deformation of a lattice (Davidov, 1980). The problem is reduced to the decision of the equation:

$$i\hbar \frac{\partial}{\partial t}\varphi_n + (\varepsilon_0 + \frac{\hbar^2}{2ma})\varphi_n + \frac{\hbar^2 e^{-W_n}}{2ma}(\varphi_{n-1} + \varphi_{n+1}) + G|\varphi_n|^2 \varphi_n = 0,$$

$$W_n = |\varphi_n|^4 Bf(\theta)$$

7.2.3.3 ELECTRIC PROPERTIES OF POLYACETYLENE

At kinetic processes in polyacetylene moves backs. The weight of soliton (M_s) is approximately equal to the weight of electron (m). Exact calculation shows, that M_s=6m. The charge and energy of soliton are homogeneous. They do not give the contribution in electroconductivity and heat conductivity. Kinetic display of presence of solitons is only spin diffusion. Participation of solitons leads strong nonlinearities in factors of carry.

Big time of the photoresponse is in polyacetylene (10^{-3} second). It connects with slow scattering of domain walls. Exponential growth of conductivity at border of zones is. Photoconductivity is absent in the *cis*-polyacetylene despite of absorption of light.

7.2.4 PROBLEMS WHICH HAVE ARISEN AT RESEARCH OF POLYACETYLENE

These problems share on two kinds: experimental and theoretical. We shall consider them more in detail.

7.2.4.1 EXPERIMENTAL PROBLEMS

1. Optical experiments are necessary for leading at different temperatures for samples of a different degree of alloying impurity that width of a crack to define, its temperature dependence and presence of conditions inside of it.
2. Character of conductivity (soliton, polaron or other) it is necessary to find out, investigate anisotropy of conductivity, photoconductivity and thermo-electro-factor.
3. The thermal capacity and its temperature dependence should be investigated in a vicinity $q = 2CF$.
4. It is necessary to investigate infrared-spectrums for studying of Konov anomaly.
5. Influence of pressure on transition metal-dielectric is necessary for investigating metal for definition of type of transition. The critical temperature of the Hardware grows with growth of pressure for Peierls change, and for Mott change falls.
6. The X-ray analysis of a monocrystal is necessary for leading to vicinities of phase transition for definition of structure.

7.2.4.2 THEORETICAL QUESTIONS

1. The basic interest represents research of dynamics in polyacetylene (quasi-one-dimensional system). To us it is not clear, whether it is possible to transfer properties of one-dimensional systems on three-dimensional with strong anisotropy.
2. Methods are necessary for developing, replacing a method of self-consistent field in approach of chaotic phases or Hartree-Foc, considering more essentially fluctuations both collective effects and leading the nonlinear equations of integrated type.
3. The theory of indignations to develop concerning the equations of integrated type. The account of fluctuations of a lattice destroys solitons and leads to non-integrated system (dynamic pinning).

These are the main questions concerning all strongly anisotropic systems. For polyacetylene it is necessary:

1. Power structure to find out in more details: presence of even lengths, what difference between *cis*- and *trans*-isomers of polyacetylene, whether is broken Peierls change at change of interaction between chains, occurrence of overlapped zones with different CF (for σ- and π-electrons), occurrence of defects.
2. Conditions to reveal when spin, also as well as charges on soliton, continuous value (fractional spin).
3. A number of formulas is necessary for receiving for kinetic factors, namely for the form of a signal of a nuclear spin induction, an electron-spin

echo, conductivity and photoconductivity. It is necessary to find out when appears paraconduction during the moment of Peierls change, connected with greater current fluctuations.

7.2.5 CONCLUSION

Almost all properties of polyacetylene are defined by an opportunity of occurrence in it solitons (domain walls, polarons, others stationary and dynamic). Mathematicians have developed methods of construction and the decision of non-linear problems for reception and researches of soliton decisions (a method of a return problem of dispersion, group methods).

The problems investigated by mathematicians, can serve as models for the description of physicochemical properties of polyacetylene and other polymers.

Properties of polyacetylene cannot be clear in terms of quasi-particles and consequently are unusual. The charge of a particle can be equal 1/3, or 1, and spin is equal 0. Properties of polyacetylene are collective and can be described by introduction of collective structures, for example, soliton-like states.

We shall not find out, what experiments should be made first of all. The main theoretical questions are the first and second of the general part and the third for polyacetylene.

7.3 THE NATURE OF KINETIC PROPERTIES OF POLYACETYLENE.

7.3.1 A STRUCTURE AND PROPERTIES OF POLYACETYLENE

The chemical formula of polyacetylene-$(CH)_x$, where C-carbon, H-hydrogen. Three external electrons of carbon form a sp^2-hybrid and give three σ-bonds. Remained electron forms π-bond. p-Clouds are focused perpendicularly to a plane in which p-bonds are located. The chain of atoms of carbon is formed as a result of polymerization. The chain has *cis*- or *trans*-form. Cis-form is formed at polymerization in usual conditions, and the *trans*-form turns out from it by thermalization or introduction of an impurity. We shall consider properties of *trans*-polyacetylene.

We shall discuss properties of polyacetylene without impurity. The length of bond C−C in a separate chain $(CH)_x$ makes approximately 1.4 Å. Distances between chains equally 3.6−4.4 Å. We can to neglect interchain interactions as a first approximation. It allows considering polyacetylene as the system consisting of separate isolated strings, bound among them. Polyacetylene is the semiconductor at a room temperature with width of a crack $2\Delta_0 = 1.4-1.8$ yM and value of a dielectric constant along chains 10−12. The crack of the order 1.6−2.0 eV turns out as a result of alternation of bonds C−C and C=C. Polyacetylene it is possible to consider as one-dimensional dielectric of Peierls. The size of displacement of atoms from balance is measured and it has appeared equal 0.03 Å.

Mobile paramagnetic centers are available in pure *trans*-polyacetylene with concentration of 1 spin on 3000 atoms of carbon. Their origin is not found out yet.

Data on electric and magnetic properties of alloy polyacetylene is. As impurity are used as donors (Na, K, NH_3), and (AsF_5, I, $FeCl_3$). Research of alloy polyacetylene has allowed making a number of the remarkable conclusions.

Residual conductivity is available in pure *trans*-polyacetylene. It is caused by presence of defects. Sharp increase of conductivity (in 10^{11} times) takes place with increase in concentration of impurity $y = N_1/N$ (N_1 – number of impurity, N – number of atoms of carbon). Saturation is reached at $y = 0.1$ and is equal 220 $(\Omega \cdot cm)^{-1}$.

Dependence on temperature has three areas over iodine:

a) $y < 0.05$. In this area $\sigma = \sigma_0 \exp(\frac{-B}{T})^{0.5}$, $B = 10^4$. One-dimensional hopping mobility takes place.

b) $0.06 < y < 0.11$ $\sigma = \sigma_0 \exp(\frac{-A}{T})^{0.25}$, $A = 10^5$. It is three-dimensional disordered an alloy.

c) $y > 0.15$. $\sigma = \sigma_0 T^v$, $v = 0.7$. It is characteristic for "dirty" metals.

The magnetic susceptibility of polyacetylene aspires to zero at y<0.05. Conductivity grows on some orders thus. Spin particles do not give the contribution to conductivity. Concentration of the paramagnetic centers falls with growth at. The magnetic susceptibility sharply increases at y<0.07. The metal behavior takes place.

Luminescence has been found out in *cis*-$(CH)_x$ at absence of a photocurrent. Luminescence is not present in *trans*-$(CH)_x$. Photoconductivity appears. Luminescence and photoconductivity is in usual semiconductors.

7.3.2 THE MODEL OF POLYACETYLENE

Dielectric properties of polyacetylene can be explained on the basis of the mechanism of Peierls. It has shown that the one-dimensional system is unstable concerning spontaneous infringement of mirror symmetry of a chain of atoms for any nonzero electron– phonon bond. Thus the crack arises in a one-electric power spectrum. The crack separates the filled and empty zones of Brillouin. One "free" π–electron on atom of carbon is available in $(CH)_x$. Doubling of the period of a lattice is preferable.

The explanation of the big number of experimental data has been received on the basis of the model using the mechanism of "internal defects" in a chain $(CH)_x$. Two approaches it is known for the description of defects.

The first approach: impurity electron or a hole will lead to infringement in alternation of communications of a chain of polyacetylene. Defect is change on 180 degrees of a phase regular polyacetylene (special case). This infringement of the order looks like a domain wall. Additional electron or a hole will borrow localized soliton-like states. The phase of lattice dimerization overturns. The elementary model describing such soliton, has Hinzburg-Landau Lagrangian.

Polyacetylene

$$L(u) = \frac{8\varepsilon_c}{(u_c\omega_\lambda)^2}\left[\frac{1}{2}\left(\frac{\partial u}{\partial t}\right)^2 - \frac{C_0^2}{2}\left(\frac{\partial u}{\partial x}\right)^2 - (\omega_\lambda u_0)^2 V(u)\right]$$

u Describes displacement of atoms of lattice,

$$V(u) = \frac{1}{8}\left[1-\left(\frac{u}{u_0}\right)^2\right]^2, u_0$$

where amplitude is dimerization of Peierls. ε_c – Stabilizing energy on atom of carbon. ω_λ – Frequency of optical phonons. c_0 – the Speed describing of a lattice dispersion. $\omega_\lambda^2(q) = \omega_\lambda^2 + c_0^2 q^2$ at $|q| = \pi/2a$. a–the constant of a lattice. Model (1) has twice degenerate the basic condition and supposes the decision in the form of soliton

$$u(x-vt) = \pm u_0 \tanh\frac{x-vt}{I}, \quad I = \frac{2c_0}{\omega_\lambda}$$

where: I characterize width of soliton. Energy of soliton excitation $E_s = 4\Delta_0/3\pi$ = 0.4 electron volt, weight of soliton M_s = 6 mm – weight of electron, I = 10a. Expression (2) will enter to nuclear structure symmetrical concerning the center of soliton. The localized electronic condition appears in the center of a crack of Peierls. This condition can be borrowed by electron or a hole. Soliton gets a charge ±e. Model (1) is in the good consent with experiment, but demands care at the analysis in the field of defect $x-vt\to 0$. The model has singularity in this area.

The second approach: the topological defects dividing areas A and B, appear even during polymerization.

The area of defect remains neutral, but has not coupled backs ½ in absence of an impurity

7.3.3 THE DISCRETE MODEL
The macroscopical model including screen and electronic members looks like *The Hamiltonian*:

$$H = -\sum_{n,s} t_{n,n+1}(C^+_{n+1,s}C_{n,s} + h.C.) + \frac{k}{2}\sum_n (y_{n+1}-y_n)^2 + \frac{M}{2}\sum y_n^2 \quad (4)$$

$C^+_{n,s}(C_{n,s})$ – the operator of a birth (destruction) of electron with spin s on n (CH)– group, y_n – configuration coordinates for everyone CH– group, describing translation along a linear skeleton of a chain. M is weight of group (CH), κ is an elastic constant. The integral of overshoot electronic clouds of π – electrons $t_{n+1,n}$ can be spread out up to the first order rather nondimerization conditions $t_{n+1,n} = t_0 - \alpha(y_{n+1} - y_n)$, where t_0 – the integral of overshoot for 1D chains, α–a constant of electron –fonon bond.

We shall note that interchained bond is excluded and Coulomb interaction of π − electrons not considered in Eq. (3). Values of parameters of Hamiltonian: $t_0 = 2.5$ electron volt, $k = 21$ electron volt/A^2, $\alpha = 4.1$ electron volt/A. Length of C–C bond is equal 1.4 F. The constant of a lattice is equal $\alpha = 1.22$ Å. Reading of energy is conducted from Fermi level.

The Hamiltonian possesses symmetry to mirror reflection $y_n \rightarrow -y_n$ at $\alpha = 0$. $H(-y_n) = H(y_n)$. Infringement of symmetry of the basic condition takes place at $\alpha \neq 0$: $H(-y_n) \neq H(y_n)$.

7.3.4 RESULTS FROM DISCRETE MODEL

We neglect kinetic energy of phonons for definition of energy and we represent static displacement of unit n in the form of
$$y_n = (-1)^n y$$
The electronic spectrum looks like in a lattice without displacement (a normal phase):
$$E = -2t_0 \cos qa$$
$qa = 2\pi n/N$, $n = 0, \pm 1, \ldots \pm N/2$, N is number of atoms in a lattice. In a lattice with displacement

$$E_{1,2} = \pm\sqrt{(\Delta_0 \sin qa)^2 + (2t_0 \cos qa)^2}$$

where $qa = qa = 2\pi n/N$, $n = 0, \pm 1, \ldots \pm N/4$. Brillouin zone has decreased twice and the crack arises $\Delta_0 = \alpha y$. The spectrum is symmetrical concerning Fermi energy.

Free energy of electrons in lattice is equal:

$$F(y,T) = -T \sum_{i=1,2;q} \ln[1+\exp(-E_i(q)/T)] + 2Nky^2$$

where: T is a temperature. Free energy on one atom of carbon looks like at T = 0 and $N \rightarrow \infty$:

$$F(y,0) = -\frac{4t_0}{\pi} E(1-z^2) + \frac{kt_0^2 z^2}{2\alpha^2}$$

$$z = \frac{2\alpha y}{t_0}, \quad E(1-z^2) = \int_0^{\pi/2} \sqrt{1-(1-z^2)\sin^2 q}\, dq$$

at
$$\Delta_0 \ll t_0 \quad E(1-z^2) = 1 + \frac{1}{2}(\frac{\ln 4}{|z|} - \frac{1}{2})z^2 + \ldots$$

We minimize F(y,0) (7) on z, we receive minima at $\pm y = \Delta_0/4\alpha$ c $\Delta_0 = 8t_0/e$ $\times \exp(-\pi k t_0/4\alpha^2)$. The local maximum is in a point y = 0 under the theorem of Peierls. The basic condition is doubly singular at low temperatures. Displacement y=0.04Å at $\Delta_0 = 0.7$ eV and α = 4.1 eV/A.

Presence of degeneration causes occurrence of topological solitons, dividing A- and B-phases. The numerical calculations lead within the limits of this model, have shown existence of decisions in the form of a domain wall. The parameter of the order is entered, connected with variables y_n:

$$\psi_n = (-1)^n y_n$$

Phases A and B can be defined
$\psi_{0n} = y_0$, a phase A;
$\psi_{0n} = -y_0$, a phase B.

We shall assume, that the domain wall is formed in a vicinity of unit n = 0. This wall divides A– and B– areas. We shall consider a birth of pair soliton–antisoliton to eliminate boundary effects. They consider far dissolved to exclude interaction between them. Model supposes the decision in the form of

$$\psi_n = y_0 \tanh(na/l)$$

Parameters of soliton have appeared the following. Width of soliton l = 7a, energy of a birth of soliton E_s = 0.42 electron volt, weight M_s = 6m.

In density of the conditions, caused of soliton, it is necessary to calculate changes for studying an electronic spectrum. Calculations give occurrence of one condition at E = 0 inside of the forbidden zone. It is connected with the advent of additional condition Φ_0 in the center of a power crack. The condition of completeness of wave functions Φ_v on each unit n means, that the integral on energy from local density $\rho_{nn}(E)$ for any unit "n" is equal to unit. We have:

$$I = \int_{-\infty}^{\infty} \rho_{nn}(E)dE = 1, \quad \rho_{nn}(E) = \sum |\phi_v(n)|^2 \delta(E - E_v)$$

The Hamiltonian is invariant concerning charging interface.

$$\rho_{nn}(E) = \rho_{nn}(-E).$$

$$I = 2\int_{-\infty}^{0} \rho_{nn}(E)dE = 2\int_{-\infty}^{0} \rho_{nn}^1(E)dE + |\phi_0(n)|^2 = 1$$

where: ρ^1 – means, that the condition of soliton is lowered.
Local deficiency of a valent zone is equal:

$$\int_{-\infty}^{0}[\rho_{nn}^1(E)-\rho_{nn}(E)]dE = \frac{1}{2}|\phi_0(n)|^2$$

where: ½ of condition is necessary on spin. One electron comes short in a zone. But a zone is neutral. Superfluous electron proves in condition Φ_0 with soliton ½. The charge of a condition is equal to zero. Neutral soliton appears with spin ½. If one charge will sit down on soliton from an impurity or its charge will leave in an impurity soliton begins to have soliton – zero, charge ±e.

Soliton picture allows to explain the basic properties of polyacetylene. Impurity conduction occurs due to solitons (with zero spin) if energy of a birth of soliton $E_s < \Delta_0$ solitons are Bose particles, can be condensed in superfluid state. Activation energy electric conductivity E_a = 0/3 electron volt it will well be coordinated with energy of bond of soliton on 0.33 electron volt. The peak 0.1 electron volt is in infrared spectrum. It will be coordinated with an oscillatory fashion of soliton. It is equal 0.07 – 0.08 electron volt. Other properties speaking of solitons exist.

Attempts to consider electrostatic interaction are. We choose the Hamiltonian of interactions in the form of:

$$H_{ee} = \frac{V}{2}\sum \rho_{n\downarrow}\rho_{n\uparrow} + \frac{1}{2}\sum u_{nm}\rho_n\rho_m, \quad \rho_n = \sum C_{ns}^+ C_{ns}$$

Model investigated numerically. Dimerization of chain increases at u_{nm} = 0 with growth V up to value V/2 = $4t_0$, and then decreases. Interaction of the first neighbors and the second also influences properties of a chain: at u_{nm+1} <V/4 amplifies dimerization, at u_{nm+1} >V/4 – it is weakened, and at u_{nm+2} – operates opposite u_{nm+1}.

7.4 CONTINUAL MODEL OF POLYACETYLENE.

7.4.1 MODEL

This model supposes the exact decision. The stationary raised conditions of such system are amplitude solitons. Quasi-classical approach we shall make on lattice variables. The Hamiltonian looks like in continual approach:

$$H = \sum_s \int dx \left\{ -iv_F(u_s^+(x)\frac{\partial}{\partial x}u_s(x) - v_s^+(x)\frac{\partial}{\partial x}v_s(x)) + [\Delta^+(x) + \eta\Delta(x)]u_s^+(x)v_s(x) + [\Delta(x) + \eta\Delta^+(x)]v_s^+(x)u_s(x) \right\} +$$
$$(2\lambda\pi v_F)^{-1} \int dx \left[2\Delta^+(x)\Delta(x) + \eta\Delta^2(x) + \eta\Delta^{+2}(x) \right],$$
$$\lambda = 4\alpha^2 a / \pi v_F k, \quad v_F = 2t_0 a, \quad \hbar = c = 1, \quad \eta = 1,0$$

$$\Lambda = \frac{4\alpha^2 a}{\pi v_F k}, \quad v_F = 2t_0 a, \quad \hbar = c = 1, \quad \eta = 1.0$$

where: u(x), v(x) – characterize electronic conditions close of Fermi momentum $\pm q_F$, $\Delta^+(x)$, $\Delta(x)$ – are connected with lattice deformation (dependence from t is excluded, adiabatic approximation). The equations follow from the Hamiltonian:

$$iu_t = -iv_F u_x + \tilde{\Delta}^+ + \frac{a^2}{2}(\tilde{\Delta}^+ v_{xx} + v_x \tilde{\Delta}_x^+ + \frac{1}{2}v\tilde{\Delta}_{xx}^+),$$
$$iv_t = iv_F v_x + u\tilde{\Delta} + \frac{a^2}{2}(\tilde{\Delta} v_{xx} + u_x \tilde{\Delta}_x + \frac{1}{2}u\tilde{\Delta}_{xx}),$$
$$\tilde{\Delta}(x,t) = \Delta(x,t) + \eta\tilde{\Delta}(x,t),$$
$$M\tilde{\Delta}_{tt} + ka^2\tilde{\Delta}_{xx} = -4k\tilde{\Delta} - 16a\alpha^2(u^+v + \eta v^+u) - 4a^2\alpha^3 \left[u^+v_{xx} + vu_{xx}^+ + \eta(uv_{xx} + v^+u_{xx}) \right]$$

This problem to solve it is difficultly. We shall make a number of approaches. We shall exclude members, proportional a^2 and above, we shall lower member $M\Delta_{tt}$ (quasi-static approximation). We have:

$$iu_t = -iv_F u_x + v\tilde{\Delta}^+,$$
$$iv_t = iv_F v_x + u\tilde{\Delta},$$
$$\tilde{\Delta} = -\frac{4\alpha^2 a}{k}\sum_{k,s}(u_k^* v_k + \eta v_k^* u_k)$$

This system has structure Dirac equations in a floor. Many of such equations have analytical decisions (Bagrov V. G.). The system (4) is invariant rather Lorentz transforms ($c = v_F$).

$$x \to x' = \frac{x - vt}{\sqrt{1 - \frac{v^2}{v_F^2}}}, \quad t \to t' = \frac{t - \frac{xv}{v_F^2}}{\sqrt{1 - \frac{v^2}{v_F^2}}}$$

We can be limited to the decision of the stationary equations, then simply to add time, to take advantage Lorentz invariance. We search decisions in the form of:

u(x, t) = e$^{-i\omega_n t}$u$_n$(x), v(x, t) = e$^{-i\omega_n t}$v$_n$(x) also we shall put η = 1 ((transition in commensurable structure). We have:

$$\omega_n u_n(x) = -iv_F \frac{\partial}{\partial x} u(x) + v_n(x)\tilde{\Delta}^+(x)$$

$$\omega_n v_n(x) = -iv_F \frac{\partial}{\partial x} v_n(x) + u_n(x)\tilde{\Delta}(x)$$

$$\tilde{\Delta}(x) = -\frac{4\alpha^2 a}{k} \sum_{n,s} (u_n^*(x)v_n(x) + \eta v_n^*(x)u_n(x))$$

7.4.2 DISCUSSION OF DECISIONS OF SYSTEM OF EQUATIONS

Four classes of decisions of this system are available. Decisions are found by various methods.

The first class has the decision in the form of a flat wave, in an electronic spectrum there is a crack $\tilde{\Delta} = \Delta_0 \exp(-1/2\lambda)$. The valent zone is filled completely. Electronic wave functions in do not look like:

$$u_k(x,t) = N_k e^{i(kx-\omega t)}, \quad v_k(x,t) = -N_k' e^{i(kx-\omega t)},$$

$$\omega = -\sqrt{\Delta_0^2 + k^2 v_F^2},$$

$$N_k = \frac{1}{\sqrt{8\pi}} (\frac{\omega - \Delta_0}{\omega})^{1/2} (\frac{\omega - \Delta_0 + kv_F}{kv_F})$$

$$N_k' = \frac{1}{\sqrt{8\pi}} (\frac{\omega - \Delta_0}{\omega})^{1/2} (\frac{\omega - \Delta_0 + kv_F}{kv_F})$$

Kinks are other class of decisions of system. These decisions look like:

$$u_0(x,t) = N_0 \sech \frac{x - v_s t + x_0}{\xi_s},$$

$$v_0(x,t) = -i\frac{N_0}{v} \sech \frac{x - v_s t + x_0}{\xi_s},$$

Polyacetylene

Where $N_0 = \sqrt{\dfrac{v\Delta_0}{4v_F}}$ width of soliton.

$$\xi_s = \dfrac{v_F}{\Delta_0}\sqrt{1-\beta^2}, \quad \beta = \dfrac{v_s}{v_F}, \quad v = \sqrt{(1+\beta)/(1-\beta)}$$

In a valent zone

$$u_k(x,t) = N_k\, e^{i\theta}\left[\tanh\dfrac{\xi}{\xi_s} + i\dfrac{\omega - kv_F}{v\Delta_0}\right],$$

$$v_k(x,t) = -\dfrac{iN_k}{v}e^{i\theta}\left[\tanh\dfrac{\xi}{\xi_s} - iv\dfrac{\omega + kv_F}{v\Delta_0}\right],$$

$$\xi = x - v_s t + x_0,$$
$$\theta = kx + \omega t + \theta_0, \quad \omega = 2t_0\delta,$$

The additional level appears at presence of soliton in the middle of a power crack of an electronic spectrum. The level can be borrowed by one, two electrons or is not borrowed absolutely. Energy of a birth of soliton is equal 0.44 electron volt, will well be coordinated with result of discrete model. It is exact result in continual model.

The third type of excitations is polaron. Analytical expression has a complex appearance. We shall not write out dynamic decisions. We receive them by application Lorentz transform. In a static case:

$$\Delta(x) = k_0 v_F (\tanh(k_0(x - x_0)) - \tanh(k_0(x - x_0)))$$

Two symmetrically located levels appear in a power spectrum at presence of such deformation $E = +\omega_0$

$$u_0(x) = N_0[(1-i)\sec h(k_0(x+x_0)) + (1+i)\sec h(k_0(x+x_0))]$$
$$v_0(x) = N_0[(1+i)\sec h(k_0(x+x_0)) + (1-i)\sec h(k_0(x-x_0))]$$

$$N_0 = \dfrac{\sqrt{k_0}}{4}, \quad k_0 v_F = \sqrt{\Delta_0^2 - \omega_0^2},$$

tanh
$$k_0 x_0 = \Delta_0 - \frac{\omega_0}{k_0 v_F}$$

Wave functions are equal $u_0 = iv_0$, $v_0 = -iu_0$. In a valent zone:

$$u_-(k,x) = N_k e^{ikx}\left[(\omega+\Delta_0-kv_F)-\gamma(1-i)t_- +\delta(1-i)t_-\right],$$
$$v_-(k,x) = -N_k e^{ikx}\left[(\omega+\Delta_0+kv_F)-\gamma(1-i)t_+ +\delta(1+i)t_-\right],$$
$$t_\pm = \tanh[k_0(x\pm x_0)], \quad \omega = \sqrt{k^2 v_F^2 + \Delta_0^2},$$
$$\gamma = \frac{k_0 v_F}{2}(1-\frac{ikv_F}{\omega-\Delta_0}), \quad \delta = \frac{k_0 v_F}{2}(1+\frac{ikv_F}{\omega-\Delta_0}),$$
$$N_k = \frac{1}{2\sqrt{2\pi}}\left[\frac{\omega-\Delta_0}{\omega(k^2 v_F^2 + k_0^2 v_F^2)}\right]^{1/2}$$

The decision of the fourth type has been received in the form of bion, when:
$\Delta(x, t) = \Delta_0[1 + \delta(x, t)]$,
Energy of superincumbent ion is equal:

$$E_R = \Delta_0 \varepsilon \frac{2\sqrt{3}}{\pi}[1-\frac{5\varepsilon^2}{9}+0(\varepsilon^4)]$$

The contribution of ion in physical processes can be essential. Studying of ion in polyacetylene is the important problem. This studying was not spent yet.

7.4.3 SOLITON EXCITATION AND PHYSICAL PROPERTIES OF POLYACETYLENE
7.4.3.1 CHARACTERISTIC WIDTH OF POLARON

$$2x_0 = \frac{2}{k_0}\arctan h(\frac{k_0 v_F}{\Delta_0 + \omega_0}) = 10.8\,A, \quad \omega_0 = \frac{\Delta_0}{\sqrt{2}}$$

Energy of excitation of polaron is equal

$$E_p = \frac{2\sqrt{2}}{\pi}\Delta_0 = 0.9\Delta_0,$$

Polyacetylene 285

Polaron the decision passes in widely dissolved pair soliton – antisoliton at $x_0 \to \infty$ position of additional levels $\omega_0 \to 0$. Dependence of parameter of a crack for kink and polaron is on Figs. 5 and 6.

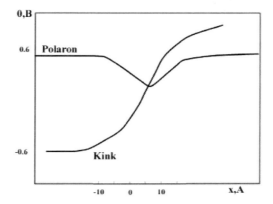

FIGURE 5 Change of a crack for kink and polaron depending on a condition.

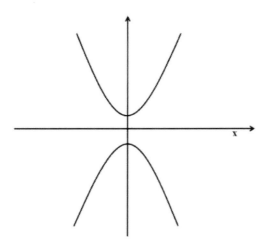

FIGURE 6 Change of a power spectrum at presence kink and polaron.

Polarons should play a greater role in the physicist of polyacetylene. Spin of system with even number of electrons should be the whole, and with odd- semi-integral value. The general number of electrons of system is kept at a birth of soliton S. Not coupled spin arises. It contradicts to Kramer theorem. Antisoliton (−S) is born simultaneously and at a great distance. It is reflected on Fig. 7. The simultaneous birth of pair S(−S) demands a condition of topological stability of line. Polaron state is more favorable and charged bearers borrow polaron level. Spin of bearer is equal ½, the charge is equal to unit. But bidirectional decision any more is not stable and breaks up on S(−S) pairs. The number of charged solitons grows with increase in number of doped electrons. Charged bearers without spin appear.

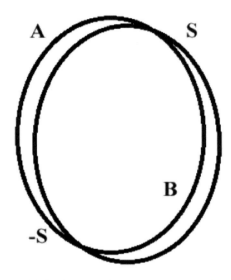

FIGURE 7 Simultaneous birth of S(−S) pair.

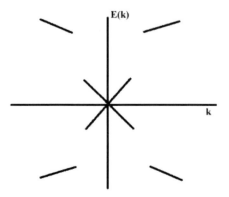

FIGURE 8 Change of an electronic spectrum depending on a degree of filling of a zone.

Formation charged or neutral soliton paires is possible depending on filling the localized levels $\pm\omega_0$. The contribution of polarons and solitons in various physical processes is well studied. Convincing acknowledgments are available in favor of soliton pictures of properties of polyacetylene.

The Soliton model continues to be specified concerning:
a) Account of Coulomb interaction,
b) Three-dimensional influence,
c) The exact decision of model is lead with any number of electrons on an elementary cell. It is important for studying of alloy polyacetylene where the deviation from a case strictly semi-occupied zone grows with growth of number of doped electrons.
The optical crack in electronic spectrum appears continuous on parameter of filling of a zone (it is equal 1 for semi-occupied zones) and is final. Change of electronic spectrum is reduced to expansion of soliton level and to formation of two symmetrical forbidden zones (Fig. 8 instead of Fig. 6).
d) Interthreaded interaction connects solitons on the next chains and soliton can get a charge not changing backs.
e) Occurrence of a nonintegral electric charge 2/3, 1/3 (it quarks) is revealed. Nobel Prize is received in 1998.

Nobel Prize on the physicist in 1999 has been given to researchers who the first observed and has explained fractional quantum Hall-effect. It is effect in which the electric current inside of two-dimensional conductive material appears, as though created of the carriers bearing a fractional charge of electron.

Standard Hall-effect is a lateral deviation of moving carriers in a magnetic field. It was revealed in 1879 *Edwin Hall*, today a basis gives for definition of a charge and density of carriers of charges in the semiconductor (electrons and holes deviate in various directions).

Klaus von Klitzing finds in 1980, that when carriers of a charge are limited inside very thin spending film (that is move in two measurements), size of a Hall current (or conductivity of a material) does not change any more smoothly from size of magnetic field at very low temperatures.

Conductivity changes from force of a field in the form of sharp steps. Conductivity changes jump on fundamental quantum size of conductivity e^2/h (where e is a charge of electron, h is Planck's constant).

Fractional quantum Hall-effect represents a deep problem as shows change in character of fundamental particles. We can tell the same about superconductivity in which electrons draw each other, and super fluidity at which atoms of super liquid any more generate viscosity.

FQHE has been noticed *Tsui and Stormer* for carry in two-dimensional electronic gas in semi-conducting heterostructure, made *Art Gossard*. The sample cooled up to 10°K, applied magnetic fields up to 30 tesla, observed transitions in spending condition with value $e^2/3h$. Carriers of a charge had a fractional charge $e/3$. The further studying have shown that charges are $2e/5$, $3e/7$ and other shares (with an odd denominator).

Laughlin has offered, that the magnetic lines of a stream getting through the sample encourage carriers of a charge to be condensed in quasi-particles. It has shown, that such quasi-particles operate as if they have fractional charges with the values noticed in experiments.

Heiblum speaks: "All from us have got used to work in the semi-conductor physics with a one-electronic picture". But studying of strongly correlated electrons in the physicist of a firm body now becomes the important field of research.

This model has the direct attitude to a number of models of relativistic quantum theory fields, to Gross–Neview model, with Fermi particles. Studying of bidimentional models is more likely formal mathematical research. Results of such research are very remarkable and have the appendix in the physicist of the condensed environments.

7.5 TWO-DIMENSIONAL MODELS OF RELATIVISTIC QUANTUM THEORY OF FIELD AND OF POLYACETYLENE

7.5.1 GROSS–NEVIEW MODEL

Gross–Neview model:

$$L = \sum_{p=1}^{N} [\bar{\psi}^p (i\partial)\psi^p] + \frac{1}{2}g^2 [\sum \bar{\psi}^p \psi^p]^2$$

Where ψ is a two-componential spinor, $\dot{g}$ is a constant.

$$\partial = \gamma_\gamma \partial^\mu, \quad \partial^0 = \frac{\partial}{\partial t}, \quad \partial^1 = \frac{\partial}{\partial x}, \quad \mu = 0.1$$

$$\gamma_0 = \sigma_3, \quad \gamma_1 = i\sigma_1,$$

This model describes N types of massless Fermi particles, cooperating in the nonlinear image. Dynamic infringement of chiral symmetry is realized in this model. Fermi particles get weight. It is similar to formation of a superconducting crack. This model is actively used in the theory of strong interactions.

The model is done renormalizable and receives properties of asymptotic freedom. This model is completely integrated. Making functional it is possible to construct:

$$Z(\eta,\bar{\eta}) = C\int d\psi d\bar{\psi}\exp(i(i\bar{\psi}\partial\psi + \frac{1}{2}g^2(\bar{\psi}\psi)^2 + \bar{\eta}\psi + \bar{\psi}\eta)) =$$

$$C'\int d\psi d\bar{\psi} d\sigma \exp(i(i\bar{\psi}\partial\psi - \frac{1}{2}\sigma^2 - g\bar{\psi}\psi\sigma + \bar{\eta}\psi + \bar{\psi}\eta))$$

And to be convinced, that Eq. (1) it is formally equivalent:

$$L = \sum_{p=1}^{N}(i\bar{\psi}^p\partial\psi^p - g\sigma\bar{\psi}^p\psi^p) - \frac{1}{2}\sigma^2,$$

Where scalar field σ has arisen which is formal object for a two-dimensional case. The equations of movement for Eq. (2) look like:

$$(i\partial - g\sigma(x))\psi^p(x) = 0,$$

$$\sigma(x) = -g\sum_{p=1}^{N}\bar{\psi}^p(x)\psi^p(x)$$

This is a condition of the self-coordination. The equation of movement becomes for stationary conditions:

$$\omega_n\psi_{1n}^p(x) = \frac{\partial}{\partial x}\psi_{2n}^p(x) + g\sigma(x)\psi_{1n}^p(x),$$

$$\omega_n\psi_{2n}^p(x) = \frac{\partial}{\partial x}\psi_{1n}^p(x) - g\sigma(x)\psi_{2n}^p(x)$$

If $\psi_1 \to (u+v)/\sqrt{2}$, to replace $\psi_{22} \to {-i(u+v)}/\sqrt{2}$, $g\sigma(x) \to \Delta(x)$, $x \to v_F x$, we shall receive the equations earlier considered for polyacetylene at N = 2.

$$\omega_n u_n(x) = -iv_F\frac{\partial}{\partial x}u(x) + v_n(x)\tilde{\Delta}^+(x)$$

$$\omega_n v_n(x) = iv_F\frac{\partial}{\partial x}v_n(x) + u_n(x)\tilde{\Delta}(x)$$

With a condition of the self-coordination

$$\tilde{\Delta} = -\frac{4\alpha^2 a}{k} \sum_{k,s} (u_k^* v_k + v_k^* u_k)$$

We shall stop on it in more detail. Dynamic infringement of symmetry means, that there is a decision σ(x)=σ₀. Well-posed interpretation a condition of the self-coordination demands performance of a parity:

$$Z(\Lambda)\sigma_0 = -q \sum \psi_n^p(x)\psi_n^p(x)$$

Z(Λ) is ultra-violet renormalization, Λ is an impulse of trimming. Summation in above equation is carried out on all conditions with energy of less zero. The sum on p is similar to the sum on backs. Z(Λ) = 1 for model of polyacetylene, that corresponds to a choice:

$$m_\psi = q\sigma_0 = 2\Lambda \exp(-\pi/Nq^2)$$

If m_ψ is fixed also Λ→∞ a constant ġ aspires to zero as 1/lnΛ. The model possesses property asymptotic freedom.

The equation supposes following decisions:
1. Flat waves at σ(x) = σ₀ which look like

$$\psi_{1k}(x) = N_k e^{ikx}, \quad \psi_{2k}(x) = \frac{N_k e^{ikx} ik}{(\omega - m_\psi)}$$

$$\omega = \sqrt{k^2 + m_\psi^2}$$

For them: $\psi_k(x)\psi_k(x) = -m_\psi/2\pi\omega$, and $\sum_{p,n,\omega<0} \bar{\psi}_n^p(x)\psi_n^p(x) = -\frac{N}{2\pi}\int \frac{m_\psi dk}{\sqrt{k^2 + m_\psi^2}}$

The condition of the self-coordination is executed.

2. "Kinks" at σ(+∞) = −σ(−∞) = ±σ₀ when σ = σ₀tanh(m_ψx + δ₀), where tanhδ₀ = (m_ψ − c₀)/(m_ψ + c₀). We come to parameters of a crack as in polyacetylene at c₀ = m_ψ. The Fermionic spectrum contains the connected condition with ω₀ = 0. Wave functions correspond to this condition:

$$\psi_{10}(x) = \psi_{20}(x) = \sqrt{\frac{m_\psi}{4}} \operatorname{sech}(m_\psi x + \delta_0)$$

and conditions with a continuous spectrum $\omega = \sqrt{k^2 + m_\psi^2}$ and wave functions:

$$\psi_1(k,x) = N_k e^{ikx}(\tanh(m_\psi x + \delta_0) - (\omega + ik)/m_\psi),$$

$$\psi_2(k,x) = N_k e^{ikx}(\tanh(m_\psi x + \delta_0) - (\omega - ik)/m_\psi),$$

$N_k = m/2\omega\sqrt{2\pi}$.

3. "Bags" is the third type of decisions. They appear under boundary conditions $\sigma(x) \to \sigma_0$, $|x| \to \infty$. Expression is the decision:
$\sigma(x) = \sigma_0 - (k_0^2/\omega_0 q)\sec h[k_0(x + x_0) + \delta_1]\gamma c\, h[k_0(x - x_0) + \delta_1]$,
$\tanh k_0 x_0 = (m_\psi - \omega_0)/k_0$, $\tanh \delta_1 = k_0(m_\psi - \omega_0) - c_0\omega_0/k_0(m_\psi - \omega_0) + c_0\omega_0$
$\omega_0 = \sqrt{m_\psi^2 - k_0^2}$

We come to the decision of type "polaron" at δ_0 in polyacetylene. The connected conditions arise in fermionic a spectrum with energy $\pm\omega_0$ and wave function

$$\psi_{10}(x) = \sqrt{\frac{k_0}{8}}(\sec h[k_0(x+x_0)+\delta_1] + \sec h[k_0(x-x_0)+\delta_1])$$

$$\psi_{20}(x) = \sqrt{\frac{k_0}{8}}(-\sec h[k_0(x+x_0)+\delta_1] + \sec h[k_0(x-x_0)+\delta_1])$$

Conditions with negative energy are available. They are similar to conditions u, v for polaron decisions in model of polyacetylene. Energy of excitation is:

$$E(n_0) = \frac{2}{\pi} N m_\psi \sin(\frac{\pi n_0}{2N})$$

The decision behaves as dissolved on infinity kink antikink pair in a limit $n_0 \to N$. Essential difference "sack" excitation from kink is a dependency ω_0 from number of filling n_0. Excitation is absent at $n_0 = 0$, $m_\psi = \omega_0$. For kink it is not essential, the condition $\omega_0 = 0$ is borrowed or not. $N = 2$ it is necessary to accept for comparison with model of polyacetylene. The condition with $n_0 = 1$ can be interpreted as presence of additional electron in a chain of polyacetylene. Bond with a phonon field is available; this excitation is considered as "polaron". Infinitely dissolved pair kink-antikink is available at $n_0 = 2$.

4. The Fourth type of decisions is the decision similar bion in model sine–Gordon. Periodic boundary conditions are imposed
$\sigma(x,t+T) = \sigma(x,t)$. If $g = 1$, $\sigma(x,t) = 1 + \xi f_2 + \eta f_4$,

$f_4 = f_4 \cos\Omega t$, $f_4 = (chkx + \cos\Omega t + b)^{-1}$, t and x constants are expressed in terms of $(q\sigma_0)^{-1}$. k, Ω, ξ, η, a, – constants. The full analogy of the considered model and continual models of polyacetylene is available only in a static case. Dynamic properties of models are various.

The Phonon field corresponds to real fluctuations of a lattice and has dynamics in model of polyacetylene. The field σ is auxiliary that is reflected by absence of a kinetic member.

7.5.2 MODEL FIE-FOUR WITH FERMI PARTICLES

$$L = \sum_{p=1}^{N} \bar{\psi}^p (i\partial - g\phi)\psi^p + \frac{1}{2}(\partial_\mu \phi)^2 + \frac{1}{2}\mu_0^2 \phi^2 - \frac{\lambda}{4}\phi^4$$

Constants have corresponding dimension. We shall note, that the kinetic member with derivative of time is available in composed for a field.

We shall exclude effects of sea Fermi from consideration and we shall consider a discrete level in an electronic spectrum. The stationary equations for model look like:

n_0 Characterizes filling a level ω_0. The index p is lowered. Decisions of system are known: kinks, "small bags," and "double bags" – excitation of polaron type.

a. Kinks are conditions with $\omega_0 = 0$, which wave functions are given by formulas

$$\phi(x) = \frac{\mu_0}{\sqrt{\lambda}} \tanh \frac{\mu_0 x}{\sqrt{2}}, \quad \psi_{10}(x) = \psi_{20}(x) = A(\operatorname{ch} \frac{\mu_0 x}{\sqrt{2}})^{-g\sqrt{2}/\sqrt{\lambda}}$$

Energy of kink does not depend from n_0 and in quasi-classical approach looks like:

$$E_k = \frac{2\sqrt{2}}{3} \frac{\mu_0^3}{\sqrt{\lambda}}$$

b. Decisions of polaron type

$$\phi(x) = f - \frac{k_0^2}{g\omega_0} \sec h[k_0(x+x_0)] \sec h[k_0(x-x_0)], \quad f = \frac{\mu_0}{\sqrt{\lambda}},$$

$$\psi_{10}(x,t) = \sqrt{\frac{k_0}{8}} e^{-i\omega_0 t} (\sec h[k_0(x+x_0)] + \sec h[k_0(x-x_0)]),$$

$$\psi_{20}(x,t) = \sqrt{\frac{k_0}{8}} e^{-i\omega_0 t} (-\sec h[k_0(x+x_0)] + \sec h[k_0(x-x_0)]),$$

$$k_0 = \frac{n_0 g^2}{8}, \quad \omega_0^2 = g^2 f^2 - k_0^2, \quad \tanh k_0 x_0 = (gf - \omega_0)/k_0$$

We define $\omega_0 = qf \cos\varphi$, $k_0 = qf \sin\varphi$, we come to bond $\sin 2\varphi(n_0) = n_0/4f^2$.

The equation has no decision at $n_0/4f^2 \rangle 1$. The equation has two decisions at $n_0/4f^2 \ll 1$ $(k_0(n_0))_+ = qf - qn_0^2/128f^3$ – "a double bag", the astable decision ($\varphi(n_0) > \pi/4$) и $(k_0(n_0))_- = qn_0/8f$ – "a fine bag", stable at $\varphi(n_0)\langle \frac{\pi}{4}$. Both decisions become equivalent at $n_0/4f^2 \to 1$. A necessary condition of stability of a condition is $\frac{d^2E}{d\varphi^2}\rangle 0$. We receive $\frac{d^2E}{d\varphi^2} = 8qf^3 \sin\varphi(n_0)\cos 2\varphi(n_0)$.

Model fie-four with Fermi particles is possible to consider as low-energy the effective theory of a field of more fundamental Gross–Neview model.

7.5.3 FINAL REMARKS

We shall discuss briefly qualitative parallels between three models considered above. All properties of solitons of polyacetylene are described by models at N=2. Crushing of the charge localized on soliton, takes place at N=1. We shall note a number of the important properties inherent in three models.

1. Dynamic infringement of symmetry of the basic condition takes place. Degeneration of the basic condition takes place as a result of spontaneous infringement of symmetry.
2. All models contain excitations in the form of topological solitons kinks, connecting vacuous vacuums. The electronic power spectrum varies at presence topological solitons a discrete level appears with zero energy. Solitons have unusual spin– charge parities.
3. All three models contain polaron the decision.

Further we shall try to investigate consequences of spontaneous infringement of symmetry in these models at final temperature of system and at presence of asymmetry of number of particles and antiparticles (electrons – holes) is as much as possible detailed. The question of crushing of the charge, localized on soliton, will be considered later.

7.6 PHASE TRANSITIONS IN BIDIMENTIONAL MODELS OF THE THEORY OF A FIELD AND POLYACETYLENE

7.6.1 INTRODUCTION

Research of systems the basic condition of which possesses the symmetry which is distinct from symmetry of Hamiltonian, is the most interesting though also a difficult problem of physics. A number of the fundamental concepts entered into the theory of many cooperating particles by Bogolyubov, underlies research of spontaneous infringement of symmetry. The model with the allocated condensate has been offered in its works for the first time. Bogolyubov's fundamental concept about "quasi-average" has huge value at studying superconductivity, superfluidity, ferromagnetics and to that similar.

The general scheme of research of phase transition consists in construction of effective potential of system at final temperature and fermionic density P(σ, T, n). The analysis of minimum P(σ, T, n) allows to receive critical values T and n_c, at which symmetry is restored. Feynman diagrams it is necessary sum up in view of replacement: zero a component of an impulse $k_0 \to i\omega_n$ and, accordingly ω_n = $(2n+1)\pi T - i\alpha(\omega_n \pm 2n\pi T)$ – Matsubara frequency. The chemical potential α is connected with presence of preservation of number of Fermi particles.

7.6.2 EFFECTIVE POTENTIAL OF GROSS–NEVIEW MODEL

We shall be limited to the account of one-loop diagrams, at construction of effective potential, and we shall use approach of an average field. We have at summation of diagrams (Fig. 9):

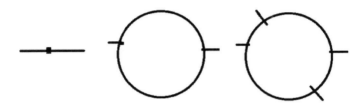

FIGURE 9 Matsubara one-loop diagrams.

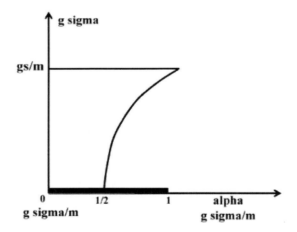

FIGURE 10 A qualitative kind of decisions.

$$P(\sigma) = \frac{1}{2}\sigma^2 - iN\sum \int \frac{dk}{(2\pi)^2} \frac{1}{n}(\frac{\lambda\sigma^2}{k^2})^n$$

N – Number of types of Fermi particles, $\lambda = Ng^2, k^2 = k_0^2 - \bar{k}_0^2$

We use decomposition and, in view of replacement, we have:

$$P(\sigma,T,\alpha) = \frac{1}{2}\sigma^2 - 2NT\int_0^\infty \frac{dk}{2\pi}\sum_{n=-\infty}^{\infty}\ln\frac{k^2+\omega_n^2+\frac{\lambda}{N}\sigma^2}{k^2+\omega_0^2}$$

We calculate the sum. We get

$$P(\sigma,T,\alpha) = \frac{1}{2}\sigma^2 - 2NT\int_0^\infty \frac{dk}{2\pi}\ln\frac{[1+\exp\frac{\alpha-\varepsilon}{T}][1+\exp\frac{\alpha+\varepsilon}{T}]}{[1+\exp\frac{\alpha-k}{T}][1+\exp\frac{\alpha+k}{T}]}$$

We investigate at zero value of chemical potential. In this case

$$\varepsilon = \sqrt{k^2 + \frac{\lambda}{N}\sigma^2}$$

where: Λ – impulse of trimming. Last member is final in previous expression. Effective potential renormalize at T=0. We use a following condition for this purpose (Kolumen–Vainberg):

$$\left.\frac{\partial^2 P(\sigma)}{\partial \sigma^2}\right|_{\sigma=\sigma_0} = 1$$

where: σ_c – the classical field. Meanwhile:

$$\ln\frac{\Lambda+\sqrt{\Lambda^2+(g\sigma)^2}}{g\sigma_0} = \frac{\Lambda}{\sqrt{\Lambda^2+(g\sigma)^2}}$$

We have a bond $2\Lambda = q\sigma_0 e$. A Dependence on parameter of trimming disappears as a result of renormalization of weights of a field in effective potential.

$$P(\sigma,\sigma_0) = \frac{1}{2}\sigma^2 + \frac{\lambda}{4}\pi\sigma^2[\ln(\frac{\sigma}{\sigma_0})^2 - 3]$$

The account one-loop correction has led to occurrence of an additional member $\sigma^2\ln\sigma^2$ in potential energy of system. This member gives the negative amendment at small values σ. This amendment dominates. Potential energy is positive and increases at great values σ. The theory is stable. The local maximum is in a point $\sigma = 0$. The condition of an extremum looks like

$$\frac{\partial P(\sigma,\sigma_0,T)}{\partial \sigma} = \sigma_M(1+\frac{\lambda}{2\pi}[\ln(\frac{\sigma_M}{\sigma_0})^2 - 2] + \frac{2\lambda}{\pi}\int_0^\infty \frac{dk}{\varepsilon}(e^{\frac{\varepsilon}{T}}+1)^{-1})$$

at T = 0;

$$\sigma_M = \pm\sigma_0 \exp(1-\frac{\pi}{\lambda})$$

Dynamic infringement of symmetry occurs. Fermi particles get weight M_F. $M_F = g\sigma_M$. Crack M_F arises in a spectrum of Fermi particles. The potential has two minima.

a) We shall consider temperature effect in above relation. The integral supposes an analytical estimation at heats:

$$\int_0^\infty \frac{dk}{\sqrt{k^2+(g\sigma)^2}}[\exp(\frac{\sqrt{k^2+(g\sigma)^2}}{T})+1]^{-1} = -\frac{1}{4}\ln(\frac{g\sigma}{\pi T})^2 - \frac{\gamma_E}{2} + 0(\sigma^4)$$

$\gamma_E = 0.577$ (Euler constant). Expression (7) becomes:

$$\frac{\partial P(\sigma,\sigma_0,T)}{\partial \sigma} = \sigma[1-\frac{\lambda}{\pi}(1+\gamma_E) + \frac{\lambda}{2\pi}\ln\frac{N}{\lambda}(\frac{\pi T}{\sigma_0})^2 + 0(\sigma^2)]$$

We receive from above expression: the minimum takes place at $\sigma = 0$, T > Tc. The maximum takes place at T < Tc.

$$T_c = \frac{1}{\pi}g\sigma_0 \exp(1-\frac{\pi}{\lambda}+\gamma_E) = \frac{\gamma M_F}{\pi}, \quad \gamma = \exp\gamma_E$$

The analogy is available with an estimation of critical temperature of the superconducting channel. The weight of Fermi particle costs instead of a crack. The last expression is connected with temperature of structural Peierls transition. Peierls transition takes place in model of polyacetylene.

b) We shall pass to the analysis at final values of chemical potential. We accept $\alpha \gg T$. In view of renormalization:

$$P(\sigma,\sigma_0,T) = \frac{1}{2}\sigma^2 + \frac{\lambda}{4\pi}\sigma^2[\ln(\frac{\sigma}{\sigma_0})^2 - 3] - \frac{N}{\pi}[\int_0^{\sqrt{k^2-(g\sigma)^2}} dk(\alpha-\varepsilon) - \frac{\alpha^2}{2}]$$

We shall consider two cases:
1. $\alpha \langle g\sigma$. We have

Polyacetylene

$$P(\sigma,\sigma_0) = \frac{1}{2}\sigma^2 + \frac{\lambda}{4\pi}\sigma^2[\ln(\frac{\sigma}{\sigma_0})^2 - 3] + \frac{N\alpha^2}{2\pi}$$

Condition of an extremum:

$$\frac{\partial P}{\partial \sigma} = \sigma[1 - \frac{\lambda}{\pi} + \frac{\lambda}{\pi}\ln\frac{\sigma}{\sigma_0}] = 0$$

Above expression does not contain chemical potential, at $\sigma = 0$ has a local maximum. This expression has minima at $\sigma_M = \pm\sigma_0 \exp(1 - \frac{\pi}{2})$.

2. The case $\alpha \rangle g\sigma$ is more interesting. We can write down;

$$P(\sigma,\sigma_0) = \frac{1}{2}\sigma^2 + \frac{\lambda}{4\pi}\sigma^2[\ln(\frac{\sigma}{\sigma_0})^2 - 3] + \frac{N}{\pi}[\frac{\alpha^2}{2} - \frac{\alpha}{2}\sqrt{\alpha^2 - (g\sigma)^2} + \frac{(g\sigma)^2}{2}\ln\frac{\alpha + \sqrt{\alpha^2 - (g\sigma)^2}}{g\sigma}]$$

Condition of an extremum:

$$\ln\frac{\alpha + \sqrt{\alpha^2 - (g\sigma)^2}}{g\sigma_M} = 0$$

The maximum takes place in $\sigma = 0$ at $\alpha \langle \frac{g\sigma}{2}$. The minimum takes place at $\alpha \rangle \frac{g\sigma}{2}$.

The decision of above expression is available in the form of:

$$\sigma^* = \sigma_M\sqrt{\frac{2\alpha}{g\sigma_M} - 1}$$

Local maximum P takes place at $\frac{g\sigma_M}{2} \langle \alpha \langle g\sigma_M$. The minimum takes place at $\alpha \langle g\sigma$. The qualitative kind of decisions is resulted on Fig. 10.

7.6.3 MODEL FIE!-FOUR WITH FERMI PARTICLES

We shall study phase transition with restoration of symmetry in this section. Potential energy of system is equal a classical case:

$$P_0(\phi) = -\frac{\mu_0^2}{2}\phi^2 + \frac{\lambda}{4}\phi^4$$

The account of quantum amendments will lead to divergence in effective potential. Divergence in case of two measurements will be logarithmic. We can eliminate these of divergence. We change weight of a field. The effective potential can be written down in case of final temperature and chemical potential in the form of:

$$P(\phi) = P_0(\phi) + \frac{1}{2}Q(T,\alpha)\phi^2$$

The factor in above expression has logarithmic divergence.

We shall define fields $\phi' = \phi - \rho$, where there is a deviation of a field from position of a minimum

$$\left.\frac{\partial P}{\partial \phi}\right|_{\phi=\rho} = 0$$

where: ρ – parameter of the order. We consider shift in initial Lagrangian. The unique one-loopback diagrams containing logarithmic divergence, are "tadpoles" and own energy of Bose particle (Fig. 11).

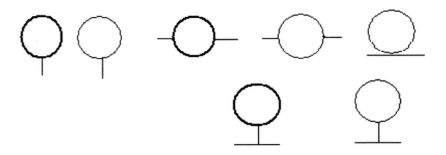

FIGURE 11 Columns, leading to divergence.

Counter-term Becomes after shift:

$$\partial L = \frac{1}{2}Q(T,\alpha)\phi'^2 + Q(T,\alpha)\rho\phi' + \frac{1}{2}Q(T,\alpha)\rho^2$$

If divergence of "tadpole" and own energy of Bose particle are reduced by counter-terms, we can define $Q(T,\alpha)$. The contribution from "tadpoles" is equal:

$$\Gamma = -i(2\pi)^2[3\lambda I_B(T,\alpha) - 2g^2 I_F(T,\alpha)]\rho,$$

where

$$I_B(T,\alpha) = \frac{1}{2\pi}\int_{-\infty}^{\infty}\frac{dk}{2\omega_B}\coth\frac{\omega_B}{2T}, \quad \omega_B = \sqrt{k^2 + m_B^2(T,\alpha)},$$

$$I_F(T,\alpha) = \frac{1}{2\pi}\int_{-\infty}^{\infty}\frac{dk}{2\omega_F}\frac{sh(\omega_F/T)}{ch(\omega_F/T)+ch(\alpha/T)},$$

$$\omega_F = \sqrt{k^2 + m_F^2(T,\alpha)}$$

We choose

$$Q(T,\alpha) = 3\lambda I_B(T,\alpha) - 2g^2 I_F(T,\alpha).$$

Counter-terms completely reduce divergence in own energy of Bose particle. We need to eliminate logarithmic divergence. It is possible to carry out by renormalization weights of bosonic field, which is to subtract a member that corresponds to replacement

$$\mu_0^2 \to \mu^2 = \mu_0^2 + Q(0) = 3\lambda I_B(0) - 2g^2 I_F(0).$$

The effective potential becomes final and looks like:

$$Q(T,\alpha) = 3\lambda I_B(T,\alpha) - 2g^2 I_F(T,\alpha)$$

The condition of an extremum of effective potential becomes

$$P(\phi) = P_0(\phi, \mu_0 \to \mu) + \frac{1}{2}Q^R(T,\alpha)\phi^2,$$

$$Q^R(T,\alpha) = 3\lambda(I_B(T,\alpha) - I_B(0)) - 2g^2(I_F(T,\alpha) - I_F(0))$$

This equation has the trivial decision $\rho = 0$. Fermi particles are massless. The weight of bosonic field is defined from the equation

$$m_B^2(T,\alpha) = -\mu^2 + 3\lambda\rho^2 + Q^R(T,\alpha)$$

Counter-term addresses in zero at zero temperature and chemical potential. The bosonic field has negative weight. It signals about spontaneous infringement of symmetry in system. From two last expressions we can write:

$$m_B^2(T,\alpha) - 2\mu^2 + 2Q^R(T,\alpha) = 0.$$

The decision is available at zero temperature and chemical potential:
$m_B^2(T,\alpha) = 2\mu^2$,

Bose particles get physical weight, $\rho(0) = \mu/\sqrt{\lambda}$, Fermi particles become massive $m_F(0) = g\mu/\sqrt{\lambda}$.

We can believe, that our results are fair everywhere, except for area near to critical temperature. The technics of temperature functions of Green allows to receive a number of interesting results and wide application finds.

The general scheme is simple enough. The effective potential, which is simple for receiving in one-loopback approach, is thermodynamic potential of system. Small deviations from position of a minimum are considered in absence. We receive thermodynamic potential in the basic condition. Kink decision the deviation of a field from kink vacuum is considered and entered. We receive thermodynamic potential again for single-kink sectors. The thermodynamic potential for one kink Ω_k is calculated as a difference.

Realization this programme leads to energy of based kink.

$$E_K = \Omega_K - T\frac{d\Omega_K}{dT} = \frac{2\sqrt{2}\mu^3}{3\lambda} - (\frac{3}{2\pi} - \frac{1}{4\sqrt{3}})\sqrt{2}\mu + \frac{\sqrt{6}}{2}\mu N(\frac{\sqrt{6}}{2}\mu) - 3\sqrt{2}\mu f$$

At $T = 0$

$$\omega_k = \sqrt{k^2 + 2\mu^2}$$

At $T \gg \mu$

$$E_k = \frac{2\sqrt{2}\mu^3}{3\lambda} - \frac{3}{\sqrt{2\pi}} + \frac{T}{2}$$

We can generalize results in case of moving soliton, find density of number of solitones.

7.6.4 MODEL OF TRANS-POLYACETYLENE

Polyacetylene is Peierls dielectric with doubling the period. Suppression of Peierls dimerization takes place with growth of temperature. The crack disappears. Presence of an external field also leads to disappearance of a mass crack. We shall construct thermodynamic potential for system of noninteracting electrons on a deformable chain of atoms

$$\Omega(\Delta,T,\alpha) = -\frac{2T}{N}\sum_i\sum_q \ln[1+\exp(\frac{\alpha-\varepsilon_i}{T})] + \frac{\Delta^2}{4\pi\lambda t_0}$$

where $\varepsilon_{1,2} = \pm\sqrt{(\Delta\sin q)^2 + (2t_0\cos q)^2}$,
where N – Number of atoms in a chain, α – chemical potential, $\lambda = 4\alpha^2 a/\pi k v_F$.
The condition of extremum looks like:

$$\frac{\partial\Omega(\Delta,T,\alpha)}{\partial\Delta} = \Delta(1-2\lambda\int_0^{kv_F}\frac{dq}{\varepsilon}\frac{sh(\varepsilon/T)}{ch(\varepsilon/T)+ch(\alpha/T)}) = 0$$

The crack remains to a constant depending on temperature up to 0.2 eV, and then quickly decreases up to zero at 0.4 eV. Such kind of the decision specifies presence of critical temperature.

Dependence on previous figure is received at temperature 300K (0.026 eV). Control calculations have shown that change of temperature within the limits of 250–350K does not influence qualitatively results. Numerical calculations will be coordinated with analytical.

The critical density is equal:
$n_c = \sqrt{2}\Delta/\pi v_F$

7.6.5 MODEL OF CIS-POLYACETYLENE

Model of *cis*-polyacetylene differs from *trans*-polyacetylene presence of seed crack. It is similar to introduction of seed fermion sludge. *Cis*-model gets a number of the important properties. The thermodynamic potential on one atom of carbon looks like:

$$\Omega = -\frac{2T}{N}\sum_q \ln([1+\exp(\frac{\alpha+\varepsilon}{T})][1+\exp(\frac{\alpha-\varepsilon}{T})])$$

The condition of an extremum is

$$\varepsilon = \sqrt{(\Delta\sin q)^2 + (2t_0\cos q)^2}$$

7.7 SOLITONS WITH FRACTIONAL CHARGES IN POLYACETYLENE

7.7.1 INTRODUCTION

To research of the unusual quantum numbers localized on topological solitones, many works are devoted. The physical basis of occurrence of fractional quantum numbers became clear.

Presence in system of objects with not trivial topology leads to change of structure of the basic condition of system. Quantification concerning new soliton vacuum also leads to unusual quantum numbers.

7.7.2 MODEL

We shall consider the mechanism of occurrence of fractional charges on an example of model with $N = 1$. Potential energy is equal

$$P_0(\phi) = -\frac{\mu_0^2}{2}\phi^2 + \frac{\lambda}{4}\phi^4 \tag{1}$$

and has minima at $\langle\phi\rangle = \rho = \pm\mu_0/\sqrt{\lambda}$.

Fermi particles get weight $m_F = g\rho$. Free Dirac equation has decisions in the form of flat waves with a continuous power spectrum $\varepsilon_k = \pm\sqrt{k^2 + m_F^2}$ decisions with positive $u_k^+(x), v_k^+(x)$ and negative $u_k^-(x), v_k^-(x)$ with fashions are allocated. Quantization is carried out by standard decomposition.

We shall enter the operator of a charge

$$Q = \psi^+(x)\psi(x)dx. \tag{2}$$

We consider Eq. (1)

$$Q = \sum(b_k^+ b_k - d_k^+ d_k),$$

$Q|0 \geq 0$. The vacuum is neutral.

It is necessary for us to solve Dirac equation at presence of soliton and in the soliton field. The discrete fashion with zero energy appears in a power spectrum. Corresponding wave function looks like:

$$\psi_0(x) = \begin{vmatrix}1\\1\end{vmatrix} A(\cosh\frac{\mu_0 x}{\sqrt{2}})^{-g\sqrt{2}/\sqrt{\lambda}}$$

Function is charge–self–conjugate. $\sigma_3\psi_0^+(x) = \psi_0(x)$.

We carry out quantification

$$\psi = a\psi_0(x) + \Sigma(e^{-iE_k t}B_k U_k^+(x) + e^{-E_k t}D_k^+ V_{k-}(x)) \qquad (3)$$

Zero mode energy and wave functions of Dirac equation at presence of soliton is allocated in the Eq. (3.) E_k, U_k^+, V_k^- − Operators $B_k^+(B_k), D_k^+(D_k)$ are certain in soliton sector $V_k^- = \sigma_3(U_k^-)^*$. We designate soliton vacuum $|S>$. $B_k|S>$, $D_k|S\geq 0$. A state $_0(x)$ is degenerate on energy, action of the operator α on a state $|S>$ generates other state with the same energy. We shall designate these states $|\pm, S>$. Then $\alpha|+, S> = |—, S>$, $\alpha^+|—, S> = |+, S>$, $\alpha|—, S> = 0$, $\alpha^+|+, S> = 0$. (4)

Plus in expression (4) concerns to the borrowed state, a minus − to free. We shall define a charge of soliton vacuum. We shall substitute Eq. (3) in Eq. (2),

$$Q = a^+a - \frac{1}{2} + \Sigma(B_k^+ B_k - D_k^+ D_k) \qquad (5)$$

$$Q|+, S> = \pm\tfrac{1}{2}|+, S>, \qquad (6)$$

Each of two soliton states bears half of charge of electron. This result already has been received at the analysis of properties of polyacetylene. Electrons have spin and crushing of the charge localized on soliton, is hidden by doubling of degrees of freedom in *trans*-polyacetylene. It has led solitones with an unusual parity a back and a charge. We shall consider two chains in the basic state (a) (for example in a phase A) and with two solitones on a final segment (b).

Simple calculation shows, that the bottom chain does not have not enough one double bond. The two-soliton state is equivalent to the basic state a minus one double bond which is formed by one electron. Two solitones give deficiency in one electron. One soliton has a charge equal ½ a charge of electron.

We shall consider model with is the triple degenerate basic state

The amplitude of displacement of n-th atom from position of balance can be written down in the form of:

$$y_n = y\cos(\tfrac{2}{3}\pi n - \theta), \qquad (7)$$

where y and θ fix displacement of atom with $n = 0$, and $\theta = \pi/6$ in a phase A, $\theta = 5\pi/6$ in a phase B and $\theta = 3\pi/2$ in a phase C. We can allocate two classes of domain walls in such system: kink type I it is formed at transition through phases A−B−C−A, kink type II corresponds to transition A−C−B−A.

We shall consider an infinite chain in a phase A at $n \to -\infty$ $\theta = \pi/6$. Three units n_1, n_2 and n_3 are allocated on a chain. Units are widely dissolved from each other. The B− phase takes place near to unit n_1. The C-phase takes place near to unit n_2.

The phase A is restored at $n=n_3$. Full change of a phase makes $\Delta\theta=2\pi$. The general charge $\Delta Q = \dfrac{-\Delta\theta y}{\pi} = -2e$ flows through Gaussian surface far from kinks. We assume identity of kinks type I. The charge of everyone kink is equal $(2/3)e$. Two kinks were in polyacetylene. Change of a phase was 2π. Similar reasonings lead to a charge localized on soliton, multiple to size e/p for system with p the degenerate basic states. These results will be coordinated with numerical calculations for chains final are long also results of method Green function.

The charge is equal $-(2/3)e$ for kinks type II. They remain anti-kinks in relation to kinks I. We can create pairs kink anti-kink S(–S) or triplets SSS, (–S) (–S) (–S) in case of –(2/3)e, because of infringements of the basic state. Some analogy with quark structure of mesones and hadrons is available. Attempt to describe baryons as solitones within the limits of Skyrme model is represented interesting.

7.7.3 SOLITONES WITH FRACTIONAL CHARGES

We shall assume that double degeneration of the basic state takes place in the basic state and soliton solutions are available $\varphi_s(x)$. It is necessary for us to know a spectrum of a problem at presence solitones and in its absence for definition of the charge localized on soliton.

$$Q = \int dx[p^s(x) - p^0(x)], \, p(x) = \sum \psi_n^+(x)\psi_n(x), \qquad (8)$$

Summation is carried out on all borrowed levels in fermionic spectrum.
Dirac equation looks like for stationary states:

$$(-i\sigma_2 \dfrac{d}{dx} + \sigma_1\varphi + \sigma_3\varepsilon)\psi_n(x) = E_n\psi_n \qquad (9)$$

or in components

$$-v_x + \varphi v + \varepsilon u = Eu,$$

$$-u_x + \varphi u - \varepsilon u = Ev, \qquad (10)$$

$$-u_{xx} + (\varphi^2 - \varphi_x)u = (E^2 - \varepsilon^2)u \qquad (11)$$

We shall note that flat waves are the decision of system (11) in absence of soliton (constant bosonic field).

Polyacetylene

$$u_k(x, t) = C$$

$$v_k(x, t) = C \qquad (12)$$

with continuous spectrum.
Discrete level $E_s = \varepsilon > 0$ with wave function is available at presence of soliton.

$$u^s(x) = N_0 \exp[-\int dz \varphi_s(z)], \quad v^s(x) = 0 \qquad (13)$$

and also continuous spectrum with wave functions

$$u^s(x) = \sqrt{\frac{E+\varepsilon}{2E}} u_k^s(x), \quad v^s(x) = -\frac{\partial_x + \varphi(x)}{\sqrt{2E(E+\varepsilon)}} u_k^s(x) \qquad (14)$$

where $u_k^s(x)$ are normalized decisions of Schrödinger equations. The concrete kind of the equations is not essential. We shall receive density of a charge, under condition of free soliton level:

$$\rho_k(x) = \frac{E+\varepsilon}{2E} |u_k(x)|^2 + \frac{1}{2E(E+\varepsilon)} |(\partial_x + \varphi) u_k(x)|^2 =$$

$$= |u_k(x)|^2 + \frac{1}{4E(E+\varepsilon)} \frac{\partial^2}{\partial x^2} |u_k(x)|^2 + \frac{1}{2E(E+\varepsilon)} \frac{\partial}{\partial x} [|u_k(x)|^2 \varphi(x)] \qquad (15)$$

Where the second equation in Eq. (12) is used. Only the first member remains in Eq. (15) in absence of kink.

$$Q = \int dx \int \frac{dk}{2\pi} [|u_k^s(x)|^2 - |u_k^0(x)|^2] + \int \frac{dk}{2\pi} \frac{1}{4E(E+\varepsilon)} [\partial_x |u_k^s(x)|^2 + 2|u_k^s(x)|^2 \varphi_s(x)] \Big|_{x=-\infty}^{x=\infty} \qquad (16)$$

The double integral is calculated in Eq. (16), property of completeness of functions is used. The integral is equal (−1). We should recollect, that $u_k^0(x)$ represents all Schrödinger modals in vacuum, and $u_k^s(x)$ is truncated from full soliton spectrum. The connected condition adding (−1) to size Q, is excluded.

We shall assume

$u_k^s(x)\big|_{x\to\infty} \to T e^{ikx}$, $u_k^s(x)\big|_{x\to-\infty} \to e^{ikx} + R e^{-ikx}$, where T and R are factors of passage and reflection accordingly, and $|T|^2 + |R|^2 = 1$.

$Q = -1/\pi \operatorname{arctg} \varphi_0/\varepsilon$ (17)

The charge localized on soliton, $Q = -1/2$ at $\varepsilon \to 0$.

Obvious kind of soliton and fermion wave function is known in polyacetylene and other models. We can directly calculate density of a charge.

Average quadratic size of fluctuation the operator of a charge of soliton concerning its average size (generally fractional) is small. Wave function of soliton is own function of the operator of a charge with fractional own value. Fermionic charge can be measured experimentally.

7.8 CHANGE OF A FRACTIONAL CHARGE OF SOLITON

Expression for a charge becomes at final temperature and chemical potential:

$$Q = (T, \alpha) = \int dx \sum [\rho_n^s(x) - \rho_n^0(x)], \qquad (16)$$

$n(\varepsilon_i - \alpha) = [\exp(\varepsilon_i - \alpha)/T + 1]^{-1}$ – accumulated distribution of Fermi. The Eq. (17) from lecture (16) will become:

$$Q(\alpha, T) = -2\varphi_0 \qquad (17)$$

We shall analyze Eq. (17). The concrete kind of potential was not considered till now. The potential became important for considering now. The effective potential of system changes at inclusion of effects of final temperature and chemical potential. Critical values temperatures (T_c) and chemical potential (α_c) appear. Symmetry of system is restored above critical values of temperature. Solitones already is not present in such system.

We shall consider a case $\alpha = 0$.

$$Q(T) = C \qquad (18)$$

We used parities $n(E) + n(-E) = 1$, $n(E) - n(-E) = -\tanh(\frac{E}{2T})$.

We come to result (19) from lecture (16) at T=0. The integral in Eq. (13) can be estimated at low temperatures:

$$Q(T) = -C \qquad (19)$$

The limit $\varepsilon \rightarrow 0$ is interesting. $Q(T) = 0$ in this limit. It means that the average charge is equal to zero at final temperature. Pairs Soliton–anti-Soliton born at final temperature. The doublet of solitones with fermion number $\pm\frac{1}{2}$ will be the basic condition in this limit. The average fermion number addresses in zero.

$T \ll \varepsilon$ $Q(T)$ aspires to the limit at $\varepsilon \neq 0$. Oscillation of size $Q(T)$ takes place at $T=\varepsilon$.

Polyacetylene

Sizes $\varphi_0 \to 0$ and $Q(T = T_c) \to 0$ at $T \to T_0$. Delocalization takes place. Width of Soliton increases indefinitely).

We shall consider Eq. (17) at final values of chemical potential. We shall be limited to limit T=0. We shall consider three areas of values of chemical potential.

1. At $\alpha > m(\alpha)$, $m(\alpha) =$ we have

$$Q(\alpha > m) = C \tag{20}$$

2. We have in the field of $\varepsilon < \alpha < m$ for a charge

$$Q(\varepsilon < \alpha < m) = C \tag{21}$$

3. We receive at $0 < \alpha < m$

$$Q = -1 + C \tag{22}$$

Dependence on chemical potential has disappeared in Eqs. (21) and (22). Received results have precise physical interpretation. Soliton level is formed of half of condition, segregated from a zone of conductivity and half from a valent zone (Fermi sea) at $\varepsilon=0$ and $T=\alpha=0$. Degeneration takes place thus. The charge of soliton is equal $\pm|e|/2$. Crushing of a charge is hidden at presence a spin at Fermi particle. Solitones have spin 1/2, but without a charge or with a charge $\pm|e|$, but without a spin. Filling of a zone of conductivity (a devastation of a valent zone) occurs at $\alpha > m(\alpha)$. The charge, localized on soliton, is "exhausted". Solitones disappear, localization of a charge is not present at critical values of temperature and chemical potential.

7.9 TRANSITION DIELECTRIC-METAL IN POLYACETYLENE

7.9.1 INTRODUCTION

We shall stop on behavior of polyacetylene at high concentration of impurity. Electric properties were stated earlier. We shall consider optical properties of alloyed polyacetylene depending on concentration of impurity in this section. The crack in an electronic spectrum disappears at concentration $y = 0.0078$. The consent with Drude theory of optical absorption, characteristic for normal metals) is not found out.

7.9.2 EXPERIMENTAL DATA

Experimental data for alloyed polyacetylene will be coordinated with behavior of simple metal with some difference in details in the field of heats. One-dimensional metal with semi-occupied zone takes place, alternation of bondes is absent.

A number of explanations of effect metallization of chains of *trans*-polyacetylene is available now. Two basic mechanisms allocate: a) soliton and b) cluster. Both of the approaches explain the experimental facts.

We shall notice, that the standard technique allows to synthesize samples of *cis*-polyacetylene. *Cis*-trans isomerization takes place at entering impurity. Lines of cis-product in Raman spectrum collapse and appear *trans*-products at 12% of an impurity of AsF_5. *Cis*-behavior prevails in the field of concentration $0.02<y<0.05$. We can exclude effect of polymerization by thermal processing of *cis*-polyacetylene. Trans-polyacetylene turns out. Three results testify against soliton picture:
1. The Spin susceptibility of *cis*-polyacetylene increases with growth of concentration of an impurity and has character of Pauli receptivity.
2. Charged not magnetic solitones are not obligatory for an explanation of conductivity in *trans*-polyacetylene.
3. The Impurity, insertion in samples, is distributed is non-uniform. "Drops" with high conductivity are formed already at concentration of 1%. Conductivity thus consists in tunneling between metal areas. Transition dielectric-metal is connected with achievement of a threshold overshoot of metal "drops". Metal continuum takes place.

7.9.3 THEORETICAL MODELS
Various interpretations of transition dielectric-metal are available according to two approaches in polyacetylene.
1. The Model of metal islands has been offered ("drops", "moustaches"). Heterogeneity of accommodation of the introduced impurity leads to formation of metal areas with rather high concentration of the impurity divided by semiconducting areas. Alternation of bondes inside of metal areas is absent. Solitones are not present there. The size of metal areas increases with growth of concentration of impurity until the threshold of overshoot will not be reached.
2. Some explanations of transition dielectric-metal are offered within the limits of soliton picture. The number of pairs Soliton–anti-Soliton in a chain grows at increase in number of atoms of impurity. Soliton sublattice it is formed, modulating a lattice of atoms (dimmer). The Narrow impurity band arises in the center of a power band. Soliton sublattice the lattice continuously creates weak exponentially damped exchange potential between solitones. Modulated lattice passes in incommensurate Peierls lattice with growth of concentration of carriers. Impurity band extends. Two Peierls cracks are formed. Conductivity is carried out on impurity band and carries Frelich disposition. We can consider effect of interchain bridge, influence of a field of doped atom and another.

7.9.4 CONCLUSION

Interest to physics of polyacetylene continuously increases. It is connected with outlook its practical use as easy metal. The theoretical description of polyacetylene leans on the model having analogue in the quantum theory of a field. It opens greater opportunities in research nontrivial hypotheses as real samples of polyacetylene suppose experimental check. Methods of the theory of a field can be used effectively also at calculations of model of polyacetylene. Certain successes are reached in this direction:

1. Bidimentional models of the theory of a field suppose existence of objects with non-trivial topology. Influence of such objects on physical properties of considered systems rather essentially. Structure of the basic condition changes at presence of solitones. Experimental data about existence solitones are for polyacetylene.
2. The Effect of spontaneous infringement of symmetry is present at many models of relativistic quantum theory fields. The opportunity of restoration of symmetry at heats and (or) density of Fermi particles has the important consequences in cosmology, astrophysics, quantum chromodynamics. Restoration of symmetry means disappearance of Peierls dimerization and promotes transition in a metal condition in polyacetylene.
3. The Opportunity of crushing of charge of Fermi particle is widely discussed in the theory of a field. Degeneration of the basic condition of system, presence topological solitones conduct to fractional charges of Fermi particles. Crushing is hidden by doubling of degrees of freedom because of presence a spin at electrons in polyacetylene. It leads solitones with an unusual parity of a charge and a spin. It finds experimental acknowledgment.

Greater work still is necessary in study of properties of polyacetylene, in the physicist of unidimensional conductors as a whole.

KEYWORDS

- **bidimentional models**
- **effect of spontaneous infringement of symmetry**
- **fermi particles**
- **polyacetylene**
- **quantum chromodynamics**
- **unidimensional conductors**

REFERENCES

1. Sladkov, A.M., Kasatochkin, V.I., Korshak, V.V., Kudryavcev, Yu. P. Diploma on discovery № 107. Bulletin of inventions, 1992, № 6.
2. Korshak, V.V., Kasatochkin, V.I., Sladkov A.M., Kudryavcev, Yu. P., Usunbaev K. About synthesis and properties of polyacetylene. Lecture Academy of Sciences the USSR, 1991, vol. 136, № 6, 1342.
3. Sladkov A.M. Carbyne – the third allotropic form of carbon. M.: Science, 2003, 152.
4. Natta, G., Pino, P., Mazzanti, G. Patent. Hal. 530753 Italy C. A. 1998, v. 52. 15128.
5. Natta, G., Mazzanti, G., Corradini P. AIII Accad. Naz. Lincei, Cl. Sci. Fis. Mat. Nat. Rend. 1998. v. 25. p. 2.
6. Watson, W.H., Memodic, W. C, Lands, L.G. 3. Polym. Sci. 1991, v. 55. 137.
7. Shirakawa, H., Ikeda, S. Polym. J. 1991, v. 2. p. 231.
8. Ito T., Shirakawa, H., Ikeda, S. J. Polym. Sci. Polym. Chem. Ed. 1974, v. 12. 11.
9. Tripathy, S. K-, Rubner, M., Emma, T. et al. Ibid. 1983, v. 44. P. C3–37.
10. Wegner, G. Macromol. Chem. 1981, v. 4. 155.
11. Schen M. A., Karasz, F.E., Chien, L.C. J. Polym. Sci.: Polym. Chem. Ed. 1983, v. 21. 2787.
12. Wnek, G.E., Chien, J.C., Karasz, F.E. et al. J. Polym. Sci. Polym. Lett. Ed. 1979, v. 17. p. 779.
13. Aldissi M. Synthetic Metals. 1984, v. 9. 131.
14. Schue, F., Aldissi Af. Colloq. Int. Nouv. Orient. Compos. Passifs. Mater. Technol. Mises Ocure. Paris. 1982, 225.
15. Chien, M.A., Karasz, F.E., Chien, J.C. Macromol. Chem. Rapid Communs. 984. v. 5. 217.
16. Chien, J.C.J. Poli. Sci. Polym. Lett. Ed. 1983, v. 21. 93.
17. Saxman, A.M., Liepins, R., Aldissi, M. Progr. Polym. Sci. 1985, v. 11. 57.
18. Chien, J.C., Karasz, F.E., Schen M. A., Hirsch, T. 4. Macromolecules. 1983, v.16. 1694.
19. Chien, J.C., Karasz, F.E., MacDiarmid, A.G., Heeger, A., J. Polym. Sci. Polym. Lett. Ed. 1980, v. 18. 45.
20. Chien, J.C. Polymer News, 1979, v. 6. 53.
21. Dandreaux, G.F., Galuin, M.E., Wnek, G.E. J. Phys. 1983, v. 44. P. C3–135.
22. Parshakov A.S. Abstract of a thesis Interaction pentachloride molybdenum with acetylene – a new method for the synthesis of nanoscale composite materials, Moscow. IONCh RAS. 2010.; E.G. Ilin, A.S. Parshakov, A.K. Buryak, D.I. Kochubei, D.V. Drobot, V.I. Nefedov. DAN, 2009, V.427, №5, 641–645.
23. Goldberg, I.B., Crowe, H.R., Newman, P.R., Heeger, A.J., Mc Diarmid, A.G. J. Chem. Phys. 1999, v. 70. № 3. 1132–1136.
24. Bernier, P., Rolland, M., Linaya, C., Disi, M., Sledz, I., Fabre, I.M., Schue, F., Giral, L. Polym. J. 1981, v. 13. № 3. 201–207.
25. Holczer, K., Boucher, J.R., Defreux, F., Nechtschein, M. Chem. Scirpta. 1981, v. 17. № 1–5. 169–170.
26. Krinichnui, V.I. Advances of chemistry. 1996, v. 65. № 1. 84.
27. Shirakawa, H., Sato, M., Hamono, A., Kawakami, S., Soga, K., Ikeda, S. Macromolec. 1980, v. 13. № 2. 457–459.

28. Soga, K., Kawakami, S., Shirakawa, H., Ikeda, S. Makromol. Chem., Rapid. Commun. 1980, v. 1. № 10. 643–646.
29. Natta, G., Pino, P., Mazzanti, G. Patent. Hal. 530753 Italy C. A. 1998, v. 52. 15128.
30. Chasko, B., Chien, J.C.W., Karasz, F.E., Mc Diarmid, A.G., Heeger, A. J. Bull. Am. Phys. Soc. 1999, № 24. 480–483.
31. Lopurev, V.A., Myachina, G.F., Shevaleevskiy O.I., Hidekel M.L. High-molecular compounds. A. 1988, v. 30. № 10. 2019–2037.
32. Kobryanskiy, V.M., Zurabyan, N.J., Skachkova, V.K., Matnishyan A.A. High-molecular compounds.B. 1985, v. 27. № 7. 503–505.
33. Berlyn A.A., Geyderih M.A., Davudov, B.E. Chem. of polyconjugate systems. M.: Chem. 1972, 272.
34. Yang, X.-Z., Chien, J.C.W. J. Polym. Sci.: Polym. Chem. Ed. 1985, v. 23. № 3. 859–878.
35. Mc Diarmid, A.G., Chiang, J.C., Halpern, M., et al. Amer. Chem. Soc. Polym. Prepr. 1984, v. 25. № 2. 248–249.
36. Gibson, H., Pochan, J. Macromolecules. 1982, v. 15. № 2. 242–247.
37. Kobryanskii, V.M. Mater. Sci. 1991, v. 27. № 1. 21–24.
38. Deits, W., Cukor, P. Rubner, M., Jopson, H. Electron. Mater. 1981, v. 10. № 4. 683–702.
39. Heeger, A.J., Mc Diarmid, A.G., Moran, M.J. Amer. Chem. Soc.Polym. Prepr. 1978, v. 19. № 2. 862.
40. Arbuckle, G.A., Buechelev, N.M., Valentine, K.G. Chem. Mater. 1994, v. 6. № 5. 569–572.
41. Korshak, J.V., Korschak, V.V., Kanischka Gerd, Hocker Hartwig Makromol. Chem. Rapid Commun. 1985, v. 6. № 10. 685–692.
42. Swager, T.M., Grubbs, R.H. Synth. Met. 1989, v. 28. № 3. D57–D62. 51.
43. Matsushita, A., Akagi, K., Liang, T.–S., Shirakawa, H. Synth. Met.1999. v. 101. № 1–3. 447–448.
44. Salimgareeva, V.N., Prochuhan Yu. A., Sannikova, N.S. and others High-molecular compounds. 1999, V.41. № 4. 667–672.
45. Leplyanin, G.V., Kolosnicin, V.S., Gavrilova A. A and others Electro chemistry. 1989, v. 25. № 10. 1411–1412.
46. Zhuravleva T. S. Advances of chemistry. 1987, v. 56. № 1. 128–147.

CHAPTER 8

FEATURES OF MACROMOLECULE FORMATION BY RAFT-POLYMERIZATION OF STYRENE IN THE PRESENCE OF TRITHIOCARBONATES

NIKOLAI V. ULITIN, TIMUR R. DEBERDEEV,
ALEKSEY V. OPARKIN, and GENNADY E. ZAIKOV

CONTENTS

8.1 Introduction 314
8.2 Experimental Part 314
8.3 Conclusion 330
Keywords 330
References 330

8.1 INTRODUCTION

The kinetic modeling of styrene controlled radical polymerization, initiated by 2,2'-asobis(isobutirnitrile) and proceeding by a reversible chain transfer mechanism was carried out and accompanied by "addition-fragmentation" in the presence dibenzyltritiocarbonate. An inverse problem of determination of the unknown temperature dependences of single elementary reaction rate constants of kinetic scheme was solved. The adequacy of the model was revealed by comparison of theoretical and experimental values of polystyrene molecular-mass properties. The influence of process controlling factors on polystyrene molecular-mass properties was studied using the model.

The controlled radical polymerization is one of the most developing synthesis methods of narrowly dispersed polymers nowadays [1–3]. Most considerations were given to researches on controlled radical polymerization, proceeding by a reversible chain transfer mechanism and accompanied by "addition-fragmentation" (RAFT – reversible addition-fragmentation chain transfer) [3]. It should be noted that for classical RAFT-polymerization (proceeding in the presence of sulphur-containing compounds, which formula is Z–C(=S)–S–R', where Z – stabilizing group, R' – outgoing group), valuable progress was obtained in the field of synthesis of new controlling agents (RAFT-agents), as well as in the field of research of kinetics and mathematical modeling; and for RAFT-polymerization in symmetrical RAFT-agents' presence, particularly, tritiocarbonates of formula R'–S–C(=S)–S–R', it came to naught in practice: kinetics was studied in extremely general form [4] and mathematical modeling of process hasn't been carried out at all. Thus, the aim of this research is the kinetic modeling of polystyrene controlled radical polymerization initiated by 2,2'-asobis(isobutirnitrile) (AIBN), proceeding by reversible chain transfer mechanism and accompanied by "addition-fragmentation" in the presence of dibenzyltritiocarbonate (DBTC), and also the research of influence of the controlling factors (temperature, initial concentrations of monomer, AIBN and DBTC) on molecular-mass properties of polymer.

8.2 EXPERIMENTAL PART

Prior using of styrene (Aldrich, 99%), it was purified of aldehydes and inhibitors at triple cleaning in a separatory funnel by 10%-th (mass) solution of NaOH, then it was scoured by distilled water to neutral reaction and after that it was dehumidified over $CaCl_2$ and rectified in vacuo. AIBN (Aldrich, 99%) was purified of methanol by re-crystallization. DBTC was obtained by the method presented in research [4]. Examples of polymerization were obtained by dissolution of estimated quantity of AIBN and DBTC in monomer. Solutions were filled in tubes, 100 mm long, and having internal diameter of 3 mm, and after degassing in the mode of "freezing-defrosting" to residual pressure 0.01 mmHg column, the tubes were unsoldered. Polymerization was carried out at 60°C. Research of polymerization's kinetics was made with application of the calorimetric method on Calvet type differential automatic microcalorimeter DAK-1–1 in the mode of immediate record of heat emission rate in isothermal conditions at 60°C. Kinetic parameters of polymerization were calculated basing on the calorimetric data as in the work [5]. The value of polymerization enthalpy $\Delta H = -73.8$ kJ $\times$ mol^{-1} [5] was applied in processing of the data in the calculations.

Molecular-mass properties of polymeric samples were determined by gel-penetrating chromatography in tetrahydrofuran at 35°C on chromatograph GPCV 2000 "Waters". Dissection was performed on two successive banisters PLgel MIXED–C 300×7.5 mm, filled by stir gel with 5 µm vesicles. Elution rate – 0.1 mL $\times$ min^{-1}. Chromatograms were processed in programme "Empower Pro" with use of calibration by polystyrene standards.

8.2.1 MATHEMATICAL MODELING OF POLYMERIZATION PROCESS

Kinetic scheme, introduced for description of styrene controlled radical polymerization process in the presence of trithiocarbonates, includes the following phases.

1. REAL INITIATION

$$I \xrightarrow{k_d} 2R(0) \cdot$$

2. THERMAL INITIATION [6]

It should be noted that polymer participation in thermal initiation reactions must reduce the influence thereof on molecular-mass distribution (MMD). However, since final mechanism of these reactions has not been ascertained in recording of balance differential equations for polymeric products so far, we will ignore this fact.

$$3M \xrightarrow{k_{i1}} 2R(1) \text{'} \quad 2M+P \xrightarrow{k_{i2}} R(1)+R(i) \text{'} \quad 2P \xrightarrow{k_{i3}} 2R(i) \cdot$$

In these three reactions summary concentration of polymer is recorded as P.

3. CHAIN GROWTH

$$R(0)+M \xrightarrow{k_p} R(1), \quad R'+M \xrightarrow{k_p} R(1), \quad R(i)+M \xrightarrow{k_p} R(i+1) \cdot$$

4. CHAIN TRANSFER TO MONOMER

$$R(i)+M \xrightarrow{k_{tr}} P(i, 0, 0, 0) + R(1) \cdot$$

5. REVERSIBLE CHAIN TRANSFER [4]

As a broadly used assumption lately, we shall take that intermediates fragmentation rate constant doesn't depend on leaving radical's length [7].

$$R(i) + RAFT(0, 0) \underset{k_f}{\overset{k_{a1}}{\rightleftarrows}} Int(i, 0, 0) \underset{k_{a2}}{\overset{k_f}{\rightleftarrows}} RAFT(i, 0) + R' \qquad (I)$$

$$R(j) + RAFT(i, 0) \underset{k_f}{\overset{k_{a2}}{\rightleftarrows}} Int(i, j, 0) \underset{k_{a2}}{\overset{k_f}{\rightleftarrows}} RAFT(i, j) + R' \qquad (II)$$

$$R(k) + RAFT(i, j) \underset{k_f}{\overset{k_{a2}}{\rightleftarrows}} Int(i, j, k) \qquad (III)$$

6. CHAIN TERMINATION [4]

For styrene's RAFT-polymerization in the trithiocarbonates presence, besides reactions of radicals quadratic termination

$$R(0) + R(0) \xrightarrow{k_{t1}} R(0)\text{-}R(0), \qquad R(0) + R' \xrightarrow{k_{t1}} R(0)\text{-}R',$$

$$R' + R' \xrightarrow{k_{t1}} R'\text{-}R', \qquad R(0) + R(i) \xrightarrow{k_{t1}} P(i, 0, 0, 0),$$

$$R' + R(i) \xrightarrow{k_{t1}} P(i, 0, 0, 0), \qquad R(j) + R(i-j) \xrightarrow{k_{t1}} P(i, 0, 0, 0).$$

are character reactions of radicals and intermediates cross termination.

$$R(0) + Int(i, 0, 0) \xrightarrow{k_{t2}} P(i, 0, 0, 0), \quad R(0) + Int(i, j, 0) \xrightarrow{k_{t2}} P(i, j, 0, 0),$$

$$R(0) + Int(i, j, k) \xrightarrow{k_{t2}} P(i, j, k, 0), \quad R' + Int(i, 0, 0) \xrightarrow{k_{t2}} P(i, 0, 0, 0),$$

$$R' + Int(i, j, 0) \xrightarrow{k_{t2}} P(i, j, 0, 0), \quad R' + Int(i, j, k) \xrightarrow{k_{t2}} P(i, j, k, 0),$$

$$R(j) + Int(i, 0, 0) \xrightarrow{k_{t2}} P(i, j, 0, 0), \quad R(k) + Int(i, j, 0) \xrightarrow{k_{t2}} P(i, j, k, 0),$$

Features of Macromolecule Formation by RAFT-Polymerization 317

$$R(m) + Int(i, j, k) \xrightarrow{k_{t2}} P(i, j, k, m)$$

In the introduced kinetic scheme: I, R(0), R(i), R', M, RAFT(i, j), Int(i, j, k), P(i, j, k, m) – reaction system's components (refer to Table 1); i, j, k, m – a number of monomer links in the chain; k_d – a real rate constant of the initiation reaction; k_{i1}, k_{i2}, k_{i3} – rate constants of the thermal initiation reaction's; k_p, k_{tr}, k_{a1}, k_{a2}, k_f, k_{t1}, k_{t2} are the values of chain growth, chain transfer to monomer, radicals addition to low-molecular RAFT-agent, radicals addition to macromolecular RAFT-agent, intermediates fragmentation, radicals quadratic termination and radicals and intermediates cross termination reaction rate constants, respectively.

TABLE 1 Signs of components in a kinetic scheme.

I	NC−C(CH₃)₂−N=N−C(CH₃)₂−CN	Int(i, 0, 0)	Ph−CH₂−S−C(S)−S−CH₂−Ph with SR(i)
R(0)	NC−C•(CH₃)₂	Int(i, j, 0)	Ph−CH₂−S−C(S)−SR(i) with SR(j)
R'	Ph•	Int(i, j, k)	R(k)S−C(S)−SR(i) with SR(j)
M	CH₂=CH−Ph	P(i, 0, 0, 0)	(polymer–polymer)
			R(0), Ph−CH₂−S−C(S)−S−CH₂−Ph, SR(i)
			R', Ph−CH₂−S−C(S)−S−CH₂−Ph, SR(i)
R(i)	~~~C•H−Ph	P(i, j, 0, 0)	R(0), Ph−CH₂−S−C(S)−SR(i), SR(j)
			R', Ph−CH₂−S−C(S)−SR(i), SR(j)
			R(j), Ph−CH₂−S−C(S)−S−CH₂−Ph, SR(i)

RAFT (0,0)	[benzyl-S-C(=S)-S-benzyl structure]	P(i, j, k, 0)	$\begin{array}{c} R(0) \\ R(k)S-\overset{|}{\underset{SR(j)}{C}}-SR(i) \end{array}$,
			$\begin{array}{c} R' \\ R(k)S-\overset{|}{\underset{SR(j)}{C}}-SR(i) \end{array}$,
			$\begin{array}{c} R(k) \\ [benzyl]-S-\overset{|}{\underset{SR(j)}{C}}-SR(i) \end{array}$
RAFT(i, 0)	[benzyl-S-C(=S)-SR(i)]	P(i, j, k, m)	$\begin{array}{c} R(m) \\ R(k)S-\overset{|}{\underset{SR(j)}{C}}-SR(i) \end{array}$
RAFT(i, j)	$R(j)S-C(=S)-SR(i)$		

The differential equations system describing this kinetic scheme, is as follows:

$$d[I]/dt = -k_d[I];$$

$$d[R(0)]/dt = 2f\, k_d[I] - [R(0)](k_p[M] + k_{t1}(2[R(0)] + [R'] + [R])) + k_{t2}\left(\sum_{i=1}^{\infty}[\text{Int}(i, 0, 0)] + \sum_{i=1}^{\infty}\sum_{j=1}^{\infty}[\text{Int}(i, j, 0)] + \sum_{i=1}^{\infty}\sum_{j=1}^{\infty}\sum_{k=1}^{\infty}[\text{Int}(i, j, k)]\right);$$

$$d[M]/dt = -(k_p([R(0)] + [R'] + [R]) + k_{tr}[R])[M] - 3k_{i1}[M]^3 - 2k_{i2}[M]^2([M]_0 - [M]);$$

$$d[R']/dt = -k_p[R'][M] + 2k_f\sum_{i=1}^{\infty}[\text{Int}(i, 0, 0)] - k_{a2}[R']\sum_{i=1}^{\infty}[\text{RAFT}(i, 0)] + k_f\sum_{i=1}^{\infty}\sum_{j=1}^{\infty}[\text{Int}(i, j, 0)] - k_{a2}[R']\sum_{i=1}^{\infty}\sum_{j=1}^{\infty}[\text{RAFT}(i, j)] - [R'](k_{t1}([R(0)] + 2[R'] + [R]) +$$

$$+k_{t2}(\sum_{i=1}^{\infty}[\text{Int}(i, 0, 0)]+\sum_{i=1}^{\infty}\sum_{j=1}^{\infty}[\text{Int}(i, j, 0)]+\sum_{i=1}^{\infty}\sum_{j=1}^{\infty}\sum_{k=1}^{\infty}[\text{Int}(i, j, k)]));$$

$$d[\text{RAFT}(0,0)]/dt = -k_{a1}[\text{RAFT}(0,0)][R]+k_f\sum_{i=1}^{\infty}[\text{Int}(i, 0, 0)];$$

$$d[R(1)]/dt = 2k_{i1}[M]^3+2k_{i2}[M]^2([M]_0-[M])+2k_{i3}([M]_0-[M])^3+k_p[M]([R(0)]+[R']-$$

$$-[R(1)])+k_{tr}[R(i)][M]-k_{a1}[R(1)][\text{RAFT}(0,0)]+k_f[\text{Int}(1, 0, 0)]-$$

$$-k_{a2}[R(1)]\sum_{i=1}^{\infty}[\text{RAFT}(i, 0)]+2k_f[\text{Int}(1, 1, 0)]-k_{a2}[R(1)]\sum_{i=1}^{\infty}\sum_{j=1}^{\infty}[\text{RAFT}(i, j)]+3k_f[\text{Int}(1, 1, 1)]-$$

$$-[R(1)](k_{t1}([R(0)]+[R']+[R])+k_{t2}(\sum_{i=1}^{\infty}[\text{Int}(i, 0, 0)]+\sum_{i=1}^{\infty}\sum_{j=1}^{\infty}[\text{Int}(i, j, 0)]+\sum_{i=1}^{\infty}\sum_{j=1}^{\infty}\sum_{k=1}^{\infty}[\text{Int}(i, j, k)])), \; i = 2,...;$$

$$d[R(i)]/dt=k_p[M]([R(i-1)]-[R(i)])-k_{tr}[R(i)][M]-k_{a1}[R(i)][\text{RAFT}(0,0)]+k_f[\text{Int}(i, 0, 0)]-$$

$$-k_{a2}[R(i)]\sum_{i=1}^{\infty}[\text{RAFT}(i, 0)]+2k_f[\text{Int}(i, j, 0)]-k_{a2}[R(i)]\sum_{i=1}^{\infty}\sum_{j=1}^{\infty}[\text{RAFT}(i, j)]+3k_f[\text{Int}(i, j, k)]-$$

$$-[R(i)](k_{t1}([R(0)]+[R']+[R])+k_{t2}(\sum_{i=1}^{\infty}[\text{Int}(i, 0, 0)]+\sum_{i=1}^{\infty}\sum_{j=1}^{\infty}[\text{Int}(i, j, 0)]+\sum_{i=1}^{\infty}\sum_{j=1}^{\infty}\sum_{k=1}^{\infty}[\text{Int}(i, j, k)])), \; i = 2,...;$$

$$d[\text{Int}(i, 0, 0)]/dt = k_{a1}[\text{RAFT}(0,0)][R(i)]-3k_f[\text{Int}(i, 0, 0)]+k_{a2}[R'][\text{RAFT}(i, 0)]-$$

$$-k_{t2}[\text{Int}(i, 0, 0)]([R(0)]+[R']+[R]);$$

$$d[\text{RAFT}(i, 0)]/dt = 2k_f[\text{Int}(i, 0, 0)]-k_{a2}[R'][\text{RAFT}(i, 0)]-k_{a2}[\text{RAFT}(i, 0)][R]+2k_f[\text{Int}(i, 1, 0)];$$

$$d[P(i, 0, 0, 0)]/dt = [R(i)](k_{t1}([R(0)]+[R'])+k_{tr}[M])+(k_{t1}/2)\sum_{j=1}^{i-1}[R(j)][R(i-j)]+$$

$$+k_{t2}[\text{Int}(i, 0, 0)]([R(0)]+[R']);$$

$$d[P(i, j, 0, 0)]/dt = k_{t2}([\text{Int}(i, j, 0)]([R(0)]+[R'])+\sum_{i+j=2}^{\infty}[R(j)][\text{Int}(i, 0, 0)]).$$

Here f – initiator's efficiency; $[R]=\sum_{i=1}^{\infty}[R(i)]$ – summary concentration of macro-radicals; t – time.

In this set, the equations for [Int(i, j, 0)], [Int(i, j, k)], [RAFT(i, j)], [P(i, j, k, 0)] and [P(i, j, k, m)] are omitted since they are analogous to the equations for [Int(i, 0, 0)], [RAFT(i, 0)] and [P(i, j, 0, 0)].

A method of generating functions was used for transition from this equation system to the equation system related to the unknown MMD moments [8].

Number-average molecular mass (M_n), polydispersity index (PD) and weight-average molecular mass (M_w) are linked to MMD moments by the following expressions:

$$M_n = (\Sigma \mu_1 / \Sigma \mu_0) M_{ST}, \quad PD = \Sigma \mu_2 \Sigma \mu_0 / (\Sigma \mu_1)^2, \quad M_w = PD \cdot M_n,$$

where $\Sigma \mu_0$, $\Sigma \mu_1$, $\Sigma \mu_2$ – sums of all zero, first and second MMD moments; $M_{ST} = 104$ g/mol – styrene's molecular mass.

8.2.2 RATE CONSTANTS

8.2.2.1 REAL AND THERMAL INITIATION

The efficiency of initiation and temperature dependence of polymerization real initiation reaction rate constant by AIBN initiator are determined basing on the data in this research, which have established a good reputation for mathematical modeling of leaving in mass styrene radical polymerization [6]:

$$f = 0.5, \quad k_d = 1.58 \cdot 10^{15} e^{-15501/T}, \text{ s}^{-1},$$

where T – temperature, K.

As it was established in the research, thermal initiation reactions' rates constants depend on the chain growth reactions rate constants, the radicals quadratic termination and the monomer initial concentration:

$$k_{i1} = 1.95 \cdot 10^{13} \frac{k_{t1}}{k_p^2 M_0^3} e^{-20293/T}, \quad L^2 \cdot \text{mol}^{-2} \cdot \text{s}^{-1};$$

$$k_{i2} = 4.30 \cdot 10^{17} \frac{k_{t1}}{k_p^2 M_0^3} e^{-23878/T}, \quad L^2 \cdot \text{mol}^{-2} \cdot \text{s}^{-1};$$

$$k_{i3} = 1.02 \cdot 10^8 \frac{k_{t1}}{k_p^2 M_0^2} e^{-14807/T}, \quad L \cdot \text{mol}^{-1} \cdot \text{s}^{-1}. \quad [6].$$

8.2.2.2 CHAIN TRANSFER TO MONOMER REACTION'S RATE CONSTANT

On the basis of the data in research [6]:

$$k_{tr} = 2.31 \cdot 10^6 e^{-6376/T}, L \cdot \text{mol}^{-1} \text{s}^{-1}.$$

8.2.2.3 RATE CONSTANTS FOR THE ADDITION OF RADICALS TO LOW-MOLECULAR AND MACROMOLECULAR RAFT-AGENTS

In research [9], it was shown by the example of dithiobenzoates at first that chain transfer to low- and macromolecular RAFT-agents of rate constants are functions of respective elementary constants. Let us demonstrate this for our process. For this record, the change of concentrations [Int(i, 0, 0)], [Int(i, j, 0)], [RAFT(0,0)] and [RAFT(i, 0)] in quasistationary approximation for the initial phase of polymerization is as follows:

$$d[\text{Int}(i, 0, 0)]/dt = k_{a1}[\text{RAFT}(0,0)][R] - 3k_f[\text{Int}(i, 0, 0)] \approx 0, \quad (1)$$

$$d[\text{Int}(i, j, 0)]/dt = k_{a2}[\text{RAFT}(i, 0)][R] - 3k_f[\text{Int}(i, j, 0)] \approx 0, \quad (2)$$

$$d[\text{RAFT}(0,0)]/dt = -k_{a1}[\text{RAFT}(0,0)][R] + k_f[\text{Int}(i, 0, 0)], \quad (3)$$

$$d[\text{RAFT}(i, 0)]/dt = 2k_f[\text{Int}(i, 0, 0)] - k_{a2}[\text{RAFT}(i, 0)][R] + 2k_f[\text{Int}(i, j, 0)]. \quad (4)$$

The Eq. (1) expresses the following concentration $[\text{Int}(i, 0, 0)]$:

$$[\text{Int}(i, 0, 0)] = \frac{k_{a1}}{3k_f}[\text{RAFT}(0,0)][R]$$

Substituting the expansion gives the following [Int(i, 0, 0)] expression to Eq. (3):

$$d[\text{RAFT}(0,0)]/dt = -k_{a1}[\text{RAFT}(0,0)][R] + k_f \frac{k_{a1}}{3k_f}[\text{RAFT}(0,0)][R].$$

After transformation of the last equation, we have:

$$\frac{d[\text{RAFT}(0,0)]}{[\text{RAFT}(0,0)]} = -\frac{2}{3}k_{a1}[R]dt.$$

Solving this equation (initial conditions: $t = 0$, $[R] = [R]_0 = 0$, $[\text{RAFT}(0,0)] = [\text{RAFT}(0,0)]_0$), we obtain:

$$\ln \frac{[RAFT(0,0)]}{[RAFT(0,0)]_0} = -\frac{2}{3} k_{a1}[R]t. \tag{5}$$

To transfer from time t, being a part of Eq. (5), to conversion of monomer C_M, we put down a balance differential equation for monomer concentration, assuming that at the initial phase of polymerization, thermal initiation and chain transfer to monomer are not of importance:

$$d[M]/dt = -k_p[R][M]. \tag{6}$$

Transforming the Eq. (6) with its consequent solution at initial conditions $t = 0$, $[R] = [R]_0 = 0$, $[M] = [M]_0$:

$$d[M]/[M] = -k_p[R]dt,$$

$$\ln \frac{[M]}{[M]_0} = -k_p[R]t. \tag{7}$$

Link rate $[M]/[M]_0$ with monomer conversion (C_M) in an obvious form like this:

$$C_M = \frac{[M]_0 - [M]}{[M]_0} = 1 - \frac{[M]}{[M]_0}, \quad \frac{[M]}{[M]_0} = 1 - C_M.$$

We substitute the last ratio to Eq. (7) and express time t:

$$t = \frac{-\ln(1 - C_M)}{k_p[R]}. \tag{8}$$

After substitution of the expression (8) by the equation (5), we obtain the next equation:

$$\ln \frac{[RAFT(0,0)]}{[RAFT(0,0)]_0} = \frac{2}{3} \frac{k_{a1}}{k_p} \ln(1 - C_M). \tag{9}$$

By analogy with introduced $[M]/[M]_0$ to monomer conversion, reduce ratio $[RAFT(0,0)]/[RAFT(0,0)]_0$ to conversion of low-molecular RAFT-agent – $C_{RAFT(0,0)}$. As a result, we obtain:

$$\frac{[RAFT(0,0)]}{[RAFT(0,0)]_0} = 1 - C_{RAFT(0,0)}. \tag{10}$$

Substitute the derived expression for $[RAFT(0,0)]/[RAFT(0,0)]_0$ from Eq. (10) to Eq. (9):

$$\ln(1 - C_{RAFT(0,0)}) = \frac{2}{3}\frac{k_{a1}}{k_p}\ln(1 - C_M). \qquad (11)$$

In the research [9], the next dependence of chain transfer to low-molecular RAFT-agent constant C_{tr1} is obtained on the monomer and low-molecular RAFT-agent conversions:

$$C_{tr1} = \frac{\ln(1 - C_{RAFT(0,0)})}{\ln(1 - C_M)}. \qquad (12)$$

Comparing Eqs. (12) and (11), we obtain dependence of chain transfer to low-molecular RAFT-agent constant C_{tr1} on the constant of radicals addition to macromolecular RAFT-agent and chain growth reaction rate constant:

$$C_{tr1} = \frac{2}{3}\frac{k_{a1}}{k_p}. \qquad (13)$$

From Eq. (13), we derive an expression for constant k_{a1}, which will be based on the following calculation:

$$k_{a1} = 1.5 C_{tr1} k_p, L \cdot mol^{-1} s^{-1}.$$

As a numerical value for C_{tr1}, we assume value 53, derived in research [4] on the base of Eq. (12), at immediate experimental measurement of monomer and low-molecular RAFT-agent conversions. Since chain transfer reaction in RAFT-polymerization is usually characterized by low value of activation energy, compared to activation energy of chain growth, it is supposed that constant C_{tr1} doesn't depend or slightly depends on temperature. We will propose as an assumption that C_{tr1} doesn't depend on temperature [10].

By analogy with k_{a1}, we deduce equation for constant k_{a2}. From Eq. (2) we express such concentration $[Int(i, j, 0)]$:

$$[Int(i, j, 0)] = \frac{k_{a2}}{3k_f}[RAFT(i, 0)][R].$$

Substitute expressions, derived for $[Int(i, 0, 0)]$ and $[Int(i, j, 0)]$ in Eq. (4):

$$d[RAFT(i, 0)]/dt = \frac{2}{3}k_{a1}[RAFT(0,0)][R] - \frac{1}{3}k_{a2}[RAFT(i, 0)][R]. \qquad (14)$$

Since in the end it was found that constant of chain transfer to low-molecular RAFT-agent C_{tr1} is equal to divided to constant k_p coefficient before expression $[RAFT(0,0)][R]$ in the balance differential equation for $[RAFT(0,0)]$, from Eq. (14) for constant of chain transfer to macromolecular RAFT-agent, we obtain the next expression:

$$C_{tr2} = \frac{1}{3}\frac{k_{a2}}{k_p}.$$

From the last equation we obtain an expression for constant k_{a2}, which based on the following calculation:

$$k_{a2} = 3C_{tr2}k_p, L \cdot mol^{-1}s^{-1}. \qquad (15)$$

In research [4] on the base of styrene and DBTK, macromolecular RAFT-agent was synthesized, thereafter with a view to experimentally determine constant C_{tr2}, polymerization of styrene was performed with the use of the latter. In the course of experiment, it may be supposed that constant C_{tr2} depends on monomer and macromolecular RAFT-agent conversions by analogy with equation (12). As a result directly from the experimentally measured monomer and macromolecular RAFT-agent conversions, value C_{tr2} was derived, equal to 1000. On the ground of the same considerations as for that of C_{tr1}, we assume independence of constant C_{tr2} on temperature.

8.2.2.4 RATE CONSTANTS OF INTERMEDIATES FRAGMENTATION, TERMINATION BETWEEN RADICALS AND TERMINATION BETWEEN RADICALS AND INTERMEDIATES

In research [4] it was shown, that RAFT-polymerization rate is determined by this equation:

$$(W_0 / W)^2 = 1 + \frac{k_{t2}}{k_{t1}} K[RAFT(0,0)]_0 + \frac{k_{t3}}{k_{t1}} K^2 [RAFT(0,0)]_0^2,$$

where W_0 and W – polymerization rate in the absence and presence of RAFT-agent, respectively, s^{-1}; K – constant of equilibrium (III), L·mol^{-1}; k_{t3} – constant of termination between two intermediates reaction rate, L·mol^{-1}·s^{-1} [11].

For initiated AIBN styrene polymerization in DBTC's presence at 80°C, it was shown that intermediates quadratic termination wouldn't be implemented and RAFT-polymerization rate was determined by equation [4]:

$$(W_0/W)^2 = 1 + 8[\text{RAFT}(0,0)]_0.$$

Since $k_{t2}/k_{t1} \approx 1$, then at 80°C $K = 8$ L·mol⁻¹ [4]. In order to find dependence of constant K on temperature, we made research of polymerization kinetics at 60°C. It was found, that the results of kinetic measurements well rectify in coordinates $(W_0/W)^2 = f([\text{RAFT}(0,0)]_0)$. At 60°C, $K = 345$ L·mol⁻¹ was obtained. Finally dependence of equilibrium constant on temperature has been determined in the form of Vant–Goff's equation:

$$K = 4.85 \cdot 10^{-27} e^{22123/T}, \text{L·mol}^{-1}. \tag{16}$$

In compliance with the equilibrium (III), the constant is equal to

$$K = \frac{k_{a2}}{3k_f}, \text{L·mol}^{-1}.$$

Hence, reactions of intermediates fragmentation rate constant will be as such:

$$k_f = \frac{k_{a2}}{3K}, \text{s}^{-1}. \tag{17}$$

The reactions of intermediates fragmentation rate constant was built into the model in the form of dependence (Eq. 17) considering Eqs. (15) and (16).

As it has been noted above, ratio k_{t2}/k_{t1} equals approximately to one, therefore it will taken, that $k_{t2} \approx k_{t1}$ [4]. For description of gel-effect, dependence as a function of monomer conversion C_M and temperature T (K) [12] was applied:

$$k_{t2} \approx k_{t1} \approx 1.255 \cdot 10^9 e^{-844/T} e^{-2(A_1 C_M + A_2 C_M^2 + A_3 C_M^3)}, \text{L·mol}^{-1}\text{s}^{-1},$$

where $A_1 = 2.57 - 5.05 \cdot 10^{-3} T$; $A_2 = 9.56 - 1.76 \cdot 10^{-2} T$; $A_3 = -3.03 + 7.85 \cdot 10^{-3} T$.

8.2.2.5 RATE CONSTANT FOR CHAIN GROWTH

We made our choice on temperature dependence of the rate constant for chain growth that was derived on the ground of method of polymerization, being initiated by pulse laser radiation [13]:

$$k_p = 4.27 \cdot 10^7 e^{-3910/T}, L \cdot mol^{-1} s^{-1}. \tag{18}$$

8.2.3 MODEL'S ADEQUACY

The results of polystyrene molecular-mass properties calculations by the introduced mathematical model are presented in Fig. 1. Mathematical model of styrene RAFT-polymerization in the presence of trithiocarbonates, taking into account the radicals and intermediates cross termination, adequately describes the experimental data that prove the process mechanism, built in the model.

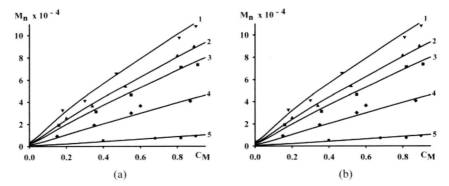

FIGURE 1 Dependence of number-average molecular mass (a) and polydispersity index (b) on monomer conversion for being initiated by AIBN ($[I]_0=0.01$ mol·L^{-1}) styrene bulk RAFT-polymerization at 60°C in the presence of DBTC (lines – estimation by model; points – experiment): $[RAFT(0,0)]_0$ = 0.005 mol·L^{-1} (1), 0.007 (2), 0.0087 (3), 0.0174 (4), 0.087 (5).

Due to adequacy of the model realization at numerical experiment it became possible to determine the influence of process controlling factors on polystyrene molecular-mass properties.

8.2.4 NUMERICAL EXPERIMENT

Research of influence of the process controlling factors on molecular-mass properties of polystyrene, synthesized by RAFT-polymerization method in the presence of AIBN and DBTC, was made in the range of initial concentrations of: initiator – 0–0.1 mol·L^{-1}, monomer – 4.35–8.7 mol·L^{-1}, DBTC – 0.001–0.1 mol·L^{-1}; and at temperatures – 60–120°C.

8.2.4.1 THE INFLUENCE OF AIBN INITIAL CONCENTRATION BY NUMERICAL EXPERIMENT

It was set forth that generally in same other conditions, with increase of AIBN initial concentration number-average, the molecular mass of polystyrene decreases (Fig. 2). At all used RAFT-agent initial concentrations, there is a linear or close to linear growth of number- average molecular mass of polystyrene with monomer conversion. This means that even the lowest RAFT-agent initial concentrations affect the process of radical polymerization. It should be noted that at high RAFT-agent initial concentrations (Fig. 3) the change of AIBN initial concentration practically doesn't have any influence on number-average molecular mass of polystyrene. But at increased temperatures, in case of high AIBN initial concentration, it is comparable to high RAFT-agent initial concentration; polystyrene molecular mass would be slightly decreased due to thermal initiation.

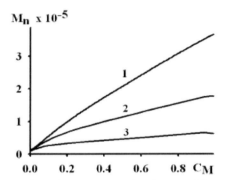

FIGURE 2 Dependence of number-average molecular mass M_n on monomer conversion C_M (60°C) $[M]_0 = 6.1$ mol·L^{-1}, $[RAFT(0, 0)]_0 = 0.001$ mol·L^{-1}, $[I]_0 = 0.001$ mol·L^{-1} (1), 0.01 (2), 0.1 (3).

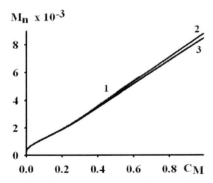

FIGURE 3 Dependence of number-average molecular mass M_n on monomer conversion C_M (60°C) $[M]_0 = 8.7$ mol·L^{-1}, $[RAFT(0, 0)]_0 = 0.1$ mol·L^{-1}, $[I]_0 = 0.001$ mol·L^{-1} (1), 0.01 (2), 0.1 (3).

Since the main product of styrene RAFT-polymerization process, proceeding in the presence of trithiocarbonates, is a narrow-dispersed high-molecular RAFT-agent (marked in kinetic scheme as RAFT(i, j)), which is formed as a result of reversible chain transfer, and widely-dispersed (minimal polydispersity – 1.5) polymer, forming by the radicals quadratic termination, so common polydispersity index of synthesizing product is their ratio. In a broad sense, with increase of AIBN initial concentration, the part of widely-dispersed polymer, which is formed as a result of the radicals quadratic termination, increase in mixture, thereafter general polydispersity index of synthesizing product increases.

However, at high temperatures this regularity can be discontinued – at low RAFT-agent initial concentrations the increase of AIBN initial concentration leads to a decrease of polydispersity index (Fig. 4, curves 3 and 4). This can be related only thereto that at high temperatures thermal initiation and elementary reactions rate constants play an important role, depending on temperature, chain growth and radicals quadratic termination reaction rate constants, monomer initial concentration in a complicated way [6]. Such complicated dependence makes it difficult to analyze the influence of thermal initiation role in process kinetics, therefore the expected width of MMD of polymer, which is expected to be synthesized at high temperatures, can be estimated in every specific case in the frame of the developed theoretical regularities.

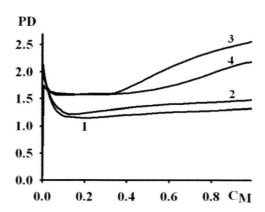

FIGURE 4 Dependence of polydispersity index PD on monomer conversion C_M (120°C) $[M]_0 = 8.7$ mol·L^{-1}, $[RAFT(0, 0)]_0 = 0.001$ mol·L^{-1}, $[I]_0 = 0$ mol·L^{-1} (1), 0.001 (2), 0.01 (3), 0.1 (4)

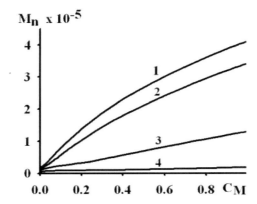

FIGURE 5 Dependence of number-average molecular mass M_n on monomer conversion C_M (120°C) $[M]_0 = 8.7$ mol·L^{-1}, $[RAFT(0, 0)]_0 = 0.001$ mol·L^{-1}, $[I]_0 = 0$ mol·L^{-1} (1), 0.001 (2), 0.01 (3), 0.1 (4)

Special attention shall be drawn to the fact that for practical objectives, realization of RAFT-polymerization process without an initiator is of great concern. In all cases at high temperatures as the result of styrene RAFT-polymerization implementation in the presence of RAFT-agent without AIBN, more high-molecular (Fig. 5) and more narrow-dispersed polymer (Fig. 4, curve 1) is built-up than in the presence of AIBN (Fig. 4, curves 2–4).

8.2.4.2 THE INFLUENCE OF MONOMER INITIAL CONCENTRATION BY NUMERICAL EXPERIMENT

In other identical conditions, the decrease of monomer initial concentration reduces the number-average molecular mass of polymer. Polydispersity index doesn't practically depend on monomer initial concentration.

8.2.4.3 INFLUENCE OF RAFT-AGENT INITIAL CONCENTRATION BY NUMERICAL EXPERIMENT

In other identical conditions, increase of RAFT-agent initial concentration reduces the number-average molecular mass and polydispersity index of polymer.

8.2.4.4 THE INFLUENCE OF TEMPERATURE BY NUMERICAL EXPERIMENT

Generally, in other identical conditions, the increase of temperature leads to a decrease of number-average molecular mass of polystyrene (Fig. 6a). Thus, polydispersity index increases (Fig. 6b). If RAFT-agent initial concentration greatly exceeds AIBN initial concentration, then the temperature practically doesn't influence the molecular-mass properties of polystyrene.

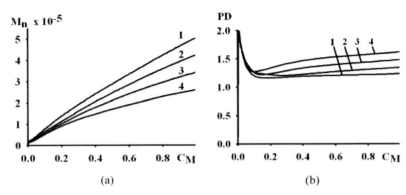

FIGURE 6 Dependence of number-average molecular mass M_n (a) and polydispersity index PD (b) on monomer conversion C_M $[I]_0 = 0.001$ mol·L^{-1}, $[M]_0 = 8.7$ mol·L^{-1}, $[RAFT(0,0)]_0 = 0.001$ mol·L^{-1}, T = 60°C (1), 90 (2), 120 (3), 150 (4).

8.3 CONCLUSION

The kinetic model developed in this research allows an adequate description of molecular-mass properties of polystyrene, obtained by controlled radical polymerization, which proceeds by reversible chain transfer mechanism and accompanied by "addition-fragmentation." This means, that the model can be used for development of technological applications of styrene RAFT-polymerization in the presence of trithiocarbonates.

Researches were supported by Russian Foundation for Basic Research (project. no. 12-03-97050-r_povolzh'e_a).

KEYWORDS

- controlled radical polymerization
- dibenzyltritiocarbonate
- mathematical modeling
- polystyrene
- reversible addition-fragmentation chain transfer

REFERENCES

1. Matyjaszewski, K.: Controlled/Living Radical Polymerization: Progress in ATRP, D.C.: American Chemical Society, Washington (2009).
2. Matyjaszewski, K.: Controlled/Living Radical Polymerization: Progress in RAFT, DT, NMP and OMRP, D.C.: American Chemical Society, Washington (2009).

3. Barner–Kowollik, C.: Handbook of RAFT Polymerization, Wiley–VCH Verlag GmbH, Weinheim (2008).
4. Chernikova, E.V., Terpugova, P.S., Garina, E.S., Golubev, V.B.: Controlled radical polymerization of styrene and n-butyl acrylate mediated by tritiocarbonates. Polymer Science, Vol. 49(A), №2, 108 (2007).
5. Stephen, Z.D. Cheng: Handbook of Thermal Analysis and Calorimetry. Volume 3 – Applications to Polymers and Plastics: New York, Elsevier, (2002).
6. LI, I., Nordon, I., Irzhak, V.I., Polymer Science, 47, 1063 (2005).
7. Zetterlund, P.B., Perrier S., Macromolecules, 44, 1340 (2011).
8. Biesenberger, J.A., Sebastian D.H.: Principles of polymerization engineering, John Wiley & Sons Inc., New York (1983).
9. Chong, Y.K., Krstina, J., Le, T.P.T., Moad, G., Postma, A., Rizzardo, E., Thang, S.H., Macromolecules, 36, 2256 (2003).
10. Goto, A., Sato, K., Tsujii, Y., Fukuda, T., Moad, G., Rizzardo, E., Thang, S.H., Macromolecules, 34, 402 (2001).
11. Kwak, Y., Goto, A., Fukuda, T., Macromolecules, 37, 1219 (2004).
12. Hui, A.W., Hamielec, A.E., J. Appl. Polym. Sci, 16, 749 (1972).
13. Li, D., Hutchinson, R.A., Macromolecular Rapid Communications, 28, 1213 (2007).

CHAPTER 9

A STUDY ON PHYSICAL PROPERTIES OF COMPOSITES BASED ON EPOXY RESIN

J. ANELI, O. MUKBANIANI, E. MARKARASHVILI, G. E. ZAIKOV, and E. KLODZINSKA

CONTENTS

9.1 Introduction 334
9.2 Basic Part 334
9.3 Results and Discussion 335
9.4 Conclusion 341
Keywords 342
References 342

9.1 INTRODUCTION

Ultimate strength, softening temperature, and water absorption of the polymer composites based on epoxy resin (type ED-20) with unmodified and/or modified by tetraethoxysilane (TEOS) mineral diatomite are described. Comparison of experimental results obtained for investigated composites shows that ones containing modified filler have the better technical parameters mentioned above than composites with unmodified filler at corresponding loading. Experimentally is shown that the composites containing binary fillers diatomite and andesite at definite ratio of them possess the optimal characteristics – so called synergistic effect. Experimental *results are explained in terms of structural peculiarities of polymer composites.*

In recent time the mineral fillers attract attention as active filling agents in polymer composites [1, 2]. Thanks to these fillers many properties of the composites are improved -increases the durability and rigidity, decrease the shrinkage during hardening process and water absorption, improves thermal stability, fire proof and dielectric properties and finally the price of composites becomes cheaper [3–5]. At the same time it must be noted that the mineral fillers at high content lead to some impair of different physical properties of composites. Therefore the attention of the scientists is attracted to substances, which would be remove mentioned leaks. It is known that silicon organic substances (both low and high molecular) reveal hydrophobic properties, high elasticity and durability in wide range of filling and temperatures [6, 7].

The purpose of presented work is the investigation of effect of modify by TEOS of the mineral –diatomite as main filler and same mineral with andesite (binary filler) on some physical properties of composites based on epoxy resin.

9.2 BASIC PART

Mineral diatomite as a filler was used. The organic solvents were purified by drying and distillation. The purity of starting compounds was controlled by an LKhM-8–MD gas liquid chromatography; phase SKTF-100 (10%, the NAW chromosorb, carrier gas He, 2m column). FTIR spectra were recorded on a Jasco FTIR-4200 device.

The silanization reaction of diatomite surface with TEOS was carried out by means of three-necked flask supplied with mechanical mixer, thermometer and dropping funnel. For obtaining of modified by 3 mass % diatomite to a solution of 50 g grind finely diatomite in 80 ml anhydrous toluene the toluene solution of 1.5 g (0.0072 mole) TEOS in 5 ml toluene was added. The reaction mixture was heated at the boiling temperature of used solvent toluene. Than the solid reaction product was filtrated, the solvents (toluene and ethyl alcohol) were eliminated and the reaction product was dried up to constant mass in vacuum. Other product modified by 5% tetraethoxysilane was produced via the same method.

A Study on Physical Properties of Composites Based on Epoxy Resin 335

Following parameters were defined for obtained composites: ultimate strength (on the stretching apparatus of type "Instron"), softening temperature (Vica method), density and water absorption (at saving of the corresponding standards).

9.3 RESULTS AND DISCUSSION

High temperature condensation reaction between diatomite and TEOS from the one side and between andesite and same modifier from the other one was carried out in toluene solution (~38%). The masses of TEOS were 3 and 5% from the mass of filler. The reaction systems were heated at the solvent boiling temperature (~110°C) during 5–6 hours by stirring. The reaction proceeds according to the following scheme:

$$\begin{array}{c}-OH\\-OH\\-OH\end{array} + Si(OC_2H_5)_4 \xrightarrow[-C_2H_5OH]{T^0C} \begin{array}{c}-O\text{-}Si(OC_2H_5)_3\\-O\text{-}Si(OC_2H_5)_2\text{-}O\sim\\-OH\end{array}$$

The direction of reaction defined by FTIR spectra analysis shown that after reaction between mineral surface hydroxyl, $-OSi(OEt)_3$ and the $-OSi(OEt)_2O-$ groups are formed on the mineral particles surface.

In the FTIR spectra of modified diatomite one can observe absorption bands characteristic for asymmetrical valence oscillation for linear °Si–O–Si° bonds at 1030 cm^{-1}. In the spectra one can see absorption bands characteristic for valence oscillation of °Si–O–C° bonds at 1150 cm^{-1} and for °C–H bonds at 2950–3000 cm^{-1}. One can see also broadened absorption bands characteristic for unassociated hydroxyl groups.

On the basis of modified diatomite and epoxy resin (of type ED-20) the polymer composites with different content of filler were obtained after careful wet mixing of components in mixer. After the blends with hardening agent (polyethylene-polyamine) were placed to the cylindrical forms (in accordance with standards ISO) for hardening, at room temperature, during 24 h. The samples hardened later were exposed to temperature treatment at 120°C during 4 h.

The concentration of powder diatomite (average diameter up to 50 micron) was changed in the range 10–60 mass %.

The curves on the Fig. 1 show that at increasing of filler (diatomite) concentration in the composites the density of materials essentially depends on both of diatomite contain and on the degree of concentration of modify agent (TEOS). Naturally the decreasing of density of composites at increasing of filler concentration is due to increasing of micro empties because of one's localized in the filler particles (Fig. 1, curve 1). The composites with modified by TEOS diatomite contain less amount of empties as they are filled with modify agent (Fig. 1, curves 2 and 3).

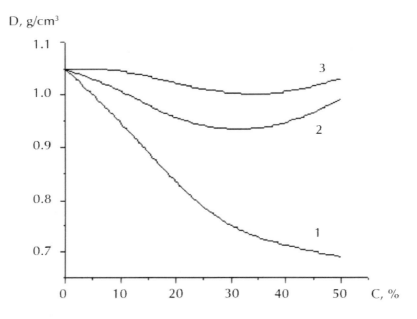

FIGURE 1 Dependence of the density of the composites based on epoxy resin on the concentration of unmodified (1), modified by 3% (2) and 5 mass % (3) tetraethoxysilane diatomite.

The dependence of ultimate strength on the content of diatomite (modified and unmodified) presented on the Fig. 2 shows that it has an extreme character. However the positions of corresponding curves maximums essentially depend on amount of modified agent TEOS. The general view of these dependences is in full conformity with well-known dependence of $\sigma - C$ [8]. The sharing of the maximum of curve for composites containing 5% of modified diatomite from the maximum for the analogous composites containing 3% modifier to some extent is due to increasing of the amount of the bonds between filler particles and macromolecules at increasing of the concentration of the filler.

A Study on Physical Properties of Composites Based on Epoxy Resin

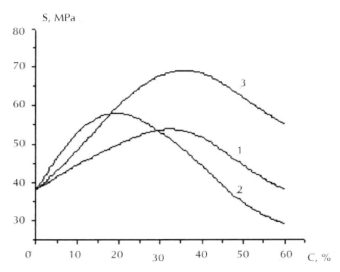

FIGURE 2 Dependence of ultimate strength of the composites based on ED-20 with unmodified (1) and modified by 3 (2), and 5 mass % (3) TEOS diatomite.

Investigation of composites softening temperature was carried out by apparatus of Vica method. Figure 3 shows the temperature dependence of the indentor deepening to the mass of the sample for composites with fixed (20 mass %) concentration of unmodified and modified by TEOS

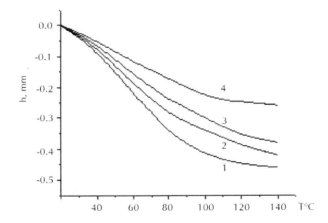

FIGURE 3 Temperature dependence of the indentor deepening in the sample for composites containing 0 (1), 20 mass % (2), 20 mass % modified by 3% TEOS (3), 20 mass % modified by 5% TEOS (4) diatomite.

Based on character of curves on the Fig. 3 it may be proposed that the composites containing diatomite modified by TEOS possesses thermo-stability higher than in case of analogous composites with unmodified filler. Probably the presence of increased interactions between macromolecules and filler particles due to modify agent leads to increasing of thermo-stability of composites with modified diatomite.

Effect of silane modifier on the investigated polymer composites reveals also in the water absorption. In accordance with Fig. 4 this parameter is increased at increasing of filler contain. However, if the composites contain the diatomite modified by TEOS this dependence becomes weak.

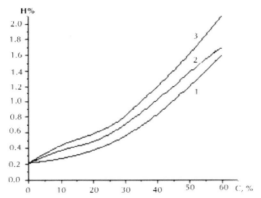

FIGURE 4 Dependence of the water-absorption on the concentration of filler in the composites based on epoxy resin containing diatomite modified by 5% (1) and 3% (2) tetraethoxysilane and unmodified (3) one.

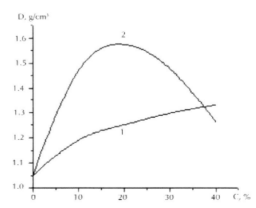

FIGURE 5 Dependence of the density on the concentration of diatomite in binary fillers with andesite. (1) unmodified and modified by 5% tetraethoxysilane (2) fillers for composites based on epoxy resin. Full concentration of binary filler in composites 50 mass %.

There were conducted the investigation of binary fillers on the properties of the composites with same polymer basis (ED-20). Two types of minerals diatomite and andesite with different ratios were used as fillers. It was interesting to establish effect both of ratio of the fillers and effect of modifier TEOS on the same properties of the polymer composites investigated above.

The curves presented on the Fig. 5 show the effect of modify agent TEOS on the dependence of the density of composites containing the binary filler diatomite and andesite on ratio of lasts when the total content of fillers is 50 mass % to which the maximal ultimate strength corresponds. The maximum of noted effect corresponds to composite, filler ratio diatomite/andesite in which is about 20/30. Probably microstructure of such composite corresponds to optimal distribution of filler particles in the polymer matrix at minimal inner energy of statistical equilibration, at which the concentration of empties is minimal because of dense disposition of the composite components. It is known that such structures consists minimal amount both of micro and macro structural defects [8].

Such approach to microstructure of composites with optimal ratio of the composite ingredients allows supposing that these composites would be possessed high mechanical properties, thermo-stability and low water-absorption. Moreover, the composites with same concentrations of the fillers modified by TEOS possess all the noted above properties better than ones for composites with unmodified by TEOS binary fillers, which may be proposed early (Figs. 6–8). Indeed the curves on the Figs. 6–8 show that the maximal ultimate strength, thermo-stability and simultaneously hydrophobicity correspond to composites with same ratio of fillers to which the maximal density corresponds.

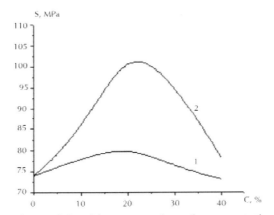

FIGURE 6 Dependence of the ultimate strength on the concentration of diatomite in binary fillers with andesite. (1) unmodified fillers and modified by 5% tetraethoxysilane (2) ones for composites based on epoxy resin Full concentration of binary filler in composites 50 mass %.

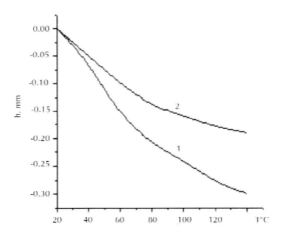

FIGURE 7 Thermo-stability of composites with binary fillers at ratio diatomite/andesite = 20/30.

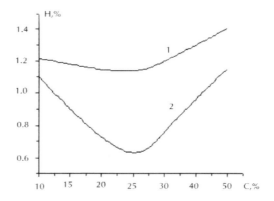

FIGURE 8 Dependence of the water-absorption of composites based on epoxy resin on the concentration of diatomite in binary fillers with andesite. (1) – unmodified and modified by 5% tetraethoxysilane (2) fillers. Total concentration of binary fillers in composites 50 mass %.

The obtained experimental results may be explained in terms of composite structure peculiarities. Silane molecules displaced on the surface of diatomite and andesite particles lead to activation of them and participate in chemical reactions between active groups of TEOS (hydroxyl) and homopolymer (epoxy group). Silane molecules create the "buffer" zones between filler and the homopolymer.

This phenomenon may be one of the reasons of increasing of strengthening of composites in comparison with composites containing unmodified fillers. The composites with modified diatomite display more high compatibility of the components than in case of same composites with unmodified filler. The modified filler has more strong contact with polymer matrix (thanks to silane modifier) than unmodified diatomite. Therefore mechanical stresses formed in composites by stretching or compressing forces absorb effectively by relatively soft silane phases, i.e. the development of micro defects in carbon chain polymer matrix of composite districts and finishes in silane part of material the rigidity of which decreases.

The structural peculiarities of composites display also in thermo-mechanical properties of the materials. It is clear that softening of composites with modified by TEOS composites begins at relatively high temperatures. This phenomenon is in good correlation with corresponding composite mechanical strength. Of course the modified filler has more strong interactions (thanks to modifier) with epoxy polymer molecules, than unmodified filler.

The amplified competition of the filler particles with macromolecules by TEOS displays well also on the characteristics of water absorption. In general loosening of micro-structure because of micro empty areas is due to the increasing of filler content. Formation of such defects in the microstructure of composite promotes the water absorption processes. Water absorption of composites with modified diatomite is lower than that for one with unmodified filler to some extent. The decreasing of water absorption of composites containing silane compound is result of hydrophobic properties of ones.

Composites with binary fillers possess so called synergistic effect- non-additive increasing of technical characteristics of composites at containing of fillers with definite ratio of them, which is due to creation of the dens distribution of ingredients in composites.

9.4 CONCLUSION

Comparison of the density, ultimate strength, softening temperature and water absorption for polymer composites based on epoxy resin and unmodified and modified by tetraethoxysilane mineral fillers diatomite and andesite leads to conclusion that modify agent stipulates the formation of heterogeneous structures with higher compatibility of ingredients and consequently to enhancing of noted above technical characteristics.

KEYWORDS

- polymer composite
- epoxy resin
- modified filler
- ultimate strength
- softening temperature
- water absorption
- synergistic effect of fillers

REFERENCES

1. Katz, H.S., Milevski, J.V. Handbook of Fillers for Plastics, RAPRA, 1987.
2. Mareri, P., Bastrole, S., Broda, N., Crespi, A. Composites Science and Technology, 1998, 58(5), pp. 747–755.
3. Tolonen, H., Sjolind, S. Mechanics of composite materials, 1996, 31(4), pp. 317–322.
4. Rothon, S.: Particulate filled polymer composites, RAPRA, NY, 2003, 205 p.
5. Lou, J., Harinath, V. Journal of Materials Processing Technology. 2004, 152(2), pp.185–193.
6. Khananashvili, L.M., Mukbaniani, O.V., Zaikov, G.E. Monograph, New Concepts in Polymer Science, "Elementorganic Monomers: Technology, Properties, Applications". Printed in Netherlands, VSP, Utrecht, (2006).
7. Aneli, J.N., Khananashvili, L.M., Zaikov, G.E. Structuring and conductivity of polymer composites. Nova Sci.Publ., New–York, 1998. 326 p.
8. Zelenev, Y.V., Bartenev, G.M. Physics of Polymers. M. Visshaya Shkola, 1978. 432 p. (in Russian).

CHAPTER 10

SYNTHESIS, STRUCTURAL PROPERTIES, DEVELOPMENT AND APPLICATIONS OF METAL-ORGANIC FRAMEWORKS IN TEXTILE

M. HASANZADEH and B. HADAVI MOGHADAM

CONTENTS

10.1	Introduction	344
10.2	Synthesis of MOFS	345
10.3	Structure and Properties of MOFS	346
10.4	Application of MOFS in Textiles	347
10.5	Conclusion	354
	Keywords	355
	References	355

10.1 INTRODUCTION

Metal-organic frameworks (MOFs) have received increasing attention in recent years as a new class of nanoporous materials. These crystalline compounds basically consist of metal ions linked by organic bridging ligands. They have found wide range of applications including gas storage, gas separation, catalysis, luminescence, and drug delivery due to their large pore sizes, high porosity, high surface areas, and wide range of pore sizes and topologies. Textile application is one of the areas MOFs started to appear recently. Interesting chemical and physical properties of MOFs make them promising candidates for future developments in textile applications. This short review intends to introduce recent progress in application of MOFs in the field of textile engineering and some of the key advances that have been made in it.

Recently the application of nanostructured materials has garnered attention, due to their interesting chemical and physical properties. Application of nanostructured materials on the solid substrate such as fibers brings new properties to the final textile product [1]. Metal-organic frameworks (MOFs) are one of the most recognized nanoporous materials, which can be widely used for modification of fibers. These relatively crystalline materials consist of metal ions or clusters (named secondary building units, SBUs) interconnected by organic molecules called linkers, which can possess one, two or three dimensional structures [2–10]. They have received a great deal of attention, and the increase in the number of publications related to MOFs in the past decade is remarkable (Fig. 1).

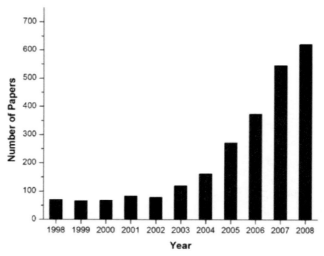

FIGURE 1 Number of publications on MOFs over the past decade, showing the increasing research interest in this topic.

These materials possess a wide array of potential applications in many scientific and industrial fields, including gas storage [11,12], molecular separation [13], catalysis [14], drug delivery [15], sensing [16], and others. This is due to the unique combination of high porosity, very large surface areas, accessible pore volume, wide range of pore sizes and topologies, chemical stability, and infinite number of possible structures [17,18].

Although other well-known solid materials such as zeolites and active carbon also show large surface area and nanoporosity, MOFs have some new and distinct advantages. The most basic difference of MOFs and their inorganic counterparts (e.g., zeolites) is the chemical composition and absence of an inaccessible volume (called dead volume) in MOFs [10]. This feature offers the highest value of surface area and porosities in MOFs materials [19]. Another difference between MOFs and other well-known nanoporous materials such as zeolites and carbon nanotubes is the ability to tune the structure and functionality of MOFs directly during synthesis [17].

The first report of MOFs dates back to 1990, when Robson introduced a design concept to the construction of 3D MOFs using appropriate molecular building blocks and metal ions. Following the seminal work, several experiments were developed in this field such as work from Yaghi and O'Keeffe [20].

In this review, synthesis and structural properties of MOFs are summarized and some of the key advances that have been made in the application of these nanoporous materials in textile fibers are highlighted.

10.2 SYNTHESIS OF MOFS

MOFs are typically synthesized under mild temperature (up to 200°C) by combination of organic linkers and metal ions (Fig. 2) in solvothermal reaction [2, 21].

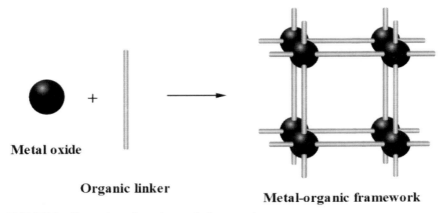

FIGURE 2 Formation of metal organic frameworks.

Recent studies have shown that the character of the MOF depends on many parameters including characteristics of the ligand (bond angles, ligand length, bulkiness, chirality, etc.), solubility of the reactants in the solvent, concentration of organic link and metal salt, solvent polarity, the pH of solution, ionic strength of the medium, temperature and pressure [2, 21].

In addition to this synthesis method, several different methodologies are described in the literature such as ball-milling technique, microwave irradiation, and ultrasonic approach [22].

Post-synthetic modification (PSM) of MOFs opens up further chemical reactions to decorate the frameworks with molecules or functional groups that might not be achieved by conventional synthesis. In situations that presence of a certain functional group on a ligand prevents the formation of the targeted MOF, it is necessary to first form a MOF with the desired topology, and then add the functional group to the framework [2].

10.3 STRUCTURE AND PROPERTIES OF MOFS

When considering the structure of MOFs, it is useful to recognize the secondary building units (SBUs), for understanding and predicting topologies of structures [3]. Figure 3 shows the examples of some SBUs that are commonly occurring in metal carboxylate MOFs. Figure 3(a–c) illustrates inorganic SBUs include the square paddlewheel, the octahedral basic zinc acetate cluster, and the trigonal prismatic oxo-centered trimer, respectively. These SBUs are usually reticulated into MOFs by linking the carboxylate carbons with organic units [3]. Examples of organic SBUs are also shown in Fig. 3(d–f).

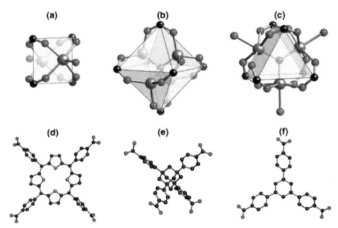

FIGURE 3 Structural representations of some SBUs, including (a–c) inorganic, and (b–f) organic SBUs. (Metals are shown as blue spheres, carbon as black spheres, oxygen as red spheres, nitrogen as green spheres).

It should be noted that the geometry of the SBU is dependent on not only the structure of the ligand and type of metal utilized, but also the metal to ligand ratio, the solvent, and the source of anions to balance the charge of the metal ion [2].

A large number of MOFs have been synthesized and reported by researchers to date. Isoreticular metal-organic frameworks (IRMOFs) denoted as IRMOF-n (n = 1 through 7, 8, 10, 12, 14, and 16) are one of the most widely studied MOFs in the literature. These compounds possess cubic framework structures in which each member shares the same cubic topology [3, 21]. Figure 4 shows the structure of IRMOF-1 (MOF-5) as simplest member of IRMOF series.

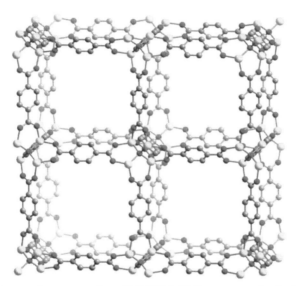

FIGURE 4 Structural representation of IRMOF-1 (Yellow, grey, and red spheres represent Zn, C, and O atoms, respectively).

10.4 APPLICATION OF MOFS IN TEXTILES

10.4.1 INTRODUCTION

There are many methods of surface modification, among which nanostructure based modifications have created a new approach for many applications in recent years. Although MOFs are one of the most promising nanostructured materials for modification of textile fibers, only a few examples have been reported to data. In this section, the first part focuses on application of MOFs in nanofibers and the second part is concerned with modifications of ordinary textile fiber with these nanoporous materials.

10.4.2 NANOFIBERS

Nanofibrous materials can be made by using the electrospinning process. Electrospinning process involves three main components including syringe filled with a polymer solution, a high voltage supplier to provide the required electric force for stretching the liquid jet, and a grounded collection plate to hold the nanofiber mat. The charged polymer solution forms a liquid jet that is drawn towards a grounded collection plate. During the jet movement to the collector, the solvent evaporates and dry fibers deposited as randomly oriented structure on the surface of a collector [23–28]. The schematic illustration of conventional electrospinning setup is shown in Fig. 5.

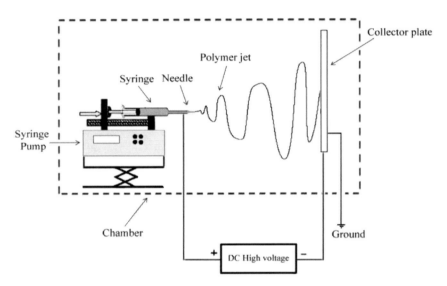

FIGURE 5 Schematic illustration of electrospinning set up.

At the present time, synthesis and fabrication of functional nanofibers represent one of the most interesting fields of nanoresearch. Combining the advanced structural features of metal-organic frameworks with the fabrication technique may generate new functionalized nanofibers for more multiple purposes.

While there has been great interest in the preparation of nanofibers, the studies on metal-organic polymers are rare. In the most recent investigation in this field, the growth of MOF (MIL-47) on electrospun polyacrylonitrile (PAN) mat was studied using in situ microwave irradiation [18]. MIL-47 consists of vanadium cations associated to six oxygen atoms, forming chains connected by terephthalate linkers (Fig. 6).

It should be mentioned that the conversion of nitrile to carboxylic acid groups is necessary for the MOF growth on the PAN nanofibers surface.

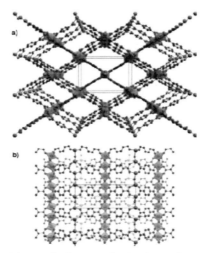

FIGURE 6 MIL-47 metal-organic framework structure: view along the b axis (a) and along the c axis (b).

The crystal morphology of MIL-47 grown on the electrospun fibers illustrated that after only 5 s, the polymer surface was partially covered with small agglomerates of MOF particles. With increasing irradiation time, the agglomerates grew as elongated anisotropic structures (Fig. 7) [18].

FIGURE 7 SEM micrograph of MIL-47 coated PAN substrate prepared from electrospun nanofibers as a function of irradiation time: (a) 5 s, (b) 30 s, (c) 3 min, and (d) 6 min.

It is known that the synthesis of desirable metal-organic polymers is one of the most important factors for the success of the fabrication of metal-organic nanofibers [29]. Among several novel microporous metal-organic polymers, only a few of them have been fabricated into metal-organic fibers.

For example, new acentric metal-organic framework was synthesized and fabricated into nanofibers using electrospinning process [29]. The two dimensional network structure of synthesized MOF is shown in Fig. 8. For this purpose, MOF was dissolved in water or DMF and saturated MOF solution was used for electrospinning. They studied the diameter and morphology of the nanofibers using an optical microscope and a scanning electron microscope (Fig. 9). This fiber display diameters range from 60 nm to 4 µm.

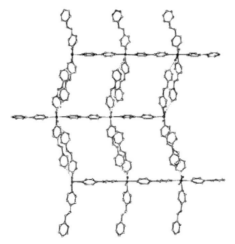

FIGURE 8 Representation of polymer chains and network structure of MOF.

FIGURE 9 SEM micrograph of electrospun nanofiber.

Synthesis, Structural Properties, Development 351

In 2011, Kaskel et al. [30], reported the use of electrospinning process for the immobilization of MOF particles in fibers. They used HKUST-1 and MIL-100(Fe) as MOF particles, which are stable during the electrospinning process from a suspension. Electrospun polymer fibers containing up to 80 wt% MOF particles were achieved and exhibit a total accessible inner surface area. It was found that HKUST-1/PAN gives a spider web-like network of the fibers with MOF particles like trapped flies in it, while HKUST-1/PS results in a pearl necklace-like alignment of the crystallites on the fibers with relatively low loadings.

10.4.3 ORDINARY TEXTILE FIBERS

Some examples of modification of fibers with metal-organic frameworks have verified successful. For instance, in the study on the growth of $Cu_3(BTC)_2$ (also known as HKUST-1, BTC=1,3,5-benzenetricarboxylate) MOF nanostructure on silk fiber under ultrasound irradiation, it was demonstrated that the silk fibers containing $Cu_3(BTC)_2$ MOF exhibited high antibacterial activity against the gram-negative bacterial strain *E. coli* and the gram-positive strain *S. aureus* [1]. The structure and SEM micrograph of $Cu_3(BTC)_2$ MOF is shown in Fig. 10.

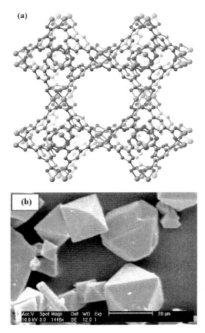

FIGURE 10 (a) The unit cell structure and (b) SEM micrograph of the Cu3(BTC)2 metal-organic framework. (Green, grey, and red spheres represent Cu, C, and O atoms, respectively).

$Cu_3(BTC)_2$ MOF has a large pore volume, between 62% and 72% of the total volume, and a cubic structure consists of three mutually perpendicular channels [32].

The formation mechanism of $Cu_3(BTC)_2$ nanoparticles upon silk fiber is illustrated in Fig. 11. It is found that formation of $Cu_3(BTC)_2$ MOF on silk fiber surface was increased in presence of ultrasound irradiation. In addition, increasing the concentration cause an increase in antimicrobial activity [1]. Figure 12 shows the SEM micrograph of $Cu_3(BTC)_2$ MOF on silk surface.

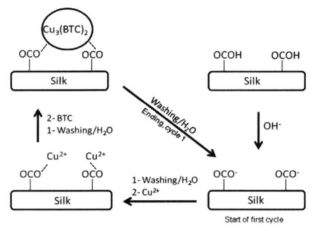

FIGURE 11 Schematic representation of the formation mechanism of $Cu_3(BTC)_2$ nanoparticles upon silk fiber.

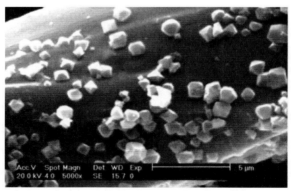

FIGURE 12 SEM micrograph of $Cu_3(BTC)_2$ crystals on silk fibers.

The FT–IR spectra of the pure silk yarn and silk yarn containing MOF (CuBTC–Silk) are shown in Fig. 13. Owing to the reduction of the C=O bond, which is caused by the coordination of oxygen to the Cu^{2+} metal center (Fig. 11), the stretching frequency of the C=O bond was shifted to lower wavenumbers (1654 cm^{-1}) in comparison with the free silk (1664 cm^{-1}) after chelation [1].

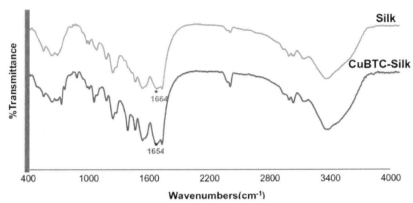

FIGURE 13 FT–IR spectra of the pure silk yarn and silk yarn containing $Cu_3(BTC)_2$.

In another study, $Cu_3(BTC)_2$ was synthesized in the presence of pulp fibers of different qualities [33]. The following pulp samples were used: a bleached and an unbleached kraft pulp, and chemithermomechanical pulp (CTMP).

All three samples differed in their residual lignin content. Indeed, owing to the different chemical composition of samples, different results regarding the degree of coverage were expected. The content of $Cu_3(BTC)_2$ in pulp samples, k-number, and single point BET surface area are shown in Table 1. k-number of pulp samples, which is indicates the lignin content indirectly, was determined by consumption of a Sulfuric permanganate solution of the selected pulp sample [33].

TABLE 1 Some characteristics of the pulp samples.

Pulp sample	MOF content[a] (wt.%)	k-number[b]	Surface area[c] (m^2 g^{-1})
CTMP	19.95	114.5	314
Unbleached kraft pulp	10.69	27.6	165
Bleached kraft pulp	0	0.3	10

[a]Determined by thermogravimetric analysis.
[b]Determined according to ISO 302.
[c]Single point BET surface area calculated at p/p0=0.3 bar.

It is found that CTMP fibers showed the highest lignin residue and largest BET surface area. As shown in the SEM micrograph (Fig. 14), the crystals are regularly distributed on the fiber surface. The unbleached kraft pulp sample provides a slightly lower content of MOF crystals and BET surface area with 165 m^2 g^{-1}. Moreover, no crystals adhered to the bleached kraft pulp, which was almost free of any lignin.

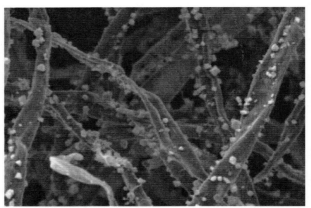

FIGURE 14 SEM micrograph of Cu$_3$(BTC)$_2$ crystals on the CTMP fibers.

10.5 CONCLUSION

New review on feasibility and application of several kinds of metal-organic frameworks on different substrate including nanofiber and ordinary fiber was investigated. Based on the researcher's results, the following conclusions can be drawn:

1. Metal-organic frameworks (MOF), as new class of nanoporous materials, can be used for modification of textile fibers.
2. These nanostructured materials have many exciting characteristics such as large pore sizes, high porosity, high surface areas, and wide range of pore sizes and topologies.
3. Although tremendous progress has been made in the potential applications of MOFs during past decade, only a few investigations have reported in textile engineering fields.
4. Morphological properties of the MOF/fiber composites were defined; the most advantageous, particle size distribution was shown.
5. It is concluded that the MOFs/fiber composite would be good candidates for many technological applications, such as gas separation, hydrogen storage, sensor, and others.

KEYWORDS

- development
- metal-organic frameworks
- structural properties
- synthesis
- textiles

REFERENCES

1. Abbasi, A.R., Akhbari, K., Morsali, A.: Dense coating of surface mounted CuBTC metal-organic framework nanostructures on silk fibers, prepared by layer-by-layer method under ultrasound irradiation with antibacterial activity. *Ultrasonics Sonochemistry*, 19, 846–852 (2012).
2. Kuppler, R.J., Timmons, D.J., Fang, Q.-R., Li, J.-R., Makal, T.A., Young, M.D., Yuan, D., Zhao, D., Zhuang, W., Zhou, H.-C.: Potential applications of metal-organic frameworks. *Coordination Chemistry Reviews*, 253, 3042–3066 (2009).
3. Rowsell, J.L.C., Yaghi, O.M.: Metal-organic frameworks: A new class of porous materials. *Microporous and Mesoporous Materials*, 73, 3–14 (2004).
4. An, J., Farha, O.K., Hupp, J.T., Pohl, E., Yeh, J.I., Rosi, N.L.: Metal-adeninate vertices for the construction of an exceptionally porous metal-organic framework. *Nature communications*, DOI: 10.1038/ncomms1618, (2012).
5. Morris, W., Taylor, R.E., Dybowski, C., Yaghi, O.M., Garcia–Garibay, M.A.: Framework mobility in the metal-organic framework crystal IRMOF-3: Evidence for aromatic ring and amine rotation. *Journal of Molecular Structure*, 1004, 94–101 (2011).
6. Kepert, C.J.: Metal-organic framework materials. in 'Porous Materials' (eds.: by Bruce, D.W., O'Hare, D. and Walton, R.I.) John Wiley & Sons, Chichester (2011).
7. Rowsell, J.L.C., Yaghi, O.M.: Effects of functionalization, catenation, and variation of the metal oxide and organic linking units on the low-pressure hydrogen adsorption properties of metal-organic frameworks. *Journal of the American Chemical Society*, 128, 1304–1315 (2006).
8. Rowsell, J.L.C., Yaghi, O.M.: Strategies for hydrogen storage in metal-organic frameworks. *Angewandte Chemie International Edition*, 44, 4670–4679 (2005).
9. Farha, O.K., Mulfort, K.L., Thorsness, A.M., Hupp, J.T.: Separating solids: purification of metal-organic framework materials. *Journal of the American Chemical Society*, 130, 8598–8599 (2008).
10. Khoshaman, A.H.: Application of electrospun thin films for supra-molecule based gas sensing. M.Sc. thesis, Simon Fraser University (2011).
11. Murray, L.J., Dinca, M., Long, J.R.: Hydrogen storage in metal-organic frameworks. *Chemical Society Reviews*, 38, 1294–1314 (2009).
12. Collins, D.J., Zhou, H.-C.: Hydrogen storage in metal-organic frameworks. *Journal of Materials Chemistry*, 17, 3154–3160 (2007).

13. Chen, B., Liang, C., Yang, J., Contreras, D.S., Clancy, Y.L., Lobkovsky, E.B., Yaghi, O.M., Dai, S.: A microporous metal-organic framework for gas-chromatographic separation of alkanes. *Angewandte Chemie International Edition*, 45, 1390–1393 (2006).
14. Lee, J.Y., Farha, O.K., Roberts, J., Scheidt, K.A., Nguyen, S.T., Hupp, J.T.: Metal-organic framework materials as catalysts. *Chemical Society Reviews*, 38, 1450–1459 (2009).
15. Huxford, R.C., Rocca, J.D., Lin, W.: Metal-organic frameworks as potential drug carriers. *Current Opinion in Chemical Biology*, 14, 262–268 (2010).
16. Suh, M.P., Cheon, Y.E., Lee, E.Y.: Syntheses and functions of porous metallosupramolecular networks. *Coordination Chemistry Reviews*, 252, 1007–1026 (2008).
17. Keskin, S., Kızılel, S.: Biomedical applications of metal organic frameworks. *Industrial and Engineering Chemistry Research*. 50, 1799–1812 (2011).
18. Centrone, A., Yang, Y., Speakman, S., Bromberg, L., Rutledge, G.C., Hatton, T.A.: Growth of metal-organic frameworks on polymer surfaces. *Journal of the American Chemical Society*, 132, 15687–15691 (2010).
19. Wong–Foy, A.G., Matzger, A.J., Yaghi, O.M.: Exceptional H_2 saturation uptake in microporous metal-organic frameworks. *Journal of the American Chemical Society*, 128, 3494–3495 (2006).
20. Farrusseng, D.: Metal-organic frameworks: Applications from Catalysis to Gas Storage. Wiley–VCH, Weinheim (2011).
21. Rosi, N.L., Eddaoudi, M., Kim, J., O'Keeffe, M., Yaghi, O.M.: Advances in the chemistry of metal-organic frameworks. *CrystEngComm*, 4, 401–404 (2002).
22. Zou, R., Abdel–Fattah, A.I., Xu, H., Zhao, Y., Hickmott, D.D.: Storage and separation applications of nanoporous metal-organic frameworks, *CrystEngComm*, 12, 1337–1353 (2010).
23. Reneker, D.H., Chun, I.: Nanometer diameter fibers of polymer, produced by electrospinning, *Nanotechnology*, 7, 216–223 (1996).
24. Shin, Y.M., Hohman, M.M., Brenner, M.P., Rutledge, G.C.: Experimental characterization of electrospinning: The electrically forced jet and instabilities. *Polymer*, 42, 9955–9967 (2001).
25. Reneker, D.H., Yarin, A.L., Fong, H., Koombhongse, S.: Bending instability of electrically charged liquid jets of polymer solutions in electrospinning, *Journal of Applied Physics*, 87, 4531–4547 (2000).
26. Zhang, S., Shim, W.S., Kim, J.: Design of ultra-fine nonwovens via electrospinning of Nylon 6: Spinning parameters and filtration efficiency, *Materials and Design*, 30, 3659–3666 (2009).
27. Yördem, O.S., Papila, M., Menceloğlu, Y.Z.: Effects of electrospinning parameters on polyacrylonitrile nanofiber diameter: An investigation by response surface methodology. *Materials and Design*, 29, 34–44 (2008).
28. Chronakis, I.S.: Novel nanocomposites and nanoceramics based on polymer nanofibers using electrospinning process–A review. *Journal of Materials Processing Technology*, 167, 283–293 (2005).
29. Lu, J.Y., Runnels, K.A., Norman, C.: A new metal-organic polymer with large grid acentric structure created by unbalanced inclusion species and its electrospun nanofibers. *Inorganic Chemistry*, 40, 4516–4517 (2001).

30. Rose, M., Böhringer, B., Jolly, M., Fischer, R., Kaskel, S.: MOF processing by electrospinning for functional textiles. *Advanced Engineering Materials*, 13, 356–360 (2011).
31. Basu, S., Maes, M., Cano–Odena, A., Alaerts, L., De Vos, D.E., Vankelecom, I.F.J.: Solvent resistant nanofiltration (SRNF) membranes based on metal-organic frameworks. *Journal of Membrane Science*, 344, 190–198 (2009).
32. Hopkins, J.B.: Infrared spectroscopy of H_2 trapped in metal organic frameworks. B.A. Thesis, Oberlin College Honors (2009).
33. Küsgens, P., Siegle, S., Kaskel, S.: Crystal growth of the metal-organic framework $Cu_3(BTC)_2$ on the surface of pulp fibers. *Advanced Engineering Materials*, 11, 93–95 (2009).

CHAPTER 11

THERMAL BEHAVIOR AND IONIC CONDUCTIVITY OF THE PEO/PAC AND PEO/PAC BLENDS

AMIRAH HASHIFUDIN, SIM LAI HAR, CHAN CHIN HAN,
HANS WERNER KAMMER, and SITI NOR HAFIZA MOHD YUSOFF

CONTENTS

11.1 Introduction 360
11.2 Experimental 361
11.3 Results and Discussion 362
11.4 Conclusions 367
Acknowledgment 368
Keywords 368
References 368

11.1 INTRODUCTION

Solution casting technique is employed to prepare the poly(ethylene oxide) (PEO)/ polyacrylate (PAc) blends. Thermal behavior and ionic conductivity of the PEO/PAc and PEO/PAc blends added with $LiClO_4$ were investigated using differential scanning calorimetry (DSC) and impedance spectroscopy (IS), respectively. Observations of a single composition-dependent glass transition temperature (T_g) which agrees closely with that calculated using the Fox equation, coupled with successive suppression of the melting temperature (T_m) and crystallinity of PEO with ascending PAc content, affirm the miscibility of the two constituents in the blend. The conductivity of salt-free PEO is enhanced with the addition of ≤ 25 wt% of PAc due to the reduced crystallinity of PEO in the blend. The T_g values of the blend at all compositions under study increase with the addition of $LiClO_4$. Ionic conductivity of the salt-added blend increases with increasing salt concentration. The amorphous phase of PEO forms the percolating pathway in the homogeneous $PEO/PAc/LiClO_4$ blends as blends with PEO content ≥ 25 wt% (PEO/PAc 75/25 blend) records slightly higher σ values at $LiClO_4$ concentration $Y > 0.02$. Enhancement in ionic conductivity in the blend is probably the result of increase charge carrier density and ionic dynamic of the PEO macromolecular chain.

The rapid development in advanced electrochemical and micro-ionic devices has attracted extensive research on polymer electrolytes, with the hope of applying these electrolytes in new generation high performance rechargeable batteries [1–5]. Over the last three decades, poly(ethylene oxide) (PEO) remained to be the focus in most of the researches on solid state batteries because of its strong solvating capability of wide variety of inorganic salt and its low glass transition temperature (T_g) [6–9].

It is well documented that the amorphous phase of PEO forms the percolating pathway for fast ion transport in PEO-salt system [10, 11]. Glass transition temperature results obtained in the previous studies [12–14] verified that no isotropic dispersion of Li^+ ion in different phases of a blend is demonstrated for both the immiscible blends of PEO/epoxidized natural rubber (ENR) and PEO/polyacrylate (PAc) with the addition of $LiClO_4$, instead, the Li^+ ion has a higher solubility in the amorphous phase of PEO as compared to ENR or PAc, respectively. Besides, $LiClO_4$ is found to be more soluble in the amorphous PAc than in ENR when equal amount of the salt is added to the immiscible PEO/ENR and PEO/PAc systems of the same blend composition [12–14]. Furthermore, with the addition of salt, the T_gs of PAc in the $PEO/PAc/LiClO_4$ blend are raised to the range of 26–42°C at which conductivity of the blend is measured. Therefore, higher charge density in the PEO amorphous phase of the heterogeneous $PEO/ENR/LiClO_4$ blend accounts for the higher ionic conductivity of the blend as compared to that in the $PEO/LiClO_4$ system [12]. On the contrary, due to a reasonable amount of the salt being locked in the glassy PAc, the reduction in charge density in the PEO amorphous phase for the immiscible PEO/PAc/LiClO$_4$ blend causes the conductivity of the blend to be lower than that of the PEO/LiClO$_4$ system [15].

In the blend preparation of PEO/PAc and PEO/PAc/LiClO$_4$ electrolyte films described in the previous study [15], solution cast free standing film was dried in a vacuum oven for 48 h at 50°C. Calorimetric analysis using differential scanning calorimetry (DSC) shows that both the salt-free and the salt-added blend systems are immiscible marked by the presence of two T_gs and a relatively constant PEO crystallinity (X^*) with increasing PAc content. However, miscible PEO/PAc and PEO/PAc/LiClO$_4$ blends are obtained in the present work when the solution cast free standing film was heated at 80°C (above the melting point of PEO) for 2 h under nitrogen atmosphere before vacuum dried for another 24 h at 50°C.

A brief description of the thermal procedure used in the preparation of the homogeneous PEO/PAc blend with and without addition of the inorganic salt, LiClO$_4$ is presented. Miscibility of the two polymer components of the blend was investigated by thermal analysis using DSC. The glass transition temperature (T_g) as well as the apparent melting temperature (T_m) and the crystallinity of the as-prepared samples of the blend were studied as functions of compositions of the blend. The effect of phase behavior on the conductivity properties of selected compositions of the blend incorporated with different salt contents is discussed here.

11.2 EXPERIMENTAL

11.2.1 MATERIALS

PEO with viscosity-average molecular weight (M_v) = 3×10^5 g mol^{-1}, was purchased from Aldrich Chemical Company and used after purification. PAc, a random copolymer with weight-average molecular weight (M_w) = 1.7x10^5 g mol^{-1} estimated by gel permeation chromatography, was supplied by the Chemistry Department, Faculty of Science, University of Malaya [16, 17]. Anhydrous LiClO$_4$ with purity $\geq$ 99% (Acrōs Organics) was vacuum dried for 24 h at 120°C prior to application. Methanol (Fisher Scientific, Leicestershire, UK) dehydrated by molecular sieves with pore diameter of 3Å (Merck, Darmstadt, Germany) was the common solvent used for both the salt-free and the salt-added PEO/PAc blends.

11.2.2 PREPARATION OF BLENDS

Free standing films of PEO/PAc and PEO/PAc/LiClO$_4$ blends were prepared by solution casting technique. Different compositions of the 4% w/w stock solutions of PEO and PAc were mixed while for the salt-added blends, different amount in mass of LiClO$_4$ (Y_s) were added to the polymer solutions. The mixtures were stirred at 50°C for 24 h and cast into Teflon dish, left to dry overnight in a fume hood. After drying at 50°C for 24 h in an oven, the thin films were heated to 80°C under nitrogen atmosphere for 2 h. Under this thermal treatment, the components of the PEO/PAc and PEO/PAc/LiClO$_4$ blends have more time to mix, hence, enhances the interactions between the blending components in the salt-free blend and the salt in the salt-added blend resulting in the formation of miscible PEO/PAc and PEO/PAc/LiClO$_4$ blends. After thermal treatment, the blend films were vacuum dried at 50°C for 24 h. The concentration of LiClO$_4$ in the blend is defined as below:

$$Y_s = \frac{\text{mass of salt}}{\text{mass of polymer}}$$

11.2.3 DIFFERENTIAL SCANNING CALORIMETRY

The values of T_g, T_m and enthalpies of fusion (ΔH_m) of as-prepared samples of the blends were performed on TA DSC Q200, calibrated with indium standard under nitrogen atmosphere. The sample was quenched from 30°C to -90°C, annealed at this temperature for 5 min before heating up to 80°C at a rate of 10°C min^{-1}.

11.2.4 IMPEDANCE SPECTROSCOPY

Ionic conductivity (σ) at 30°C of the as-prepared samples of PEO/PAc/LiClO$_4$ blends was determined from ac-impedance measurements using a Hioki 3532–50 Hi–Tester over the frequency range between 50 Hz and 1 MHz. Thin films of the polymer electrolytes were sandwiched between two stainless steel block electrodes with a surface area of 3.142 cm^2. The bulk resistance (R_b) of the electrolyte were extracted from the impedance spectrum of the sample at the point of intersection between the semicircle and the real impedance axis (Z_r). Ionic conductivity (σ) was calculated from the bulk resistance (R_b) by adopting the equation $\sigma = L/(AR_b)$, where L and A represent, respectively, the thickness and the active area of the electrode.

11.3 RESULTS AND DISCUSSION

11.3.1 THERMAL ANALYSIS

The T_g for neat PAc and PEO extracted from the heating cycle of DSC traces are 16 and −54°C, respectively. A single, composition dependent T_g is observed with increasing PAc added to PEO in the salt-free PEO/PAc blends as shown in Fig. 1, indicating that the salt-free blend is miscible. The monotone dependence of T_g on composition of the binary PEO/PAc blend as depicted in Fig. 1 concurred with values calculated from the Fox equation as given in Eq. (1) [18, 19]. The Fox equation is defined as

$$\frac{1}{T_g} = \frac{W_1}{T_{g1}} + \frac{W_2}{T_{g2}} \qquad (1)$$

where W_1, T_{g1} and W_2, T_{g2} refer to the weight fractions and T_gs of PEO and PAc, respectively.

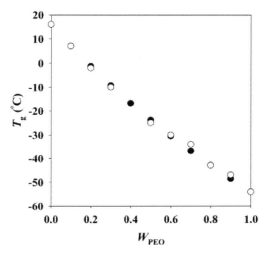

FIGURE 1 Variations of glass transition temperature of PEO/PAc blend as functions of weight fractions of PEO (W_{PEO}). Symbols (●) and (○) denote experimental value and values calculated from Fox equation, respectively.

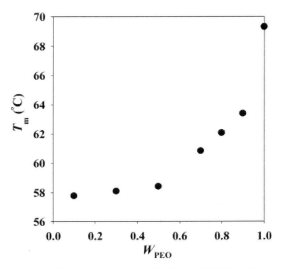

FIGURE 2 Apparent melting temperature of PEO in PEO/PAc blends as functions of weight fraction of PEO.

Figure 2 presents the variations of the apparent melting temperature (T_m) of PEO in the as-prepared samples of the PEO/PAc blend as a function of W_{PEO}. The T_m values, extracted from the first heating cycle of the DSC thermogram, descend rapidly with the addition of 10–60 wt% of PAc, then decrease gradually when more than 60 wt% of PAc is added. From the thermodynamic viewpoint, the miscibility of the PEO/PAc is affirmed by the suppression of the T_m of PEO in the presence of PAc.

PEO crystallinity (X^*) in PEO/PAc blend is calculated from the enthalpy of fusion (ΔH_m) after Eq. (2):

$$X^* = \left\{ \frac{\Delta H_m}{\Delta H_{ref}^o \times (W_{PEO})} \right\} \times 100\% \qquad (2)$$

where $\Delta H_{ref}^o = 188.3$ J g^{-1} is the enthalpy of fusion of 100% crystalline PEO [20]. Figure 3 demonstrates a successive decrease in the crystallinity of PEO with ascending PAc content in the blend showing the miscibility of the PEO/PAc blend. The presence of PAc in the homogeneous PEO/PAc blend greatly hinders the migration of the crystallizable material of PEO to the crystal growth front, thus, slows down the PEO crystallization rate, leading to a reduction in PEO crystallinity.

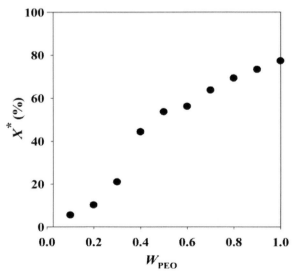

FIGURE 3 Crystallinity (X^*) of PEO in PEO/PAc blend versus W_{PEO}.

The addition of low concentrations of LiClO$_4$ ($Y_s \leq 0.02$) causes a sharp and linear increase in the T_gs of both PEO and PAc as shown in Fig. 4. However, the T_g of PAc remains relatively constant while that of PEO ascends gradually with salt concentration $0.02 < Y_s \leq 0.12$. Similar linear relationship between the T_gs of both PEO and PAc and the initial concentrations of LiClO$_4$ from $Y_s = 0$ to ~ 0.15 were also observed in the previous study by Sim et al. [15]. The difference is that the T_gs of PEO and PAc both level-off at salt concentration $Y_s \geq 0.12$ and 0.15, respectively.

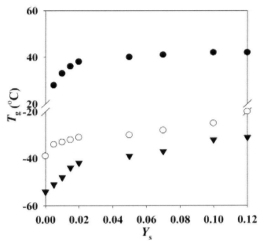

FIGURE 4 Values of T_g of (●) neat PAc, (○) PEO/PAc 75/25 blend and (▼) PEO as functions of concentrations of LiClO$_4$.

It is interesting to note that the T_gs of PEO and PAc recorded at $Y_s = 0.12$ are −31 and 42°C as compared to −37 and 25°C reported in the previous study [13]. The higher T_gs of PEO and PAc obtained at the initial salt concentration in the present work are the result of enhanced solvation of LiClO$_4$ by the polymers as the polymer films were subjected to prolong interaction at temperature above the melting point of PEO (at 80°C for 2 h under N$_2$ atmosphere) as described in the experimental section. The plateau observed for PAc at $Y_s \geq 0.05$ implies that no further Li$^+$–PAc coordination is formed whereas coordination of Li$^+$ ion to the ether oxygen of PEO continues, leading to a gradual increase in its T_g. This shows that the lithium salt has a higher preference in the amorphous phase of PEO than that of PAc. Under this thermal treatment, the T_g values of the homogenous PEO/PAc 75/25 blend are observed to increase linearly with the addition of LiClO$_4$ from $Y_s = 0$ to 0.12.

11.3.2 CONDUCTIVITY ANALYSIS

Ionic conductivity (σ) of the electrolyte films of the salt-free PEO/PAc blends and the blends added with 12 wt% ($Y_s = 0.12$) LiClO$_4$ are presented in Fig. 5. In the salt-free electrolyte system, neat PEO records a conductivity of 1.9×10^{-9} S cm^{-1} as compared to PAc with a lower conductivity of 3.5×10^{-11} S cm^{-1} in close proximity to 1.1×10^{-11} S cm^{-1} reported in the previous study [14]. The conductivity of the PEO/PAc blend as shown in Fig. 5 increases by 2–4 orders of magnitude with salt concentration, $Y_s = 0.12$ especially for blends with PEO content ≥ 50 wt% suggesting that the amorphous phase of PEO in the blend forms the percolating pathway for ion transport.

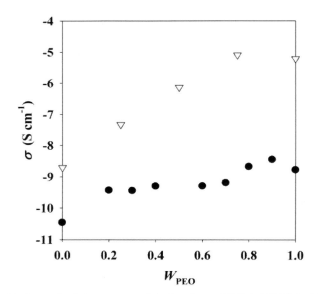

FIGURE 5 Ionic conductivity versus weight fractions of PEO in PEO/PAc blend with (●) $Y_s = 0$ and (∇) $Y_s = 0.12$.

Both the salt-free and the salt-added PEO/PAc blends with PEO content 75 wt% attained conductivity higher than that of neat PEO. On the contrary, none of the compositions in the previous heterogeneous PEO/PAc system reported in [15] recorded higher σ values than neat PEO. The improved conductivity obtained as compared to reference [15] is attributed to the extended time in the thermal treatment during sample preparation, which allowed the polymer and the salt to interact and achieve equilibrium. This thermal treatment not only enables the formation of the miscible PEO/PAc blends but more so enhances conductivity of the neat polymers and their blends.

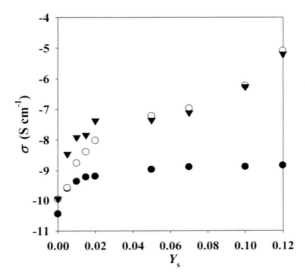

FIGURE 6 Semilogarithm plots of σ versus Y_s, for (●) neat PAc, (o) PEO/PAc 75/25 blend and (▼) neat PEO.

Figure 6 shows the dependence of conductivity on the concentration of LiClO$_4$ incorporated into 1 gram dry weight of the neat polymers and the PEO/PAc 75/25 blend. With ascending salt content from $Y_s = 0$ to 0.02, there is a marked increase in σ followed by a level-off for PAc but a gradual increase in σ for PEO at $Y_s > 0.02$. This variation trend in σ observed for both the neat polymers correlate closely to that of T_g (c.f. Fig. 4), which in turn reflects the mode of complexation of the Li$^+$ ion to the monomer units of the macromolecular chains. Increase in the charge carrier density leads to higher ion transport through the percolating pathway of PEO whereas ion-transport through the percolation network in PAc appears to reach a maximum at $Y_s > 0.02$. In addition to the neat polymer, the homogeneous PEO/PAc 75/25 blend records slightly enhanced σ values at $Y_s > 0.02$ implying a synergistic effect on the charge carrier density and ionic dynamic in PEO brought about by the miscible blending of a small amount of PAc which increases the amount of amorphous phase leading to improved ion transport through the percolating pathway.

11.4 CONCLUSIONS

Films of salt-free PEO/PAc blends prepared by solution casting technique are found to be miscible after being subjected to 2 h of thermal treatment at 80°C under nitrogen atmosphere. The presence of a single, composition dependent T_g

and the suppression of the T_m of PEO in the as-prepared PEO-based blend by the addition of PAc affirm the miscibility of the PEO/PAc blend.

Addition of LiClO$_4$ from $Y_s = 0$ to 0.02 causes a marked increase in the T_g and conductivity of neat PEO and neat PAc. Enhancement in the conductivity of the two neat polymers is the result of an increase in the charge carrier density. At $Y_s >$ 0.02, both the T_g and the conductivity of PAc remains constant whereas those for PEO increase gradually. PEO/PAc blends with PEO content ≥ 75 wt% with and without the addition of LiClO$_4$ recorded higher σ values than neat PEO. Enhancement in the conductivity of the blend is due to improvement in the ion mobility attributed by an increase in Li$^+$ ion solvation and the addition of PAc increases the amorphous phase volume of PEO, which forms the percolating pathway for ion-transport in the blend system.

ACKNOWLEDGMENT

This study is supported by Fundamental Research Grant from Ministry of Higher Education, Malaysia, FRGS/2/2010/ST/UI/TM/02/4.

KEYWORDS

- **conductivity**
- **electrochemical and micro-ionic devices**
- **lithium batteries**
- **PEO/PAc and PEO/PAc blends**
- **polymer electrolyte**
- **thermal behavior**

REFERENCES

1. Armand, M.B.: Polymer electrolytes. in Annual Reviews Inc. 16, 245–261 (1986).
2. Eds.: Mac Callum, J.R., Vincent, C.A.: Polymer Electrolyte Reviews. Elsevier, Amsterdam (1987/1989).
3. Owen, J.R., Laskar, A.L., Chandra, S.: Superionic solids and solid state electrolytes–Recent trends. Academic Press, New York (1989).
4. Dias, F.B., Plomp, L., Veldhuis, J.B.J.: Trends in polymer electrolytes for secondary lithium batteries. Journal of Power Sources, 88, 169–191 (2000).
5. Yu, X.Y., Xiao, M., Wang, S.J., Zhao, Q.Q., Meng, Y.Z.: Fabrication and characterization of PEO/PPC polymer electrolyte for lithium-ion battery. Journal of Applied Polymer Science, 115, 2718–2722 (2010).
6. Watanabe, M., Oohashi, S.I., Sanui, K., Ogata, N., Ohtaki, Z.: Morphology and ionic conductivity of polymer complexes formed by segmented polyether poly(urethane urea) and lithium perchlorate. Macromolecules, 18, 1945–1950 (1985).

7. Rhodes, C.P., Frech, R., Local structures in crystalline and amorphous phases of diglyme–LiCF$_3$SO$_3$ and poly(ethylene oxide)–LiCF$_3$SO$_3$ systems: Implications for the mechanism of ionic transport. Macromolecules, 34, 2660–2666 (2001).
8. Caruso, T., Capoleoni, S., Cazzanelli, E., Agostino, R.G., Villano, P., Passerini, S.: Characterization of PEO–Lithium Triflate polymer electrolytes: Conductivity, DSC and Raman investigations. Ionics, 8, 36–43 (2002).
9. Gitelman, L., Israeli, M., Averbuch, A., Nathan, M., Schuss, Z., Golodnitsky, D.: Modeling and simulation of Li-ion conduction in poly(ethylene oxide). Journal of Computational Physics, 227, 1162–1175 (2007).
10. Berthier, C., Gorecki, W., Minier, M., Armand, M.B., Chabagno, J.M., Rigaud, P.: Microscopic investigation of ionic conductivity in alkali metal salts-poly(ethylene oxide) adducts. Solid State Ionics, 11, 91–95 (1983).
11. Reddy, M.J., Chu, P.P.: Optical microscopy and conductivity of poly(ethylene oxide) complexed with KI salt. Electrochimica Acta, 47, 1189–1196 (2002).
12. Chan, C.H., Kammer, H.W.: Properties of Solid Solutions of Poly(ethylene oxide)/Epoxidized Natural Rubber Blends and LiClO$_4$. Journal of Applied Polymer Science, 110, 424–432 (2008).
13. Sim, L.H., Chan, C.H., Kammer, H.W.: Selective localization of lithium perchlorate in immiscible blends of poly(ethylene oxide) and epoxidized natural rubber. IEEE conference proceedings: International Conference on Science and Social Research (CSSR 2010). Kuala Lumpur, Malaysia, 499–503 (2011).
14. Sim, L.H., Chan, C.H., Kammer, H.W.: Melting behavior, morphology and conductivity of solid solutions of PEO/PAc blends and LiClO$_4$. Materials Research Innovations, 15, S71–S74 (2011).
15. Sim, L.H., Gan, S.N., Chan, C.H., Kammer, H.W., Yahya, R.: Compatibility and conductivity of LiClO$_4$ free and doped polyacrylate-poly(ethylene oxide) blends. Materials Research Innovations, 13, 278–281 (2009).
16. Gan, S.N.: Water-reducible Acrylic Copolymer Dipping Composition and Rubber Products Coated with Same. PI20055440, Malaysia (2005).
17. Sim, L.H., Gan, S.N., Chan, C.H., Yahya, R.: ATR–FTIR studies on ion interaction of lithium perchlorate in polyacrylate/poly(ethylene oxide) blends. Spectrochimica Acta – Part A: Molecular and Biomolecular Spectroscopy, 76, 287–292 (2010).
18. Fox, T.G.: Influence of diluent and of copolymer composition on the glass temperature of a polymer system. Bull. Am. Phys. Soc., 1, 13, 123–135 (1956).
19. Brostow, W., Chiu, R., Kalogeras, I.M., Vassilikou–Dova, A. Prediction of glass transition temperatures: Binary blends and copolymers. Materials Letters, 62, 3152–3155 (2008).
20. Cimmino, S., Pace, E.D., Martuscelli, E., Silvestre, C.: Syndiotactic polystyrene: crystallization and melting behavior. Polymer, 32, 1080–1083 (1991).

CHAPTER 12

CORRELATION BETWEEN THE STORAGE TIME OF THE NRL AND THE EFFICIENCY OF PMMA GRAFTING TO NR

YOGA SUGAMA SALIM, NUR AZIEMAH ZAINUDIN, CHIN HAN CHAN, and KAI WENG CHAN

CONTENTS

12.1 Introduction 372
12.2 Experimental 373
12.3 Results and Discussion 377
12.4 Conclusions 381
Acknowledgments 382
Keywords 382
References 382

12.1 INTRODUCTION

Natural rubber latex (NRL) is often stored for a certain period of time before the grafting of poly(methyl methacrylate) (PMMA) to natural rubber (NR). It is well known that the properties of NRL change as a function of storage time. This paper describes the influence of storage time of NRL on its mechanical stability, followed by the effect on grafting efficiency of PMMA onto the NR backbone in NRL. The mechanical stability time (MST) of NRL decreases dramatically after 30 days of storage, implying that there is an increase in volatile fatty acid content leading to lower MST. Quality control tests (i.e., total solid content (TSC), alkalinity, pH, viscosity and MST) show that the alkalinity, viscosity and pH of NRL slightly fluctuate as a function of time in both NRL that fails the MST test and the NRL after addition of potassium oleate. Both NRLs having low and high MST at fixed period of storage time were used for the grafting of PMMA onto the NR in NRL. Correlation between storage time of NRL in low and high MST and the grafting efficiencies was studied. Results suggest that there is no significant change in the grafting efficiency, which ranges from 84–88% for all samples, as well as MMA monomer conversion to PMMA for NRL with low and high MST at fixed storage time. The grafting of PMMA onto NRL backbone is confirmed using Fourier-transform infrared (FTIR) spectroscopy.

Hevea brasiliensis tree has been the object of research for the past few decades. The dry phase of latex [known as natural rubber (NR)] from the *Hevea* tree consists of 93–95 wt% *cis*-1,4-poly(isoprene) [1, 2]. NR possesses excellent physical properties such as high mechanical strength, excellent flexibility, and resistance to impact and tear [3, 4], but the unsaturated and non-polar nature of the chain makes it susceptible to flame, chemicals, solvents, ozone and weather [5, 6]. Investigations of NR modified by graft-copolymerization with a second monomer such as vinyl benzene [7, 8], methyl methacrylate (MMA) [9–11], acrylonitrile [12, 13], phosphonate [12, 14] have been reported. Among these, NR grafted with MMA [methyl-grafted (MG) rubber or NR-*g*-PMMA] has been marketed in Malaysia since 1950s under trade name "Heveaplus MG." There are three grades of MG rubber widely available: MG 30, MG 40 and MG 49 (the number denotes wt% of MMA content in the grafted NR) [9, 15]. The chemical structure of NR-*g*-PMMA is illustrated in Fig. 1.

In practical applications, NR-*g*-PMMA serves as adhesives [16], reinforcing agents and impact modifiers for thermoplastics [6, 11, 17–19]. Before the synthesis of NR-*g*-PMMA, natural rubber latex (NRL) is often stored for a certain period of time. Santipanusopon and Riyajan [20] studied the influence of ammonia content as a function of storage period of NRL on the properties of NRL including alkalinity, magnesium content and viscosity. They suggest that the magnesium content of NRL decreases while the alkalinity and viscosity of NRL increases as a function of storage time within 45 days. Furthermore, Sasidharan and coworkers [21] aged the NRL up to 300 days, and their properties were monitored at different storage intervals of 0, 20, 40, 60, 120, 180, 240, and 300 days. The MST values of all samples increase during the first 120 days of NRL storage and then begin to decrease gradually. MST is a measure of resistance against mechanical influences and it tends to increase the number of collisions between particles that are likely to coacervate the NRL. This parameter correlates to the volatile fatty acid content in NRL. Influence of storage time on properties of NRL has been numerously reported, but to our best knowledge, none have investigated the influence of storage time on grafting efficiency of PMMA onto NRL backbone. In this study, correlation between the storage time of the NRL and the efficiency of PMMA grafting to NR is discussed.

12.2 EXPERIMENTAL

12.2.1 MATERIALS

NRL with high ammonia content (alkalinity = 250 mEq) containing 60 wt% dried rubber content (DRC) was supplied by Thai Rubber Latex Ltd. (Rayong, Thailand). Reagent grade MMA monomer (purity ~ 99 wt%) was purchased from Rohm GmbH & Co. (Darmstadt, Germany) and used without further purification. The stabilizer oleic acid (Palm Oleo Sdn. Bhd., Malaysia), the activator tetraethylenepentamine (TEPA, purity ~ 99.9 wt%, Hunstman, Woodsland, USA), tetramethyl thiuramdisulphide (TMTD, conc. ~ 50 wt%, Flexy Sys, Belgium), tert-butyl hydroperoxide (TBHP, conc. ~ 70 wt%), and potassium oleate (conc. ~ 20 wt%) were used as received. Other reagents were commercially available in reagent grade and were used without further purification.

12.2.2 STORAGE PERIOD

A total of 12 kg of NRL was kept in 6 containers for 30, 60, 90, 120, 150, and 180 days at 30°C. At the specified storage time, 2 kg NRL was tested for quality control tests (TSC, alkalinity, MST, pH, viscosity), described in Section 12.2.5. Numerical subscripts in NRL refer to the numbers of storage days, for example, NRL_{30} and NRL_{90} refers to 30 days and 90 days of storage period, respectively. For NRL that passed the MST test (MST value of 800 s and above), the NRL would proceed to grafting reaction. However, data of the NRL that passed the MST test is not shown. Instead, we focus on NRL that failed the MST tests to observe the effects of storage period. For NRL that failed the MST tests, two experiments were conducted: (1) the NRL that failed the MST test was added with potassium oleate (coded as NRL_R). Approximately 20 wt% of potassium oleate was added drop-wise into every 1000 g of NRL. After 30 min of stirring at 25°C, the NRL containing potassium oleate was left overnight and the MST value was measured again on the following day. Once the MST value had been adjusted to a minimum specification (800–1000 s), it was then used for the synthesis of NR-g-PMMA. (2) the NRL that failed the MST test was used directly without any treatment for the synthesis of NR-g-PMMA (coded as NRL_F). The aim of the latter experiment is to study the effect of low MST on grafting efficiency of PMMA to NR.

12.2.3 SYNTHESIS OF NR-G-PMMA WITH 40 WT% OF PMMA CONTENT

Various quantities of NRL (424 g), distilled water (269 g) and ammonia solution (14.45 wt%, 43 g) were added into 1 L reactor. After 10 min of stirring at 250 rpm at 25°C, roughly 259 g MMA monomer and suitable amount of oleic acid (as a stabilizer) and TBHP were added slowly to the mixture. The mixture was maintained under stirring for 10 min before it was left 2–4 hours to obtain maximum percent grafting. After 2–4 hours, stirring was started again and suitable amount of TEPA (as an activator agent) was fed to the reactor. Addition of TEPA increased the temperature to approximately 65°C within 15–30 min. The mixture was left overnight at 25°C without stirring. Insoluble components of the reacted product (NR-g-PMMA containing ungrafted NR and free PMMA) were removed using 60 μm filter mesh. Appropriate amount of TMTD was added to the filtrated NR-g-PMMA to retard further grafting. The crude NR-g-PMMA latex was subjected to quality control tests before extraction and drying process. The ungrafted NR and free MMA in NR-g-PMMA were removed by Soxhlet extraction. Approximately 2.0 ± 0.5 g of non-purified NR-g-PMMA was extracted using light petroleum ether for 24 hours at 40°C. The residue inside the thimble was dried to a constant weight at 70°C, which was further extracted with acetone for 24 hours at 50°C and dried to a constant weight at 70°C. The flow chart for the analyses of NRL and NR-g-PMMA is shown in Fig. 2.

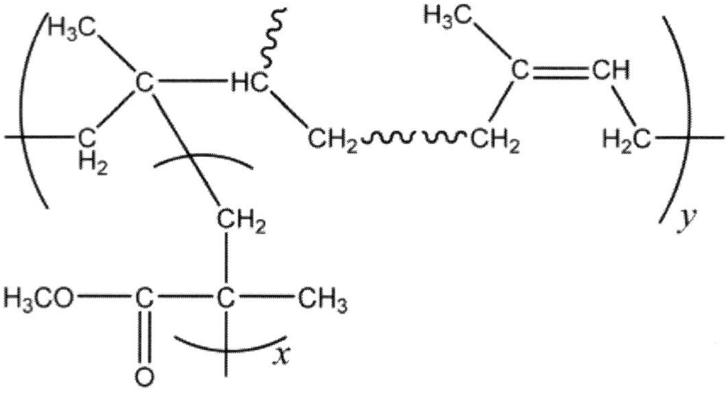

FIGURE 1 Chemical structure of NR-g-PMMA, where x and y refer to the wt% of PMMA and NR, respectively.

FIGURE 2 Flow chart for analyses of NRL and NR-g-PMMA.

12.2.4 FREE NR, FREE PMMA, GRAFTING EFFICIENCY AND PERCENTAGE OF MMA MONOMER CONVERSION

Free NRL and free PMMA can be calculated based on weight differences before and after the extraction interval, as shown in Eqs. (1) and (2). Mass difference between the initial and final mass of NR-g-PMMA was then used to calculate grafting efficiency (GE) according to Eq. (3) [6].

$$\text{Free NRL (wt\%)} = \frac{W_1 - W_2}{W_1} \times 100\% \tag{1}$$

$$\text{Free PMMA (wt\%)} = \frac{W_2 - W_3}{W_1} \times 100\% \tag{2}$$

Where, W_1 is the dry weight of non-purified NR-g-PMMA, W_2 is the weight of residue inside the thimble after Soxhlet extraction with light petroleum ether, and

W_3 is the weight of residue inside the thimble after Soxhlet extraction by acetone [6].

$$\text{GE (\%)} = \left[100\% - \text{free NRL (wt\%)} - \text{free PMMA (wt\%)}\right] \quad (3)$$

Free PMMA was used to calculate the percentage of MMA monomer conversion by using Eq. (4):

$$\text{Monomer conversion (CV\%)} = \left[100\% - \text{free PMMA (wt\%)}\right] \quad (4)$$

12.2.5 QUALITY CONTROL TESTS

The total solid content (TSC) was determined according to ISO 124:2011. In this test, NRL was placed on a weighing dish and dried at 100°C for 2 hrs. The dried sample was kept in desiccators. It is worth noting that the TSC of neat NRL was diluted from 60% to 55% with ammonium solution for the MST measurement. The dilution of NRL may increase or decrease the stability of NRL, depending on the types of NRL [22]. The TSC values were calculated according to formula given in Eq. (5). The alkalinity was determined according to ISO 125:2003, in which 5 g of NRL in 250 mL conical flask was added to 200 mL of distilled water. Subsequently, 2–3 drops of methyl red (an indicator) was added drop-wise into the mixture, and was titrated with 0.1 N of Sulfuric acid until the indicator turned pink against the white background of a slightly coagulated NRL. Viscosity tests were conducted with a Brookfield DV–I+ viscometer according to ISO 1652:2011. The stirrer of the viscometer was immersed into the NRL after it had been attached to the viscometer. The viscosity value was measured in centipoises (cPs) using a proper speed that allowed the stirrer to rotate until a stable reading was attained. The MST was determined according to ISO 35:2004. The sample was agitated at 14,000 rpm using latex testing machine (Klaxon, Secomak Ltd.) until end point was reached. This is indicated by a visual formation of aggregate and a change in the sound of stirring speed. The sound of agitation becomes loud as the sample becomes thick. All the results of the above-mentioned tests are mean values from two replicates.

$$\text{Total solid content (TSC) (\%)} = \frac{C-A}{B-A} \times 100\% \quad (5)$$

A = Weight of weighing dish (g), B = Weight of weighing dish with NRL (g), C = Weight of weighing dish with dried NR (g).

12.2.6 FTIR CHARACTERIZATION

FTIR sample analysis was carried out using Attenuated Total Reflection (ATR) on Perkin Elmer Spectrum One spectrometer. FTIR spectra were recorded in the transmittance mode over the range of 4000–600 cm^{-1} by averaging 16 scans at maximum resolution of 2 cm^{-1} in all cases.

12.3 RESULTS AND DISCUSSION

12.3.1 MST, TSC AND ALKALINITY OF NRL AS A FUNCTION OF STORAGE TIME

Figure 3 shows the MST tests of neat NRL before and after addition of potassium oleate as a function of storage time. The MST of NRL shows great reduction from 30–180 days of storage. Conversely, Sasidharan and coworkers [21] found that the MST of NRL gradually decreases after 120 days of storage. This trend can be explained by the fact that MST correlates closely to volatile fatty acid content in NRL. Low MST in NRL means higher volatile fatty acid content in the NRL. It is suggested that the increase in volatile fatty acid content of NRL would affect the stability of NRL [22, 23]. The MST value of approximately 800–1000s is set to be the lowest standard specification for the synthesis of NR-g-PMMA; thus in this study, the amount of potassium oleate added to the system must be sufficient to increase the MST value to the expected level. Other quality control tests such as TSC seems to be constant at 61.5 ± 0.2%, while alkalinity slightly fluctuates from 354 to 417 mEq as a function of storage time (Table 1).

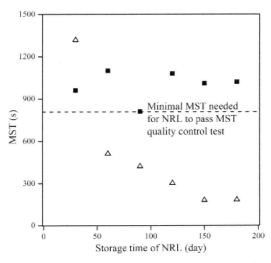

FIGURE 3 MST of NRL aged for 30 to 180 days (Δ: NRL; ■: aged NRL that failed the MST after addition of potassium oleate).

TABLE 1 TSC and alkalinity of NRL as a function of storage time.

Sample Code	TSC (%)	Alkalinity (mEq)
NRL_{30}	61.5	417.6
NRL_{60}	61.5	390.2
NRL_{90}	61.2	409.6
NRL_{120}	61.3	358.7
NRL_{150}	61.5	354.5
NRL_{180}	61.6	407.2

12.3.2 TSC, ALKALINITY, MST, AND VISCOSITY OF NR-G-PMMA LATEX

Tables 2 and 3 summarize the quality control tests and physical appearance of NR-g-PMMA. The former shows the NRL that failed the MST test was used for synthesis without addition of potassium oleate (coded as NRL_F), and the latter shows the NRL that passed the MST test after addition of potassium oleate was used for the synthesis of NR-g-PMMA (coded as NRL_R). Results show that the TSC and alkalinity values do not differ from each another in NRL_F and NRL_R, however the viscosities of NR-g-PMMA, after addition of potassium oleate to the NRL_R increase (Table 3). Higher viscosity of NR-g-PMMA may be caused by the poor colloidal stability of NRL during storage [24] or by the formation of homopolymer PMMA through crosslinking by hydrogen bonding when potassium oleate is added [20]. Physical appearance shows a thickening of NR-g-PMMA, with a drop in meniscus of the NRL in water indicated after 1800s, in both NRL_F and NRL_R samples. Above 1800s, NR-g-PMMA solidifies and may harm the testing machine. The MST values of NR-g-PMMA investigated in this study are similar to that observed by Kalkornsurapranee et al. [6].

TABLE 2 Quality control tests and physical appearance of NR-g-PMMA with 40wt% of PMMA content for the aged NRL that failed MST test before the synthesis of NR-g-PMMA.

Sample code	NR-g-PMMA					
	TSC (%)	Alkalinity (mEq)	MST (s)	Viscosity (cP s)	pH	Physical appearance of NR-g-PMMA latex after MST test
NRL_{30}	—	—	—	—	—	
$NRL_{60/F}$	51.1	348.1	>1800	155.0	10.6	NR-g-PMMA latex thickens up and it is stable

TABLE 2 *(Continued)*

$NRL_{90/F}$	51.1	372.3	> 1800	68.5	10.6	NR-g-PMMA latex thickens up and it is stable
$NRL_{120/F}$	51.0	357.8	> 1800	194.0	10.6	NR-g-PMMA latex thickens up and it is stable
$NRL_{150/F}$	51.1	342.9	> 1800	122.5	10.6	NR-g-PMMA latex thickens up and it is stable
$NRL_{180/F}$	51.4	376.3	> 1800	158.5	10.6	NR-g-PMMA latex thickens up and it is stable

Subscript 'F' indicates the NRL that failed MST test and was used for the synthesis of NR-g-PMMA without addition of potassium oleate, and subscripted numbers indicates the storage time in days.

TABLE 3 Quality control tests and physical appearance of NR-g-PMMA after adding potassium oleate in the NRL, which failed MST test before the synthesis of NR-g-PMMA.

Sample code	NR-g-PMMA					
	TSC (%)	Alkalinity (mEq)	MST (s)	Viscosity (cP s)	pH	Physical appearance of NR-g-PMMA latex after MST test
NRL_{30}	—	—	—	—	—	
$NRL_{60/R}$	50.5	379.4	> 1800	220.0	10.6	NR-g-PMMA latex thickens up and it is stable
$NRL_{90/R}$	51.0	386.8	> 1800	259.0	10.6	NR-g-PMMA latex thickens up and it is stable
$NRL_{120/R}$	50.9	401.7	> 1800	157.5	10.6	NR-g-PMMA latex thickens up and it is stable
$NRL_{150/R}$	51.3	376.3	> 1800	209.1	10.6	NR-g-PMMA latex thickens up and it is stable
$NRL_{180/R}$	51.2	374.6	> 1800	83.5	10.6	NR-g-PMMA latex thickens up and it is stable

Subscript 'R' indicates the NRL that passed MST test after addition of potassium oleate and was used for the synthesis of NR-g-PMMA, and subscripted numbers indicates the storage time in days.

TABLE 4 Effect of storage time of NRL on grafting efficiency and monomer conversion of NR-*g*-PMMA.

Sample code	Free NR (wt%)	Free PMMA (wt%)	Grafting efficiency (%)	MMA Conversion (wt%)
NRL$_{30}$	3.3	8.6	88.1	91.4
NRL$_{60/F}$	2.0	9.4	88.7	90.6
NRL$_{90/F}$	3.1	8.9	88.0	91.1
NRL$_{120/F}$	2.0	9.8	88.1	90.2
NRL$_{150/F}$	1.6	10.5	87.9	89.5
NRL$_{180/F}$	1.2	10.3	88.5	89.7
NRL$_{60/R}$	1.7	10.9	87.3	89.1
NRL$_{90/R}$	2.3	10.5	87.2	89.5
NRL$_{120/R}$	2.3	10.7	87.0	89.3
NRL$_{150/R}$	2.3	10.3	87.3	89.7
NRL$_{180/R}$	2.6	11.4	86.0	88.6

12.3.3 FREE PMMA, FREE NR, GRAFTING EFFICIENCY AND MMA MONOMER CONVERSION

The NR-*g*-PMMA products obtained from the polymerization were extracted and characterized to determine free NR, free PMMA, monomer conversion and grafting efficiency. Table 4 shows the effect of storage NRL on grafting efficiency and monomer conversion. The grafting efficiency and monomer conversion was calculated after the synthesis of NR-*g*-PMMA using Eq. (1) to Eq. (4). The grafting efficiencies of all samples range from 86 to 88 wt%, while monomer conversion ranges from 89 to 91 wt%. Free NR and free PMMA in all samples slightly fluctuate after 30 days of storage, with 1.2–3.1 wt% and 8.6–11.4 wt%, respectively. Results strongly suggest that there is no significant difference in NR-*g*-PMMA for samples obtained from different sources and different times.

12.3.4 CHARACTERIZATION OF THE NR-G-PMMA WITH 40 WT% PMMA CONTENT WITH FOURIER TRANSFORM INFRARED

Figure 4 shows representative IR spectra of NR-*g*-PMMA from the NRL that passed the quality control tests after 120 days ageing (sample code NRL$_{120/R}$). All other spectra of NR-*g*-PMMA samples from the storage study show similar adsorption band. The characteristic peaks of the saturated aliphatic sp^3 C–H bonds are observed at 2853 cm^{-1} and 2952 cm^{-1}, which corresponds to $v_{as}(CH_3)$ and $v_s(CH_2)$, respectively. A strong sharp peak located at 1727 cm^{-1} is attributed to the symmetrical stretching mode of C=O (carbonyl). The C=C stretching, CH_3 symmetrical deformation, and CH_2 twisting modes of NR are observed at 1660, 1376 and 1242 cm^{-1}, respectively. A strong peak due to O–CH_3 is located at 1446 cm^{-1}. The doublet peak of C–O stretching mode of PMMA can be observed at 1100–1210 cm^{-1}, with sharp and strong maximum peak at 1147 cm^{-1}. A peak at 987 cm^{-1} observed in the spectrum could be assigned to C–O–C symmetrical of PMMA [23]. FTIR results confirmed the occurrence of grafting polymerization of PMMA onto NRL.

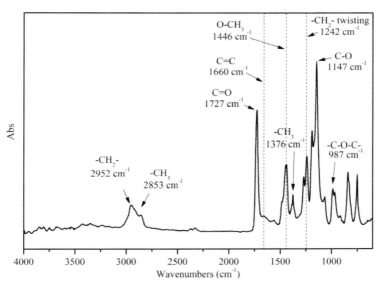

FIGURE 4 FTIR spectrum of NR-g-PMMA with 40 wt% PMMA content for NRL$_{120/R}$.

12.4 CONCLUSIONS

It is shown that the optimum storage time of NRL is approximately 30 days as the MST values drop exponentially after 30 days. Meanwhile, TSC and alkalinity of NRL are not affected by the storage time of NRL. Nevertheless, the viscosity of NR-g-PMMA latex with the aged NRL treated with potassium oleate is higher than that without treatment. The GE and monomer conversion are not affected by the storage period of NRL. Current work focuses on the effect of stabilizers on the colloidal stability of ammonia-preserved NRL.

ACKNOWLEDGMENTS

This work has been supported by Research Intensive Faculty Grant [600–RMI/DANA 5/3/RIF(636/2012)] from University Teknologi MARA (UiTM), Shah Alam, Malaysia. We are grateful for invaluable help from NR division in Synthomer (M) Sdn. Bhd., Kluang, Johor, Malaysia.

KEYWORDS

- effect of stabilizers
- efficiency of PMMA Grafting to NR
- mechanical fatigue limit
- NRL
- physical properties of natural rubber
- storage time

REFERENCES

1. Chen, H.Y. (1962) Determination of *cis*-1,4 and *trans*-1,4 contents of polyisoprenes by high resolution nuclear magnetic resonance. Anal Chem 34, 1793–1795.
2. Whelan, T. (1994) Polymer technology dictionary, Technology and engineering. Chapman & Hall, London.
3. Nasir, M., Teh, G.K. (1988) The effects of various types of crosslinks on the physical properties of natural rubber. Eur Polym J, 24, 733–736.
4. Kongparakul, S., Prasassarakich, P., Rempel, G.L. (2008) Catalytic hydrogenation of methyl methacrylate-*g*-natural rubber (MMA-*g*–NR) in the presence of homogeneous osmium catalyst $OsHCl(CO)(O_2)(PCy_3)_2$. Appl Catal, A., Gen 344, 88–97.
5. Lake, G.J., Lindley, P.B. (1965) The mechanical fatigue limit for rubber. J Appl Polym Sci 9, 1233–1251.
6. Kalkornsurapranee, E., Sahakaro, K., Kasesaman, A., Nakason, C. (2009) From a laboratory to a pilot scale production of natural rubber grafted with PMMA. J Appl Polym Sci 114, 587–597.
7. Minoura, Y., Mori, Y., Imoto, M. (1957) Vinyl polymerization XXI. Polymerization of styrene in the presence of natural rubber. Die Makromolekulare Chemie 24, 205–221.
8. Mays, J.W. (1990) Synthesis of "simple graft" poly(isoprene-*g*-styrene) by anionic polymerization. Polym Bull 23, 247–250.
9. Thiraphattaraphun, L., Kiatkamjornwong, S., Prasassarakich, P., Damronglerd, S. (2001) Natural rubber-*g*-methyl methacrylate/poly(methyl methacrylate) blends. J Appl Polym Sci 81, 428–439.
10. George, V., Britto, I.J., Sebastian, M.S. (2003) Studies on radiation grafting of methyl methacrylate onto natural rubber for improving modulus of latex film. Radiat Phys Chem 66, 367–372.

11. Kalkornsurapranee, E., Sahakaro, K., Kasesaman, A., Nakason, C. (2010) Influence of reaction volume on the propertries of natural rubber-g-methyl methacrylate. J Elastomers Plast 42, 17–34.
12. Arauj'o PHH, Sayer, C., Poco, J.G.R., Giudici, R. (2002) Techniques for reducing residual monomer content in polymers, A review. J Polym Eng Sci 42, 1442–1468.
13. Bhattacharya, A., Misra, B.N. (2004) Grafting, a versetiles means to modify polymers technique, factors and applications. Prog Polym Sci 29, 767–814.
14. Derouet, D., Intharapat, P., Quang, N.T., Gohier, F., Nakason, C. (2008) Graft copolymers of natural rubber and poly(dimethyl (acryloyloxymethyl) phosphonate) (NR-g-PDMAMP) or poly(dimethyl (methacryloyloxyethyl) phosphonate) (NR-g-PDMMEP) from photopolymerization in latex medium. Eur Polym, J., 45, 820–836.
15. Kamisan A.S., Kudin, T.I.T., Ali, A.M.M., Yahya, M.Z.A. (2011) Polymer gel electrolytes based on 49% methyl-grafted natural rubber. Sains Malaysiana 40, 49–54.
16. Rezaifard, A.H., Hodd, K.A., Tod, D.A., Barton, J.M. (1994) Toughening epoxy resins with poly(methyl methacrylate)-grafter-natural rubber, and its use in adhesive formulations. Int, J., Adhes Adhes 14, 153–159.
17. Keskkula, H., Kim, H., Paul, D.R. (2004) Impact modification of styrene-acrylonitrile copolymers by methyl methacrylate grafted rubbers. Polym Eng Sci 30, 1373–1381.
18. Keskkula, H., Paul, D.R., McCreedy, K.M., Henton, D.E. (1987) Methyl methacrylate grafted rubbers as impact modifiers for styrenic polymers. Polym 28, 2063–2069.
19. Charmondusit, K., Seeluangsawat, L. (2009) Recycling of poly(methyl methacrylate) scrap in the styrene–methyl methacrylate copolymer cast sheet process. Resour Conserv Recy 54, 97–103.
20. Santipanusopon, S., Riyajan, S.A. (2009) Effect of field natural rubber latex with different ammonia contents and storage period on physical properties of latex concentrate, stability of skim latex and dipped film. Phys Procedia 2, 127–134.
21. Sasidharan, K.K., Joseph, R., Palaty, S., Gopalakrishnan, K.S., Rajammal, G., Pillai, P.V. (2005) Effect of the vulcanization time and storage on the stability and physical properties of sulphur-prevulcanized natural rubber latex. J Appl Polym Sci 97, 1804–1811.
22. Dawson, H.G. (1949) Mechanical Stability Test for Hevea Latex. Anal Chem 21, 1066–1071.
23. Allen P.W., Merrett, F.M. (1956) Polymerization of methyl methacrylate in polyisoprene solutions. J Polym Sci 22, 193–201.
24. Blackley, D.C. (1966) High polymer Science Lattices (Vol 1 & 2). McLaren & Sons Ltd, New York.

CHAPTER 13

A STUDY ON COMPOSITE POLYMER ELECTROLYTE

TAN WINIE, N. H. A. ROSLI, M. R. AHMAD, R. H. Y. SUBBAN, and C. H. CHAN

CONTENTS

13.1 Introduction 386
13.2 Experimental 387
13.3 Results 388
13.4 Discussion 390
13.5 Conclusions 395
Acknowledgments 395
Keywords 395
References 396

13.1 INTRODUCTION

Hexanoyl chitosan that exhibited solubility in THF was prepared by acyl modification of chitosan. Atactic polystyrene was chosen to blend with hexanoyl chitosan. $LiCF_3SO_3$ was employed as the doping salt. Untreated and HNO_3 treated TiO_2 fillers were dispersed in hexanoyl chitosan-polystyrene–$LiCF_3SO_3$ electrolyte at 4 wt.% concentration. We observed better filler dispersion in the matrix for the acid treated system. The resulting composite electrolyte films were characterized for the electrical and tensile properties. Untreated TiO_2 improved the electrolyte conductivity while HNO_3 treated TiO_2 decreased the conductivity. A model based on interaction between Lewis acid-base sites of TiO_2 with ionic species of $LiCF_3SO_3$ has been proposed to understand the conductivity mechanism brought about by the fillers. The conductivity enhancement by untreated TiO_2 is attributed to the increase in the number and mobility of Li^+ cations. HNO_3-treated TiO_2 decreased the conductivity by decreasing the anionic contribution. An enhancement in the Young's modulus and toughness was observed with the addition of TiO_2 and greater enhancement is found for the treated TiO_2. This is discussed using the percolation concept.

Polymeric electrolytes are the fastest growing and most widely investigated electrolyte system ever since the proposition and recognition of its potential application in solid-state electrochemical systems [1]. Solid polymer electrolyte (SPE) is formed by dissolving an alkali metal salt in a polymer. To date, various combinations of salts and polymers forming polymer-salt complexes have been investigated. A major drawback of these polymer-salt complexes, however, is the low ionic conductivity at ambient temperature.

In general, factors affecting the ionic conductivity of a SPE are amorphous phase, number and mobility of charge carriers. It has been shown that ionic transport takes place only in the amorphous phase [2]. The number and mobility of charge carriers are governed by the interactions between the salt and polymer matrix. Thus, two primary strategies have been adopted. The first one is to suppress crystallinity in polymer system by co-polymerization [3–6], cross-linked polymer networks [7, 8], comb formation [9, 10] and plasticizer addition [11, 12]. In these cases, the conductivity enhancement is achieved by increasing the polymer chain mobility. The second strategy to increase the number of charge carriers is by increasing salt concentration and using highly dissociable salt [13]. However, the mechanical property is often scarified by increasing salt concentration [14]. For practical applications of SPEs in various electrochemical devices, it is important for the SPE to retain good mechanical property.

One promising way to improve both the mechanical and conductivity properties of SPE is the incorporation of inorganic fillers such as TiO_2, SiO_2 and Al_2O_3 to form a kind of composite polymer electrolyte (CPE). Based on those reported in the literature, it is still not clear on the role played by the fillers as the conductivity enhancer. For example, some researchers suggested that the conductivity enhancement in CPE could not be attributed to the enhanced polymer segmental motion as no appreciable change in T_g is observed [15–17]. On the other hand, some studies have shown that T_g is affected due to the addition of fillers [18,19]. Studies by Wieczorek et al. [20] suggested that the conductivity increase is due to the increase in the number of ions. Best and co-workers [16] in their study of some CPEs have suggested that the conductivity enhancement brought about by the fillers does not come from the change in the number of ions but from the increase in the ions mobility. XRD studies on hexanoyl chitosan:polystyrene–$LiCF_3SO_3$–TiO_2 have suggested that the increase in conductivity is attributable to the decrease in percentage of crystallinity [21].

It has been shown that the surface properties of fillers could affect the conductivity performance, but limited attention has been devoted to their influence on the tensile properties. In the present work, we attempt to correlate the surface properties of fillers to the electrical and tensile properties of a CPE based on hexanoyl chitosan:polystyrene- $LiCF_3SO_3$. Surface acid-base investigation of the fillers has been carried out. The variation in conductivity was discussed on the basis of number and mobility of ions. The tensile properties, specified in terms of Young's modulus and toughness were compared and discussed using the percolation concept.

13.2 EXPERIMENTAL

Hexanoyl chitosan that exhibited solubility in tetrahydrofuran (THF) were prepared by acyl modification of chitosan [22]. Polystyrene (M_w of 280,000) used in this work has an atactic chain configuration and commercially available through Sigma–Aldrich. Lithium trifluromethanesulphonate ($LiCF_3SO_3$) with purity >96% from Aldrich was dried for 24 h at 120°C prior to use. TiO_2 having particle size of 30–40 nm was acid-treated by stirring in diluted HNO_3 solution (242 mL, 0.83 v/ v% in de-ionized water) for 8 h at 80°C, rinsed by de-ionized water until the filtrates were neutralized and then dried for 12 h at 100°C. The amounts of the acidic site at the surface of TiO_2 untreated and acid-treated were determined by titration method. They are represented in terms of number of mole of H^+ and were found to be 0.06 and 0.74 mmolg^{-1} for the untreated and acid-treated, respectively. Untreated and HNO_3 treated TiO_2 were then used as the fillers.

To prepare the CPE, required amount of TiO_2 was added to hexanoyl chitosan:polystyrene (90:10) blend and $LiCF_3SO_3$ dissolved in THF. The solutions were stirred at room temperature until TiO_2 particles are dispersed homogeneously before pouring into separate glass Petri dishes. They are left to evaporate at room temperature for the films to form. For the impedance measurement, the film was sandwiched between two stainless steel electrodes. Impedance of the films was measured using HIOKI 3532–50 LCR Hi-tester impedance spectroscopy in the frequency range from 100 Hz to 1 MHz at different temperatures from 273 to 333 K. The ionic conductivity of the film was calculated using the equation

$$\sigma = \frac{t}{R_b A} \quad (1)$$

where t is the thickness of the film and A is the film-electrode contact area. The bulk resistance, R_b was obtained from the complex impedance plot. The dielectric constant, ε_r is related to the measured real, Z_r and imaginary parts, Z_i of impedance as follows:

$$\varepsilon_r = \frac{Z_i}{\omega C_o \left(Z_r^2 + Z_i^2\right)} \quad (2)$$

where $C_o = \varepsilon_o A/t$, ε_o is the permittivity of free space, symbols A and t have their usual meaning. $\omega = 2\pi f$, f being the frequency in Hz.

The Li⁺ transference numbers, τ_{Li+} in the CPEs were determined by a combination of ac impedance and dc polarization methods. The film was sandwiched between two lithium metal electrodes and the measurement was performed as described in Ref. [23]. The τ_{Li+} was calculated using the following equation

$$\tau_{Li+} = \frac{I_{ss}(\Delta V - I_o R_o)}{I_o(\Delta V - I_{ss} R_{ss})} \quad (3)$$

where I_o is the initial current, I_{ss} is the steady-state current. ΔV is the applied voltage bias, R_o and R_{ss} is the initial and final interfacial resistance, respectively.

Tensile properties were studied by using Instron 3366 tensile tester. Tensile stress-strain tests were conducted at crosshead speed of 1 mm/min. The films were cut into a 0.5 mm × 7 mm × 25 mm rectangular strip samples. The gauge length was controlled to be 24 mm. The Young's modulus was determined from the slope of the initial linear part of stress-strain curve. The toughness is characterized by the area under the curve.

13.3 RESULTS

Figure 1 shows the temperature dependence of ionic conductivity for hexanoyl chitosan:polystyrene–LiCF$_3$SO$_3$ without and with TiO$_2$. Untreated TiO$_2$ improved the conductivity while HNO$_3$ treated TiO$_2$ decreased the electrolyte conductivity. Room temperature conductivity achieved for the untreated and HNO$_3$ treated TiO$_2$ system was 2.27 10^{-4} S cm^{-1} and 6.33 10^{-5} S cm^{-1}, respectively. Activation energies for ionic conduction, E_a were obtained from the slope of the plots in Fig. 1 and are summarized in Table 1. E_a is the energy required for an ion to begin migration from one donor site to another. This ion migration results in conduction. It can be observed that high conducting sample exhibits low value of E_a. This indicates that ions in high conducting sample require lower energy to begin migration. The experimental results of τ_{Li+} are presented in Table 1. The increase in Li$^+$ mobility is reflected in the increase value of τ_{Li+}. The τ_{Li+} for the samples differs as follows: HNO$_3$-treated < TiO$_2$ free < untreated TiO$_2$.

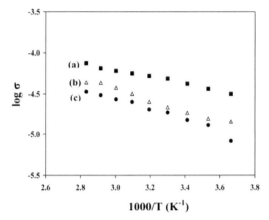

FIGURE 1 Temperature dependence of ionic conductivity for hexanoyl chitosan:polystyrene–LiCF$_3$SO$_3$ with (a) 4 wt% untreated TiO$_2$; (b) TiO$_2$ free and (c) 4 wt% HNO$_3$-treated TiO$_2$.

TABLE 1 Room temperature conductivity, activation energy and Li$^+$ transference number for hexanoyl chitosan: polystyrene–LiCF$_3$SO$_3$ electrolyte.

	σ_{RT} (S cm^{-1})	E_a (eV)	τ_{Li+}
TiO$_2$-free	7.21 ′ 10^{-5}	0.11	0.41
Untreated TiO$_2$	2.27 ′ 10^{-4}	0.05	0.46
HNO$_3$-treated TiO$_2$	6.33 ′ 10^{-5}	0.13	0.40

Figure 2 presents the frequency dependence of dielectric constant for electrolyte system without and with TiO$_2$. The dielectric constant is found to increase in the order: TiO$_2$ free < HNO$_3$ treated < untreated TiO$_2$. We have deduced that increment in the number of free ions is reflected in the increment in dielectric constant [24].

In our previous work [25], the tensile properties of hexanoyl chitosan is found to increase with addition of polystyrene. The introduction of LiCF$_3$SO$_3$, on the other hand, deteriorated the tensile properties of hexanoyl chitosan:polystyrene blend. The deteriorated tensile properties are overcome with the addition of TiO$_2$. As seen in Table 2, the Young's modulus increases from 11.77 MPa to 21.53

MPa and 22.90 MPa with addition of untreated and HNO_3 treated TiO_2, respectively. Similar enhancement trend in toughness was observed. This shows that the enhancement in Young modulus and toughness brought about by the treated TiO_2 is greater than the untreated TiO_2. The HNO_3 treatment of TiO_2 improves the dispersion of filler particles in polymer matrix. In comparison with the case of untreated TiO_2, the particles tend to agglomerate. These observations suggest that the surface state of TiO_2 influences the electrical and tensile properties of the present electrolyte system, which will be discussed in the later part of this paper.

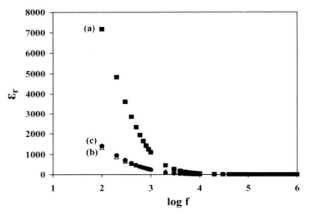

FIGURE 2 Frequency dependence of dielectric constant, ε_r for hexanoyl chitosan:polystyrene–$LiCF_3SO_3$ with (a) 4 wt% untreated TiO_2; (b) TiO_2 free and (c) 4 wt% HNO_3 treated TiO_2.

TABLE 2 Young's modulus and toughness for hexanoyl chitosan:polystyrene–$LiCF_3SO_3$ at 4 wt.% TiO_2.

	Young's Modulus (MPa)	Toughness (MPa)
TiO_2-free	11.77	0.025
Untreated TiO_2	21.53	0.036
HNO_3-treated TiO_2 (wt.%)	22.90	0.063

13.4 DISCUSSION

The conductivity mechanism by TiO_2 fillers in hexanoyl chitosan:polystyrene–$LiCF_3SO_3$ system is proposed in Fig. 3. Blend of hexanoyl chitosan and polystyrene is immiscible [25]. FTIR results showed that the Li^+ ions of $LiCF_3SO_3$ interacted with the donor atoms (nitrogen and oxygen atoms) of hexanoyl chitosan [26] and no interaction between $LiCF_3SO_3$ and polystyrene [25] as there is no complexation site for the salt in the structure of polystyrene. DSC results revealed that T_g of hexanoyl chitosan in the blend increases with increasing salt concentration whereas the T_g of polystyrene in the blend remains constant (results not shown here). We thus propose that $LiCF_3SO_3$ salt to be located in the hexanoyl chitosan phase. The hexanoyl chitosan–Li^+ interactions are electrostatic in nature. A Li^+ ion can hop from one donor site to another leaving a vacancy which will be filled by another Li^+ from a neighboring site, as illustrated in Fig. 3a. In the absence of TiO_2, this Li^+ motion is facilitated by the segmental motion of polymer chain.

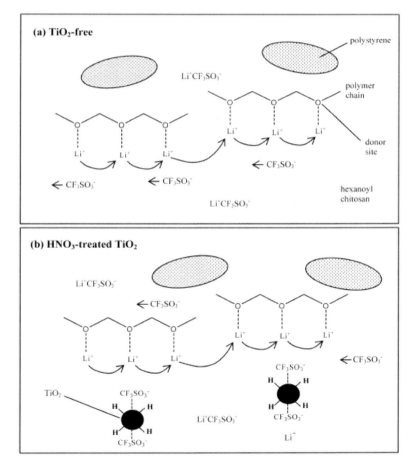

FIGURE 3 *(Continued)*

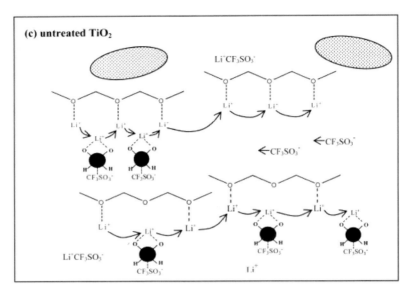

FIGURE 3 Conductivity mechanism in hexanoyl chitosan:polystyrene–$LiCF_3SO_3$–TiO_2 composite polymer electrolytes: (a) TiO_2-free; (b) HNO_3-treated TiO_2 and (c) untreated TiO_2.

No evidence is yet to show that TiO_2 fillers are located in the hexanoyl chitosan phase. Relevant studies on localization of TiO_2 in immiscible blends of hexanoyl chitosan and polystyrene are underway and are being conducted separately. Conductivity in CPE is due to the free ions from the salt. $LiCF_3SO_3$ is located in the hexanoyl chitosan phase. Thus, the conductivity variation as a result of interaction between $LiCF_3SO_3$ salt and TiO_2 suggests that the filler particles are most probably located in the same polymer phase with $LiCF_3SO_3$. Figure 3b shows the electrolyte system with HNO_3-treated TiO_2. The HNO_3-treated TiO_2 fillers have more acidic sites (0.74 mmolg^{-1}) as compared to untreated TiO_2 (0.06 mmolg^{-1}). The acidic site consists of OH group. For the simplicity of discussion, we denote the treated TiO_2 grain surface consists of solely OH groups and are marked with H in the figure. It is shown that anion of salt has larger affinity towards filler surface acid site than cation [27]. $CF_3SO_3^-$ anions dissociated from the salt are bonded with the H of OH groups. The anions are now immobilized on TiO_2 grains, which would, as compared to the filler-free system where anions migration is assisted to some extent by the segmental motion of polymer chain. The cation migration along the polymer chain remains unaffected. Immobilization of anions decreases the anionic contribution to the conductivity.

Conductivity mechanism in the case of untreated TiO_2 is shown in Fig. 3c. Each untreated TiO_2 grain surface is to have equal number of acidic and basic sites. The basic site is due to the oxygen of TiO_2. Function of the acidic site will be compensated by the basic site. Li^+ cations interacted with the O atoms of surface basic groups of TiO_2. The bonds between Li^+ and TiO_2 are similar to the bonds between Li^+ with donor atoms of hexanoyl chitosan i.e. bonds are subjected to breaking and forming during Li^+ transport process. This provides additional conduction pathway for Li^+ cations, which would, otherwise, be moving only along the polymer chain. This improved Li^+ transport is reflected in a higher value of τ_{Li^+}.

In addition, TiO_2 particles may also act as transit sites for Li^+ ions to make several small jumps from one donor site in polymer chain to another. The opportunity to make several small jumps implies that the Li^+ ions need not acquire a lot of energy as is required to make one bigger jump to the next donor site. Thus more Li^+ ions can hop with ease. This explains why the activation energy for ionic conduction is reduced.

The interaction between Lewis acid-base sites of TiO_2 with ionic species of salt helps in dissociation of $LiCF_3SO_3$ and yields greater number of free ions. It is important to note that the number of free ions represented by dielectric constant is indiscriminate of speciation. The conductivity of HNO_3 treated TiO_2 system is lower than that of TiO_2 free system despite the higher number of free ions in HNO_3 treated TiO_2 system (see Fig. 2). This is attributed to the anions immobilization as discussed previously. Anions immobilization decreases the effective average mobility of carrier ions. Slight increase in the number of free ions alone is not sufficient to satisfy the requirement of conductivity enhancement.

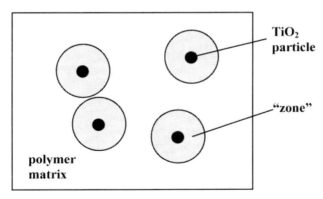

FIGURE 4 "Zone" around each TiO_2 particle.

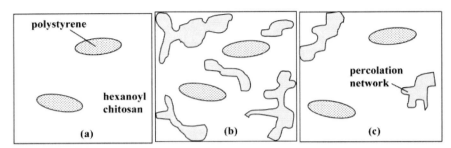

FIGURE 5 Formation of percolation networks within polymer matrix: (a) TiO_2-free; (b) HNO_3-treated TiO_2 and (c) untreated TiO_2.

The role of treated TiO_2 in modulus enhancement is discussed on the basis of the percolation concept proposed by He and Jiang [28]. Each individual filler particle is surrounded by a "zone" as presented in Fig. 4. The distances between well dispersed particles are closer. Hence, these zones tend to overlap, leading to the creation of percolation networks (see Fig. 5). Here, the continuous percolation networks develop more easily as compared to the agglomerated untreated TiO_2 particles at the same filler content. Better filler dispersion thus, leads to a greater modulus (i.e. stiffer sample film). Similar observation is reported by Svehlova and Poloucek [29].

It is generally accepted that the conductivity enhancement in CPE is attributed to the enhanced cation transport on the filler grain boundaries. However, this conductivity enhancement as a result of percolation mechanism is inconsistent with our findings. The percolation networks formed in HNO_3-treated TiO_2 system did not help in increasing the conductivity. This is because the OH functional groups on TiO_2 grains did not form the cation conduction pathway. Instead, they act as the anion trapper.

The toughness of a material is a measure of its ability to resist crack growth. The principle of the toughening process in the presence of TiO_2 is proposed as illustrated in Fig. 6. The tip of the crack is blunted by the percolation network formed by the fillers and stopped from spreading. Aggregation of untreated filler results in poorer development of continuous percolation networks in the polymer matrix. This increases the probability of crack growth (see Fig. 6c). This explains the treated TiO_2 sample film is tougher than the untreated TiO_2 film.

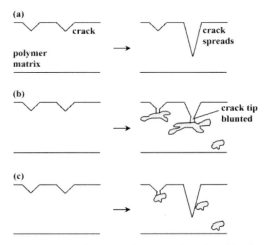

FIGURE 6 Toughening process in composite electrolyte: (a) TiO_2-free; (b) HNO_3-treated TiO_2 and (c) untreated TiO_2.

13.5 CONCLUSIONS

Untreated TiO_2 improved the electrolyte conductivity while HNO_3-treated TiO_2 decreased the conductivity. The conductivity enhancement by untreated TiO_2 comes from the increase in the number and mobility of Li^+ cations. HNO_3-treated TiO_2 decreased the conductivity by decreasing the anionic contribution. The addition of TiO_2 enhances the tensile properties of hexanoyl chitosan:polystyrene–$LiCF_3SO_3$ and greater enhancement is found for the treated TiO_2.

ACKNOWLEDGMENTS

The authors wish to thank Universiti Teknologi MARA, Malaysia for supporting this work through grant DANA 5/3 Dst (420/2011).

KEYWORDS

- characterization of polymer
- composite
- conductivity enhancement
- electrolyte
- polymer
- treated TiO_2

REFERENCES

1. Gray, F.M.: Polymer Electrolytes. The Royal Society of Chemistry, UK (1997).
2. Berthier, C., Gorecki, W., Minier, M., Armand, M.B., Chabagno, J.M., Rigand, P.: Microscopic investigation of ionic conductivity in alkali metal salts-poly (ethylene oxide) adducts. Solid State Ionics, 11(1), 91–95 (1983).
3. Fonseca, C.P., Neves, S.: Characterization of polymer electrolytes based on poly(dimethyl siloxane-co-ethylene oxide). J. Power Sources, 104, 85–89 (2002).
4. Rajendran, S., Mahendran, O., Kannan, R.: Characterisation of $[(1-x)$PMMA-xPVdF$]$ polymer blend electrolyte with Li^+ ion. Fuel, 81, 1077–1081 (2002).
5. Park, Y.W., Lee, D.S.: The fabrication and properties of solid polymer electrolytes based on PEO/PVP blends. J. Non–Crystalline Solids, 351, 144–148 (2005).
6. Yuan, F., Chen, H.Z., Yang, H.Y., Li, H.Y., Wang, M.: PAN–PEO solid polymer electrolytes with high ionic conductivity. Mat. Chem. Phys., 89, 390–394 (2005).
7. Kobayashi, N., Kubo, N., Hirohashi, R.: Control of ionic conductivity in solid polymer electrolyte by photo irradiation. Electrochim. Acta, 37, 1515–1516 (1992).
8. Watanabe, M., Nishimoto, A.: Effects of network structures and incorporated salt spesies on electrochemical properties of polyether-based polymer electrolytes. Solid State Ionics, 79, 306–312 (1995).
9. Ding, L.M.: Synthesis, characterization and ionic conductivity of solid polymer electrolytes based on modified alternating maleic anhydride copolymer with oligo (oxyethylene) side chains. Polymer, 38(16), 4267–4273 (1997).
10. Ding, L.M., Shi, J., Yang, C.Z.: Ion-conducting polymers based on modified alternating maleic anhydride copolymer with oligo (oxyethylene) side chains. Synth. Met., 87, 157–163 (1997).
11. Rajendran, S., Mahendran, O., Kannan, R.: Lithium ion conduction in plasticized PMMA–PVdF polymer blend electrolytes. Mat. Chem. Phys., 74, 52–57 (2002).
12. Tan Winie, Ramesh, S., Arof, A.K.: Studies on the structure and transport properties of hexanoyl chitosan-based polymer electrolytes. Physica B: Condensed Matter, 404, 4308–4311 (2009).
13. Murata, K., Izuchi, S., Yoshihisa, Y.: An overview of the research and development of solid polymer electrolyte batteries. Electrochim. Acta, 45, 1501–1508 (2000).
14. Fan, L., Dang, Z., Nan, C.W., Li, M.: Thermal, electrical and mechanical properties of plasticized polymer electrolytes based on PEO/P(VDF–HFP) blends. Electrochim. Acta, 48, 205–209 (2002).
15. Krawiec, W., Scanlon, L.G., Jr., Fellner, J.P., Vaia, R.A., Vasudevan, S., Giannelis, E.P.: Polymer nanocomposites: a new strategy for synthesizing solid electrolytes for rechargeable lithium batteries. J. Power Sources, 54, 310–315 (1995).
16. Best, A.S., Adebahr, J., Jacobsson, P., MacFarlane, D.R., Forsyth, M.: Microscopic interactions in nanocomposite electrolytes. Macromolecules, 34(13), 4549–4555 (2001).
17. Chung, S.H., Wang, Y., Persi, L., Croce, F., Greenbaum, S.G., Scrosati, B., Plichita, E.: Enhancement of ion transport in polymer electrolytes by addition of nanoscale inorganic oxides. J. Power Sources, 97–98, 644–648 (2001).

18. Capiglia, C., Mustarelli, P., Quartarone, E., Tomasi, C., Magistris, A.: Effects of nanoscale SiO_2 on the thermal and transport properties of solvent-free, poly(ethylene oxide) (PEO)-based polymer electrolytes. Solid State Ionics, 118, 73–79 (1999).
19. Kim, Y.W., Lee, W., Choi, B.K.: Relation between glass transition and melting of PEO-salt complexes. Electrochim. Acta, 45, 1473–1477 (2000).
20. Wieczorek, W., Raducha, D., Zalewska, A., Stevens, J.R.: Effect of salt concentration on the conductivity of PEO-based composite polymeric electrolytes. J. Phys. Chem. B, 102(44), 8725–8731 (1998).
21. Rosli, N.H.A., Chan, C.H., Subban, R.H.Y., Tan Winie: Studies on the structural and electrical properties of hexanoyl chitosan/polystyrene-based polymer electrolytes. Physics Procedia, 25, 215–220 (2012).
22. Zong, Z., Kimura, Y., Takahashi, Yamane, M.H.: Characterization of chemical and solid state structures of acylated chitosans. Polymer, 41, 899–906 (2000).
23. Bruce, P.G., Vincent, C.A.: Steady state current flow in solid binary electrolyte cells. J. Electroanal. Chem., 225, 1–17 (1987).
24. Tan Winie, Arof, A.K.: Dielectric behavior and AC conductivity of $LiCF_3SO_3$ doped H-chitosan polymer films. Ionics, 10, 193–199 (2004).
25. Tan Winie, Rosli, N.H.A., Hanif, N.S.M., Chan, C.H., Ramesh, S.: Polymer electrolytes based on blend of hexanoyl chitosan and polystyrene. Submitted for publication in elsewhere.
26. Tan Winie, Arof, A.K.: FT–IR studies on interactions among components in hexanoyl chitosan-based polymer electrolytes. Spectrochim. Acta A, 62, 677–684 (2006).
27. Kumar, B., Scanlon, L.G.: Polymer-ceramic composite electrolytes: conductivity and thermal history effects. Solid State Ionics, 124, 239–254 (1999).
28. He, D., Jiang, B.: The elastic modulus of filled polymer composites. J. Appl. Polym. Sci., 49, 617–621 (1993).
29. Svehlova, V., Poloucek, E.: Mechanical properties of talc-filled polypropylene. Influence of filler content, filler particle size and quality of dispersion. Angew. Makromol. Chem., 214(3762), 91–99 (1994).

CHAPTER 14

A STUDY ON SOLID POLYMER ELECTROLYTES

SITI NOR HAFIZA MOHD YUSOFF, SIM LAI HAR,
CHAN CHIN HAN, AMIRAH HASHIFUDIN,
and HANS–WERNER KAMMER

CONTENTS

14.1 Introduction 400
14.2 Experimental 401
14.3 Results and Discussions 402
14.4 Conclusions 408
Acknowledgment 409
Keywords 409
References 409

14.1 INTRODUCTION

Solid solutions of epoxidized natural rubber with 25 and 50 mol % epoxidation, ENR-25 and ENR-50, respectively, added with LiClO$_4$ were prepared by solution casting technique. Glass transition temperature (T_g) values obtained using differential scanning calorimetry (DSC) and the ionic conductivity evaluated from bulk resistance (R_b) determined using the impedance spectroscopy point towards higher solubility of the lithium salt in ENR-50 when the ratio of the mass of salt to the mass of polymer (Y) 0.15. This ramification correlates with spectroscopic results demonstrated in FTIR spectra. Ionic conductivity (σ) is observed to increase with ascending values of Y. When Y 0.15, ENR-50 exhibits higher ionic conductivity than ENR-25 but the σ values of ENR-25 increase sharply with increasing salt content to above that of ENR-50 when Y 0.20. Higher ion mobility, better salt molecule-chain segment correlation and higher charge carrier diffusion rate account for the higher σ value for ENR-50 at 0.00 < Y 0.15. However, restricted ion transport for ENR-50 and relatively flexible segmental motion for ENR-25 at Y 0.20 cause the conductivity of ENR-25 to be higher than that of ENR-50. Therefore, conductivity of the epoxidized natural rubber is primarily governed by the segmental motion of the elastomer rather than charge carrier density, since the discovery of ion-conducting polymer by Fenton et al. [1] followed by the application of polymer electrolyte in lithium batteries by Armand et al. [2], solid polymer electrolyte (SPE) has been widely studied especially on the enhancement of ionic conductivity. To date, SPE has become the focus of extensive research in pursue for a new generation of power source to cater for the latest development in electrochemical devices. Polymer electrolyte is a complex formed by dissolving an inorganic salt in a polymer matrix with polar groups acting as an immobile solvent. It is generally accepted that the charge carrier density and ion mobility are the two important parameters contributing to the ionic conductivity of a SPE [3–5]. For SPE with a semi-crystalline polymer matrix like poly(ethylene oxide) (PEO), ion mobility is attributed to the segmental motion of the amorphous phase of the macromolecular chain. Epoxidized natural rubber (ENR) is derived from natural rubber by converting different percent of the C=C bonds on the macromolecular backbone to the polar epoxy groups. ENR has good potential to be polymer host in SPE because of their distinctive characteristic such as low glass transition temperature (T_g), soft elastomeric characteristics at room temperature [5] and good electrode-electrolyte adhesion. Furthermore, the highly flexible macromolecular chain and the polar epoxy oxygen provide excellent segmental motion and coordination sites for Li$^+$ ion transport, respectively, in ENR-based polymer electrolytes [4, 6]. However, Chan and Kammer [7], in their study on the properties of PEO/ENR/lithium perchlorate (LiClO$_4$) solid solutions concluded that the solubility of the ionic salt is comparatively higher in PEO than in ENR. Therefore, the conductivity mechanism as well as the role of the oxirane ring in the dissociation of ionic salt in ENR-salt complex will be investigated in detail. Other than ENR with 25 and 50 mol percent of epoxidation, ENR-25 and ENR-50, respectively, deproteinized natural rubber (DPNR) is used as a polymer reference [8].

14.2 EXPERIMENTAL

14.2.1 MATERIAL

Epoxidized natural rubber (ENR-25 and ENR-50) purchased from Malaysian Rubber Board (Sungai Buloh, Malaysia) was used after purification. The chemical structures of the two elastomers were shown in Fig. 1. Deproteinized natural rubber (DPNR) was supplied by Green HPSP (M) Sdn Bhd (Petaling Jaya, Malaysia). $LiClO_4$ was purchased from Acrōs Organic Company (Geel, Belgium) and the solvent tetrahydrofuran (THF) was purchased from Merck.

FIGURE 1 Chemical structures of (a) ENR-25, (b) ENR-50.

14.2.2 SAMPLE PREPARATION

Thin films of polymer electrolytes with ENR-25, ENR-50 and DPNR as the polymer host were prepared by solution casting technique. Appropriate amounts of the rubber and $LiClO_4$ were dissolved in THF by stirring with a magnetic stirrer at 50°C until a homogeneous solution was obtained. The $LiClO_4$ content (Y) added varied in a range from 0.01–0.30 of 1 gram dry weight of the polymer. It can be represented by Eq. (1)

$$Y = \frac{\text{mass of salt}}{\text{mass of polymer}} \qquad (1)$$

The electrolyte solution was cast into a Teflon dish and left to dry overnight at room temperature to form a thin film. The electrolyte films were dried in an oven for 24 hours at 50°C before heated for another 24 hours in nitrogen atmosphere at 80°C. This was to ensure a good interaction between the salt and the elastomer. The free standing film was further dried in a vacuum oven at 50°C for another 24 hours before keeping in desiccators for further characterization.

14.2.3 IMPEDANCE MEASUREMENTS

The ionic conductivity measurements were performed using the HIOKI 3532-50 LCR Hi-Tester interfaced to a computer. The samples were scanned at frequencies ranging from 50 Hz to 1 MHz at room temperature. The thin film samples were sandwiched between two stainless steel block electrodes of 20 mm in diameter. The conductivity (σ) of each sample was calculated using the equation $\sigma = t/(R_b A)$, where t is the thickness of the sample, R_b is the bulk resistance and A is the cross-section area of the film. The R_b value is the intersection point between the semicircle and the x-axis in a cole–cole plot. The average of the thickness (t) was calculated from four measurements of the thickness of the thin film using Mitutoyo Digimatic Calliper (Japan).

14.2.4 DIFFERENTIAL SCANNING CALORIMETRY

Differential scanning calorimeter, TA Q200 DSC calibrated with indium standard was used to study the thermal properties of the sample. For estimation of the T_g, approximately 10–12 mg of the sample was used for each analysis. The sample was cooled down to –90°C and was heated up to 80°C at a rate of 10°C min^{-1}.

14.2.5 FOURIER TRANSFORM INFRARED SPECTROSCOPY

Infrared spectra of all samples were obtained using the Attenuated Total Reflectance (ATR) method on Perkin Elmer Spectrum 1 spectrometer at room temperature with frequency range of 4000–650 cm^{-1}. For each sample, 32 scans were taken at maximum resolution of 2 cm^{-1} using the Ge crystal plate.

14.3 RESULTS AND DISCUSSIONS

14.3.1 GLASS TRANSITION TEMPERATURE

The glass transition temperature (T_g) of a polymer, observed as an endothermic shift from the baseline of a thermogram, is determined using differential scanning calorimetry (DSC). It is governed mainly by the heating and cooling rates applied in the DSC run. The T_g values of the modified natural rubber (MNR), extracted from the second heating runs, are presented in Fig. 2. Among all the three salt-free rubber samples, DPNR records the lowest T_g value of –65°C compared to –42°C and –21°C for ENR-25 and ENR-50, respectively. It is well documented that an increase in the epoxidation level of ENR will cause a reduction in the free volume of the chain phases leading to stiffening of the molecular chain structure [9].

Figure 2 depicts that the T_g values of the ENR samples increase with ascending LiClO$_4$ from $Y = 0.00$ to 0.30 while that of the DPNR remains relatively constant in the same range of salt content. Similar result on ENR was also reported by Idris et al. [4]. Solubility of LiClO$_4$ in the ENR samples through the complexation between Li$^+$ ion and the epoxy oxygen of the oxirane group results in the stiffening of the polymer chain. ENR-50 with more epoxy oxygen in its macromolecular backbone as coordination sites for the Li$^+$ ion experience a larger increase in T_g values from –21 to 8°C as compared to ENR-25 when the salt content increases from $Y = 0.00$ to 0.30. Meanwhile, weak interaction between Li$^+$ ion and the unsaturated ethylene group (C=C) of DPNR accounts for the lower solubility of the

A Study on Solid Polymer Electrolytes

salt in the sample leading to relatively constant T_g values obtained with ascending salt content.

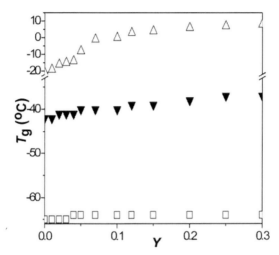

FIGURE 2 Glass transition temperature of the rubber samples as a function of Y, (□) DPNR, (▼) ENR-25, and (Δ) ENR-50.

14.3.2 INFRARED SPECTROSCOPY

ATR- FTIR spectroscopy is applied to investigate the effect of the ion-dipole interactions between ENR-25, ENR-50 and DPNR with the lithium salt.

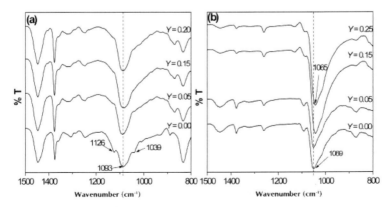

FIGURE 3 FTIR spectra in the region of 1500–800 cm^{-1} for (a) ENR-25 and (b) ENR-50 added with different content of LiClO$_4$ (Y).

The triplet centered at 1093 cm^{-1} with two shoulders at 1039 and 1126 cm^{-1} as shown in Fig. 3a is assigned to the C–O–C stretching mode of the oxirane group in ENR-25. Addition of LiClO$_4$ from $Y = 0.00$ to 0.20 results in the disappearance of the two shoulders, indicating the coordination of Li$^+$ to the oxygen atom of the oxirane group. However, no significant shifting of the center peak at 1093 cm^{-1} is observed. On the other hand, addition of $Y = 0.01$ to 0.15 of LiClO$_4$ to ENR-50 as shown in Fig. 3b, causes an immediate downshifting of the C–O–C stretching mode from 1069 to 1065 cm^{-1}. No further shifting is observed with increasing salt content from $Y = 0.15$ to 0.30. The downshifting of the C–O–C vibration mode observed in ENR-50 and not in ENR-25 implies that more Li$^+$ ions form complexes with ENR-50 than with ENR-25. This result concurs closely with that observed in the T_g values discussed in the T_g section.

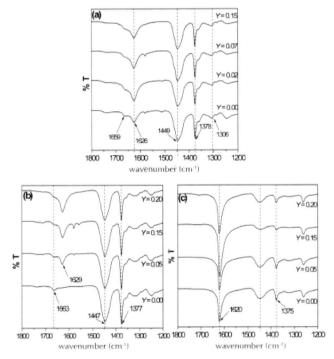

FIGURE 4 FTIR spectra for (a) DPNR, (b) ENR-25 and (c) ENR-50 doped with different salt content in the range of 1800–1200 cm^{-1}.

The unsaturated ethylene groups (C=C) which absorb moderately at 1659 cm^{-1} and strongly at 1626 cm^{-1} as shown in Fig. 4a, are the only nucleophilic sites in DPNR available for ion-dipole interactions with the Li$^+$ ions. No significant shifting but merging of the two absorbance peaks to form a broad peak at 1626 cm^{-1} for the C=C vibration mode is observed suggests weak ion-dipole interaction between the C=C group and Li$^+$ ion. This result correlates with the relatively constant T_g observed in DPNR when increasing salt content is added. Addition of epoxy oxygen to the C=C bond of natural rubber causes a shifting in the absorbance bands of the remaining C=C groups to 1663 cm^{-1} for ENR-25 and 1620 cm^{-1} for ENR-50 as shown in Figs. 4b and 4c, respectively. In Fig. 4b, one observes that the broad absorbance peak at 1663 cm^{-1} for ENR-25 continues to broaden with increasing salt content whereas the new peak at 1629 cm^{-1} which is formed on the addition of salt increases progressively in intensity. Examining the chemical structure of ENR-25, one can see that the C=C bonds experience different electron environment in the neighborhood of the epoxy groups. Therefore, coordination between Li$^+$ ions and the epoxy groups causes the vibration mode of the C=C bonds to downshift to different wavenumbers according to their positions relative to the epoxy group. The absorbance band at 1629 cm^{-1} is ascribed to the one next to the epoxy group while the very broad peak in the vicinity of 1663 cm^{-1} is assigned to the rest of the C=C bonds further away from the epoxy groups. It is noteworthy that apart from the increase in the Li$^+$-epoxy oxygen interaction, the increase in intensity with ascending salt content is also partly due to the overlapping of the C=C vibration mode at 1629 cm^{-1} and the internal vibration mode of ClO$_4^-$ anion which absorbs at the same wavenumber [10]. With the C=C bonds alternate with the epoxy groups on the macromolecular backbone of ENR-50, the vibration mode of C=C gives a sharp and intense absorbance at 1620 cm^{-1} for the neat ENR-50 as shown in Fig. 4c. No change in the intensity and peak position of the C=C vibration mode observed in ENR-50 suggests that the Li$^+$ ions prefer to coordinate with epoxy oxygen than with the C=C bonds. Furthermore, it is noted in Fig. 4c that there are broadening and reduction in intensity in the bending modes of CH$_2$ scissoring, CH$_2$ wagging and CH$_2$ twisting at 1447, 1377 and 1309 cm^{-1}, respectively [11]. This observation points towards stronger interactions at the C–O–C of ENR-50 which influence the electron environment of the neighboring methylene (CH$_2$) groups, and thus closely relates to the higher increase in T_g observed with ascending LiClO$_4$ for ENR-50.

14.3.3 CONDUCTIVITY

Figure 5 demonstrates the semi-logarithm plot of σ versus Y for DPNR, ENR-25 and ENR-50 at room temperature. It is noted that the ionic conductivity of all the three rubber samples increase with ascending salt content due to an increase in the number of charge carriers. Being non-crystalline elastomers, the segmental motion of the macromolecular chains promotes ion transport of Li$^+$ ions within the polymer [12].

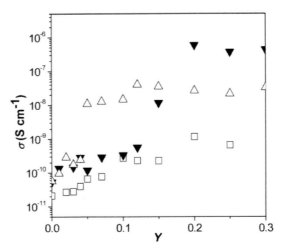

FIGURE 5 Ionic conductivity (σ) versus Y of MNR polymer electrolytes at 30°C with salt content ranging from 0.00 to 0.30; ($\square$) DPNR, ($\blacktriangledown$) ENR-25, and ($\triangle$) ENR-50, respectively.

Among the three NR-based polymer electrolytes investigated, DPNR records the lowest σ values at all salt concentrations while ENR-50 exhibits the highest σ values at low salt content from $Y = 0.00$ to 0.15. However, as the $LiClO_4$ content increases from $Y = 0.12$, ENR-25 encounters a marked increase in σ value such that when $Y \geq 0.20$, the σ value of ENR-25 is approximately 2 orders and 3 orders in magnitude higher than that of ENR-50 and DPNR, respectively. Furthermore, it is noted that both the ENR polymer electrolytes demonstrate a sharp increase in σ value (~2–3 orders in magnitude) at a certain range of $LiClO_4$, follow by a plateau with further increase in salt content. The significant jump in σ values which occurs at $Y = 0.05$–0.12 for ENR-50 and $Y = 0.12$–0.20 for ENR-25 is most probably the result of the formation of a stable percolation network in the polymer electrolytes due to polarization relaxations. Meanwhile, ENR-50 records a maximum σ value of 3.7 10^{-8} S cm^{-1} at $Y = 0.12$ whereas ENR-25 has a maximum σ value of 5.7 10^{-7} S cm^{-1} at $Y = 0.20$.

As mentioned in the discussion on T_g, flexibility and the segmental motion of the polymer chain decrease with increasing epoxidation level and salt concentration. At relatively high salt content, the macromolecular backbone of ENR-50 experiences increasing stiffness caused by higher degree of epoxidation and increasing Li$^+$-polymer complex formation [13]. In addition, ENR-25 with an optimal number of epoxy group, has an advantage over ENR-50 because it not only possesses coordination sites for Li$^+$ ions but is also able to maintain the good elastomeric characteristics of natural rubber at relatively high salt content. Therefore, ENR-50 as shown in Fig. 5 exhibits higher ionic conductivity at salt content $Y \geq$

0.15 due to higher ion mobility. However, with salt content increases to $Y \geq 0.20$, restricted segmental motion in ENR-50 and a relatively steady ion transport in the stabilize percolation network of ENR-25 account for the higher σ value for the latter electrolyte system.

At low salt concentration, the dependence of conductivity on $LiClO_4$ content (Y) can be described using the power law [14] as shown in Eq. (2).

$$\sigma_{DC} = N_A e(\alpha\mu) \frac{\rho_{MNR}}{M_{salt}} (Y)^x \qquad (2)$$

where N_A represents Avogadro's number, e is the elementary charge, μ and α denote the ion mobility and the degree of dissociation, respectively. The product of degree of dissociation and mobility ($\alpha\mu$) is referred to as the ion mobility for the following discussions. The exponent x which is determined experimentally gives the extent of correlations between the salt molecules and the MNR segments. Meanwhile, ρ_{MNR} and M_{salt} represent the density of the MNR and the molar mass of the salt molecule, respectively.

A double logarithmic plot of σ versus Y at the range of low salt content from $Y = 0.00$ to 0.15 is shown in Fig. 6. The exponent x and ion mobility ($\alpha\mu$) can be extracted from the slope and the y-intercept of the regression functions, respectively after Eq. (2) in Fig. 6. Molecular characteristics adopted for the determination of $\alpha\mu$ are M_{salt} (M_{LiClO4}: 106.5 g mol^{-1}) and ρ_{MNR} (ρ_{DPNR}: 0.920 g cm^{-3}, ρ_{ENR-25}: 0.971 g cm^{-3} and ρ_{ENR-50}: 1.027 g cm^{-3}).

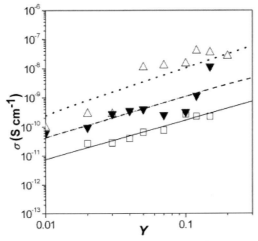

FIGURE 6 The dependence of ionic conductivity (σ) on salt concentration at 30°C, with salt content ranging from 0.00–0.15 for (□) DPNR, (▼) ENR-25, and (△) ENR-50 with solid, dashed and dotted curves, respectively, representing linear regression after Eq. (2).

The value of diffusion coefficient (D) of the charge carrier can be evaluated by adopting the Nernst's relationship as given in Eq. (3). This equation reflects the dependence of D on the absolute temperature (T) and the ion mobility ($\alpha\mu$).

$$D = \frac{k_B T \alpha\mu}{e} \qquad (3)$$

where k_B is the Boltzmann constant ($k_B = 1.381 \times 10^{-23}$ m^2 kg s^{-2} K^{-1}) and T is the absolute temperature ($T = 303$ K). All the parameters determined from the regression functions after Eqs. (2) and (3) at room temperature and the salt content ranging from $Y = 0.00$ to 0.15 are listed in Table 1. The values of $\alpha\mu$ and the exponent x for ENR-50 are the highest among all the three rubber samples, and this explains its highest σ value at $Y \geq 0.15$ as shown in Figs. 5 and 6. Besides, LiClO$_4$ has the highest diffusion rate in ENR-50 at low salt content. On the contrary, DPNR displays the lowest values for ion mobility, exponent x and D which agree closely with its lowest σ value and the weak interaction between the polymer chain and the salt as described in the T_g section.

TABLE 1 Regression functions after Eq. (2), ion mobility ($\alpha\mu$); exponent x and diffusion coefficient (D) of MNR after Eq. (3) for all the MNR with LiClO$_4$ at salt content ranging from $Y = 0.00$ to 0.15.

	DPNR	ENR-25	ENR-50
Regression function	$\sigma = 3.8\times10^{-9}\ Y^{1.35}$	$\sigma = 2.9\times10^{-8}\ Y^{1.41}$	$\sigma = 4.1\times10^{-7}\ Y^{1.58}$
Correlation, R	0.967	0.844	0.908
Exponent x	1.35	1.41	1.58
$\alpha\mu$ (cm^2 V^{-1} s^{-1})	4.6×10^{-12}	3.3×10^{-11}	4.4×10^{-10}
D (cm^2 s^{-1})	1.2×10^{-13}	8.5×10^{-13}	1.1×10^{-11}

14.4 CONCLUSIONS

Solid polymer electrolytes of ENR-25/LiClO$_4$, ENR-50/LiClO$_4$ and DPNR/LiClO$_4$ with various concentrations of LiClO$_4$ were successfully prepared using solution casting technique. The effect of LiClO$_4$ on the conductivity, thermal properties and polymer-salt interaction of ENR-25 and ENR-50 were investigated by AC impedance spectroscopy, differential scanning calorimetry and ATR–FTIR, respectively. Ascending T_g values with increasing salt content which is higher in ENR-50 as compared to ENR-25 and DPNR correlates with increased ion-dipole interaction for ENR-50 observed in the FTIR spectra. At LiClO$_4$ content $Y \geq 0.15$, ENR-50 exhibits higher ionic conductivity than ENR-25 as a result of higher ion mobility, better correlation between salt molecules and chain segments and higher diffusion rate of charge carrier. However, at higher salt content Y 0.20, reduction in segmental motion leading to restricted ion transport for ENR-50 as compared to good ion mobility due to flexible chain movement for ENR-25 results in higher ionic conductivity in the latter electrolyte system. However, the conductivity of ENR-25 is too low for any practical application. Blending ENR with a second polymer consisting of polar groups like PEO and incorporation with inorganic fillers can be applied to enhance the conductivity of the MNR.

ACKNOWLEDGMENT

The authors would like to express our gratitude to the Research Management Institute (RMI), University Teknologi MARA for awarding the grant 600–RMI/ST/Dana 5/3/Dst (426/2011).

KEYWORDS

- **ATR–FTIR**
- **conductivity of the MNR**
- **differential scanning calorimetry**
- **electrolytes**
- **polymer-salt interaction**
- **solid polymer**
- **spectroscopy**

REFERENCES

1. D.E. Fenton, J.M. Parker, and P.V. Wright, "Complexes of alkali metal ions with poly(ethylene oxide), "*Polymer,* vol. 14, pp. 589–594, 1973.
2. M.B. Armand, M.J. Duclot, and P. Rigaud, "Polymer solid electrolytes: Stability domain, "*Solid State Ionics,* vol. 3–4, pp. 429–430, 1981.
3. F. Latif, M. Aziz, N. Katun, A.M.M. Ali, and M.Z. Yahya, "The role and impact of rubber in poly(methyl methacrylate)/lithium triflate electrolyte, "*Journal of Power Sources,* vol. 159, pp. 1401–1404, 2006.
4. R. Idris, M.D. Glasse, R.J. Latham, R.G. Linford, and W.S. Schlindwein, "Polymer electrolytes based on modified natural rubber for use in rechargeable lithium batteries, "*Journal of Power Sources,* vol. 94, pp. 206–211, 2001.
5. F. Latif, M. Aziz, A.M.M. Ali, and M.Z.A. Yahya, "The Coagulation Impact of 50% Epoxidised Natural Rubber Chain in Ethylene Carbonate–Plasticized Solid Electrolytes, "*Macromolecular Symposia,* vol. 277, pp. 62–68, 2009.

6. M.L. Hallensleben, H.R. Schmidt, and R.H. Schuster, "Epoxidation of poly(cis-1,4-isoprene) microgels, "*Die Angewandte Makromolekulare Chemie,* vol. 227, pp. 87–99, 1995.
7. C.H. Chan and H.W. Kammer, "Properties of solid solutions of poly(ethylene oxide)/epoxidized natural rubber blends and LiClO$_4$, "*Journal of Applied Polymer Science,* vol. 110, pp. 424–432, 2008.
8. W. Klinklai, S. Kawahara, T. Mizumo, M. Yoshizawa, Y. Isono, and H. Ohno, "Ionic conductivity of highly deproteinized natural rubber having epoxy group mixed with alkali metal salts, "*Solid State Ionics,* vol. 168, pp. 131–136, 2004.
9. I.R. Gelling, "Modification of natural rubber latex with peracetic acid, "*Rubber Chemistry and Technology,* vol. 58, pp. 86–96, 1985.
10. S. Rajendran, O. Mahendran and R. Kannan, "Ionic conductivity studies in composite solid polymer electrolytes based on PMMA, "*J. Phys. Chem. Solids,* vol. 63, pp. 303–307, 2002.
11. S.C. Ng and L.H. Gan, "Reaction of natural rubber latex with performic acid, "*European Polymer Journal,* vol. 17, pp. 1073–1077, 1981.
12. A.M.M. Ali, M.Z.A. Yahya, H. Bahron, and R.H.Y. Subban, "Electrochemical studies on polymer electrolytes based on poly(methyl methacrylate)-grafted natural rubber for lithium polymer battery, "*Ionics,* vol. 12, pp. 303–307, 2006.
13. R.H.Y. Subban, A.K. Arof, and S. Radhakrishna, "Polymer batteries with chitosan electrolyte mixed with sodium perchlorate, "*Materials Science and Engineering: B,* vol. 38, pp. 156–160, 1996.
14. N.H.A. Nasir, C.H. Chan, H.-W. Kammer, L.H. Sim, and M.Z.A. Yahya, "Ionic conductivity in solutions of poly(ethylene oxide) and lithium perchlorate, "*Macromolecular Symposia,* vol. 290, pp. 46–55, 2010.

CHAPTER 15

MODIFICATION OF PC–PHBH BLEND MONOLITH

YUANRONG XIN and HIROSHI UYAMA

CONTENTS

15.1 Introduction 412
15.2 Experimental Section 413
15.3 Results and Discussion 415
15.4 Conclusion 418
Acknowledgment 419
Keywords 419
References 419

15.1 INTRODUCTION

A polyethylenimine (PEI)-modified blend monolith with porous structure was prepared as an effective adsorbent to remove copper ion (Cu^{2+}) in aqueous media. The polycarbonate (PC) and poly(3-hydroxybutyrate-*co*-3-hydroxyhexanoate) (PHBH) blend monolith was selected as matrix, which was fabricated *via* non-solvent induced phase separation (NIPS). PEI, a chelating agent to bind metal ions, was covalently connected onto the surface of the blend monolith by aminolysis reaction. The adsorption capacity of the PEI-modified blend monolith for Cu^{2+} was evaluated. The adsorption has been examined under various conditions such as solution pH, adsorption time, and Cu^{2+} concentration. The maximum capacity for Cu^{2+} is 55 mg/g. The present PEI-modified blend monolith has large potential for wastewater treatment.

Removal of toxic metal ions has become an urgent issue due to their serious contamination in waste effluents of many industrial processes. Copper ions are one of these toxic metal ones which could result in severe health problems. Excessive uptake of copper ions may cause damage to heart, kidney, liver, pancreas and brain [1–5]. Typical treatments of copper-containing wastewater are adsorption, chemical precipitation, ion exchange, solvent extraction, reverse osmosis and membrane separation [6–10]. Among them, adsorption technique is considered to be an attractive method because of its low cost, simple operation, potential recovery of metals, and regeneration of the adsorbent by suitable desorption process.

Monoliths, materials with open-cellular three-dimensional continuous structure, have attracted considerable attention due to their good permeability, fast mass transfer property, high stability and easy modification [11–15]. During the last decade, polymer-based monoliths have become greatly significant. The unique features such as tunable bulk and surface properties and surface functionalizations afford wide applications in various industrial fields [16–18]. These monolithic materials could provide a convenient and effective route to adsorb copper ions in an aqueous solution due to their large surface area.

Recently, we have developed a novel approach to fabricate polymer monoliths by thermally induced [19, 20] and non-solvent induced [21] phase separation techniques using the polymer itself as precursor. The fabrication process was very convenient and clean for both techniques. In the case of non-solvent induced phase separation (NIPS), a monolith was obtained by addition of a non-solvent to a homogeneous polymer solution. An appropriate selection of solvent/non-solvent and their mixed ratio enabled formation of the monolith with uniform structure. In the present work, a blend monolith composed of polycarbonate (PC) and poly(3-hydroxybutyrate-*co*-3-hydroxyhexanoate) (PHBH) fabricated *via* NIPS method was utilized as matrix for adsorption of copper ion (Cu^{2+}) [22] (Fig. 1). PC is one of the most widely used thermoplastics due to its properties such as impact resistance, heat resistance and dimensional stability [23–26]. Hence PC monolith is an ideal candidate to be used as solid substrate bearing high mechanic strength. PHBH is a kind of microbial polyesters naturally produced by microorganisms from biomass. Due to the segment of 3-hydroxyhexanoate (HH) unit in the molecular structure [27–30], PHBH could afford improved flexibility and ductility to the blend monolith.

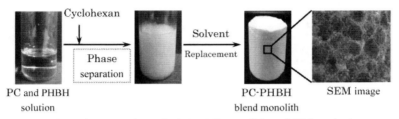

FIGURE 1 Fabrication procedure of PC–PHBH monolith *via* NIPS method.

In order to provide the affinity towards metal ions in polymer monoliths, the attachment of functional groups on the surface of monolith is required. Branched polyethylenimine (PEI) is a water-soluble polycation consisting of primary, secondary and tertiary amine functional groups in 1/2/1 molar ratio [2–3, 31–34]. PEI is widely used for removal of toxic metal pollutants due to the excellent chelation ability towards heavy various metal ions.

In this study, the PC–PHBH blend monolith [22] was modified with PEI by aminolysis reaction for immobilization of the PEI chain *via* covalent bond on the monolith surface. The PEI-modified blend monolith was utilized as adsorbent to remove Cu^{2+} in an aqueous solution.

15.2 EXPERIMENTAL SECTION

15.2.1 MATERIALS

PC (M_n=2.3×10^4, M_w/M_n=1.4) was purchased from Sigma–Aldrich Co. PHBH with 11 mol% HH content (M_n=4.8×10^4, M_w/M_n=1.1) was supplied by Kaneka Co. PEI with molecular weight of 1x10^4 was purchased from Wako Pure Chemical Industries. Ltd. All reagents were used as received without further purification.

15.2.2 MEASUREMENTS

Scanning electron microscopic (SEM) images were recorded on a Hitachi S-3000N instrument at 15 kV. A thin gold film was sputtered on the samples. FT–IR measurement was carried out by a Perkin Elmer Spectrum One System B2 spectrometer with a universal ATR sampling accessory. Metal concentrations in an aqueous solution were analyzed by induced coupled plasma-atom emission spectrometry (ICP–AES, an ICP-7510 Shimadzu sequential plasma spectrometer). pH of the aqueous solution was measured by a Horiba Compact pH meter B-211.

Elemental analysis was performed by a Yanaco CHN corder instrument (MT-5 type, Yanagimoto Mfg. Co., Ltd., Japan).

15.2.3 SYNTHESIS OF PEI-MODIFIED BLEND MONOLITH

PC–PHBH blend monolith was prepared according to the literature [22]. The PEI-modified blend monolith was synthesized through aminolysis reaction (Scheme 1). Briefly, PC–PHBH blend monolith (0.15 g, PC/PHBH=80/20 (wt)) was immersed into ethanol solution of PEI (10 mL, 0.020 g/mL), followed by shaking gently at 20°C. After 36 h, the monolith was rinsed by ethanol for 3 times and dried in vacuo.

15.2.4 ADSORPTION OF CU^{2+}

The PEI-modified blend monolith was utilized to adsorb Cu^{2+} in aqueous solutions (Scheme 1). The modified blend monolith (0.15 g) was immersed into $CuCl_2$ solution (7.5 mL) in the concentration ranging from 0.010 to 2.0 mg/mL with different pH (1.3~4.8), and the solution was shaken at 300 rpm at 20°C. pH of $CuCl_2$ solution was adjusted by 0.1 mol/L HCl or 0.1 mol/L NaOH. The adsorption procedures for other metal ions, potassium (K^+), sodium (Na^+), nickel (Ni^{2+}) and cobalt (CO_2^+) ions, were similar to those of Cu^{2+}.

The concentration of metal ions in an aqueous solution was determined by ICP–AES. The amount of metal ions adsorbed by the blend monolith was calculated using the following equation:

$$qt = (C_0 - C_t)V/m \qquad (1)$$

where qt (mg/g) is the metal adsorption capacity of the blend monolith, C_0 and C_t are the metal concentrations (mg/mL) initially and at a given time, respectively, V is the solution volume (mL), and m is the weight (g) of the PEI-modified blend monolith.

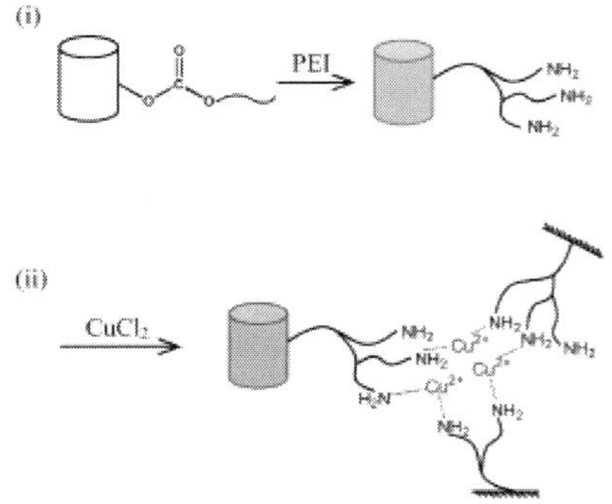

SCHEME 1 Synthesis of PEI-modified blend monolith (i) and its adsorption for Cu^{2+} (ii).

15.3 RESULTS AND DISCUSSION

15.3.1 MODIFICATION OF BLEND MONOLITH BY PEI

The PC–PHBH blend monolith was modified by aminolysis with PEI. Figure 2 shows the SEM images of the blend monolith before and after the reaction. Figure 2a shows that inside the blend monolith, PC and PHBH phases were well distributed into each other and the latter was of the round circle-shape with larger diameter, which dispersed uniformly in the continuous matrix of the former consisting of smaller pores. The sizes of the skeleton and smaller pore of the blend monolith were in the range of 0.2–0.7 and 0.9–2.3 µm, respectively. The morphology of the monolith after the aminolysis was hardly changed (Fig. 2b).

FIGURE 2 SEM images of original PC–PHBH blend monolith (a), after PEI modification (b), and after Cu^{2+} adsorption (c).

Figure 3 shows the FT–IR spectra of the unmodified (a) and modified PC (b) monoliths. For the modified monolith, the new peaks appear at 3280 and 1680 cm^{-1}, which are assigned to the stretching vibrations of N–H and amide bonds, respectively. These data strongly suggest that the aminolysis took place on the surface of the PC–PHBH blend monolith. The nitrogen content of the blend monolith determined by elemental analysis was 3.4 wt%. The PEI content in the modified monolith was estimated to be roughly 10 wt%.

15.3.2 ADSORPTION OF METAL IONS BY MONOLITH

At first, the effect of pH on the adsorption of Cu^{2+} by the blend monolith was investigated in the pH range from 1.3 to 4.8 (Fig. 4). The experiment at pH over 4.8 was not carried out because of the copper hydroxide precipitation. The adsorption capacity was strongly dependent on the pH of the solution, and the modified monolith had a maximum adsorption amount at pH of 4.8. At pH 1.3, all the amino groups of PEI on the surface of the blend monolith were protonated and repel positive metal ions, resulting in the very weak chelation ability of the monolith. At higher pH, on the other hand, the amino groups of PEI were partly or mostly deprotonated and thereby the adsorption capacity of the monolith increased. 9995632114.

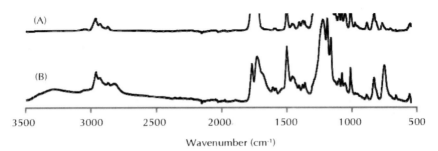

FIGURE 3 FT-IR spectra of PC-PHBH monolith before (a) and after (b) PEI modification.

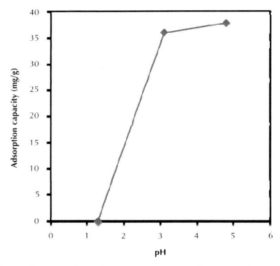

FIGURE 4 Effect of pH on the adsorption capacity of PEI-modified PC-PHBH blend monolith (initial Cu^{2+} concentration; 1.0 mg/mL; adsorption time: 14 h).

Figure 5 shows time-course in the adsorption of Cu^{2+} by the PEI-modified monolith at pH 4.8. The rapid increase of the adsorbed amount of Cu^{2+} was found from 0.5 to 12 h; afterwards, the amount gradually increased. After 14 h, the adsorption came to the equilibrium. The inside structure of the PEI-modified monolith was hardly changed through adsorption of Cu^{2+} (Fig. 2c).

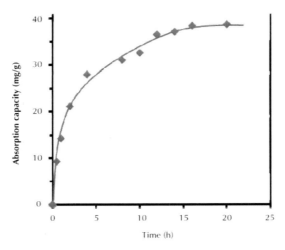

FIGURE 5 Time-course of Cu^{2+} adsorption by PEI-modified PC-PHBH blend monolith (initial Cu^{2+} concentration: 1.0 mg/mL; pH=4.8).

Figure 6 represents the relationship between the initial Cu^{2+} concentration and adsorption capacity of the PEI-modified blend monolith. The initial concentration varied from 0.010 to 2.0 mg/mL. The adsorption capacity increased from 0.24 to 55 mg/g as a function of the initial concentration. This may be because in the higher initial concentration, Cu^{2+} has higher driving force to overcome mass transfer resistance, leading to the higher adsorption capacity [6].

The adsorption of various metals by the PEI-modified monolith was examined (Fig. 7). K^+ and Na^+ ions were not adsorbed on the monolith, which is due to the very low affinity of the PEI chains towards these monovalent cations. For divalent cations of Cu^{2+}, Ni^{2+}, and CO_2^+, the adsorption capacities were 38, 16 and 11 mg/g, respectively. These data indicate that the PEI-modified blend monolith has preferable adsorption towards Cu^{2+}. The chelating order ($Cu^{2+} > Ni^{2+} > CO_2^+$) may be related to the electronegativity and size of these metals ions; metal ions of stronger electronegativity and smaller ionic radii are more easily diffused into the interior of the blend monolith and thus have larger affinity with the PEI chain of the blend monolith.

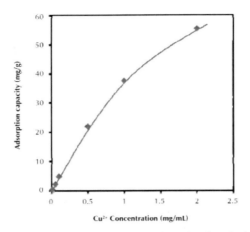

FIGURE 6 Relationship between initial concentration of Cu^{2+} and adsorption capacity of PEI-modified PC-PHBH blend monolith (adsorption time: 14 h; pH=4.8).

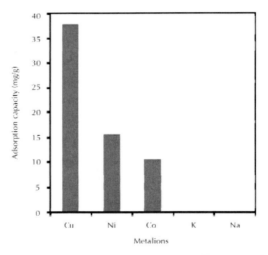

FIGURE 7 Adsorption of various metal ions by PEI-modified PC-PHBH blend monolith to (initial metal concentration: 1.0 mg/mL; adsorption time: 14 h; pH=4.8).

15.4 CONCLUSION

In this paper, the PEI-modified PC–PHBH blend monolith with continuous interconnected structure was fabricated successfully *via* NIPS and aminolysis reaction. The resulting monolith has good chelating ability for Cu^{2+}. The maximum adsorption was found at pH 4.8. The chelating order of the monolith was Cu^{2+} > Ni^{2+} > CO_2^+ and no chelation was found for Na^+ and K^+.

Monoliths have characteristic properties suitable for applications of water treatment such as large surface area and high mass transfer. Furthermore, the immobilization of functional molecules on the monolith will expand their industrial applications. Therefore, the PEI-modified PC–PHBH blend monolith has large potential for various applications including removal of toxic metal ions in wastewater treatment.

ACKNOWLEDGMENT

This study is financially supported by a Grant-in-Aid for Scientific Research from the Japan Society for the Promotion of Science (No. 24655208) and the New Energy and Industrial Technology Development Organization (NEDO) of Japan. We gratefully acknowledge the gift of PHBH from Kaneka Co.

KEYWORDS

- adsorption
- aqueous media
- magnetic nanoparticles
- modification
- PC–PHBH blend
- porous structure

REFERENCES

1. Chen, J., Zeng, F., Wu, S., Su, J., Zhao, J., Tong, Z.: A facile approach for cupric ion detection in aqueous media using polyethylenimine/PMMA core-shell fluorescent nanoparticles. Nanotechnology, 20, 1–7 (2009).
2. Goon, I.Y., Zhang, C., Lim, M., Gooding, J.J., Amal, R.: Controlled fabrication of polyethylenimine-functionalized magnetic nanoparticles for the sequestration and quantification of free Cu^{2+}. Langmuir, 26, 12247–12252 (2010).
3. Wu, A., Jia, J., Luan, S.: Amphiphilic PMMA/PEI core-shell nanoparticles as polymeric adsorbents to remove heavy metal pollutants. Colloid and Surface A: Physicochem. Eng. Aspects, 384, 180–185 (2011).
4. Molinari, R., Gallo, S., Argurio, P.: Metal ions removal from wastewater or washing water from contaminated soil by ultrafiltration-complexation. Water Res., 38, 593–600 (2004).
5. Liu, M., Zhao, H., Quan, X., Chen, S., Yu, H.: Signal amplification *via* cation exchange reaction: an example in the ratiometric fluorescence probe for ultrasensitive and selective sensing of Cu(II). Chem. Commun., 46, 1144–1146 (2010).

6. Aluigi, A., Tonetti, C., Vineis, C., Tonin, C., Mazzuchetti, G.: Adsorption of copper(II) ions by keratin/PA6 blend nanofibers. Eur. Polym. J., 47, 1756–1764 (2011).
7. Sun, X.F., Liu, C., Ma, Y., Wang, S.G., Gao, B.Y., Li, X.M.: Enhanced Cu(II) and Cr(VI) biosorption capacity on poly(ethylenimine) grafted aerobic granular sludge. Colloids and Surfaces B: Biointerfaces, 82, 456–462 (2011).
8. Yin, C.Y., Aroua, M.K., Daud, W.M. A.W.: Fixed-bed adsorption of metal ions from aqueous solution on polyethylenimine-impregnated palm shell activated carbon. Chem. Eng. J., 148, 8–14 (2009).
9. Chen, Y., Pan, B., Zhang, S., Li, H., Lv, L., Zhang, W.: Immobilization of polyethylenimine nanoclusters onto a cation exchange resin through self-crosslinking for selective Cu(II) removal. J. Hazard. Mater., 190, 1037–1044 (2011).
10. Chen, Y., Pan, B., Li, H., Zhang, W., Lv, L., Wu, J.: Selective removal of Cu(II) ions by using cation-exchange resin-supported polyethylenimine (PEI) nanoclusters. Environ. Sci. Technol., 44, 3508–3513 (2010).
11. Buchmeiser, M.R.: Polymeric monolithic materials: syntheses, properties, functionalization and applications. Polymer, 48, 2187–2198 (2007).
12. Svec, F.: Porous polymer monoliths: amazingly wide variety of techniques enabling their preparation. J. Chromatogr. A, 1217, 902–924 (2010).
13. Courtois, J., Byström, E., Irgum, K.: Novel monolithic materials using poly(ethylene glycol) as porogen for protein separation. Polymer, 47, 2603–2611 (2006).
14. Yang, H., Liu, Z., Gao, H., Xie, Z.: Synthesis and catalytic performances of hierarchical SAPO-34 monolith. J. Mater. Chem., 20, 3227–3231 (2010).
15. Unger, K.K., Skudas, R., Schulte, M.M.: Particle packed columns and monolithic columns in high-performance liquid chromatography-comparison and critical appraisal. J. Chromatogr. A, 1184, 393–415 (2008).
16. Wei, S., Zhang, Y.L., Ding, H., Liu, J., Sun, J., He, Y., Li, Z., Xiao, F.S.: Solvothermal fabrication of adsorptive polymer monolith with large nanopores towards biomolecules immobilization. Colloids and Surfaces A: Physicochem. Eng. Aspects, 380, 29–34 (2011).
17. Nordborg, A., Hilder, E.F.: Recent advances in polymer monoliths for ion-change chromatography. Anal. Bioanal. Chem., 394, 71–84 (2009).
18. Svec, F., Fréchet, J.M. J.: Continuous rods of macroporous polymer as high-performance liquid chromatography separation media. Anal. Chem., 64, 820–822 (1992).
19. Okada, K., Nandi, M., Maruyama, J., Oka, T., Tsujimoto, T., Kondoh, K., Uyama, H.: Fabrication of mesoporous polymer monolith: a template-free approach. Chem. Commun., 47, 7422–7424 (2011).
20. Nandi, M., Okada, K., Uyama, H.: Functional mesoporous polymer monolith for application in ion-exchange and catalysis. Func. Mater. Lett., 4, 407–410 (2011).
21. Xin, Y., Fujimoto, T., Uyama, H.: Facile fabrication of polycarbonate monolith by non-solvent induced phase separation method. Polymer, 53, 2847–2853 (2012).
22. Xin. Y., Uyama, H.: Fabrication of polycarbonate and poly(3-hydroxybutyrate-co-3-hydroxyhexanoate) blend monolith via non-solvent induced phase separation method. Chem. Lett., in press.
23. Woo, B.G., Choi, K.Y., Song, K.H., Lee, S.H.: Melt polymerization of bisphenol–A and diphenol carbonate in a semibatch reactor. J. Appl. Polym. Sci., 80, 1253–1266 (2001).

24. Haba, O., Itakura, I., Ueda, M., Kuze, S.: Synthesis of polycarbonate from dimethyl carbonate and bisphenol–A through a non-phosgene process. J. Polym. Sci. Part A: Polym. Chem., 37, 2087–2093 (1999).
25. Fukuoka, S., Tojo, M., Hachiya, H., Aminaka, M., Hasegawa, K.: Green and sustainable chemistry in practice: development and industrialization of a novel process for polycarbonate production from CO_2 without using phosgene. Polym. J., 39, 91–114 (2007).
26. Okuyama, K., Sugiyama, J., Nagahata, R., Asai, M., Ueda, M., Takeuchi, K.: Direct synthesis of polycarbonate from carbon monoxide and bisphenol A catalyzed by Pd-carbene complex. Macromolecules, 36, 6953–6955 (2003).
27. Asrar, J., Valentin, H.E., Berger, P.A., Tran, M., Padgette, S.R., Garbow, J.R.: Biosynthesis and properties of poly(3-hydroxybutyrate-*co*-3-hydroxyhexanoate) polymers. Biomacromolecules, 3, 1006–1012 (2002).
28. Iwata, T.: Strong fibers and films of microbial polyesters. Macromol. Biosci., 5, 689–701 (2005).
29. Yang, Y., Ke, S., Ren, L., Wang, Y., Li, Y., Huang, H.: Dielectric spectroscopy of biodegradable poly(3-hydroxybutyrate-*co*-3-hydroxyhexanoate) films. Eur. Polym. J., 48, 79–85 (2012).
30. Abe, H., Ishii, N., Sato, S., Tsuge, T.: Thermal properties and crystallization behaviors of medium-chain-length poly(3-hydroxyalkanoate)s. Polymers, 53, 3026–3034 (2012).
31. Wang, X., Min, M., Liu, Z., Yang, Y., Zhou, Z., Zhu, M., Chen, Y., Hsiao, B.S.: Poly(ethylenimine) nanofibrous affinity membrane fabricated *via* one step wet-electrospinning from poly(vinyl alcohol)-doped poly(ethylenimine) solution system and its application. J. Membrane Sci., 379, 191–199 (2011).
32. Chen, Z., Deng, M., Chen, Y., He, G., Wu, M., Wang, J.: Preparation and performance of cellulose acetate/polyethylenimine blend microfiltration membranes and their applications. J. Membrane Sci., 235, 73–86 (2004).
33. Maketon, W., Ogden, K.L.: Synergistic effects of citric acid and polyethylenimine to remove copper from aqueous solutions. Chemosphere, 75, 206–211 (2009).
34. Bessbousse, H., Rhlalou, T., Verchère, J.F., Lebrun, L.: Removal of heavy metal ions from aqueous solutions by filtration with a novel complexing membrane containing poly(ethylenimine) in a poly(vinyl alcohol) matrix. J. Membrane Sci., 307, 249–259 (2008).

INDEX

A

Abrasion force, 226
Absorbent media, 34
AC impedance spectroscopy, 408
Acidic media, 234
Acidic site, 387, 392, 393
Acrōs organic company, 401
Actuation strain, 228–231, 234
Adding oleic acid, 227
Addition-fragmentation, 2, 26, 314, 330
Adsorption for metal ions, 414
    Cobalt, 414
    Nickel, 414
    Potassium, 414
    Sodium, 414
Aerated concrete, 69
Aibn initiator, 9, 320
Aldrich chemical company, 361
Alkalinity, 372, 373, 376–381
Aluminum hydroxide bayerite, 69
Aluminum hydroxide boehmite, 69
Aluminum oxide, 69
Aluminum powder, 69, 70, 71
Ambipolar diffusion, 116
Amorphous analogue, 108
Amorphous phase, 254, 360, 361, 365–368, 386, 400
Amount of pozzolans, 54, 63
Amplitude solitons, 280
Amplitude, 116, 227, 280, 303
Anhydrous toluene, 334
Anionic contribution, 386, 392, 395
Anions immobilization, 393
Anisotropic membranes, 31, 32
Anisotropic techniques, 32
    drawing, 33, 186
    electrospinning in the laboratory, 32
    electrospinning, 30, 32, 33
    high porosity, 32
    high surface area, 32
    industrial scale, 32
    interfacialpolymerization, 32
    nanofibrous membranesflexibility, 32
    phase inversion, 32
    phase separation, 33, 34, 412
    plasma deposition, 32
    selfassembly, 33
    small pore, 32, 34, 45
    solution coating, 32
    template synthesis, 33
Annealing techniques, 44
    objective-based simulated annealing, 44
    pixel-based simulated annealing, 44
Anodic current pulse, 231
Antoine Lavoisier, 190
Applications in robotics, 228
Aqueous solutions, 414
Arburg Allounder, 109
Area per mass ratio, 30
Area-to-volume ratios, 45
Area-weighted, 41
Aromatic Cycle, 265
Artificial muscles, 228
Asobis, 2, 314
Atactic polystyrene, 386
Atmosphere of argon, 258
Atom statistic model, 212
Atom valence orbitals, 120, 121
Atomic scale, 144
Atomic structure, 138, 219
Atomic-power microscopy, 259
ATR- FTIR spectroscopy, 403
Attenuated total reflection, 377
Autocorrelation function, 44
Avogadro number, 110
Axial electric field, 199

B

Backscattered electron images, 41
Ball-milling technique, 346
Baseline of a thermogram, 402
Bead-on-string structure, 237
Bead-spring model, 204
Binary hooking, 110
Binary images, 39, 42
Biomedical agents, 34
Blend monolith, 412–419
Blend nanofiber, 237
Blend solution, 234,235, 236, 243
Blending in melt, 109
Blends ratios, 244
Blind pore, 36, 38, 39, 40
Bogoliubov, 270
Bond line, 117
Boundary elements, 41
Braille screen, 228
Brillouin zone, 278
Brittle manner, 66
Brookfield viscometer, 235
Bubble point method, 38
Buckingham formula, 178
Buffer zones, 340
Bulk resistance, 362, 388, 400, 402

C

Calcium hydroxide, 69
Calcium silicates, 69
Calculation consistency, 109
Calculation of porosity, 42
Calorimetric data, 3, 315
Calorimetric method, 315
Calvet type, 315
Camber propeller, 228
Capillary tip, 33, 239
Carbon nanostructures, 226, 227
Carbyne, 253,254, 255, 2546, 310
Cellulose tubular membrane, 234
Cement-based materials, 41
    Mortars, 41
    Plain cement paste, 41
    Silica fume, 41
Centrifugal force, 226

Chain growth, 5, 10, 23, 315, 317, 320
Chain transfer mechanism, 26, 314
Channel length, 142
Chemical engineering, 40, 114, 190
Chemical particles, 130, 219, 220, 224,
Chemical potential, 295, 298–307
Chemical species, 31
Chemithermomechanical pulp, 353
Chiral symmetry, 288
Chirality, 137, 138, 156, 163, 169
    Armchair, 137, 139
    Chloroform, 34
    Helical, 137
    Zigzag, 137,139
Chromatogram, 3, 153, 315
Clapeyron equation, 147
Clayey sand, 81, 84, 93
Clementi, 121
Closed pores, 37
Cluster network density, 110
Cluster of layers, 31
Cobalt, 141, 219, 414
Coefficient of slippage, 172
Cole–cole plot, 402
Concrete mix, 62, 63
Conductivity of the MNR, 409
Conical flask, 376
Constant of termination, 14, 324
Construction industries, 51, 52
Contact angle, 37, 38, 154
Contact of teeth, 102
Continuum hypothesis, 143, 146
Controlled radical polymerization, 2, 3, 26,
    314, 315, 330
Conversion of monomer, 11, 322
Cooling rates, 402
Copper ions causes, 412
    Brain, 412
    Heart, 412
    Kidney, 412
    Liver, 412
    Pancreas, 412
Couette flow, 172
Coulomb force, 196
Coulomb interaction, 278, 287
Coulomb's law, 196, 197

Counter electrode, 187
Covalent bond, 413
Cross section morphology, 31
Crushing, 220, 226, 227, 309
Crystalline graphite, 131, 136
Crystallinity degree, 109
Crystallinity of PEO, 360, 364
Cubic topology, 347
CuBTC–silk, 353
Cured specimens, 96, 102, 104
Curing age, 71
Curing period, 90, 100
Cyclic voltammetry, 230
Cyclooctatetraene, 267

D

Darcy's friction factor, 143
Dark green solution, 234
Dark-brown powder, 254
Dark-yellow-green, 258
De broglie wavelength, 147
Deborah number, 208
Deep blue emeraldine base, 234
Dialysis tubing, 234
Diameter of electrospun, 236, 237, 240, 243
Dibenzyltritiocarbonate, 2, 314
Dimagnesium dibromine, 254
Dirac equation, 281, 302, 303, 304
Direct dependence, 121
Direct shear, 82, 102
Discrete model, 283
Displacement curves, 66
Distilled aniline, 234
Distorting pore structure, 40
Dithiobenzoates, 10, 40, 321
Domain walls, 271, 273, 275, 303
Domino effect, 161–165
Dominoes diameter, 163
Donor Atoms, 390, 393
Doped complexes, 264
Dried rubber content, 373
Droplets connected, 244
Drops of methyl red, 376
Drude theory, 307

Dry density, 71, 75, 78, 82, 84, 244
Dry fibers deposits, 33
DSC thermogram, 364
Dual mode lever system, 229
Ductile failure, 63
Durability tests, 82
Dynamics of interactions, 117

E

Earnshaw's theorem, 196
Effect of hydrolysis, 69
Effect of polyamide, 84
Elastic dumbbell, 200, 204
Electric arc, 130, 200, 227, 255
Electric charge, 191, 192, 196, 287
Electrical resistivity, 141, 189
Electrically heated set, 58
Electrochemical actuation cycle, 228
Electrochemical cycling, 230
Electrodynamics, 125, 193, 194, 204
Electrokinetic, 142
Electrolyte films, 361, 366, 386, 401
Electron-fonon bond, 277
Electron compton wavelength, 124
Electron density, 121
Electron in hydrogen atom, 124
Electrokinetic, 142
    Electro-osmosiselectrophoresis, 142
    Potential of sedimentation, 142
    Streamy potential, 142
Electron kinetic energy, 123
Electron shells, 157
Electron transport probability, 116
Electron volt, 120, 277–280, 283
Electron-vibrational interaction, 272
Electron-volts, 120
Electronic angular momentum, 116
Electronic clouds, 277
Electrospun nanofibrous mat, 39
Element's angels, 194
Elementary clusters, 219
Elementary reactions, 23, 328
Elengational flow, 237
Elution rate, 3, 315
Empower pro, 3, 315

Energy conversion, 31
Energy of soliton excitation, 277
ENR samples, 402
Enthalpies of fusion, 362
Epoxidized natural rubber, 360, 400, 401
Epoxy resin, 334–336, 338–401
Ethereal solution, 266
Euclidean space, 109
Euler constant, 296
Excitation of polaron, 284, 292

F

Fermi energy, 278
Fermi particles, 288, 292–294, 296, 299–302, 309
    Astrophysics, 309
    Cosmology, 309
    Quantum chromodynamics, 309
Fermionic charge, 306
Fermionic spectrum, 290, 304
Feynman diagrams, 293
Fiber bonding, 34
Fiber bridging, 66
Fibers on dilation, 102, 103
Filaments, 131, 135, 186
Film-electrode, 388
Filtration, 30-35, 41, 263
Fisher scientific, 361
Flat waves, 290, 302, 304
Flexural strength tests, 57
Flow porosimetry, 30, 38, 39, 40
Food industry, 31
Forbidden zone, 272, 279, 287
Forces of repulsion, 157, 159
Form of bion, 284
Four probe method, 236
Fourier theory, 116
Fourier-transform infrared, 372
Fox equation, 360, 363
Frameworks, 38, 108, 343–351, 354
Free atoms, 116
Free dirac equation, 302
Free motion, 117
Free stroke, 230, 231
Freeze-thaw, 103

Freezing-defrosting, 117, 314
FTIR sample analysis, 377
FTIR spectra, 334, 335, 377, 400, 403, 404, 408
Fullerene molecule, 131, 134, 164
Fullerene tubules, 219, 220

G

Galerkin, 143
Gamma high voltage research, 235
Gas foaming, 34
Gas of electrons, 269
Gas separation, 32, 344, 354
Gas storage, 344, 345
Gas to escape, 38
Gauge length, 388
Gaussian distribution, 44
Gaussian field, 44
Gaussian method, 44
Gel permeation chromatography, 361
Gel-effect, 16, 325
Glass petri dishes, 387
Glass transition temperature, 360, 361, 363, 400, 402
Global thresholding, 39
Globular anthraxolite, 221
Globules and fibrils, 130
Graft-copolymerization, 372
Grafting efficiency, 372, 373, 374, 375, 380
Grafting reaction, 373
Graphite plane, 132, 137, 138
Graphite sheet, 134, 138
Grasp electron, 271
Greenberg, 143
Grignard reaction, 254
Grinding, 220, 224, 226, 227
Gross–neview model, 288, 293, 294
Ground waste glass, 54, 55
Grounded collector, 33, 235
Gun trunk, 164

H

Haake twinscrew extruder, 109
Hagen–poiseuille equation, 166

Hamiltonian, 120, 277, 278, 293
Harmonic oscillations, 116
Harmonic waves, 116
Heiblum speaks, 288
Hemispherical caps, 131, 136
Heterogeneous borders, 146
Heterogeneous PEO/ENR, 360
Hevea tree, 372
Heveaplus MG, 372
Hexanoyl chitosan-polystyrene, 386
Hi-tester impedance, 387
Hinzburg-landau lagrangian, 276
Hollow tube, 132
Honeycombs, 132
Hooke's law, 199, 201
Horiba compact, 413
Hydration, 58, 69
Hydrogen atom, 124, 126, 155
Hydrogen gas, 69
Hydrology fields, 40
Hydrophobic surfaces, 141, 154
Hypodermic needle, 235

I

Idzhima, 130
Image analysis techniques, 41, 45
Immobilization substrates, 34
Impact of resolution, 44
Impedance spectroscopy, 360, 387, 400, 408
Impulse of trimming, 290, 295
Indirect tensile, 82, 83, 92, 94
Inorganic salt, 360, 361, 400
Inorganic SBUs, 346
    Square paddlewheel, 346
    Zinc acetate cluster, 346
Instron, 109, 335, 388
Insulating matrix, 243
Interface blocks, 264
Interfacial regions, 108, 112
Intrinsic features, 40
    Geometric, 40, 144, 260
    Topological, 40, 44, 271, 293, 309
Inversion techniques, 34
Ion migration, 388

Ion-dipole interactions, 403, 405
Ionic radius, 121
IR losses, 332
IR radiation, 222
Irradiation time, 349
Isobutirnitrile, 2, 314
Isoreticular metal-organic frameworks, 347
Isotropic dense membranes, 32
Isotropic mobility, 202
Iteration method, 179

J

Jet deposition, 189, 196
Jets dry, 235

K

Kalkornsurapranee, 378
Kaolinite layer, 86
Kaolinite powder, 86
Kayser, 124
Kernels with electrons, 269
Kinetic measurements, 15, 135, 325
Kinetic modeling, 2, 314
Kinetic parameters, 314, 315
Kinetic scheme, 2, 3, 5, 6, 7, 23, 315, 317, 328
Kink anti-kink, 304
Kink vacuum, 300
Kinks, 292, 293, 304
    Double bags, 292
    Small bags, 292
Knudsen number, 144, 152
Kolumen−vainberg, 295
Kramer theorem, 285

L

Lagrangian equation, 118
Laminar layer of liquid, 176, 177
Laplace equation, 37
Large pore volume, 352
Laser pyrolysis, 220
Laser sublimations of carbon, 255
Latin carboneum, 253
Latter stiffness, 108

Lattice structure, 131
Lementary portions-quants, 123
Lennard–jones potential, 145, 146, 161
Lewis acid-base, 386, 393
Ligand ratio, 347
Light petroleum ether, 374, 375
Light weight composites, 34
Linear filter transform, 44
Linear low density polyethylene, 109
Linear spectrum, 270
Link rate, 12, 322
Liquid chromatography, 334
Liquid extrusion flow porometry, 37
Liquid jet, 33, 204,240, 348
Literature, 34, 58, 130, 174, 263, 386, 414
Lithium metal electrodes, 388
Lithium salt, 403
Logarithmic divergence, 298, 299
Longitudinal axis, 131
Low density polyethylene, 108, 109, 110
Low stiffness, 37
Low strains, 228
Low voltage, 228
Low-density polyethylenes, 110
Low-molecular raft-agent, 5, 12, 13, 14, 322, 323
Lower wavenumbers, 353
Luminescence, 276, 344
Luttinger catalyst, 266
Luttinger–tomonug, 270

M

Macro-dimensional cones, 221
    Complex fibers, 221
    Fibrous vetcillite, 221
    Spheroids, 221
Macromolecular backbone, 400, 402, 405, 406
Macromolecular entanglements, 110
Macromolecular raftagent, 5
Macromolecules, 108, 204, 336, 338
Macroradicals, 9, 319
Macroscopic properties, 30, 40, 44, 143
Magnetic constant, 125
Magnetting tendency, 157

Malaysian rubber board, 401
Manganese carbonyls, 219
Mark dowlex, 109
Marker image, 42
Mask image, 42
Mass styrene, 9, 320
Mass-spectrometry, 257
Materials, 335, 341, 344, 354
    Husk ash, 52, 60, 63
    Portland cement, 52, 54, 61, 63, 67, 69, 81
    Rice, 52
    Silica fume, 52, 54, 60, 63, 69, 77
    Standard sand, 52, 54
    Tap water, 52
    Tow particle size, 52
Mathematical modeling, 2, 9, 314, 320
Matrix polymer elasticity, 108
Matrix polymer, 108, 109
Matsubara frequency, 294
Maximum dry density, 82, 84, 85
Mechanical crushing, 220, 226
Mechanical stability time, 372
Mechanical tests, 108, 109, 111
Medial axis, 43
Melamine content, 72, 78
Melamine plastic, 74, 78, 79, 80
Melt flow index, 109
Melting point, 109, 365
Membrane image, 30
Membrane porosity, 35
Membrane technologies, 31
Mercury intrusion porosimetry, 37, 38, 40, 42
Metal spatula, 57
Metal-organic frameworks, 344, 347, 348, 351, 354
Metal/carbon nanocomposite, 227
Methylene, 405
Micro tomography, 43
Micro-ionic devices, 360
Microflow substances, 142
    Electrokinetic, 142
    Hydraulic, 142
Microfluidics, 142, 143, 144
Microscopic rods, 135

Microscopic techniques, 37
  Electron microscopy, 37, 39, 78
  Light microscopy, 37, 43
  X-ray diffraction, 37
Microstructure grains, 44
Migration, 142, 389
Mineral diatomite, 334
Minimal polydispersity, 23, 338
Miscible PEO, 360–362, 366
Mitutoyo digimatic caliper, 402
Mixture of aggregates, 61
Mixture of cyclohexane, 266
Modulus of elasticity, 64, 65, 90
Moist chamber, 82
Molding machine, 109
Molecular cutoff, 234
Molecular-mass distribution, 3, 315
Monofilament fibers, 52
Monomer initial concentration, 10, 23, 25, 320, 329
Montmorillonite, 108, 109, 112
Morphing technique, 43
Morphology of fibers, 244
Morphology tools, 45
Mortar mixes, 58
Mortar specimens, 57, 58
MST tests, 373, 377
Multi-vicissitude of methods, 226
Multi-walled nanotubes, 132, 136
Multifunctional membranes, 34
Multiphase flow, 40, 44
Multiphase model, 43

N

Nanocluster, 108, 109, 110, 219, 225
Nanofiber mat mass, 42
Nanofiber sheets, 38
Nanofiltration, 30, 31, 33
Nanomaterials, 30, 130, 169, 220, 245
Nanoporous membranes, 30, 45
Nanoscience, 130, 131, 221
Nanostructures, 130, 132, 218, 220, 225, 227
  Bamboo-like tubules, 220
  Gigantic fullerenes, 220

Nanobands, 220
Nanowires, 220
Onions, 220, 227
Russian dolls, 220
Welded nanotubes, 220
Nanotechnology, 130, 222
Nanotube length, 135, 175
Naphthalene, 264
Narrow channels, 157
Natural rubber, 360, 372, 400–402, 405, 406
Natural science theories, 116
Negative work, 118, 120, 126
Neo-hookean equation, 202
Nernst's relationship, 408
New era pump systems, 235
Newton and Lagrange, 117
Nickel pentadienyls, 219
Nickel, 141, 215, 414
Nitrogen atmosphere, 361, 362, 367, 401
Non-reacting gas, 38
Non-wetting liquid, 37
Nonlinear equations, 207
Nonwoven image, 39
Novel image analysis, 42
NR-g-pmma products, 380
NRL backbone, 372, 373
Nucleus, 120, 121, 219
Numerical relaxation, 216

O

OH functional groups, 394
One-loop back approach, 300
One-loop diagrams, 294
Optical micrograph, 95
Optimum moisture, 82, 102
Optimum water content, 83, 76
Orbital energy, 120
Orbital radius, 120, 124
Organic bridging ligands, 344
Organic matrix clusters, 258
Organoclay, 107, 108, 111, 112
Oscillator energy, 123
Ostwald-de waele, 200
Oxirane group, 402, 404

Ozonation hydrolysis, 253
Ozonation of polycumulene, 253

P

Paint application, 188
Palladium complexes, 265
Pani matrix actuation, 232, 233
Parabolic form, 173
Parabolic shape, 170, 183
Parquet, 271
Particle leaching, 34
PC–PHBH blend, 413, 414–419
Peak solitons, 273
Pearl necklace, 351
PEI-modified monolith, 416, 417
Peierls instability, 269, 272
Pentagons, 131, 135, 136
PEO macromolecular, 360
PEO/PAC and PEO/PAC blends, 360
Percentage error, 124
Percolating pathway, 360, 366, 367, 368
Percolation threshold, 243, 244
Perimeter-area scaling law, 41
Periodic function shape, 116
Perkin elmer spectrum, 413
Permeability of biomembranes, 116
Petroleum engineering, 40
pH of solution, 346
Phase of polymerization, 10, 40, 321
Phases of sinusoids, 116
Photomicrograph, 30, 43, 44, 237, 259
Pico-sized chemical, 224, 225
Pixel-movement, 44
Planck's constant, 116, 124, 147, 287
Planetary ball mills, 227
Plank's constant, 121
Plasma sources, 220
Plasma-atom emission, 413
Plasmachemical synthesis, 252
Poiseuille flow, 152, 156, 172, 183
Poisson's ratio, 109, 166
Poly(ethylene oxide), 360, 400
Polycarboxylate, 69
Polycumulene, 253, 254,
Polydispersity index, 9, 19–25, 326, 328

Polyethylene, 108, 1009, 110, 266, 335
Polyethylenimine, 412, 413
Polymer blends, 234
Polymer casting film, 34
Polymer density, 110
Polymer electrolyte, 360, 362, 386, 400, 401, 406, 408
Polymer nanofiber mats, 30
Polymer rupture, 96
Polymer skin, 240
Polymer-based monoliths, 412
Polymer-salt interaction, 408
Polymeric glycol, 254
Polymerization enthalpy, 3, 315
Polystyrene standards, 3, 315
Pore compressibility, 41
Pore network structure, 41, 43
Pore space skeleton, 43
Pore throat diameters, 40
Porosity of mortar, 74, 78
Porous membranes, 29, 30, 37, 45
Post-synthetic modification, 346
Potassium oleate, 372, 373, 377–381
Potential energies, 118, 119
Power-law fluid, 199
Pozzolanic materials, 52, 54, 62
Pozzolans, 54, 61, 63
Presence of aibn, 24, 54, 329
Pressure on mercury, 37
Propargyl alcohol, 265
Protective clothing, 34
Proton-electron, 125
Psa method, 44, 125
Pulp fibers, 353
Pulse laser radiation, 325

Q

Quadratic termination, 4, 5, 23, 316, 317, 324
Quality control tests, 372, 373, 374, 377–380
Quantitative models, 108, 190
Quantum
    amendments, 297
    conceptualizations, 116

field theory, 193
mechanical data, 121
mechanics, 117
number, 116, 120, 124
Quark structure, 304
  Hadrons, 304
  Mesons, 304
Quartz rock, 221
Quasi-classical concepts, 121
Quasi-one-dimensional system, 268–270, 274,
Quasi-solid, 175, 176
Quasistationary, 10,175, 321

R

Radial density profile, 155, 156, 175
Radial velocity profile, 156, 170
RAFT-agents, 2, 10, 314, 321
Raman spectrum, 308
Rate constants, 5, 10, 23, 317, 324
Rate of cooling, 58
Ratio of tensile strength, 93
Real and thermal initiation, 9, 320
Reception of films, 256, 263
Rectangular specimens, 82
Rectangular strip, 388
Recycled materials, 52, 105
Regions of interactions, 118
Rehabilitation glove, 228
Reinforced mortars, 54
Reinforcement degree, 108
Relative fraction, 109, 112
Relative motion, 117, 118, 119
Replicates, 376
Repulsion of the molecules, 145, 146
Repulsive forces, 145
Research of polyacetylene, 273
  Anisotropy of conductivity, 274
  Degree of alloying impurity, 274
  Metal-dielect, 274
  Vicinity, 274
  X-ray analys is monocrystal, 274
Rest energy, 124
Reverse osmosis, 31, 32, 412

Reversible addition-fragmentation chain transfer, 314, 330
Rifluoromethylbenzene, 263
Rigidity loss, 101
Riyajan, 372
Rocks microstructures, 44
Rubber whit course, 64
Rubberized concrete mixtures, 62
  Rubcrete, 63, 65
Russian foundation, 27, 63, 330
Rydberg equation, 27, 124

S

Salt content, 361, 367, 400, 402–409
Salt molecule-chain, 400
Salt-free blend, 362
Salt-free PEO, 360, 362, 366, 367
Sandstones, 41
  Berea, 41, 44
  Fontainebleau, 41
Santipanusopon, 372
Saturated aliphatic, 381
Scanning electron microscopy, 39, 78
Schrödinger equations, 305
Science of chemistry, 191
Science of nanomaterials, 130, 222
Secant modulus, 90, 91
Second law states, 193
Secondary building units, 344, 346
Seed fermion sludge, 301
SEM images, 41, 42, 413, 415
SEP dimensionality, 120
Shear box, 102
Shell of ligands, 219
Sigma–Aldrich, 387, 413
Silica fume, 41, 52, 54, 60, 61, 63, 68, 71, 73, 78
Silvery-white strike, 256
Sine–gordon, 291
Single wall nanotubes, 132, 229
Single-kink sectors, 300
Skyrme model, 304
Sladkov two-stage method, 254
Slaked lime, 69
Slippage of the fluid, 152

no slippage, 152
  with slippage flow, 152
  complete slippage, 152
Slump tests, 63
Soccer ball, 135
Soft elastomeric, 400
Soft rocks, 102
Soft X-ray, 222
Soil-cement mixtures, 83, 85
Soil-cement specimens, 85, 100
Solid polymer electrolytes, 408
Soliton model, 287
Solvothermal reaction, 345
Soxhlet extraction, 374, 375, 376,
Sp-hybrid carbon, 264
Sp2-hybrid, 275
Space connectivity, 40, 44
Space geometry, 40
Spatial correlation, 41, 44
Spatial distribution, 43
Spatial-energy parameter, 116, 120, 122, 126
Spectrum EPR, 259, 260
Speed heterogeneity, 166
Speed of the drum, 227
Sphere, 135, 219, 221
Spherical form, 262
Split tensile, 62, 63, 64, 66
Spraying, 188, 224, 244, 266
SPT numbers, 81
Spun fibrous scaffolds, 34
Sputter coater, 236
Stabilized clayey, 84
Stabilized soil, 87, 91, 103
Statistical flexibility indicator, 110
Steel block electrodes, 362, 402
Stirring speed, 376
Stochastic reconstruction, 43
Strain energy, 66, 162
Stream nitrogen plasma, 255
Stream of inert gas, 266
Stress-strain diagram, 100
Stretched chains, 108
Structural heterogeneity, 155
Structural mechanism, 108
Structure of a zone, 272

Styrene bulk RAFT-polymerization, 19, 20, 326
Styrene controlled radical polymerization, 2, 3, 314, 315
Styrene, 216, 314, 315, 316, 320, 324, 326, 328
Styrene's molecular mass, 9, 320
Sulfuric acid, 234, 376
Super plasticizer, 69
  Premia, 69
Suprasegmental level, 108, 113
Surface tension force, 33, 208
Surface tension of mercury, 37
Surface texture, 63, 67
Symmetry of grapheme, 134
Synergistic effect, 334, 341, 367
Syringe, 33, 235, 348
System total energy, 117

T

Tadpoles, 298
Taylor's cone, 237
Teflon dish, 361, 401
Teflon® tape, 229
Tensile
  force, 217
  loads, 63
  properties, 386–390, 395
  specimens, 109
Test programme mix
  fibers mixer, 57
  three-quarters, 57
Tester instron, 109
Tetraethoxysilane, 334, 341
Tetrahydrofuran, 3, 34, 254, 266, 315, 387, 401
Textile fibers, 345, 347, 354
Thermogravimetric curves, 267
Thermosetting plastic, 67
Thick nonwoven fabrics, 36
Three-necked flask, 334
Thresholded 2D image, 44
Throat diameter, 39, 40
Tier fibers, 66
Tire chips rubber, 63

Tissue engineering scaffolds, 34
Topological solitones, 271, 302, 309
Toxic metal ions, 412, 419
Trance-polyacetylene, 261
Trans-polyacetylene, 308
Tricalcium silicate, 69
Trigonal prismatic oxo-centered trimer, 346
Trithiocarbonates, 3, 4, 18, 23, 26, 315, 326, 330
Trivalent molybdenum, 259
Trivial topology leads, 302
Truncated gaussian methods, 44
Tsui and Stormer, 287
Two-dimensional analysis, 30, 45
Types of pores, 36
    Blind pore, 36
    Closed pore, 36

U

Ultra-Violet renormalization, 290
Unbleached kraft pulp, 353, 354
University of Malaya, 361
University teknologi MARA, 382, 409
Unsaturated ethylene groups, 404

V

Vacuous vacuums, 293
Valence orbitals, 120, 121
Valent zone, 280, 282–284, 307
Van der waals forces, 145, 159, 161
Vanadic catalyst, 263
Vanadium cations, 348
Vant-goff's equation, 15, 325
Vinylacetylene, 264, 265
Viscoelastic
    dumbbell, 204
    liquids, 190
    fluid, 176, 178

Visual formation, 376
Volatile fatty acid, 372, 373, 377
Volatile organic solvent, 34
Voronoi diagram, 43
Voxel, 43

W

Waste fibers, 81
    Acrylic, 81, 82, 84, 85, 92
    Polyamide, 81, 82, 84, 85, 92, 95
Waste polypropylene fibers, 52
Waste tires, 62, 63, 65
Waste to wealth, 52
Water treatment, 31, 419
Wave number, 124, 125, 353, 405
Weak subgrade areas, 103
Web density, 42
Weight of,
    average molecular mass, 9, 320
    bosonic field, 299
    cement, 54, 79, 80,
    electron, 273, 277
    solution, 273, 277
Wet-dry soil cement, 103
Wetting liquid, 38
White sediments, 255
Wound dressings, 34
Woven fabrics, 36

Y

Yield stress, 109

Z

Zero mode energy, 303
Ziegler catalyst system, 265

To Daniel (從善山) and Himiko (南夠姬)

Contents

Foreword by Francisco Rios	x
Acknowledgements	xv

1   Introduction: Latina/o Teacher Career Pathway   1

2   Growing Up, K-12 Schooling, and Working as
Paraprofessionals   25

3   Becoming Teachers of Color   53

4   Teaching at Public Schools: Workplace Environment   87

5   Teaching With Warmth and Demands   134

6   Conclusion: Latina/o Teachers Advancing in
the Profession   165

Index   178

Foreword

I've been fortunate to have had several teachers of color who impacted the way I've come to understand teaching and learning. In talking with other peers, colleagues, and students, I realize that I was also unique in even having any teachers of color at all!

Ms. Dukes, an African-American educator in my middle school, was what has been called a warm-demander. She had the highest of expectations for us to learn language arts (we *were* going to learn English!) while at the same time showing that grace, care, and intellectual prowess can go hand in hand. From Ms. Dukes, I was taught that teaching and learning are an academic, an intellectual activity.

While in Ms. Saucedo's art classroom, a Latina teacher, I felt the pride in her Mexican heritage that exuded from her as she taught us about the master artists of Mexico: Rivera, Siqueros, Orozco, and Kahlo. But we also learned about the murals being painted in our own community that told the story of *nuestro pueblo*. **From Ms. Saucedo, I was taught that teaching and learning are a cultural activity.**

I took my first Chicano Studies class from Mr. Acosta. I would see Mr. Acosta at almost all of our community's events: baseball games, theatrical presentations, and political rallies. This Latino/Chicano teacher wanted us to know about the various Latino communities' experiences with oppression and marginalization but also about our communities' resistance and resilience. **From Mr. Acosta, I was taught that teaching and learning are a political activity.**

It wasn't until I was in the 11th grade that I was first encouraged to go to college. Mr. Vidal, a *Cubano*, was teaching a speech communications course and singled me out to share his belief that I was "college material." Beyond the belief, he developed a relationship not only with me but also with my family as we explored, together, what going to college—and being successful—would entail. **From Mr. Vidal, I was taught that teaching and learning are a relational activity.**

I was, indeed, most fortunate!

Foreword xi

There has been a clarion call for more teachers of color for our nation's K-12 schools. Even at the federal level, the Department of Education held a national summit on teacher diversity and issued the report titled "The State of Racial Diversity in the Educator Workforce" (U.S. Department of Education, 2016a). This follows what communities of color have been calling for—and, at times, demanding—regarding the diversification of the teaching workforce over a much longer time period including during the civil rights movement.

The most typical reason given for the need for more teachers of color is due to the mismatch between an increasingly diverse K-12 student population (over 50% of all students; U.S. Department of Education, 2016b) and the nearly nonexistent increase in the percentage of teachers of color (18%, only up from 16% 20 years earlier; U.S. Department of Education, 2013). More specifically, advocates for increasing the diversity of the teacher workforce point to their function as role models for students of color, the likelihood that a greater match between a teacher and student social identities would lead to increased academic achievement, and to counter negative stereotypes white students have of people of color (U.S. Department of Education, 2016a). Consider the following from then Secretary of Education John B. King, Jr:

> Without question, when the majority of students in public schools are students of color and only 18 percent of our teachers are teachers of color, we have an urgent need to act. We've got to understand that all students benefit from teacher diversity. We have strong evidence that students of color benefit from having teachers and leaders who look like them as role models and also benefit from the classroom dynamics that diversity creates. But it is also important for our white students to see teachers of color in leadership roles in their classrooms and communities. The question for the nation is how do we address this quickly and thoughtfully?
> (U.S. Department of Education, 2016a).

Beyond these important advantages rests a contested story. To begin, we recognize that the dearth of teachers of color has its historical roots, especially exacerbated after the landmark 1954 Supreme Court ruling in *Brown versus Topeka School Board*. Epps (2002), for example, estimated that nearly 37,000 teachers of color, just in the south alone, lost their positions in the nearly 20 years post-Brown decision. Before they lost their jobs, these teachers were recognized as essential to the material, intellectual, social, and cultural development of students of color (Fairclough, 2004). In addition, Fairclough found that teachers of color, pre-Brown, were vital to the vibrancy of the communities where they worked and were actively engaged in the pursuit of educational equity as part of the larger social justice project. Indeed, while many teachers of color recognized the significant lack of access to opportunities that existed during the segregation period, they also worried

xii *Foreword*

about what would happen to the young black and brown students from their communities as they were integrated into predominantly white schools.

As important, educational researchers have begun to extend the understanding of the significant and substantial advantages teachers of color bring to the teaching profession. In a comprehensive review of this research, Quiocho and Rios (2000) found that teachers of color are more likely to, among other things, use their cultural assets to motivate students, see culture/language as assets, implement culturally responsive, anti-racist pedagogy, promote school climate for diversity, assure community resources are affirmed, and develop mediation skills for "bringing others" along. They also found that teachers of color, in the main, use their political assets to advocate for students of color, work with ethnic minority students in professionally nurturing context, hold positive yet realistic expectations for students of color, hold self/colleagues/ institution accountable, identify and name racist practices in education, and offer a counter-hegemonic ideology. Nieto (2017) has extended our understanding of teachers of color as socio-cultural mediators.

In sum, there is a compelling need for teachers of color. At the same time, our efforts to diversify the teacher workforce seem to be too slow and too weak to match the need.

I am pleased and honored to provide this forward to Yukari Takimoto Amos's *Latina Bilingual Education Teachers: Examining Structural Racism in Schools*. This book provides a deeper exploration of an important pathway for Latina bilingual teachers: from being paraprofessionals, to earning their teacher credential, to becoming teachers. In doing so we get a glimpse inside and alongside a common pathway for Latina, bilingual teachers specifically and teachers of color more generally.

We appreciate the opportunity to get an in-depth and longitudinal view of what one particular group of teachers of color—namely Latina bilingual teachers—experience in their pursuit of a career teaching. In this way, we get to see how K-12 schooling, teacher education programs, and K-12 teaching are gendered and raced within one particular academic area: bilingual education. But given this focus, we are also able to see how race, class, gender, and bilingualism intersect and interact. We also get to see how institutional racism plays itself out in these two teachers' lives.

We also get to hear their stories, their narratives and counter narratives, their *testimonios*. *Testimonios* are consistent with the long Latin American genre of sharing spiritual narratives aimed at raising critical consciousness (Chamberlin & Thompson, 1998) both to expose actions designed to oppress as well as to highlight the ways in which individuals and groups enact resistance to that oppression. It is purposefully designed to create an emotional response on the part of those listening as well as to elicit an urgency to respond. Jara and Vidal (1986) described this as a *narración de urgencia*. The

Foreword xiii

very act of writing these *testimonios*—what Nash (2004) would describe as scholarly professional narratives—provides these participants with a way to heal, to find meaning, and to seek wholeness (Shelhorn, 2015).

As this book demonstrates, Latina educators encounter many obstacles along the way. These include cultural barriers, academic barriers, and institutional barriers. With respect to cultural barriers, we gain a glimpse into how a white racial frame held by the white professionals profoundly influences the experiences these teachers of color face as they enter into the teaching workforce. White privilege (McIntosh, 1988) and white fragility (DiAngelo, 2011), along with color and power-evasiveness, and white supremacy (Leonardo, 2004) is a central characteristic of many white people as they interact with people of color.

We would be surprised—though no less angry—if the academic and institutional barriers obstacles were not present given that most K-12 schools were structured for white, male students. Likewise, most teacher education programs are, in the main, structured for white teacher education candidates. Especially salient is the new narrative around "increasing" academic rigor in teacher education but defining that academic rigor narrowly to mean passing standardized assessments that measure English language proficiency above all else.

Once they receive their certification and are employed as teachers, their experience includes both hyperinvisibility and hypervisibility. With respect to hyperinvisibility, they are seen from a deficit ideology, experience micro-aggressions, and find that their cultural and linguistic capital is of very little value. With respect to hypervisibility, they find that they are expected to do "extra" work given their cultural and linguistic capital.

In short, what is especially helpful in this book is that we get to see how institutional and cultural racism (and the enactment of white racial superiority, more broadly) are experienced, tangibly and concretely, in these Latina Bilingual teachers' lives. We can more fully understand the depth of the challenges around diversity in schooling and we can connect this to the long history of educational oppression experienced by LatinX people. Borrowing from Dumas (2016), teachers of color are not the problem; rather they are humans suffering from problems created by anti-blackness, anti-brownness, anti-indigeneity, anti-immigration, and anti-bilingualism. As Dumas (2016) states, "radical improvements are impossible without a broader, radical shift in the racial order" (p. 17).

Important in these stories is also the resilience that these Latina teachers demonstrate. They are committed to making a difference in the lives of the young people they teach. They recognize that this is as much about community uplift as it is about what occurs in their particular classrooms. And they find great solidarity by working with the Latina/o parents and caregivers that they meet along the way. It is these desires which drive them forward and are as central to their story as are the pains they experience along the way.

xiv *Foreword*

At end, what is most important is that we see teachers of color as truly human, who have a right to pursue their dreams unimpeded because they are human. And we desperately need them in our nation's schools.

Francisco Rios, Ph.D.

References

Chamberlin, M., & Thompson, P. (1998). *Narrative and genre*. New York: Routledge.

DiAngelo, R. (2011). White fragility. *The International Journal of Critical Pedagogy*, *3*(3).

Dumas, M. J. (2016) Against the dark: Antiblackness in education policy and discourse. *Theory Into Practice*, *55*(1), 11–19.

Epps, E. (2002). Race and school desegregation: Contemporary legal and educational issues. *Urban Education Journal*, *1*(1), 388–393.

Fairclough, A. (2004). The costs of Brown: Black teachers and school integration. *The Journal of American History*, *91*(1), 43–55.

Jara, R., & Vidal, H. (Eds.). (1986). *Testimonio y Literatura*. Minneapolis, MN: University of Minnesota Institute for the Study of Ideologies and Literatures.

Leonardo, Z. (2004). The color of supremacy: Beyond the discourse of "white privilege". *Educational Philosophy and Theory*, *36*(2), 137–152.

McIntosh, P. (1988). *White privilege: Unpacking the invisible knapsack*. Retrieved from https://files.eric.ed.gov/fulltext/ED355141.pdf?utm_campaign=Revue%20newsletter&utm_medium=Newsletter&utm_source=revue#page=43

Nash, R. J. (2004). Liberating scholarly writing: The power of personal narrative. New York: Teachers College Press.

Nieto, S. (2017). Becoming sociocultural mediators: What all educators can learn from bilingual and ESL teachers. *Issues in Teacher Education*, *26*(2), 129.

Quiocho, A., & Rios, F. (2000). The power of their presence: Minority group teachers and schooling. *Review of Educational Research*, *70*(4), 485–528.

Shelhorn, B. J. (2015). Review of *How Stories Heal: Writing Our Way to Meaning in the Academy* by R. J. Nash & S. Viray. *Educational Review*, *22*. Retrieved from http://edrev.asu.edu/index.php/ER/article/review/1866/507

U.S. Department of Education. (2013). "Number and percentage of teachers in public and private elementary and secondary schools, by selected teacher characteristics: Selected years, 1987–88 through 2011–12," Table 209.10. National Center for Education Statistics, Schools and Staffing Survey (SASS), "Public School Teacher Data File," 1987–88 through 2011–12; "Private School Teacher Data File," 1987–88 through 2011–12; and "Charter School Teacher Data File," 1999–2000. Retrieved from https://nces.ed.gov/programs/digest/d13/tables/dt13_209.10.asp

U.S. Department of Education. (2016a). *The state of racial diversity in the educator workforce*. Washington, DC: Office of Planning, Evaluation and Policy Development, Policy and Program Studies Service.

U.S. Department of Education. (2016b). "Enrollment and percentage distribution of enrollment in public elementary and secondary schools, by race/ethnicity and region: Selected years, fall 1995 through fall 2026," Table 203.50. National Center for Education Statistics, Common Core of Data (CCD), "State Nonfiscal Survey of Public Elementary and Secondary Education," 2004–05 and 2014–15; and National Elementary and Secondary Enrollment by Race/Ethnicity Projection Model, 1972 through 2026. Retrieved from https://nces.ed.gov/programs/digest/d16/tables/dt16_203.50.asp

Acknowledgements

My first meeting with Erika and Raquel—the two participants in this research study and whose stories this book narrates—began when I started my career as a college professor. They were part of a handful of Latina/o teacher candidates in my university's teacher education program, despite the proximity of the institution to one of the biggest Latina/o communities in the state. I was one of the few professors of color in the university's teacher education program. Although our ethnicity and race were different, we shared the same extreme minority status at a predominantly white space and its effects—presumed incompetence, isolation among peers, and marginalized feelings. In a way, I was seeing myself through them. My biggest thanks go to them for allowing me to investigate their journey for such a long time.

I would like to thank Geneva Gay and James A. Banks for giving me superb training in the fields of multicultural education and culturally responsive teaching when I was a graduate student at the University of Washington. Without their teaching and advice my understanding of race, ethnicity, culture, and racism and how these affect students of color would not have been sufficient to write this book. When I moved to Hong Kong, Francisco Rios remained kind and encouraging, although my work slowed and it took longer to complete the book. I am thankful for his patience. Sharing a similar research interest—teachers of color, Rita Kohli gave me fruitful feedback for the book. I thank for her intellectual collegiality.

The Office of Graduate Studies at Central Washington University awarded me research appointment time to write this book. As a faculty member at a teaching-intense university the released time given to me was more than generous. I have completed this book in Hong Kong during sabbatical from CWU. Matthew Friberg, a Routledge editor, was flexible and patient. He gave me one due date extension after another until I settled down in Hong Kong, but never gave up on me. As a first-time author of a book, I appreciate the guidance he gave me throughout.

I am deeply thankful for my husband, Daniel Miles Amos. He edited and proofread each chapter, gave me more feedback than I wanted, and frequently asked me to rewrite large sections of text. We had good discussions about Erika and Raquel's experiences and I incorporated many ideas from

xvi *Acknowledgements*

those conversations into the book. He deserves a long tip of the hat for his superb editing skills.

This book is derived in part from the following three articles I published.

Amos, Y. T. (2013). Becoming a teacher of color: Mexican bilingual paraprofessionals' journey to teach. *Teacher Education Quarterly*, *40*(3), 51–73.

Amos, Y. T. (2016a). "Are you gonna take my job away?": Latina bilingual education teachers' relationships with white colleagues in the white racial frame. *Urban Education*. doi: 10.1177/0042085916656900.

Amos, Y. T. (2016b). Wanted and used: Latina bilingual education teachers at public schools. *Equity and Excellence in Education*, *49*(1), 41–56.

Y. T. A.
Hong Kong
November 2017

1 Introduction
Latina/o Teacher Career Pathway

This book explores how individuals of color are affected by structural racism when they attempt to make an upward mobility in a white dominated organization called school. To accomplish this purpose, the book traces two Latina individuals who currently teach bilingual education as certified teachers at four different life stages: when they were pupils at K-12, when they were paraprofessionals, when they were teacher candidates at a university, and after they started teaching at public school.

Latinas/os are the largest minority group in the U.S. and comprise of approximately 17.6% of the nation's total population (U.S. Census Bureau, 2016). In the year 2014–15 alone Latinas/os added 1.2 million people to the nation's population. This number is nearly half of the approximately 2.5 million people added in the same year (U.S. Census Bureau, 2016). It is estimated that Latinas/os will constitute 28.6% of the nation's population in 2060 (U.S. Census Bureau, 2016). Although Latinas/os encompass all people groups from Latin America, 63.4% of Latinas/os in the U.S. are of Mexican origin (U.S. Census Bureau, 2016).

Latina/o's population growth is certainly reflected in the nation's schools as well. Racial and ethnic minorities are a growing segment of the K-12 student population in the U.S. with approximately 54% being students of color in 2013 (National Center for Education Statistics, 2016). Among the population of students of color, the number of Latina/o students enrolled in elementary and secondary schools increased from 19% in 2003 to 25% in 2012 (NCES, 2016), reaching approximately 12.5 million students. The 2050 estimate claims that Latina/o students will comprise approximately 29% of all students in U.S. K-12 public schools (NCES, 2016). In 2013, the percentage of children under 18 living in poverty was highest for African American children (28%), followed by Latina/o children (27%) and Asian children (14%) (Musu-Gillette, Robinson, McFarland, KewalRamani, Zhang, Wilkinson-Flicker, 2016). Among 4.9 million English learners in the nation's public schools (10% of all students), 3.8 million students (8% of all students) spoke Spanish as their native language (NCES, 2017).

With the increased Latina/o student population it would be reasonable to expect that there has been increased number of Latina/o teachers as well.

2   Introduction

However, this is not the reality. In general, the proportion of students of color in schools has remained far greater than that of teachers of color. For example, in the 2011–12 academic year Latina/o teachers consisted of only 7.8% of all teachers, only slightly greater than the number of African American teachers (6.8%) (Ingersoll, Merrill, & Stuckey, 2014). The population of both Latina/o teachers and teachers of color in general has not caught up with student demographic change. In 2004 the National Collaborative on Diversity in the Teaching Force reported that about 40% of schools had no minority teachers. In the 2011–12 school year only 17.3% of all elementary and secondary teachers were people of color (Ingersoll et al., 2014).

Research suggests that increasing the racial diversity of classroom teachers benefits the increasingly diverse student population because teachers of color tend to have high expectations of students of color, use culturally relevant teaching, serve as advocates for students of color, and confront racism through teaching (Basit & McNamara, 2004; Villegas & Irvine, 2010; Villegas & Lucas, 2004), all of which directly contribute to students of color's academic achievement. Nevertheless, teachers of color remain a minority population in an increasingly diverse public school system. With the dominance of white teachers unlikely to change soon, the shortage of teachers of color is a grave concern to anyone involved in teaching.

Latinas/os Becoming Teachers

A close examination of the process of Latinas/os becoming teachers informs us why we constantly suffer a shortage of teachers of color in general and Latina/o teachers in particular.

Academic Performance of Latina/o Students at K-12

Becoming a teacher requires one to meet several criteria. Ocasio (2014) asserts that a candidate needs to achieve a certain level of education as well as holding aspirations to continue going to school, earn a specialized degree, and attain a teaching position. Meeting all these criteria can be made possible only with solid academic standing.

Although the achievement gap between the dominant white group and Latinas/os in reading and mathematics has narrowed, the gap still exists. In 2013, at Grade 12, the white-Latina/o gap in reading was 22 points, while in mathematics, it was 21 points (Musu-Gillette et al., 2016), and this gap has been consistent throughout K-12. Rumberger and Angulano (2004) contend that Latina/o students are disadvantaged even before they enter kindergarten, suggesting that the achievement gap between Latinas/os and other groups begins early. The fact that a significant number of Latina/o students are English learners and kindergarten has the highest concentration of English learners (Musu-Gillette et al., 2016) may partially contribute to an early achievement gap.

Introduction   3

At high school, according to Swail, Cabrera, and Lee (2004), Latina/o students are more likely to be held back in school, take lower forms of mathematics, change high schools more than twice, leave high school before graduation, and earn a general equivalency diploma. For example, in 2013, only 10% of Latina/o students earned the highest math course credit in calculus compared to 18% of white students (Musu-Gillette et al., 2016). Lower math achievement is said to be a predictor of a reduced likelihood of completing high school and limited admission to postsecondary institutions (Crisp & Nora, 2010; Swail et al., 2004). Lower grades and lower standardized test scores are also frequently seen among Latina/o students (Becerra, 2012). From 1990 to 2013, the high school dropout rate among 16- to 24-year old Latina/o students dramatically decreased, from 32 to 12%. Nevertheless, this rate is still significantly higher than that of African American students (7%) and white students (5%) (Musu-Gillette et al., 2016).

How Latina/o students are taught and treated at K-12 seems to be related to the alarming statistics listed above. Latina/o students are more likely to learn from under-trained and noncredentialed teachers (Darling-Hammond, 2010), attend overcrowded schools, and experience minimal support in addressing their unique learning styles (Clark, Ponjuan, Orrock, Wilson, & Flores, 2013). Studies further found that Latina/o students are often not encouraged to take college preparation courses, lack support and guidance from counselors or ill-advised by them, and tracked into lower level courses throughout their schooling (Vela-Gude, Cavazos, Johnson, Fielding, Campos, & Rodriguez, 2009). Since student feelings of belonging in school directly correlate to academic competence and expectations (Hernández, Robins, Widaman, & Conger, 2016), a lack of support and low expectations manifested by teachers, counselors, and administrators at school naturally inhibit Latina/o high school students from moving on to higher education. Simply said, many Latina/o students lack preparation for college (Downs, Martin, Fossum, Martinez, Solorio, & Martinez 2008; Swail et al., 2004), and without a bachelor's degree none can become a teacher.

College Enrollment and Retention, and Teacher Education

Academic achievement and college preparedness directly reflect the college enrollment rate for any groups: 42% of 18- to 24-year-old whites enrolled in college in 2013, while 34% of Latinas/os in the same age group did (Musu-Gillette et al., 2016). On top of Latinas/os' lower college enrollment rates, their retention is subject to the revolving door syndrome, a cycle where Latina/o students enter higher education, drop out, and are subsequently replaced by new Latina/o students (Castellanos & Jones, 2003). Although Latina/o students' enrollment in college has increased, more than half of them enroll in community colleges. In fact, Latinas/os are more likely than any other racial and ethnic group to choose a community college rather than a four-year institution (Krogstad, 2016). Although community colleges

4   Introduction

are thought to serve as a gateway to enrollment at four-year institutions, Latina/o students have low transfer, retention, and graduation rates at four-year institutions (Carolan-Silva & Reyes, 2013). A teaching certificate usually requires a bachelor's degree. Thus, neither the lower retention and graduation rate nor the lower transfer rate to a four-year institution contributes to increased numbers of Latina/o teachers.

In order to obtain a teaching license, one usually needs to go through a teacher education program at college. Unfortunately, the number of students of color in these programs is disproportionately low (Darling-Hammond, Chung, & Frelow, 2002; Irizarry & Raible, 2011). In 2012–13, 7% of white students received a bachelor's degree in education, while only 4% of African American students, 4% of Latina/o students, and 2% of Asian students received the same (Musu-Gillette et al., 2016). Many factors account for the underrepresentation of people of color in teacher education programs. Gordon (1994) listed three main reasons why students of color do not choose teaching: negative educational experiences, cultural and community concerns, and social and economic obstacles. The work of Hollins and Guzman (2005) highlights the sociocultural obstacles faced by students of color, noting the basic fact that most teacher candidates of color are prepared in programs where the majority of their peers are white. This reality frequently leaves students of color feeling uncomfortable and out of place within teacher education programs (Gomez, Rodriguez, Agosto, 2008; Salinas & Castro, 2010). In this sense, Robinson, Paccione, and Rodriguez (2003) were accurate: the issue of retention of students of color through teacher education programs is indeed problematic.

The fact that students of color frequently feel uncomfortable and sometimes experience hostility from their white counterparts in teacher education programs is well-documented. These studies highlight the alienating and unsupportive environments where teachers of color across different races experience a disconnection from the larger program community and a sense of "not seeing themselves" in teacher education (Brown, 2014). A sense of alienation emerges in white dominated teacher training program where students of color feel stereotyped, feel the weight of prejudice, and are reminded of the existence of racism. Teacher candidates of color are frequently made to feel invisible. They describe their white peers as displaying racial stereotypes and prejudices, being socially blind, unable to recognize or acknowledge the reality of racism as it exists on many sociocultural levels in the United States (Frank, 2003; Gomez et al., 2008; Nguyen, 2008). As a result, teacher candidates of color describe not feeling safe to speak up in predominantly white teacher education programs. The negative interactions with white peers sometimes intensify in multicultural education classes where many white students tend to express strong resistance. Sheets and Chew (2002), for example, found that Chinese American teacher candidates in a multicultural education class were disgusted, angered, and disturbed by comments made by white students and eventually became silent under the pressure of majority disapproval. In Amos's (2016) study, four female teacher candidates of color (including two Latinas) were afraid that they might be labeled as the ones who spoke up against white students and were fearful of the possibility of retaliation and ostracism from their white peers.

As Lewis, Chesler, and Forman (2000) suggest, white students' behaviors such as colorblindness and racial and ethnic stereotyping have a negative impact on students of color. Clearly, this situation is not encouraging for students of color who aspire to become teachers. The frustration students of color frequently experience with white peers at predominantly white teacher education programs adds psychological burdens to students who need to concentrate on their college-level studies. At worse, it prevents them from pursuing careers in the teaching profession.

Paraprofessionals

One promising strategy for increasing the number of teachers of color in general and Latina/o teachers in particular is to tap the paraprofessional pool. Paraprofessionals, frequently called teacher aides, perform a variety of tasks. They may work as media center aides, computer lab assistants or interpreters for students with limited English skills. They may also work as teacher assistants providing instructional support services under the supervision of a teacher. Nationwide, paraprofessionals account for 12% of full-time staff at school (Hoffman & Sable, 2006), and the majority of instructional paraprofessionals are assigned to the areas of special education, Title I, ESL/bilingual, and library/media (NCES, 2007).

The route to a teaching career via a paraprofessional career is an important one, especially for Latina/o teacher candidates (Ocasio, 2014). Considering the higher poverty rate among Latinas/os (23.5%), with 62% of them holding a high school diploma or less in 2014 (Stepler & Brown, 2016), many Latinas/os who aspire to become teachers simply lack the financial means or education necessary to pursue a bachelor's degree. Becoming a paraprofessional is often a path these Latinas/os choose because it usually does not require a degree beyond high school. With the increasing number of Spanish-speaking students entering public schools, Latina/o paraprofessionals' Spanish language skills are highly desired. Ocasio (2014) concludes that paraprofessional positions allow Latinas/os a step on the path towards becoming a teacher, and that the paraprofessional route towards becoming a teacher offers opportunities for growth and support in the Latina/o teacher career pathway.

Indeed, Latina/o paraprofessionals have advantages in the form of sociocultural capital: they usually come from the same ethnic communities as their students, they are familiar with Latina/o children's cultural experiences, serve as role models with their ability to shift trajectory, and have personal insight into the experience of learning English as a second language (Monzó & Rueda, 2001; Bonner, Pacino, & Stanford, 2011). These

6   *Introduction*

Latina/o paraprofessionals' utmost strength is summarized as their ability to provide "an important source of expertise that many suggest is critical for students of diverse backgrounds" (Rueda, Monzó, & Higaereda, 2004, p. 54). Chopra, Sandoval-Lucero, Aragon, Bernal, de Balderas, & Carroll (2004) further observed that close relationships with students and their parents provided the basis for the paraprofessionals to act as connectors between parents and teachers, parents and community services, students and teachers, students and their parents, and students and their peers. Because of "their extensive school experiences and personal insight into the lives of students from minority communities" (Villegas & Clewell, 1998, p. 124), Latina/o paraprofessionals are considered to be strong teacher candidates.

Research, however, has also documented that paraprofessionals are under-appreciated and regarded lower in status in predominantly white workplaces. Weiss (1994), for example, reported that teachers unnecessarily dominated paraprofessionals, which created an unhealthy relationship between them. A theme of paraprofessionals and teachers not being equals was also observed by Rueda and Monzó (2002) who closely documented the hierarchical structure of social relations between teachers and paraprofessionals in school cultures. These hierarchical relations produced unhealthy conflict as well. Chopra et al. (2004) found that teachers sometimes became jealous and felt threatened when paraprofessionals developed rapport with children, and frequently disvalued their work, and demeaned their important contributions to the academic growth and personal development of students.

Despite their useful knowledge and skills paraprofessionals of color are not necessarily appreciated and valued. The negative work environment many paraprofessionals of color experience may prevent them from pursuing a teaching degree, a situation not conducive to increasing the number of teachers of color.

Work Environment of Teachers of Color

Even if they successfully navigate unsupportive and unwelcoming school and work environments, teachers of color are hard to keep. Ingersoll and May (2010) lament that once they become teachers, teachers of color change schools and abandon the profession at higher rates than white teachers. In 2008–2009, for example, teacher of color's turnover rate was 19.3% while that of white teachers was 15.6% (Ingersoll & May, 2010). The gap is alarming considering that the number of teachers of color is significantly lower than that of white teachers. In sum, an increased number of teachers of color being employed does not necessarily contribute to an increased number of them actually remaining in the profession.

There are a number of reasons why teachers of color leave the profession. Schools' demographic characteristics are highly important to teachers of color's initial employment. However, they are not significantly correlated

Introduction 7

to their later decision to stay or leave (Ingersoll & May, 2010). Instead, Ingersoll and May (2010) found that organizational conditions in school, such as decision-making influence and the degree of individual classroom autonomy were more strongly related to teachers of color's departure. Achinstein and Ogawa (2011) conclude that the main reason for teachers of color's low retention rate is stressful work conditions. Stress can specifically impact persistence attitudes (Johnson, Wasserman, Yildirim, & Yonai, 2014; Wei, Ku, & Liao, 2011) at a workplace. Furthermore, Quiocho and Rios's (2000) comprehensive review identified that teachers of color's feeling of alienation from white colleagues and administrators mostly contributed to the stressful work conditions.

Schools may not recognize the strengths teachers of color possess (Feuerverger, 1997) due to stereotypes and prejudice. Experiencing a lack of warmth from white colleagues and administrators, teachers of color may start harboring feelings of alienation (Basit & McNamara, 2004; Kohli, 2016; Matias, 2016; Rogers-Ard, Knaus, Epstein, & Mayfield, 2012). Furthermore, the clash of values between teachers of color who believe in equity and social justice and white teachers and administrators who want the maintenance of status quo may become problematic (Achinstein & Ogawa, 2011; Arce, 2004; Feuerverger, 1997; Ochoa, 2007; Urrieta, 2009). Ingersoll and Connor's (2009) analysis also indicated that teachers of color lacked support from colleagues and administrators. There seems to be a differential treatment between white teachers and teachers of color when they make an effort to create an inclusive and multicultural environment (Feuerverger, 1997; Klassen & Carr, 1997). Sometimes teachers of color vigorously fight against schools' unequal distribution of resources and tracking (Ochoa, 2007) while receiving little support from colleagues and administrators.

Studies of bilingual education teachers found similar negativity as well. More than 30 years ago, Ada (1986) contended that the mistaken perception of bilingual education teachers as nonprestigious made it difficult for them to build peer support networks. Ada's (1986) observation is still relevant in the present. The perceived lower status of bilingual education teachers (Creese, 2002; Reyes & Kleyn, 2010) and tensions between bilingual education teachers and subject teachers (Valdez, 2001; Varghese, 2006) have been consistently reported. Varghese (2006) revealed that Latina/o bilingual education teachers felt they were marginalized by mainstream teachers who held prejudicial viewpoints towards bilingual education. The negativity towards bilingual education seems to be directly related to the negativity towards English learners. Bigelow (2010), for example, showed that Somali English learners were perceived as uneducable because of their national origin and English learner status. These negative attitudes are frequently expressed in insensitive, disrespectful, and racist comments and behaviors, which cause conflict between mainstream teachers and bilingual education teachers who advocate for English learners' education.

8   *Introduction*

Unfavorable and discriminatory treatment, conflicts with colleagues and administrators, and lack of administrative support all lead to a sense of marginalization and alienation. As Vance, Miller, Humphreys, and Reynolds (1989) argued a long time ago, alienation from colleagues and administrators at the school site is a source of stress and may cause teachers of color to leave the profession. In summary, teachers of color seem to experience a greater sense of marginalization and alienation at a workplace in relation to their white counterparts, and opportunities for collegial work appear to be minimal.

Teachers of color are not alone with regard to their feelings of alienation. Professionals of color in other professions also find themselves feeling alienated. Chávez (2011) found that Latina/o lawyers felt that they were unwelcomed in predominantly white mainstream organizations, experienced a lack of support in the firms, and had to deal with subtle but demeaning stereotypes and discrimination from both coworkers and clients. Hochschild (1996) and Tuan (1999) also discovered similar findings regarding African American and Asian professionals respectively. Tuan's (1999) study of Asian American professionals in the fields of law and medicine reveals that they are still commonly seen as foreigners and considered to be honorary whites as opposed to real Americans. The fact that professionals of color similarly experience alienation regardless of their chosen professions indicates that race does seem to matter at predominantly white workplaces.

Through Critical Lenses

Existing studies indicate that the teacher pathway for Latinas/os is covered with obstacles both academically and socially, and that the work environment for Latina/o paraprofessionals and teachers is also stressful. It appears that race and ethnicity play a salient role in affecting their academic performance and causing unpleasant and negative interactions with white classmates and colleagues. Furthermore, differences other than race, such as language, seem to intersect with race and increase the conditions for conflict. In an attempt to explore the role race plays in the teacher development pathway for teachers of color in general and Latina/o teachers in particular, this book traces two Latinas in four different life stages: when they were pupils at K-12, when they were paraprofessionals, during their teacher education program, and after they started teaching bilingual education at public school as certified teachers.

As the number of Spanish-speaking Latina/o students at public school has dramatically increased and with many being English learners who need significant academic assistance to master both the English language and content areas in English, the demand for Spanish-speaking teachers has also

Introduction 9

risen. Restuccia (2013) describes the urgent need for Spanish-speaking teachers to accommodate the growing number of Latina/o immigrant students. The U.S. Department of Education (2015) lists bilingual education and ESL (English as a second language) together as one of the six high-need subjects nationwide. Goldberg and Proctor's (2000) survey shows that approximately three quarters of urban school districts indicated an immediate need for bilingual education teachers. This book is particularly interested in examining how the increasing demand for Spanish-speaking teachers must consider the roles played by ethnicity and race within a racially and socially stratified school system, and makes the teacher development pathway and the teaching profession for the Latinas/os who speak Spanish fluently stressful and inherently charged with conflict. I look at the different life stages of the two Latinas here within the context of the relationship between the majority whites and minority groups using the ideas of white racial frame, microaggressions, critical race theory, and whiteness studies.

White Racial Frame

White racial frame examines how whites feel about themselves and racial others and behave at the unconscious level, and how such feelings and behaviors affect people of color. Feagin (2013) defines the white racial frame as below:

> there is in North America and elsewhere a dominant, white-created racial frame that provides an overarching and generally destructive worldview . . . this powerful frame has provided the vantage point from which white Americans have constantly viewed North American society. Its centrality in white minds is what makes it a dominant frame throughout the country and, indeed, in much of the Western world and in numerous other areas. Over time, this powerful frame has been elaborated by or imposed on the minds of most Americans, becoming thereby the country's dominant "frame of mind" and "frame of reference" in regard to racial matters.
>
> (p. 10)

Because the white racial frame operates at the unconscious level in whites' mind, practices that are actually racially-based have become common sense and are embedded in nearly every aspect of our daily lives. In other words, the white racial frame is embedded in society's structures as a structural advantage and a built-in bias, directly connected to institutionalized power and privileges that benefit whites (Chubbuck, 2004; Lipsitz, 2006), thus providing a basis for structural racism and whiteness.

A central theme of the white racial frame is in-group superiority and out-group inferiority. Jensen (2005) says, "White people are so often expressing, through their behavior as well as words, a feeling of superiority" (p. 68).

10 *Introduction*

This superiority assumes that "whites are typically more American, moral, intelligent, rational, attractive, and/or hard-working than other racial groups" (Feagin, 2013, p. 94). It is entwined in the history of colonialism throughout the world, guided by powerful emotions and deep negative feelings towards racial others, and expressed in racial hatred and arrogance, greed, gratification, and a desire to dominate others. According to Feagin (2013), the dominant framing of Latinas/os is the image of people who are in "lower" in the racial hierarchy. Solórzano (1997) states that Chicanas/os have been and are sometimes seen in television, film, and print as "dumb," "stupid," and "slow," and these stereotypes are usually not socially condoned. Weyant's (2005) experiment indicates that people in the U.S. hold a stereotype that Latinas/os are less intelligent than whites, at least at an implicit level.

The white racial frame uses language to further enhance in-group superiority. Hill (2008) argues that whites intensely monitor the speech of racialized populations such as Latinas/os and African Americans for signs of linguistic disorder. For example, Latinas/os' Spanish accent in English signifies disorder, while whites' heavy English accent in Spanish and mock Spanish are perfectly acceptable (Hill, 2008). The obvious hierarchy between English and Spanish is a result of the hegemony of English that reinforces the ideological superiority of the English language where all other languages are rendered as subordinate (Macedo, Bessie, & Panayota, 2003). The attempt to dismantle bilingual education buttressed by the official English movement is an example of whites' rejection of "Spanish as a language that is valid in public space" (Hill, 2008, p. 123).

The white racial frame also includes the emotions of "fears and anxieties, conscious and unconscious" (Feagin, 2013, p. 14). The existence of fear is underlined by a loss-gain binary perspective. Milner (2008) illustrates it below:

> The ability, will, and fortitude of whites to negotiate and make difficult decisions in providing more equitable policies and practices might mean that they lose something of great importance to them, including their power, privilege, esteem, social status, linguistic status, and their ability to reproduce these benefits and interests to their children and future generations.
>
> (p. 334)

This perspective fundamentally points to "the impact of a threat to the social status of whites" (Dixson & Rousseau, 2005, p. 19) and excludes any possibility that power be shared between the dominant and subordinate groups.

Historically, whites rationalized their fears by blaming people of color. When the economy entered a depression both in the 1870s and the late 1920s, whites racially framed Chinese and Japanese American workers respectively as allegedly taking job opportunities from whites (Feagin,

Introduction 11

2013). These days, this nativist sentiment specifically targets the persistent inflow of Latina/o immigrants to the U.S. and has turned into racist nativism (Pérez Huber, 2011). In Huntington's (2004) view, Latinas/os pose a threat to white institutions because they grant too much value to their native Spanish tongue, whereas earlier immigrants from Europe readily accepted English as their primary language.

Another way for whites to deal with fear is to ignore or reject "when important but inconvenient facts are presented that do not fit this dominant frame" (Feagin, 2013, p. 15). For example, one Latina lawyer in Chávez's (2011) study mentioned, "Sadly, this often leads to condescension towards us rather than respect for the extraordinary things we've overcome to get where we are. Moreover, many colleagues deeply resent our achievements" (p. 81). Another Latina lawyer in her study stated:

> If you succeed, you are "uppity" and must be taught your proper "place." If things go badly, they offer tea, sympathy, and plenty of bad advice, but secretly delight in your failure—because you are "proof" of the inherent inferiority of those of "your kind."
>
> (p. 83)

Because in-group superiority and out-group inferiority are the central theme of the white racial frame, when people of color act like serious intellectuals and thwart expectations of presumed incompetence, it stirs emotions of fear, threat, and anxiety among whites and is met with attitudes such as resentment, sarcasm, and hostility.

Microaggressions

The various themes in the white racial frame are frequently displayed by whites in the form of microaggressions. Sue, Capodilupo, Torino, Bucceri, Holder, Nadal, & Esquilin (2007) define microaggressions as "brief and commonplace daily verbal, behavioral, or environmental indignities, whether intentional or unintentional, that communicate hostile, derogatory, or negative racial slights and insults toward people of color" (p. 271). Microaggressions are often delivered carelessly without thought and do not have malice intentions to harm. However, they are considered insulting by most people of color, while whites often do not see how they can be offensive (Villalpando, 2003). For example, the Latina/o professionals received a message at a workplace that they were only hired because of their race and ethnicity, their achievements were attributed solely to affirmative action, and were not seen as qualified based on their work experience or accomplishments (Chávez, 2011; Rivera, Forquer, & Rangel, 2010). Microaggressions at a workplace can also include "more subtle acts such as interrupting, ignoring or questioning the validity of the contributions" (Rollock, 2012, p. 519) of the individuals. Faculty of color frequently receives inadequate mentoring. An African American faculty in Constantine,

12   *Introduction*

Smith, Redington, and Owens's (2008) study mentioned, "You just don't get mentored and supported [enough] by other faculty" (p. 352). Delgado Bernal and Villalpando (2002) illustrated a struggle experienced by a Latina faculty member to legitimize her research that was considered lacking academic rigor by white colleagues.

According to Sue and Constantine (2007), people of color describe microaggressions as a pattern of being overlooked, under-respected, and devalued because of their race. The persistent casual deployment of microaggressions is understood as "an additional, unspoken aspect of daily life" (Rollock, 2012, p. 519), and a cycle of contending and resisting take place daily. Continuous "othering" and questioning of social belongingness (Kohli & Solórzano, 2012; Sue, 2010) and ongoing second-guessing their intellectual capabilities and legitimacy (Giles & Hughes, 2009; Smith, Allen, & Danley, 2007), however, physically and emotionally drain people of color. The cumulative effect of daily microaggressions is major part of the hostile professional environment they face.

Whiteness Studies and Critical Race Theory (CRT)

Critical race theory (CRT) postulates that race is no longer rooted in essentialism, but rather is viewed as a social construct (Delgado & Stefancic, 2001; Feagin, 2014). This departure from defining race strictly in a biological term allows us to see racism not about the construction of negative racial images, attitudes, and identities, but about "the creation, development, and maintenance of white privilege, economic wealth, and sociopolitical power" (Feagin, 2014, p. 21). As a social system in which resources and rights are favorably given to whites and unjustly denied to people of color, whiteness is known as a function of power and privilege (Roberts, 2009), thus is best thought of as a form of treasured property (Harris, 1993). Due to its powering and privileging functions, whiteness can be seen to provide material and symbolic privilege to whites, those passing as white, and sometimes honorary whites (Thompson, 2009).

Because of its significant value as treasured property, whiteness yields high status, and that status needs to be protected. According to Harris (1993):

> The right to exclude was the central principle, too, of whiteness as identity, for mainly whiteness has been characterized, not by an inherent unifying characteristic, but by the exclusion of others deemed to be "not white." The possessors of whiteness were granted the legal right to exclude others from the privileges inhering in whiteness; whiteness became an exclusive club whose membership was closely and grudgingly guarded.
>
> (p. 1736)

For example, Harris (1993) asserts that in *the Regents of University of California v. Bakke* (1978) decision, the Supreme Court validated that the

Introduction 13

special admissions plan only for students of color had violated neutrality, thus ultimately protected Bakke's property interest in whiteness. Despite Brutt-Griffler's (2002) observation that "English has become a world language to the extent that it has been stripped of any simplistic association with Anglo-American and Western culture" (pp. vii–viii), the ultimate authority over the English language still rests with "native speakers who are tacitly assumed to be white, and of a certain social class and educational level (Nero, 2006, p. 28). No matter how well they speak English, nonnative speakers of color are automatically excluded from an abstracted notion of an idealized speaker of English (Norton, 1997).

In a social justice oriented inquiry that attempts to expose racism, CRT problematizes how whiteness is perpetuated through "normality" and "invisibility." CRT premises that racism is "a normal fact of daily life in U.S. society that is neither aberrant nor rare" (Taylor, 2009, p. 4) and that it is pervasive, omnipresent, and ingrained in society. Because racism is so embedded in the structure of our society, it becomes normalized and invisible to the beneficiaries. Therefore, from the beneficiaries' standpoint, racism is a common/everyday experience that is often taken for granted (Lopez, 2003). The purpose of CRT is to unmask and expose racism that is hidden from people's consciousness and to examine "how educational theory and practice are used to subordinate certain racial and ethnic groups" (Solórzano & Yosso, 2001, p. 2).

de Plevitz (2007), for example, found that apparently benign and race-neutral policies and practices may unwittingly be having an adverse impact on indigenous students' education. Delgado Bernal and Villalpando (2002) described how the knowledge of faculty of color was deliberately marginalized by the dominant Eurocentric epistemology in higher education. Thompson (2009) claims that the history of whites in the U.S. counts as just plain old American history, whereas the history of people of color is a special case that not all Americans need to know.

Solórzano and Yosso (2009) contend that CRT acknowledges "the intercentricity of racialized oppression—the layers of subordination based on race, gender, class, immigration status, surname, phenotype, accent, and sexuality" (p. 133). Simply said, race intersects with other forms of differences in determining inequality and inequity. According to Andersen (1996), "Class, gender and race matter, and they matter because they structure interactions, opportunities, consciousness, ideology and the forms of resistance that characterize American life . . . They matter in shaping the social location of different groups in contemporary society" (p. ix). Parker and Lynn (2002), for example, explain that "in the case of Black women, race does not exist outside of gender and gender does not exist outside of race" (p. 12) and conclude that racism sustains and rearticulates sexism. To fully understand how discrimination and oppression are perpetuated in educational settings, it is imperative that CRT employ intersectionality because racism is intertwined with other oppressive forms in a complex way that creates unequal power relations among different individuals.

14 *Introduction*

With regard to intersectionality and the two Latinas in focus, it is also indispensable that this study employ Latina/o critical race theory (LatCrit). It examines experiences unique to the Latina/o community, such as immigration status, language, ethnicity, and culture. LatCrit enables us to better articulate the experiences of Latina/os specifically, through a more focused examination of the unique forms of oppression this group encounters. Pérez Huber (2010, 2011) further developed a framework from LatCrit specific to the intersections of racism and nativism called racist nativism. It is directly tied to definitions of whiteness and is defined as:

the assigning of values to real or imagined differences in order to justify the superiority of the native, who is perceived to be white, over that of the nonnative, who is perceived to be people and immigrants of color, and thereby defend the native's right to dominance.

(p. 81)

Sanchez (1997) explains that there is an inextricable link between race and immigration status, contextualized by the historical racialization of immigrants of color and the current moment of increased anti-immigrant sentiment. Since racist nativism operates purely based on perceptions and creates false perceptions of people of color as "nonnative," Latinas/os are racialized as nonnatives regardless of actual immigration status and continued to be excluded and discriminated against, which rationalizes white dominance.

In an attempt to expose racism, CRT privileges the voices of people of color. It recognizes that "the experiential knowledge of people of color is legitimate, appropriate, and critical to understanding, analyzing, and teaching about racial subordination" (Solórzano & Yosso, 2009, p. 133). Therefore, CRT often uses storytelling, counterstories, narrative, autobiography, and parable as a way to illuminate the unique experiences of people of color. Fernández (2002), for example, presented a counterstory of a successful Latina/o college student who resisted inadequate schooling by sometimes choosing alternative activities or practices over attending school. Ninth grade poor and working-class African American and Latina/o youth in Knight, Norton, Bentley, and Dixton's (2004) study retold their counterstories of family involvement as a major influence on their college-going processes and stressed the inequitable structures that hindered access to college-going resources.

Through the critical lenses mentioned above, this book investigates how structural racism affected the two Latinas at the different life stages.

Investigating Latina Bilingual Education Teachers

This book is based on a five-year-long longitudinal study in which I traced two Latina bilingual females, Erika and Raquel, from the entry into a teacher education program to the end of the third year of their teaching as certified teachers. Both women were paraprofessionals before, but currently

Introduction 15

teach as fully credentialed teachers at public schools. All the names used in this book are pseudonyms.

The selection of Erika and Raquel as research participants was both intentional and coincidental. I first met both women in the teacher education program at the university I currently teach (Public Northwest University, PNU). Originally, my intention was to explore nontraditional teacher candidates of color's experiences in a predominantly white teacher education program because I was interested in investigating how the life experiences of such students, particularly the experiences of racism, influence how they teach. I also intentionally targeted bilingual teacher candidates of color because I wanted to investigate how language complicates and intersects with racism. Due to my university's proximity to larger Latina/o communities, Latinas/os are the largest minority group at my university's teacher education program—approximately 13%. When I narrowed down potential participants as nontraditional Latina/o Spanish speakers, three candidates emerged from the pool. By accident, all three candidates were females and had had experiences as paraprofessionals before. Although all three candidates agreed to participate in the research study initially, the third participant withdrew due to her pregnancy. As soon as obtaining their teaching licenses, both Erika and Raquel were employed and began teaching in the field of bilingual education. Therefore, investigating bilingual teacher candidates of color was intentional, while analyzing Latinas and earlier experiences while employed as paraprofessionals was accidental.

Among the Latina/o teachers nationwide, 78% are females (Feistritzer, 2011). Therefore, an examination of the experiences of Latina teachers will contribute to our understanding of the experiences of Latina/o teachers more generally, and to our overall understanding of the experiences of teachers of color as well. Further, the book's focus on Latina bilingual education teachers with low socioeconomic backgrounds and the experiences as paraprofessionals will strengthen CRT's tenet of intersectionality by analyzing how race and other difference, such as class, gender, and language will intertwine in creating inequity.

In order to gain deep understandings of Erika and Raquel's experiences, I conducted semi-structured interviews with them. Both women were officially interviewed three times separately: 1) during the spring quarter of the first year of their teacher education program in which they talked about their upbringing, their life experiences as paraprofessionals, and their experiences in the teacher education program; 2) after completing the first year of teaching as certified teachers in which they reflected upon what they experienced in the teacher education program and what they were experiencing as first-year teachers; 3) in the middle of their third year of teaching in which they discussed what they had been feeling as novice teachers. Each interview lasted approximately two hours. All the interviews were tape-recorded and transcribed. Although both women's stories are mere reflections of their subjective perspectives, they are valuable for revealing

16   *Introduction*

their "perceptions of truth, fairness, and justice" (Taylor, 2009, p. 8). Their stories challenge the dominant ideology that knowledge is always neutral and value free. It serves to legitimize the experiential knowledge of people of color. I also observed both women's teaching at their schools for three hours each time (twice for Erika, and once for Raquel). For the duration of the research, I frequently conversed with both women in person and communicated with them through email. Fieldnotes were taken after each observation and communication. Documents relevant to the book's purpose, such as hand-made worksheets, photos, school policies, and state regulations were also collected.

For the data analysis, I employed the constant comparison method. According to Charmaz (2005), this method involves comparing different people, data from the same individuals at different points in time, incident with incident, data with category, and one category with another until a pattern or theory can be formulated. First, I coded each woman's transcript for words or phrases that stood out related to the study's research questions. Observational fieldnotes and relevant documents were analyzed at this time. Then, these codes were compared with each other within each woman and sorted into common groupings. The common groupings in each woman's transcript were subsequently compared and contrasted against each other to create larger categories. The comparative method was particularly useful because even in the same categories, the two women's experiences and feelings were derived from different factors. When the categories were integrated with the aforementioned critical lenses, overall patterns emerged with regard to how both women were treated at school and how they felt about and responded to those treatments.

A Note on the Politics of Research

For the first two years of the research, I was one of the professors from whom Erika and Raquel took classes (they each took three classes in total from me). It is undeniable that a power hierarchy existed between us when they were teacher candidates. After they graduated from PNU with a teaching license, our relationship became more equal, and we eventually became professionals and trusted friends who shared the same concerns about the education of students of color.

Chapman (2007) notes that the role of the researcher has "a personal dimension that cannot be severed from the researcher's professional interests or personal identity" (p. 158). As an Asian American female, I have had experiences similar to both women I was studying. As a racial and language minority female, what Erika and Raquel felt resonated with me. Like them, I taught language minority students of color as an ESL specialist at public schools before I became a university professor and witnessed how negatively my ESL students were treated by teachers and administrators. Therefore, I could have unconsciously focused on similar incidents to mine

Introduction 17

in the data and made my experiences overlap theirs in my interpretation of the data. In other words, the degree of affinity I had with Erika and Raquel could have threatened the validity of the study.

The above possibility made me constantly reflect upon how my own beliefs and experiences impacted the ways I interpreted the data. To overcome this possibility, I employed several strategies. To enhance validity, I engaged with Erika and Raquel for a long time—five years, triangulated the data by observing, interviewing, and collecting relevant documents. Most importantly, I shared my interpretations with them when I analyzed the data recommended by Glesne (2015). Obtaining their reactions verified that I reflected their perspectives and helped me develop new ideas and interpretations. When both women disagreed with my interpretations, the disagreements were discussed and negotiated with them and revised. To increase reliability, I followed Shenton's (2004) advice and used peer scrutiny. I had my colleagues review all coding of data and my personal writing styles to confirm whether or not my interpretations were not biased. For example, my colleagues pinpointed my preferences for certain kinds of evidence, interpretations, and explanations, and offered alternatives. Their comments challenged my assumptions as "closeness to the project frequently inhibits his or her ability to view it with real detachment" (Shenton, 2004, p. 67).

On the other hand, the similar experiences I had with Erika and Raquel could be considered as the patterns that teachers of color experience while at work in predominantly white educational institutions. In this sense, focusing on our similarities may have served a generative purpose by exposing systemic patterns experienced by teachers of color in general.

Stories to Follow

In this chapter, I have discussed the purpose of the research and related literatures. I have also included the description of the critical lenses I employed to guide my research and the research process. For the rest of the book, I will focus on the stories Erika and Raquel narrated to me.

In Chapter Two, Erika and Raquel recall how they were raised and how they were treated at K-12 schools. In addition, their stories include what they experienced as paraprofessionals and provide us with detailed descriptions of how they were treated by the white teachers with whom they worked. In particular, their stories focus on how the hierarchical relationship between them and the white teachers influenced their decisions to become teachers. Furthermore, both women also reveal why they became paraprofessionals in the first place. The existing literature does not focus on the reasons for an entry to and a departure from the paraprofessional world. These reasons are, however, essential to understand how institutional racism affects individuals of color's life choices.

In Chapter Three, Erika and Raquel focus on detailing their academic struggles and negative interactions they had with their white peers and

18   *Introduction*

instructors as teacher candidates in a predominantly white teacher education program at PNU. Their stories in this chapter highlight the intersectionality of structural racism: in other words, how being poor Latina, having experiences as paraprofessionals, and being able to speak Spanish, in particular, all lead to more institutional obstacles.

In Chapter Four, both women describe the feelings of alienation and marginalization they felt as novice bilingual education teachers during the first three years of teaching. They narrate in detail how the feelings of alienation gradually emerged among them as their schools systematically exploited their Spanish language skills and cultural competence. That is, Erika and Raquel endured heavy workloads with no extra compensation. Their white colleagues hazed them through put-downs and microaggressions, and overtly acted as though they were incompetent teachers by implying they were only hired because of their Spanish language skills. In reality, both women's skills as educators became a threat to many senior white teachers. Neither woman received much support from fellow teachers and administrators. In fact, Erika was continually harassed and negatively evaluated by her vice principal.

Chapter Five covers a period of time when Erika and Raquel are more up-beat, having clearly established themselves as competent professionals. They provide a detailed account of how their cultural knowledge, Spanish language skills, and understanding of effective methods of teacher-student bonding and discipline positively affected their Latina/o students' academic success and engagement in school. In this chapter, their competence as teachers becomes crystal clear to the reader.

Chapter Six reviews the major findings of my study by summarizing the key points of Chapters Two through Five followed by recommendations for policy and practice. Each key point is followed by theoretical discussions of how structural racism negatively affected Erika and Raquel. It also describes, more generally, the paradox between a growing interest in recruiting and preparing paraprofessionals of color and students of color for teaching and the harsh reality they face during teacher education and employment as teachers. The chapter ends with a conclusion of the book.

The investigation of Latina/o bilingual education teachers through the critical lenses is timely and important. Further, studies of paraprofessionals of color are lacking in the field. In addition, a study that analyzes the same individuals of color's experiences at different life stages for a long period of time is rare. This book combines these timely and sparsely researched topics and attempts to make a comprehensive understanding of how teachers of color's personal and professional lives are negatively affected by structural racism. By focusing on structural racism, this book highlights that even in the current post-racial era, racism still exists and professionals of color encounter obstacles as they attempt to be included in organizations dominated by whites. The book attempts to unmask and analyze racist practices that are hidden from people's consciousness within teacher education programs and public schools.

References

Achinstein, B., & Ogawa, R. (2011). *Change(d) agents: New teacher of color in urban schools*. New York: Teachers College Press.

Ada, A. F. (1986). Creative education for bilingual teachers. *Harvard Educational Review*, 56(4), 386–394.

Amos, Y. T. (2016). Voices of teacher candidates of color on white evasion: "I worried about my safety!". *International Journal of Qualitative Studies in Education*, 29(8), 1002–1015.

Andersen, M. (1996). Forward. In E. N.-L. Chow, D. Wilkinson, & M. B. Zinn (Eds.), *Race, class, and gender: Common bonds, different voices* (pp. ix–xii). Thousand Oaks, CA: Sage.

Arce, J. (2004). Latino bilingual teachers: The struggle to sustain an emancipatory pedagogy in public schools. *International Journal of Qualitative Studies in Education*, 17(2), 227–246.

Basit, T. N., & McNamara, O. (2004). Equal opportunities or affirmative action?: The induction of minority ethnic teachers. *Journal of Education for Teaching*, 30(2), 97–115.

Becerra, D. (2012). Perceptions of educational barriers affecting the academic achievement of Latino K–12 students. *Children & Schools*, 34(3), 167–177.

Bigelow, M. H. (2010). *Mogadishu on the Mississippi: Language, racialized identity, and education in a New Land*. Malden, MA: Wiley-Blackwell.

Bonner, P. J., Pacino, M. A., & Stanford, B. H. (2011). Transition from paraprofessionals to bilingual teachers: Latino voices and experiences in education. *Journal of Hispanic Higher Education*, 10(3), 212–225.

Brown, K. D. (2014). Teaching in color: A critical race theory in education analysis of the literature on preservice teachers of color and teacher education in the US. *Race Ethnicity and Education*, 17(3), 326–345.

Brutt-Griffler, J. (2002). *World English: A study of its development*. Clevedon, UK: Multilingual Matters.

Carolan-Silva, A., & Reyes, J. R. (2013). Navigating the path to college: Latino students' social networks and access to college. *Educational Studies*, 49(4), 334–359.

Castellanos, J., & Jones, L. (2003). Latina/o undergraduate experiences in American higher education. In J. Castellanos & L. Jones (Eds.), *The majority in the minority: Expanding the representation of Latina/o faculty, administrators and students in higher education* (pp. 1–13). Sterling, VA: Stylus Publishing.

Chapman, T. K. (2007). Interrogating classroom relationships and events: Using portraiture and critical race theory in education research. *Educational Researcher*, 36(3), 156–162.

Charmaz, K. (2005). *Grounded theory: Methods for the 21st century*. London: Sage.

Chávez, M. (2011). *Everyday injustice: Latino professionals and racism*. Lanham, MA: Rowman & Littlefield.

Chopra, R. V., Sandoval-Lucero, E., Aragon, L., Bernal, C., de Balderas, H. B., & Carroll, A. D. (2004). The paraprofessional role of connector. *Remedial and Special Education*, 25(4), 219–231.

Chubbuck, S. M. (2004). Whiteness enacted, whiteness disrupted: The complexity of personal congruence. *American Educational Research Journal*, 41(2), 301–333.

Clark, M. A., Ponjuan, L., Orrock, J., Wilson, T., & Flores, G. (2013). Support and barriers for Latino male students' educational pursuits: Perceptions of counselors and administrators. *Journal of Counseling & Development*, 91(4), 458–466.

20   Introduction

Constantine, M. G., Smith, L., Redington, R. M., & Owens, D. (2008). Racial microaggressions against black counseling and counseling psychology faculty: A central challenge in the multicultural counseling movement. *Journal of Counseling and Development, 86*(3), 348–355.

Creese, A. (2002). The discursive construction of power in teacher partnerships: Language and subject specialists in mainstream schools. *TESOL Quarterly, 36*(4), 597–616.

Crisp, G., & Nora, A. (2010). Hispanic student success: Factors influencing the persistence and transfer decisions of Latino community college students enrolled in developmental education. *Research in Higher Education, 51*(2), 175–194.

Darling-Hammond, L. (2010). *The flat world and education: How America's commitment to equity will determine our future.* New York, NY: Teachers College Press.

Darling-Hammond, L., Chung, R., & Frelow, F. (2002). Variation in teacher preparation: How well do different pathways prepare teachers to teach? *Journal of Teacher Education, 53*(4), 286–302.

Delgado Bernal, D., & Villalpando, O. (2002). An apartheid of knowledge in academia: The struggle over the "legitimate" knowledge of faculty of color. *Equity and Excellence in Education, 35*(2), 169–180.

Delgado, R., & Stefancic, J. (2001). *Critical race theory: An introduction.* New York: New York University Press.

de Plevitz, L. (2007). Systemic racism: The hidden barrier to educational success for indigenous school students. *Australian Journal of Education, 51*(1), 54–71.

Dixson, A. D., & Rousseau, C. K. (2005). And we are still not saved: Critical race theory in education ten years later. *Race, Ethnicity, and Education, 8*(1), 7–27.

Downs, A., Martin, J., Fossum, M., Martinez, S., Solorio, M., & Martinez, H. (2008). Parents teaching parents: A career and college knowledge program for Latino families. *Journal of Latinos and Education, 7*(3), 227–240.

Feagin, J. R. (2013). *The white racial frame: Centuries of racial framing and counterframing* (2nd ed.). New York: Routledge.

Feagin, J. R. (2014). *Racist America: Roots, current realities, and future reparations* (3rd ed.). New York: Routledge.

Feistritzer, C. E. (2011). *Profile of teachers in the U.S. 2011.* Washington, DC: National Center for Education Information.

Fernández, L. (2002). Telling stories about school: Using critical race and Latino critical theories to document Latina/Latino education and resistance. *Qualitative Inquiry, 8*(1), 45–65.

Feuerverger, G. (1997). "On the edges of the map": A study of heritage language teachers in Toronto. *Teaching and Teacher Education, 13*(1), 39–53.

Frank, A. M. (2003). If they come, we should listen: African American education majors' perceptions of a predominantly white university experience. *Teaching and Teacher Education, 19*(7), 697–717.

Giles, M., & Hughes, R. (2009). CRiT walking place, race and space in the academy. *International Journal of Qualitative Studies in Education, 22*(6), 687–696.

Glesne, C. (2015). *Becoming qualitative researchers: An introduction* (5th ed.). Boston: Pearson.

Goldberg, P. E., & Proctor, K. M. (2000). *Teacher voices 2000: A survey on teacher recruitment and retention.* New York: Scholastic.

Gomez, M. L., Rodriguez, T. L., & Agosto, V. (2008). Who are Latino prospective teachers and what do they bring to U.S. schools? *Race and Ethnicity Education, 11*(3), 267–283.

Gordon, J. (1994). Why students of color are not entering teaching: Reflections from minority teachers. *Journal of Teacher Education, 45*(5), 346–353.

Introduction 21

Harris, C. I. (1993). Whiteness as property. *Harvard Law Review*, *106*(8), 1707–1791.

Hernández, M. M., Robins, R. W., Widaman, K. F., & Conger, R. D. (2016). School belonging, generational status, and socioeconomic effects on Mexican-origin children's later academic competence and expectations. *Journal of Research on Adolescence*, *26*(2), 241–256.

Hill, J. H. (2008). *The everyday language of white racism*. Hoboken, NJ: Wiley-Blackwell.

Hochschild, J. L. (1996). *Facing up to the American dream: Race, class, and the soul of the nation*. Princeton, NJ: Princeton University Press.

Hoffman, L., & Sable, J. (2006). *Public elementary and secondary students, staff, schools, and school districts: School year 2003–04* (NCES 2006–307). U.S. Department of Education. Washington, DC: National Center for Education Statistics.

Hollins, E., & Guzman, M. T. (2005). Research on preparing teachers for diverse populations. In M. Cochran-Smith & K. M. Zeichner (Eds.), *Studying teacher education: The report of the AERA panel on research and teacher education* (pp. 477–548). Mahwah, NJ: Lawrence Erlbaum Associates.

Huntington, S. (2004). The Hispanic challenge. *Foreign Policy*, March/April, 30–45.

Ingersoll, R. M., & Connor, R. (2009, April). *What the national data tell us about minority and Black teacher turnover*. Paper presented at the annual meeting of the American Educational Research Association, San Diego, CA.

Ingersoll, R., & May, H. (2010). *Recruitment, retention, and the minority teacher shortage*. CPRE Research Report #RR-69. Philadelphia, PA: Consortium for Policy Research in Education.

Ingersoll, R., Merrill, E., & Stuckey, D. (2014). *Seven trends: The transformation of the teaching force, updated*. CPRE Research Report #RR-80. Philadelphia, PA: Consortium for Policy Research in Education, University of Pennsylvania. Retrieved from www.cpre.org/7trends

Irizarry, J., & Raible, J. (2011). Beginning with El Barrio: Learning from exemplary teachers of Latino students. *Journal of Latinos & Education*, *10*(3), 186–203.

Jensen, R. (2005). *The heart of whiteness: Confronting race, racism, and white privilege*. San Francisco: City Lights.

Johnson, D. R., Wasserman, T. H., Yildirim, N., & Yonai, B. A. (2014). Examining the effects of stress and campus climate on the persistence of students of color and white students: An application of Bean and Eaton's psychological model of retention. *Research in Higher Education*, *55*(1), 75–100.

Klassen, T. R., & Carr, P. R. (1997). Different perceptions of race in education: Racial minority and white teachers. *Canadian Journal of Education*, *22*(1), 67–81.

Knight, M. G., Norton, N. E., Bentley, C. C., & Dixton, I. R. (2004). The power of Black and Latina/o counterstories: Urban families and college-going processes. *Anthropology and Education Quarterly*, *35*(1), 99–120.

Kohli, R. (2016). Behind school doors: The impact of hostile racial climates on urban teachers of color. *Urban Education*. Advance online publication. doi:10.1177/0042085916636653.

Kohli, R., & Solórzano, D. (2012). Teachers, please learn our names! Racial micro-aggressions and the K-12 classroom. *Race, Ethnicity, & Education*, *15*(4), 441–462.

Krogstad, J. M. (2016). *5 facts about Latinos and education*. Retrieved from www.pewresearch.org/fact-tank/2016/07/28/5-facts-about-latinos-and-education/

Lewis, A. E., Chesler, M., & Forman, T. A. (2000). The impact of "colorblind" ideologies on students of color: Intergroup relations at a predominantly white university. *Journal of Negro Education*, *69*(1/2), 74–91.

22   Introduction

Lipsitz, G. (2006). *The possessive investment in whiteness: How white people profit from identity politics* (Revised & Expanded Ed.). Philadelphia: Temple University Press.

Lopez, G. R. (2003). The (racially neutral) politics of education: A critical race theory perspective. *Educational Administration Quarterly, 39*(1), 68–94.

Macedo, D., Bessie, D., & Panayota, G. (2003). *The hegemony of English.* Boulder, CO: Paradigm.

Matias, C. E. (2016). *Feeling white: Whiteness, emotionality, and education.* Rotterdam, The Netherlands: Sense Publishers.

Milner, H. R. IV. (2008). Critical race theory and interest convergence as analytic tools in teacher education policies and practices. *The Journal of Teacher Education, 59*(4), 332–346.

Monzó, L. D., & Rueda, R. S. (2001). Professional roles, caring and scaffolds: Latino teachers' and paraeducators' interaction with Latino students. *American Journal of Education, 109*(4), 438–472.

Musu-Gillette, L., Robinson, J., McFarland, J., KewalRamani, A., Zhang, A., & Wilkinson-Flicker, S. (2016). *Status and trends in the education of racial and ethnic groups 2016* (NCES 2016–007). Washington, DC: U.S. Department of Education, National Center for Education Statistics. Retrieved from http://nces.ed.gov/pubsearch.

National Center for Educational Statistics. (2007). *Issue brief: Description and employment criteria of instructional paraprofessionals.* Retrieved from https://nces.ed.gov/pubs2007/2007008.pdf

National Center for Educational Statistics. (2016). *The condition of education: Racial/ethnic enrollment in public schools.* Retrieved from https://nces.ed.gov/programs/coe/indicator_cge.asp

National Center for Educational Statistics. (2017). *The condition of education: English language learners in public schools.* Retrieved from https://nces.ed.gov/programs/coe/indicator_cgf.asp

National Collaborative on Diversity in the Teaching Force. (2004). *Assessment of diversity in America's teaching force: A call to action.* Washington, DC: Author.

Nero, S. (2006). An exceptional voice: Working as a TESOL professional of color. In A. Curtis & M. Romney (Eds.), *Color, race, and English language teaching: Shades of meaning* (pp. 23–36). Mahwah, NJ: Lawrence Erlbaum Associates.

Nguyen, H. T. (2008). Conceptions of teaching by five Vietnamese American preservice teachers. *Journal of Language, Identity and Education, 7*(2), 113–136.

Norton, B. (1997). Language, identity, and the ownership of English. *TESOL Quarterly, 31*(3), 409–429.

Ocasio, K. M. (2014). *Nuestro camino:* A review of literature surrounding the Latino teacher pipeline. *Journal of Latinos and Education, 13*(4), 244–261.

Ochoa, G. L. (2007). *Learning from Latino teachers.* San Francisco: Jossey-Bass.

Parker, L., & Lynn, M. (2002). What race got to do with it? Critical race theory's conflicts with and connections to qualitative research methodology and epistemology. *Qualitative Inquiry, 5*(1), 7–22.

Pérez Huber, L. (2010). Using Latina/o critical race theory (LatCrit) and racist nativism to explore intersectionality in the educational experiences of undocumented Chicana college students. *Educational Foundations, 24*(1–2), 77–96.

Pérez Huber, L. (2011). Discourse of racist nativism in California public education: English dominance as racist nativist microaggressions. *Educational Studies, 47*(4), 379–401.

Quiocho, A., & Rios, F. (2000). The power of their presence: Minority group teachers and schooling. *Review of Educational Research, 70*(4), 485–528.

Regents of University of California v. Bakke, 438 U.S. 265 (1978).

Introduction 23

Restuccia, D. (2013, August 8). Why demand for Spanish-speaking teacher is increasing. *The Huffington Post.* Retrieved from www.huffingtonpost.com/2013/11/25/spanish-speaking-teachers_n_4338605.html

Reyes, S. A., & Kleyn, T. (2010). *Teaching in 2 languages: A guide for K-12 bilingual educators.* Thousand Oaks, CA: Corwin.

Rivera, D. P., Forquer, E. E., & Rangel, R. (2010). Microaggressions and the life experience of Latina/o Americans. In D. W. Sue (Ed.), *Microaggressions and marginality: Manifestation, dynamics, and impact* (pp. 59–83). Hoboken, NJ: Wiley-Blackwell.

Roberts, N. S. (2009). Crossing the color line with a different perspective on Whiteness and (anti)racism: A response to Mary McDonald. *Journal of Leisure Research, 41*(4), 495–509. Robinson, J. J., Paccione, A., & Rodriguez, F. (2003). A place where people care: A case study of recruitment and retention of minority-group teachers. *Equity and Excellence in Education, 36*(3), 202–212.

Rogers-Ard, R., Knaus, C. B., Epstein, K. K., & Mayfield, K. (2012). Racial diversity sounds nice; Systems transformation? Not so much: Developing urban teachers of color. *Urban Education, 48*(3), 451–479.

Rollock, N. (2012). Unspoken rules of engagement: Navigating racial microaggressions in the academic terrain. *International Journal of Qualitative Studies in Education, 25*(5), 517–532.

Rueda, R., & Monzó, L. D. (2002). Apprenticeship for teaching: Professional development issues surrounding the collaborative relationship between teachers and paraeducators. *Teaching and Teacher Education, 18*(5), 502–521.

Rueda, R., Monzó, L. D. & Higaereda, I. (2004). Appropriating the sociocultural resources of Latino paraeducators for effective instruction with Latino students: Promise and problems. *Urban Education, 39*(1), 52–90.

Rumberger, R. W., & Angulano, B. A. (2004). *Understanding and addressing the California Latino achievement gap in early elementary school.* Retrieved from the University of California Linguistic Minority Research Institute website: http://escholarship.org/uc/item/65d6v84n

Salinas, C., & Castro, A. J. (2010). Disrupting the official curriculum: Cultural biography and the curriculum decision making of Latino preservice teachers. *Theory and Research in Social Education, 38*(3), 428–463.

Sanchez, G. (1997). Face the nation: Race, immigration and the rise of nativism in late twentieth century America. *International Migration Review, 31*(4), 1009–1030.

Sheets, R. H., & Chew, L. (2002). Absent from the research, present in our classrooms: Preparing culturally responsive Chinese American teachers. *Journal of Teacher Education, 53*(2), 127–141.

Shenton, A. K. (2004). Strategies for ensuring trustworthiness in qualitative research projects. *Education for Information, 22*(2), 63–75.

Smith, W. A., Allen, W. R., & Danley, L. L. (2007). "Assume the position . . . you fit the description": Campus racial climate and the psychosocial experiences and racial battle fatigue among African American male college students. *American Behavioral Scientist, 51*(4), 551–578.

Solórzano, D. G. (1997). Images and words that wound: Critical race theory, racial stereotyping, and teacher education. *Teacher Education Quarterly, 24*(3), 5–19.

Solórzano, D. G., & Yosso, T. J. (2001). From racial stereotyping and deficit discourse toward a critical race theory. *Multicultural Education, 9*(1), 2–8.

Solórzano, D. G., & Yosso, T. J. (2009). Critical race methodology: Counter-storytelling as an analytical framework for educational research. In E. Taylor, D. Gilborn, & G. Ladson-Billings (Eds.), *Foundations of critical race theory in education* (pp. 131–147). New York: Routledge.

Stepler, R., & Brown, A. (2016). *Statistical portrait of Hispanics in the United States.* Retrieved from www.pewhispanic.org/2016/04/19/statistical-portrait-of-hispanics-in-the-united-states/

24   Introduction

Sue, D. W. (Ed.). (2010). *Microaggressions and marginality: Manifestation, dynamics, and impact.* Hoboken, NJ: Wiley-Blackwell.

Sue, D. W., Capodilupo, C. M., Torino, G. C., Bucceri, J. M., Holder, A. M. B., Nadal, K. L., & Esquilin, M. (2007). Racial microaggressions in everyday life. *American Psychologist, 62*(4), 271–286.

Sue, D. W., & Constantine, M. G. (2007). Racial microaggressions as instigators of difficult dialogues on race: Implications for student affairs educators and students. *College Student Affairs Journal, 26*(2), 136–143.

Swail, W. S., Cabrera, A. F., & Lee, C. (2004). *Latino youth and pathway to college.* Washington, DC: Pew Hispanic Center Study.

Taylor, E. (2009). The foundations of critical race theory in education: An introduction. In E. Taylor, D. Gilborn, & G. Ladson-Billings (Eds.), *Foundations of critical race theory in education* (pp. 1–13). New York: Routledge.

Thompson, A. (2009). *Summary of whiteness theory.* Retrieved from www.pauahtun.org/Whiteness-Summary-1.html

Tuan, M. (1999). *Forever foreigners or honorary whites?: The Asian ethnic experience today.* New Brunswick, NJ: Rutgers University Press.

Urrieta, L. Jr. (2009). *Working from within: Chicana and Chicano activist educators in whitestream schools.* Tucson, AZ: The University of Arizona Press.

U.S. Census Bureau. (2016). *FFF: Hispanic heritage month 2016.* Retrieved from www.census.gov/newsroom/facts-for-features/2016/cb16-ff16.html

U.S. Department of Education. (2015). *Teacher shortage areas nationwide listing 1990–1991 through 2015–2016.* Washington, DC: Author.

Valdez, E. O. (2001). Winning the battle, losing the war: Bilingual teachers and post-proposition 227. *The Urban Review, 33*(3), 237–253.

Vance, B., Miller, S., Humphreys, S., & Reynolds, F. (1989). Sources and manifestations of occupational stress as reported by full-time teachers working in a BIA school. *Journal of American Indian Education, 28*(January), 21–31.

Varghese, M. M. (2006). Bilingual teachers-in-the-making in Urbantown. *Journal of Multilingual and Multicultural Development, 27*(3), 211–224.

Vela-Gude, L., Cavazos, J., Johnson, M., Fielding, C., Campos, L., & Rodriguez, I. (2009). "My counselors were never there": Perceptions from Latino college students. *Professional School Counseling, 12*(4), 272–279.

Villalpando, O. (2003). Self-segregation or self-preservation?: A critical race theory and Latina/o critical theory analysis of Chicana/o college students. *International Journal of Qualitative Studies in Education, 16*(5), 619–646.

Villegas, A. M., & Clewell, B. C. (1998). Increasing teacher diversity by tapping the paraprofessional pool. *Theory Into Practice, 37*(2), 121–130.

Villegas, A. M., & Irvine, J. J. (2010). Diversifying the teaching force: An examination of major arguments. *Urban Review, 42*(3), 175–192.

Villegas, A. M., & Lucas, T. (2004). Diversifying the teacher workforce: A retrospective and prospective analysis. In M. A. Smylie & D. Miretky (Eds.), *Developing the teacher workforce: 103rd yearbook of the national society for the study of education, part I* (pp. 70–104). Chicago: University of Chicago Press.

Wei, M., Ku, T.-Y., & Liao, K. Y.-H. (2011). Minority stress and college persistence attitudes among African American, Asian American, and Latino students: Perception of university environment as a mediator. *Cultural Diversity & Ethnic Minority Psychology, 17*(2), 195–203.

Weiss, M. S. (1994). Marginality, cultural brokerage, and school aides: A success story in education. *Anthropology & Education Quarterly, 25*(3), 336–346.

Weyant, J. M. (2005). Implicit stereotyping of Hispanics: Development and validity of a Hispanic version of the implicit association test. *Hispanic Journal of Behavioral Sciences, 27*(3), 355–363.

2 Growing Up, K-12 Schooling, and Working as Paraprofessionals

Both women whose stories this book traces are Americans of Mexican descent. Erika was born in Dallas, Texas, with a father who was born in Mexico and a mother who was born in the U.S. When she was only a couple weeks old, the family moved to a rural and agricultural area of the Pacific Northwest where she spent most of her life. Both of her paternal and maternal grandparents and many relatives remained in Texas. Raquel, on the other hand, was born in San Pedro, Mexico. When she was just one or two years old, her family immigrated to a rural and agricultural area of the Pacific Northwest, in close proximity to where Erika grew up. In terms of personality, Erika and Raquel were close to opposites. Erika articulated what she felt and wanted clearly, which caused her to be sometimes perceived as argumentative, confrontational, and occasionally combative by her classmates and colleagues at work. Raquel was quiet, reserved, and rarely expressed her opinions in public. Despite these differences, both women's experiences were strikingly similar at three different life phases: growing up, K-12 schooling, and working as paraprofessionals.

Erika's Story

Growing up

Erika grew up poor and her upbringing was not as easy as she wished to be. She believes that her family's educational background partially was the cause. Erika spoke about it in quiet tones.

> My mother's schooling was very choppy. She went to school in Texas, but she got pulled out of school a lot to do farmwork, so she didn't graduate. She can't read or write in Spanish. Her Spanish is choppy and so is her English. My dad has no education at all, neither in Mexico nor America. He can't read or write, neither in Spanish nor English.

At home only Spanish was spoken. This situation made Erika fluent in Spanish orally. With regard to the ability to read and write in Spanish,

26   *Growing Up, Schooling, and Working*

however, her parents were reluctant to engage Erika in the process, reflecting on their lower level literacy skills in Spanish. Luckily, her paternal grandmother in Texas contributed to Erika's literacy development in Spanish and nurtured important cultural knowledge of Spanish literacy by demanding that they regularly exchange letters in Spanish. Erika smiled: "Grandma would call us all the time and say, 'You need to learn to read and write in Spanish. Otherwise, how am I going to write letters to you?' So I learned it basically on my own through her." Thus, Erika's linguistic skills in Spanish, both oral and written, developed in the environment where the affirmation of the Spanish language was provided by the collaboration between her parents and immediate family members. Erika was deeply thankful for her grandmother's insistence, in particular, because the ability to read and write in Spanish would later play a key role in obtaining a job she came to love.

The community where Erika grew up is a poor, rural area which employs many Mexican farmworkers for a variety of agricultural businesses. These farmworkers' labor enables people to enjoy high-quality, low-cost vegetables and fruits year-round. However, they are the lowest-paid, least-protected, and unhealthiest workers in the U.S. More than half of them nationwide are undocumented and their average level of education is completion of eighth-grade (National Center for Farmworker Health, 2012). Among these farmworkers, approximately 500,000 of them are estimated to be under the age of 18 (National Farm Worker Ministry, 2017). Growing up in a poor agricultural community and seeing her own parents doing farmwork, it was expected that Erika help her family labor in the fields. As a result, Erika began helping her family at her young age. She explained how hard it was to work in the field as a child:

> I went out at three or four o'clock in the morning. Before, they used to have steel pipes which irrigated the trees. They were long pipes. We had to pick up those pipes, unhitch them, take them over a couple rows, and then hitch them back up to irrigate. We had to do that at four o'clock in the morning with my Dad and my brother before going to school.

By the time other children went to school in the morning, Erika had already finished her morning work in the field. Although feeling exhausted, her parents never allowed Erika and her siblings to skip school.

As she got older, the time spent in the field also increased, and more of her school friends began harder physical labor—picking fruit. Erika felt peer pressure to follow her girlfriends' footsteps:

> I'd say, "Dad, let me go work. I want to pick. My friends did it, so I want to know what it's like." Finally, he let me go pick. It was awful. He's like, "Do you still want to do it?" I'm a kind of a person when I start something, I'll finish it. So I finished that whole season because at that time, our school would give us a week off of school to do harvesting.

Growing Up, Schooling, and Working   27

So during that week I would go and then on weekends, I would go pick again. It was just hard.

Clearly, Mexican students at her school, both directly and indirectly, encouraged each other to engage in the farmwork to help their families. This type of peer pressure seems to have been strong considering the fact that picking fruit and vegetables in the field is indeed a hard labor. The type of peer pressure Erika felt while growing up seems to have positively contributed to the development of her strong sense of work ethic and self-reliance, affirmed her ethnic identity, and strengthened social capital among Mexican students.

After picking fruit in the field, the task of sorting them at a warehouse was waiting. This experience was simply "awful":

> Just standing at the conveyer belt, picking up apples. They don't tell you which ones are bad, they're watching you, they'll yell at you, "Get that one!" They don't train you. They just throw you out there. And then a bathroom is just awful. They go like, "Okay, here's your break!" There are only so few stalls for all these people. And when you have to go to the bathroom, you have to get permission from the supervisor and they time you.

Hard physical labor and an inhumane treatment of workers are reported to be directly linked to health problems, both physical and mental. Thus, it is reasonable for Erika or anyone who experienced it to contend, "I will never do it for a living!" However, interestingly, Erika showed appreciation for such experiences when she obtained a teaching position: "My bilingual ability definitely helped me get a job. But it was more to do with my whole experience, such as doing farmwork." Child labor certainly contributes to supporting and sustaining families and households, but it also leads to a form of civic spirit in social processes, which provides them with own learning and development (Gutiérrez & Rogoff, 2003; Orellana, 2001). In this learning opportunity outside of school, Erika recognized that farmwork experience had become one of the important aspects of how she defined who she was and how she understood who Latina/o students were.

K-12 Schooling

Farmwork among children usually comes at the expense of schooling. Child farmworkers are frequently forced to miss school, and infrequent attendance makes it harder for them to catch up with schoolwork. Unlike her parents who attended school either sporadically or not at all, Erika regularly attended school because her parents insisted on the importance of being educated, which supports several findings that Latina/o parents maintain consistently high aspirations for their children's future (Ceja, 2004;

28   Growing Up, Schooling, and Working

Gándara & Contreras, 2009; Hill & Torres, 2010). Not being educated themselves, Erika's parents instilled in her a culture of possibility in which a link between parents' current occupational status and their children's future academic attainment would break someday. The fact that her family did not migrate like other Mexican families really helped the situation: Erika's family was one of five Mexican families who stayed in the same town while the majority of other Mexican families migrated out.

Her schooling experience, however, was not positive. Erika summarized her K-12 schooling: "I don't have a good memory of school. I was one of those kids who slipped through the cracks. I was seen but not heard. I just got shoved along." At the elementary and middle level, Erika received indifferent instruction—the detachment of teachers from student learning, low expectations, and unimaginative teaching (Segura, 1993). She was simply passed on from grade to grade irrespective of her academic performance.

Erika's biggest academic problem was reading comprehension, and this problem still persists. Erika described her reading skills:

> When I read textbooks I know my reading is not good because I don't comprehend. It takes me a long time to finish reading a chapter. Long, long time. I have the same problem with reading in Spanish. I really did struggle in reading. I still do.

Although recognizing her struggles in reading as a problem, Erika's teachers all simply attributed her difficulty to her speaking Spanish: "Oh, it's a language thing. You speak Spanish." This simple and scant analysis of Erika's achievement illustrates a widely-believed perception among many teachers that "language difference is the primary cause of educational difficulties for Latina/o students, and if that problem were addressed, the education crisis would be resolved" (Gándara & Contreras, 2009, p. 121). This perception erroneously blames students themselves for speaking a different language other than English, and selfishly neglects any analysis of how teachers' indifferent instructions and negative attitudes, subtle and systemic, help reinforce academic difficulties many students of color experience. Fundamentally, it justifies endogenous deficit thinking which posits that students fail in school because of their internal deficiencies (Valencia, 2011).

Despite her teachers' belief that her native language was the cause of her difficulties in reading English, Erika was never referred to the ESL service. Erika now wonders: "I think maybe the reason I wasn't looked at or identified is because my mother spoke English, although choppy. You know, so that could have been." Her teachers never formally assessed Erika's language skills in English nor in Spanish. Apparently they assumed that Erika's English language skills were adequate because her mother's was. In the absence of the ESL service, the help Erika received was in the form of haphazard pull-out time as she described, "I was always in a pull-out because of my low reading level. At middle school, I had a Title I reading class. That's all." And while in a pull-out, Erika did not seem to receive a coherent curriculum to assist her with reading comprehension. Erika's schooling experiences highlight how seriously indifferent teachers were to her academic needs. Despite that appropriate assessment, identification, and placement in the ESL service was required by the law (*Lau v. Nicholas,*[1] 1974), teachers did not take time to professionally assess her English language skills and analyze why Erika struggled in reading. Her teachers' indifference shows their low expectations and lack of interest in Erika's educational development, which, when accumulated, amounted to "a form of oppression—that is, the cruel and unjust use of authority and power to keep a group of people in their place" (Valencia, 2010, p. 9).

In reality, Erika was desperate for attention and help from the teachers. Her voice sounded self-piteous when she asserted with a defiant look, "Nobody cared about me. Nobody paid attention to me." Feeling left behind and alone, and witnessing her problems in reading escalating as she got older, Erika began questioning whether or not she had a learning disability and increasingly doubted her academic capability. Erika's negative self-image suggests how difficult it is for students who receive indifferent instruction to examine how schools and teachers systemically operate to prevent poor students and students of color from learning, and how easy it is for them to start blaming themselves instead. The negative self-image Erika developed at school was, however, convenient for her school and teachers because it diverted attention and scrutiny from what they did not provide her. It was when she started to work as a paraprofessional that Erika clearly understood that the barriers to learning poor students and students of color encounter were systemically embedded in normalized school practices. Her following comment signifies how uncaring the teachers were to Erika: "Lots of people say, 'Oh, one of my teachers really inspired me.' I can't say that at all." Uncaring attitudes by the teachers transmitted to Erika that she was unimportant and unworthy.

At high school, indifferent instruction took a hostile turn with the implementation of a "No-Spanish" policy. This policy was a direct response to the gradual surge of Spanish-speaking Mexican students at school and in the surrounding communities. Erika explained the process as below:

> My math teacher told us not to speak Spanish. He said, "If you can't speak English, you need to get out of my classroom." The same was true pretty much throughout the school. So, it was a lot of that kind of resentment towards us. If we spoke Spanish, it was derogatory. They really didn't want us to speak Spanish.

The "No-Spanish" policy was enforced not only in classes but also on school grounds. As a result, Erika and other Spanish-speaking Mexican students were chastised for conversational use of Spanish among themselves. This is an example of a language suppression which characterizes "the

30   *Growing Up, Schooling, and Working*

perceived inferiority of the Spanish language and of the students who speak it" (Pérez Huber, 2011, p. 394). It is deeply rooted in deficit thinking where students' educability—capacity to learn—intersects with race, class, and culture. Rather than emphasizing the language assets the Spanish-speaking Mexican students brought to school and taking an advantage what they had and teaching where they were, Erika's school bluntly and blatantly refused to effectively educate them. The "No-Spanish" policy effectively solidified the hegemony of English and the subordination of the Spanish language, the Mexican students, and other native speakers of Spanish. In other words, the Spanish-speaking Mexican students were forced wear a "stigmatized" label.

What was paradoxical was the teachers' encouragement of students who spoke English as a native language for learning Spanish in the school enrichment program. The contradiction in which Spanish-speaking students were prohibited from speaking Spanish, while English-speaking students were encouraged to do so underlines the dominant group's selfishness and racism in advancing the interests of white children. A similar contradiction has been researched with regard to dual-language programs.[2] Describing a school district's decision to implement a dual-language program in English and Spanish, Milner (2008) observed:

> The district and school were willing to negotiate and provide the resources necessary for the "non-English speakers" to "learn English" because the majority white students would, of course, benefit from the various racial, ethnic, cultural, and linguistic backgrounds that would be present and represented in the school.
>
> (p. 333)

Palmer (2010) also discovered that in a dual-language program within a predominantly Latina/o and African American community, white middle-class children were in the majority, while only some Latina/o students and almost no African American students were included. In addition, Valdés (1997) raised the concern that the one advantage language minority children might have may be usurped by English speakers who become bilingual in dual-language programs. The popularity of dual-language programs among the dominant group and the formidable attacks on bilingual education programs clearly evidence whose interests are considered the utmost priority, and thus served first. These double-standards were obvious to Spanish-speaking Mexican students like Erika, but her teachers did not notice their own contradictions.

Not only were Spanish-speaking Mexican students but their parents too were adversely affected by the "No-Spanish" policy. Erika regretfully recounted: "A lot of parents wouldn't have their children speak Spanish at home or anywhere because of the policy." Studies report that low-income parents and immigrants with limited education tend to believe that schools and teachers are responsible for teaching their children and feel that school

Growing Up, Schooling, and Working 31

personnel are professionals, and thus leave educational concerns solely to teachers (Lareau, 1987; Ochoa, 2007). Strictly following the policy presented by the school as imperative, many parents unintentionally chose an assimilationist path, and minimized Mexicanness by not speaking Spanish. Erika witnessed more and more Mexican parents ceasing to speak Spanish to their children, and eventually these children lost the ability to speak Spanish. Strict adherence to this school policy by the Mexican parents was conveniently used to help reinforce a racist practice.

The "No-Spanish" policy negatively affected a social dimension, which was most detrimental: it effectively reinforced segregation, hostility, and identity shift not only between white students and Mexican students but also between the Spanish-speaking Mexican students and the non-Spanish-speaking Mexican students. Erika described: "There were three groups defined by language at school. We were very much separated even at such a small school. Each group stuck together." The intra-ethnic hostilities between Mexicans are neither a new nor unique phenomenon and have been reported in past studies (Ochoa, 2007; Tovar & Feliciano, 2009). The tensions could be attributed to differences in class, generation in the U.S., and language ability. The most detrimental effect of the intra-ethnic hostilities is a loss of resistant capital—an important capital to resist oppression as a collective group. As the same ethnic group, these Mexican students could have been able to tap the collective resources, such as their collective experiences, knowledge, and support networks (Valenzuela, 1999) to strengthen their collective identity. However, when students segregate themselves, both by force and choice, and engage in practices of division, the force of resistant capital will naturally weaken, which further strengthens and naturalizes the dominant group's dominance.

At Erika's school, the institutional force of the "No-Spanish" policy successfully accomplished what it had purported: furthering the dominance of white English-speaking Americans through intra-Mexican segregation and the suppression and eradication of the Spanish language. To Erika, however, the policy's adverse effect went beyond the tangible aspect. It weakened the Mexican peer network which was strong before high school. Peer relations could be initiated and controlled by students themselves. However, in a school setting schools themselves may "play a substantial and ongoing role in determining the types of peer relations that occur" (Gibson, Gándara, & Koyama, 2004) through rewards and sanctions to manipulate students. Erika sighed, "It really made a difference, you know, how we were set up by the school, particularly the school was small in size. They've lost part of who they are." By describing the effect of the "No-Spanish" policy as instituted and manipulated by school, Erika displayed her deep understanding how the marginalization of Mexicans and the weakening of their peer network took place at school. Language is the vehicle of culture, thus an indispensable aspect. Fundamentally, to Erika, the loss of language was a loss of cultural and ethnic being as Latinas/os.

32 Growing Up, Schooling, and Working

Working as a Paraprofessional

Erika managed to complete high school, but without adequate academic skills. Deeply disappointed with the indifferent instruction she received and feeling unprepared, unguided, and unencouraged to go to college by her teachers, Erika started taking classes randomly at a nearby community college without much thought. At this time, Erika's goal was extremely vague: she just wanted to obtain a business degree for a nonspecific reason. This, however, lasted only a quarter or so. Erika was soon married and stopped attending community college.

Employment at school came purely by accident. Desperate for a job, Erika, by chance, found an employment as a secretary at a local elementary school. The negative consequence of indifferent instruction at K-12 was evident when Erika explained, "Without a degree and with only a high school diploma, a secretary was my best qualification at that time." Erika's skills immediately after high school were adequate only for semi-skilled jobs such as clerical jobs. This is when Erika realized the negative effect of her K-12 schooling: it would continue to haunt her and restrict her career choices.

Four years later, she left her hometown for a big city. There, she again found an employment as an assistant in the human resources department at a local public school and eventually was transferred to an elementary school in the same school district as a secretary. Erika stated that she had bitter memories of her school years, and it was totally by chance and simply a means to make a living that she found herself working at school. Ironically, her Spanish language skills that her high school had attempted to suppress and eradicate came in handy at school in this new city. The school was particularly interested in Erika's ability to read and write in Spanish. Thanking her grandmother for strongly insisting on amassing this particular linguistic capital, Erika stated with a big smile: "I loved it (working at school) with all my heart!" She was 21 years old.

U.S. Department of Education (2004) defines paraprofessionals as employees who provide instructional support. More specifically, paraprofessionals' duties include: (1) providing one-on-one tutoring if such tutoring is scheduled at a time when a student would not otherwise receive instruction from a teacher, (2) assisting with classroom management, such as by organizing instructional materials, (3) providing instructional assistance in a computer laboratory, (4) conducting parental involvement activities, (5) providing instructional support in a library or media center, (6) acting as a translator, or (7) providing instructional support services under the direct supervision of a highly qualified teacher. Paraprofessionals are not allowed to provide planned direct instruction, or introduce new skills, concepts, or academic content to students. Despite the restrictions with regard to direct instruction, the work of paraprofessionals in all areas of education has shifted from mostly clerical tasks to greater instructional, diagnostic, and counseling responsibilities (Ernst-Slavit & Wenger, 2006; Wenger, Lubbss, Lazo, Azcarraga, Sharp, & Ernst-Slavit, 2004).

Growing Up, Schooling, and Working 33

Although hired as a secretary, Erika soon assumed a paraprofessional role. To her surprise, she found herself loving being with children. Alongside her official responsibilities as a secretary, Erika frequently served as a translator for the school: "I did a lot of parent teacher conference translations for the principal and teachers." Her duties also included small group instruction, particularly in reading, caretaking for recess and lunch duty, summer school, etc. Erika, in particular, displayed great ability as a teacher to students in constant trouble. Most of these students were minority or Mexican male students. Erika described:

> The kids who got into trouble were sent to the principal's office. They wouldn't work for teachers, but they would always work for me. When they were at the office with me, I was able to somehow get them to do their work.

That students of color feel more ease with paraprofessionals than with teachers was also observed in Chopra, Sandoval-Lucero, Aragon, Bernal, de Balderas, & Carroll's (2004) study. According to Chopra et al. (2004), these students consider teachers as disciplinarians and paraprofessionals as supporters, and this perception works as an advantage for paraprofessionals. In other words, students of color believe that paraprofessionals are more accessible to them, compared to teachers and other school personnel.

When asked why these students worked for her, but not for other teachers, Erika answered:

> I really think my race and ethnicity meant a lot to those students. When they see that my family is the same as theirs, we are on a whole another different level. I think the kids relate to that because I'm very . . . yeah . . . we come from the same background. So I can relate to their family get together, I can relate to what they eat, I can relate to the Godmother dancing with the bell, I can relate to all of those things.

Classrooms are a dynamic environment where students and teachers interact with each other, and how these interactions take place influence student performance both positively and negatively. The conventional wisdom among educators is that students of color are more likely to excel educationally when matched with teachers who share their race and ethnicity (Dee, 2004). Such teachers may provide effective role models (Ladson-Billings, 1994; Maylor, 2009) and their perceptions and expectations of students may influence student access to future educational opportunities and also shape the learning environment in meaningful ways (Dee, 2005).

Sharing the same ethnicity, relational trust—a level of mutual understandings of obligations and expectations held by and for others in relationship (Bryk & Schneider, 2002)—between Erika and the troubled students she worked with was strong and solid. In particular, Erika's success with

34   Growing Up, Schooling, and Working

these students could be mostly attributed to the fact that she provided them with what she desperately wanted but did not receive during her own K-12 schooling—a caring attitude with high expectations. Her teaching motto reflected her caring attitude: "I would always say, 'You can do this. Don't use anything as an excuse.' Kids know I am not judging them. They also know not to mess around with me." Although sounding authoritarian, Erika held high expectations of students she worked with. It was obvious that Erika's style made a difference among students who would refuse to work with other teachers who were probably similar to the teachers Erika herself had in the past—uncaring and indifferent. Weiss (1994) observed that paraprofessionals are often "more tolerant of disruptive student behavior" (p. 340) and "unlike some teachers, never give up on a problem child" (p. 340). Knowing each student intimately and having strong cultural connections to the students, Erika was motivated to provide quality instruction to those students who were not in the mainstream. As many Latina/o paraprofessionals do, Erika, as a mother herself, seems to have intuitionally relied on her own experience as a mother and successfully engaged the students in school work (Abbate-Vaughn, 2007). Committed herself to her wider ethnic community's well-being, Erika was their mother in a broader understanding of kinship and cultural representation (Chopra et al., 2004). In this family-like nurturing environment, high expectations Erika demanded of students and her belief in their potential seems to have increased trust between them. Her talent as a paraprofessional began to blossom as she attempted to nurture familial capital with the students.

Working with the troubled students made Erika ponder why they were constantly in trouble to begin with. Erika's analysis was:

> Once a child is frequently disciplined, people are watching out and so any little thing they do, they're called on it, and so I think once you get tagged as a problem maker, people are always watching you and any little thing you do, they catch you on it.

The school at which Erika worked accepted an overflow of students from different schools in lower socioeconomic communities, and these students were getting bussed in. In the midst of a crowded school with so many students and with so little attention to each individual student, the majority of the boys Erika worked with were Mexicans. Erika witnessed the reality of poverty among inner-city Mexican students. Many of them came to school unfed in the morning and wearing the same unwashed clothes over and over. Some of them had head lice. It was a real eye opener for Erika who was raised poor but in a rural town to see the reality of inner-city students' life in poverty. Having just had her first child and constantly comparing her own child with these troubled Mexican students, Erika gradually got so involved with the kids who were sent to the office for disciplinary reasons. She came home crying, "What can I do? Nobody understands what they have to go

Growing Up, Schooling, and Working 35

through. These kids come to school and are in trouble all the time!" As her students became more like her own children and her sense of motherhood increased towards the students, Erika became more concerned about the students' adverse conditions, both in school and life. Simultaneously, she became increasingly frustrated with her own inability to help these students. Erika began to wonder: "If I were a teacher, I would be able to provide the kinds of instructions I want and they deserve."

It was around this time that Erika noticed how deeply negative many teachers were towards Mexican students. What bothered her the most was their synonymous use of two terms, "illegal immigrants" and "Mexicans." Erika said with anger:

> They believe that illegal immigrants are someone who are lazy and come and live off welfare. Those are the people who wake up at four o'clock in the morning, picking apples in this valley! I think about my Dad. I think he's had one speeding ticket so far and he was so humiliated at that, you know. He's worked in the fields his whole life. My Mom works for a preschool now and she's a bus driver. Yeah, she gets up real early. But just a lot of those kinds of remarks make me really angry.

Although Erika herself had negative experiences and felt uncared for throughout her K-12 schooling, witnessing how teachers perceived and treated Mexican students from the teachers' side made her realize that her life mirrored that of the students in many ways and provided her with richer and more sophisticated understanding of the reproductive and institutional force of miseducation of Mexican students at school. Her sense of motherhood towards the students being at the highest at this time, Erika was reminded of her bitter memory of her own K-12 schooling where she was unwanted and uncared for. This overlap was clear between hers and her students' in Erika's mind: teachers were not giving the students she worked with the kinds of education they deserved. They were reproducing another set of students like she had been without holding themselves accountable for their failures as teachers.

After going through two bitter divorces and becoming a single mother with two little children, Erika returned to her hometown to be near with her parents. Her desire to be a role model for Mexican students had grown intensely by this time. Erika declared, "Our kids need a role model. They need people like them who have grown up like them to become someone. Someone like your teacher who upholds education. If you're a teacher and you can influence them." It is not uncommon that bilingual paraprofessionals who witness inequitable practices towards Latina/o students at school view certification "as a means to improve their own ability to work effectively with students within the school setting" (Wenger et al., 2004, p. 105). Erika's aspiration to become a certified teacher was based on her desire to challenge school practices in which Mexican students were institutionally

36 *Growing Up, Schooling, and Working*

neglected, and encouraged to slip between the crack without care or concern from their teachers and administrators. However, the idea of going to university for a teaching license intimidated her because she was not certain about her ability to handle university classes. Even with such high aspiration, Erika was still dominated by the lingering ill-effect of her own K-12 schooling. Erika was simply "not ready to go back to school academically." Erika's fear of university studies resonates many paraprofessionals' lack of confidence as university students when they consider a career change to become certified teachers (Flores, Keehn, & Pérez, 2002).

Considering what was best for her children, Erika went back to the same community college she had previously attended and eventually obtained an associate's degree. The degree gave her a chance to further obtain an emergency teaching certificate which officially allowed her to substitute without a teaching license. Erika began substituting at public schools in the town where she was born and soon this became her daily work and source of income.

Erika's instructional skills coupled with her caring attitudes and high expectations towards Mexican students evoked admiration from students. The students loved her and begged her to come back to teach them again. Although teachers at the schools she worked at were negative about Mexican students in general, they were surprisingly supportive about Erika's aspiration to become a teacher and made comments such as, "You really need to go back to school. You're really good with these kids, Erika." After working as a paraprofessional for seven years, Erika had grown personally and professionally and acquired knowledge, experience, and the ability to analyze the educational conditions of her students. Upon reflecting on the day she submitted an application to the teacher education program, Erika admitted, "I was not ready academically. But I was well-prepared with real life situations. I was ready to work for those kids I cried for." This comment highlights Erika's determination to change the school life of Latina/o students—her strong resistance capital.

Raquel's Story

Growing up and K-12 Schooling

Raquel has been living in the same rural and agricultural town since she immigrated from Mexico. This small town has been heavily populated by Mexicans, most of who are involved in farmwork. As a predominantly Mexican town, the town's mayor is a Mexican man.

Her parents' English skills were minimal, thus Spanish was always spoken at home. Unlike Erika's parents, Raquel's parents were literate in Spanish. Thus, Raquel learned basic literacy skills in Spanish through her parents and older siblings. Similar to Erika, the acquisition and maintenance of the Spanish language skills, particularly literacy skills, developed in a tight-knit

Growing Up, Schooling, and Working   37

family collaboration. Raquel still remembers her parents struggling to complete paperwork in English that they were required to submit to the federal government when she was at elementary school. Later Raquel learned that her parents were undocumented when they immigrated to the U.S., and that the paperwork was for naturalization purposes.

Growing up poor, Raquel and her family worked in a warehouse sorting fruit and vegetables. Raquel depicted this experience as "a fun family time" although she added that she would not do it for a living. She said, "You know, what's strange? I kinda miss it. My Mom made the food, bean tacos or meat tacos, for lunch. It was a hard work, but we had a good family time." Like Erika, Raquel had a positive view of her experiences of farmwork.

Children who do farmwork learn while working. They need to pick and sort fruit and vegetables, which require not only physical labor but also mental work for selecting the right ones to pick. Through farm activities, children develop observational skills by being modeled by their parents and learn the importance of perseverance by repeating the same tasks over and over (Sanderson, Dukeshire, Rangel, & Garbes, 2010). In this sense, Erika and Raquel's positive comments about their farmwork experiences reflects their pride and assertion of their "funds of knowledge" (González, Moll, & Amanti, 2005) which the dominant group may not recognize at all as valid learning experiences but which are wholeheartedly valued by the Mexican community. The fact that both women were proud of their farmwork experiences indicates that these experiences were indeed valid and meaningful to them, and showcases their strong resistant capital against those claimed otherwise.

Although low-income immigrant parents are often viewed "as being indifferent to their children's schooling, failing to encourage their children's achievement, and, in general, placing low value on education" (Orozco, 2008, p. 22), Raquel's parents emphasized the importance of education from an early age. She described her parents with a yearning look: "That's how I was brought up. You know, my parents talked about how important education was. I knew it was very important from early on." Despite her high expectation for being educated, similar to Erika Raquel's schooling experience was not positive either. Her concise summary illustrates its intense negativity: "The education I had is affecting my learning pretty much."

Raquel received a pull-out type of bilingual education from kindergarten until second-grade, but she pondered its effectiveness: "I was doing a lot of simple activities like, 'Point to this. Color this.' Very easy stuff that really didn't teach me the skills I needed to read and write in English properly." Although research concludes that "well-designed, long-term bilingual programs best promote the cognitive and academic development for academic success" (Izquierdo, 2011, p. 160), the bilingual education Raquel received was of low-quality. Gándara and Contreras (2009) contend that the problem of English learners' underachievement is more likely related to

38   Growing Up, Schooling, and Working

the quality of education that these students receive, regardless of the language of instruction.

Raquel attributed a lack of quality in her bilingual education program to the teachers who taught her. Her bilingual education program was only taught by bilingual paraprofessionals, not certified teachers. This is a common practice in rural and semi-urban districts to save money (Wenger et al., 2004). Raquel was fond of her bilingual teachers because "Mexican teachers and Filipino teachers in the program really pushed me. They knew exactly what I was lacking and tried to help me improve my skills." However, these minority bilingual paraprofessionals were not able to transform their deep knowledge about Raquel's academic weaknesses into effective teaching due to their lack of training and preparation which certified teachers would have had. Studies concur that paraprofessionals, irrespective of their areas, such as bilingual education and special education, in general, are rarely allotted preparation time for teaching, material preparation, and training necessary to conduct area-specific testing and evaluation (Chopra et al., 2004; Downing, Ryndak, & Clark, 2000; Wenger et al., 2004). For example, teachers across grade levels identified it problematic when paraprofessionals' command and modeling of writing and oral language was deemed less than acceptable (e.g., errors in grammar and spelling) and their content knowledge was not sufficient (Giangreco, Broer, & Edelman, 2002). However, it was after she became a paraprofessional that Raquel realized how unprepared her bilingual paraprofessionals had been despite their good intentions, efforts, and caring attitudes towards her.

Studies show that students of color, low-income students, and low-performing students, particularly in urban and poor rural areas, are disproportionately taught by novice and less qualified teachers (Darling-Hammond & Bransford, 2005; Duarte, 2000). According to Darling-Hammond (2010), there are growing numbers of racially segregated, apartheid-like schools like the one Rachel experienced, which serve more than 90% of students of color. In these schools the majority of teachers are inexperienced and uncertified. This situation not only adversely affects the academic performance of students of color, it also reinforces the social reality that these students are treated by society as lower-status and not given priority.

From the third-grade, Raquel was "always pulled out for something, migrant, ESL, or speech, always put into a lower level reading group, and never learned essential skills in content areas." In these classes, instruction was remedial rather than challenging. By the third-grade, her conversational English was good, but she still had a hard time making grammatically correct sentences. Raquel recalled, "I would have my words mixed up and still have that problem. I always think, 'Wait a second, that's not what I'm trying to say. It's not coming out the way I want to say it!'" Basically, Raquel was keenly aware of her lower academic standing even when she was in elementary school. She knew that she was getting by, but never fully comprehended how to learn, in particular, how to read, what steps were needed for better

Growing Up, Schooling, and Working 39

comprehension, and why reading was necessary. Unfortunately, her parents could not help Raquel at home because almost all of the materials she brought back home were written in English. Even when written in Spanish, her parents had a hard time helping Raquel due to their lower educational level. Similar to Erika, Raquel began to see herself as having a learning disability.

Raquel attributed her lack of essential skills to the frequent pull-out time at school: "I was usually pulled out for something when the teacher was teaching essential skills, if it was in spelling, reading, or math." Raquel's analysis resonates Ovando, Collier, and Combs' (2011) warning that problems with pull-out programs are lost time in students' access to the full curriculum.

When Raquel was in the sixth-grade, her grades were: "In reading, always C's and D's, English always C's and D's, some B's once in a while. I had really low scores in reading and writing. Math was pretty good. I was OK in math, not great." One day an upsetting event that had harsh and long-lasting consequences for Raquel occurred. Her memory was vivid on this matter:

> We had to take a standardized test. I remember a teacher saying, "Oh, don't worry about it. Just mark whatever you want." So I listened to my teacher and marked whatever I wanted. When I got my test back, I got the score, low, you know, 10% of something of what I should have gotten. Later, Mr. Smith, the principal, said, "Raquel, since you were so low on the test, you need to be in remedial math and reading classes. After you go through these classes, we'll see if you can move up." So I'm like, "Wait a second! I thought this wasn't a serious test. I just marked them randomly." But I didn't dare to tell that to Mr. Smith.

Still furious about this incident, Raquel confessed, "I felt cheated, but I was scared to complain. You don't talk against teachers, you know. It's something that you don't mention because they're usually right and you're wrong." Apparently deficit thinking operated in this test incident where Raquel was never expected to perform well from the beginning. Raquel continued, "Teachers didn't know what it was like to have a second language, so they didn't help me. They probably just assumed that I wasn't smart enough." Race, class, and language seem to have intersected against her and this intersection seems to have justified her teachers' assumption that Raquel had profound cognitive deficits.

Extremely afraid and embarrassed to discuss this incident with her parents, Raquel intentionally kept it secret to the parents. She hoped that she would be moved up soon so that she would not have to reveal to her parents. Despite the fact that she received A's in the remedial classes, Raquel was never taken out of them. She was never moved up, and two years later Raquel felt obliged to confess what had happened to her parents. When she confessed, her father became enraged and fiercely scolded her: "You were the only one who didn't pay attention to what you were supposed to be doing! You're stupid!" Since Raquel had expected that her parents would

40 Growing Up, Schooling, and Working

feel the way she felt, her father's indignation at her was difficult to swallow. Raquel felt, "Ok, well, now I have to deal with what I did. My fault. I need to tell when I should listen to a teacher and when I shouldn't." Rather than protesting her unreasonable placement which was based on invalid test scores, both Raquel and her family blamed themselves instead. Similar to Erika's friends' parents who adhered to the school's "No Spanish" policy, Raquel's family did not question the school's decision. To her parents, school and teachers were the absolute authority. Raquel explained, "For Mom and Dad, it was hard because they didn't know the language. They had never been to school here. So of course, to them, educators are the ones who know the best." The absence of protest from the parents, unfortunately, normalized the school's operation and decision as valid. However, this normalized practice—not attempting to involve Latina/o parents in their children's education illustrates how indifferent the school was to Latina/o parents and students. Unfortunately, Raquel's parents' lack of involvement in the decisions the school made for her reinforced the existing system of social stratification.

Because of the above incident, Raquel stayed in remedial classes from the sixth through eighth-grade despite her desperate hope that she would soon exit this program. This never happened. When Raquel entered high school, her counselor informed her that she would again take remedial math. Raquel was already aware that once in a remedial class it is difficult to change tracks and catch up with regular class peers because the curriculum in remedial classes are generally at a slower pace and watered down. Further, Raquel' sense of teachers regarding her as a lower level student had already been solidified. For example, no matter how many times Raquel asked the same question, "What do I have to do to improve my grade?" her teachers' answers were always the same: "You are doing fine. Don't worry about it." Raquel's anger towards these teachers was intense:

> It was really hard. Sometimes I thought, 'Maybe this is not for me. I shouldn't try." I didn't understand what was expected of, what I was missing. Whenever I heard them, "You're fine," I was like, "That's not what I'm asking! I need to know what I'm missing because I'm not getting the grade I want. I'm not understanding what I need. But after a while, I started shutting myself down, you know.

Similar to Erika, Raquel's teachers' indifference lowered her self-esteem, and she began to doubt about her academic capability. Unfortunately, a combination of her teachers' indifference and her own self-doubt contributed to her never getting the help that would have raised her academic performance. Raquel finally concluded that school had not been and would not be "guiding her into the right direction."

It seems that Raquel had a deep understanding that there is a causal relationship between perceived low educability and academic performance, and that teachers and schools systemically denied her access to quality

Growing Up, Schooling, and Working 41

education. Contrary to the myth held by many white teachers that Mexican American students, particularly students of low socioeconomic background, do not value education (Valencia, 2002), Raquel valued educational quality, and resented the inferior education her schools provided.

Lack of academic rigor in curricula is frequently seen at schools that serve predominantly students of color (Kozol, 1991). Fernández (2002) discovered that the school her participants attended did not offer a rigorous academic curriculum, and the school's policies focused on discipline instead. The remedial math class Raquel was told to repeat by a school counselor was in fact ridiculously easy for her age: all she did was simple addition and subtraction. Not only was the curriculum embarrassingly easy and much less demanding than what Raquel wished to receive, some of her teachers were overtly prejudiced. Her math teachers did not hesitate to show their low expectations: "Oh, you can't do this? Girls can't learn." On another occasion, one of her English teachers, Mrs. Rogers, marked her down without explanation. Mrs. Rogers told the class to present a drama in which the class would pretend as if they had had won something. Raquel thoroughly prepared for the drama and presented with her group members. Mrs. Rogers gave her group an F grade without explaining how this grade was calculated. Raquel was enraged: "I don't know what I did to her. What else was it she wanted? It's so unfair!"

Raquel was desperate to attend regular classes. She felt that she needed to know more. Each day, her determination not to go back to simple adding and subtracting increased. Usually reserved, but sometimes extremely bold, Raquel swore to herself one day: "I got myself into this trouble. I have to get myself out of here. What have I got to lose?" With a strong will, Raquel forged her counselor's signature and successfully got herself into regular math. In other words, Raquel used her resistance capital to resist the inequitable practices. Chuckling, she stated:

> Regular math was not as difficult as I thought it was going to be. I practiced a lot! My brother tutored me when I needed the help. He was an excellent tutor. Sometimes the school offered after-school tutoring. Of course, I went to the tutoring and got some help. Teachers never encouraged me to go to the tutoring session though.

Raquel's comment above and her action of forging the counselor's signature illustrates her resilience and her determination for more meaningful educational opportunities. With help from her brother and her own efforts, Raquel graduated from high school with an average of B's and C's.

Raquel commented:

> There're a lot of us who want to be teachers. We need more teachers like us. But, we're always outsiders within a school, you know. You feel, "Well, what's a school gonna do for me?" You have this negative about

42   *Growing Up, Schooling, and Working*

a school. I struggled a lot at school, but they never helped me. Instead, I had negative experiences with lots of teachers. In my view, I would be the same parent who's like, "School just blames my kids." I think I would have been one of those mothers who are always against school. It's easy to be like that from the experience I had in school.

A shortage of Latina/o teachers, in Raquel's mind, was a result of the systemic negativity they frequently have towards school and teachers. And her story illustrates that her analysis is reasonable.

Working as a Paraprofessional

After graduating from high school, Raquel began working full-time at a warehouse at which she had been already working since childhood. Every day with her mother she picked cherries in the field and sorted peaches and nectarines at the warehouse. She had no intention to go to college: she felt that her public schools had not prepared nor guided her to do so. Raquel just accepted that working at a warehouse and earning a minimal wage was the fate of people like her—a poor Mexican female who lacked academic skills, and that would be her job until she retired.

While working at a warehouse, however, she applied for a teacher's assistant position at local schools because the job description read, "Spanish speakers preferred." When she applied, Raquel thought that obtaining this position would be a longshot considering her low academic standing. An opportunity to become a paraprofessional came out of the blue when Raquel had almost forgotten about the fact that she had even applied for the position. One day, Mr. Smith whom she had known as her middle school principal called her and asked, "Can you come to a job interview?" Raquel immediately answered, "Sure." The interviewed went well. The next time Mr. Smith called Raquel, she was offered a job as a teacher's assistant. Raquel was elated: "I shouted to myself, 'Hooray! Beats going to cherries and apples. I don't need to pick and sort them for the rest of my life!' I know my Spanish had a lot to do with it." It is ironic that Mr. Smith desperately needed Raquel's Spanish language skills which he and other teachers regarded as a sign of deficiency while she was at school. Raquel was 19 years old at this time.

As a paraprofessional, Raquel did everything teachers asked her to do: making copies, collecting and filing papers, sending out parent letters, and translating information for conferences. Like Erika, Raquel also ended up actually teaching Mexican students by conducting small group sessions in almost all content areas in Spanish. Ironically, this teaching opportunity provided Raquel with an important "reeducation" time. Raquel relished that as a paraprofessional, she learned many things she had missed in her K-12 schooling and stated with a smile, "I loved it. I learned a lot. They taught me a lot." Raquel explained how uneducated and ignorant she was:

Growing Up, Schooling, and Working 43

I didn't know many little things. I didn't know there are consonant and vowel sounds in English and Spanish. I didn't know that the word "hot" has the "o" sound and didn't understand there are a long "o" and short "o," you know. I didn't know how to read effectively, either. I didn't know that when you read a sentence and don't know a word, you should go on and then come back. That way, you may be able to find out the meaning of the word in a context. Nobody taught me these little things when I was in school.

It was a reciprocal learning process between her students and herself. Every day Raquel sat next to the students listening to their teachers teach in English. She took notes on many things she had not learned before, and taught them back to the students in Spanish. Raquel's observation skills had been already well-developed through doing farmwork. However, around this time, her observation skills were heightened: she carefully observed what teachers would do and listened to what they said and watched how they taught. In other words, Raquel did not waste what she had learned from farmwork. Rather, she was resilient and used her observation skills to sharpen her teaching skills.

Raquel remembered the students always coming up to her saying cheerfully, "You know more than me," and her responding back to them in a warm but convincing voice, "Trust me. You guys probably know more than me," and both laughed together. Raquel admitted that what she had learned as a paraprofessional made her a parent who could more effectively educate her own children. The idea that a good parent makes a good teacher seems to be an overarching theme for Latina/o bilingual paraprofessionals who make an effort to provide good teaching (Abbate-Vaughn, 2007).

Without this valuable time to reeducate herself and recover from her past miseducation, Raquel admitted that she would not have been able to enter a community college and especially not a university and begin preparation to become a teacher. Raquel summarized this time as "I learned what I was missing. It helped me come back to school." Raquel successfully made up for the years' loss during her K-12 schooling. This reeducation time provided Raquel with the knowledge and skills needed to succeed in college. It also gave her a sense of self-empowerment as she taught students whose lives that mirrored her own.

Although having a fun and meaningful time reeducating herself with Mexican students, Raquel realized that nothing had changed as to how Mexican students were treated at school. First of all, most teachers lacked rapport with Mexican students due to the language barrier. Most students did not speak English well enough. Because these teachers had never been in a situation where they needed to learn content while simultaneously learning a new language, Raquel felt that they did not understand what the students were going through. A lack of understanding seems to have correlated with a fear of dealing with the unknown among the teachers. Raquel saw

44 *Growing Up, Schooling, and Working*

some teachers adamantly refusing to deal with Mexican students. In addition, most teachers would "be very strict and treat kids as if they were dumb" just like in her own K-12 days. Consequently, the students learned to be quiet and not speak. Raquel witnessed a mirror-like treatments of her own in the Mexican students she worked with. She said with a deep sigh, "The students had the same problems as I."

Raquel described one teacher:

> She made this comment, "I don't know how to teach these kids! I don't know what planet they are from!" I couldn't believe she said that, in front of me. I was like, "Hello! Aren't we from the same planet?" She pretty much thought that Mexican children were aliens! And offensive enough, this teacher didn't even bother to check how I was feeling listening to her negative comment about my people!

Raquel attributed the Mexican students' low academic achievement to ignorance and negative attitudes on the part of teachers. She noted that some teachers "simply do not want to help students and do not even hesitate to hide their attitudes and say, 'What are you doing here? You need to go back to where you came from.'" These teachers, according to Raquel, would "make them feel as if they are dumb and keep saying the same phrase sarcastically: 'Oh, you can't do this, can you?'" With such negative attitudes it is difficult for teachers to build good, trusting relationships with students. Without trusting relationships, students naturally disengage themselves in learning from teachers they do not trust. Unfortunately, this cycle reinforces itself within the school setting and worsens if not interrupted. Raquel made another sigh, "It was really hard. I saw discrimination right in front of my face every day." Witnessing deeply rooted negativity towards Mexican students from teachers, Raquel began to understand how these negative perceptions and practices were accepted uncritically by teachers and administrators, in other words, normalized at school. Raquel's consciousness was heightened by seeing the phenomenon through a different angle—from a teacher's perspective.

As her sense of normalized discrimination at school became more and more acute, Raquel began to see a limit for what she could do to help Mexican students. Raquel illustrated this sense of powerlessness in detail:

> I saw a little boy who just came from Mexico in a fourth-grade class. His teacher was teaching vowel sounds as if he were in kindergarten. She stayed on the same letter for like 20 minutes! Not only the pace was ridiculously slow, but also the teacher intimidated him. He was about to cry in class. I sat next to him and asked in Spanish, "Let's figure out what you don't understand." But his stress level was way up. The teacher would ignore him as if he had been invisible. Sadly enough, his classmates were not willing to help this boy, either. When I asked one

Growing Up, Schooling, and Working 45

classmate to help the boy, he did. But after a while, he moved away and turned his back at the boy. When I was about to leave the room, the boy asked me in Spanish and in a sobbing voice, "Where are you going? Why can't you stay here?" I couldn't say anything to him.

Raquel wanted to criticize the teacher for her patronizing and low-quality teaching. However, as a paraprofessional, she was not allowed to critique teachers. She felt that "as a paraprofessional, there is always a time when you can't do anything because you are at a certain level. So, you can't really say much." Rueda and Monzó (2002) observed that a hierarchical structure of social relations between teachers and paraprofessionals are embedded in school cultures in which teachers have direct authority and control over paraprofessionals. Under this hierarchical structure, Raquel was positioned below the teachers.

Raquel's sense of powerlessness exacerbated when most teachers had difficulty sharing power with paraprofessionals. According to Raquel:

I was invited to a group meeting where teachers discussed the Latina/o students' academic needs. One teacher asked my opinion, so I honestly stated, "These kids are lacking the basic skills as far as their phonics knowledge go. It was clear what they need. They don't know their letter sounds in Spanish nor in English." The teacher was a little upset and said, "Oh, we can't teach phonics. We are already in the second-grade." I nodded but continued, "Okay, but this is what they are missing. Isn't our job to give what they need so that they can move on? Or because they have already missed, that's the end?" All I remember afterwards is a dead and awkward silence.

During this awkward silence, Raquel observed teachers looking angry with their faces turning red. She analyzed their anger:

They thought I was intruding. I knew they didn't want to hear it from me. It was like, "What do you know? You are just a para. You've never been to university. You don't know how to teach." I could tell just by watching their body language. But I thought, "Well, Okay, if I don't say anything, nothing's gonna change." So I said it. They were the ones who invited me in and wanted my opinions. Then they got upset because of what I said. But most likely they were mad because I spoke up.

Raquel was reminded that as a paraprofessional she would always be considered to be lower in the hierarchy. Raquel's story is strikingly similar to some teachers in Chopra et al. (2004)' study in which teachers refused to value paraprofessionals' work and rebuffed, "We don't want your input. I'm the teacher; I'm certified; I'm the one with degree" (p. 228). Most paraprofessionals express hope that their contribution would be truly

46   *Growing Up, Schooling, and Working*

understood and valued by teachers and administrators (Giangreco, Broer, & Edelman, 2002). However, Lucero (2010) claims, "As minority language speakers, their contribution may not be recognized as legitimate in larger conversations about education" (p. 140). The desire to maintain the existing hierarchical structure between teachers and paraprofessionals on the part of many teachers make it impossible for paraprofessionals to be considered worthy. This situation further perpetuates the lower-status afforded to paraprofessionals within school settings.

Avoiding confrontations as usual, Raquel never revealed her frustrations directly towards the teachers and kept it to herself. After this incident, Raquel never spoke up at meetings. At home, however, she thoroughly discussed it with her parents, let her frustrations out, and made up her mind: "I'm not going to stay at this level. I'm going to find a way to move myself up. In order to help Mexican students, I need to go back to school." Her sense of powerlessness as a paraprofessional gave her a stronger aspiration to become a teacher. Ironically, teachers who always regarded Raquel as lower in status also suggested to her: "You have to go back to school. You have it in you." After ten years of teaching as a paraprofessional, Raquel was finally confident and determined:

> I will give students the help they need. They know they can go to college. They know they can learn to read and write, you know. In order to help my students, I need to go back to school. I'll be able to do something to make a difference.

Raquel finally put an end to her life as a paraprofessional with a $12 per hour salary and no benefits. She entered a community college and later transferred to a teacher education program. Raquel summarized:

> My siblings would say, "10 years, you wasted your life!" I didn't waste my life. Actually I learned what I had missed. It helped me to come back to school. It helped me boost my confidence and gave me a direction for what I should be doing.

Paraprofessionals to Teacher Candidates

Paraprofessionals, estimated 500,000 individuals in the U.S., are unique but crucial figures in pre-12 classrooms. They are frequently recruited from local communities and reflect their demographics. In major urban and rural areas, 60 to 75% of paraprofessionals are from racial and language minority groups (Burbank, Bates, & Schrum, 2009). This percentage is in sharp contrast to the percentage of teachers of color—only 18% (7% African Americans, 8% Latinas/os, 3% others) (U.S. Department of Education, 2016).

It is important to consider why there is such a wide racial and ethnic gap between paraprofessional and teacher forces. In the school year 2013–14,

Growing Up, Schooling, and Working 47

78% of Latina/o students graduated from high school (National Center for Educational Statistics, 2017a), but their dropout rate (10.6%) was higher than that of whites (5.2%) and African Americans (7.4%) (NCES, 2017b). Only 52% of them immediately enrolled in higher education (NCES, 2017c). In addition, Latinas/os are more enrolled in two-year colleges than four-year colleges (NCES, 2017d). Financial conditions are the key to understanding this point. Approximately 30% of Latina/o parents hold less than a high school education, the highest among any group (NCES, 2017d). The low educational level of Latina/o parents directly related to the poverty rate of Latina/o children—32% in 2014 (NCES, 2017e). Erika and Raquel's parents, as we saw, struggled in English, had low educational levels, and were employed as agricultural workers. As a result, both women were needed to work in the fields with their parents. Without securing a solid financial source, such as scholarships, it is difficult to attend school for four more years after high school, and scholarships require stellar academic performance. This adverse situation hinders many from becoming teachers in a process that requires at least four years of college study.

Although this "class ceiling" certainly plays a role, Erika and Raquel's story reminds us what students learn and how they are treated at school make a difference in their academic achievement. Most importantly, it will determine their career choice after high school. Their K-12 schooling indicates the normalized pervasiveness of deficit thinking towards Latina/o students which inaccurately depicts Latina/o students as less intelligent, lazy and not valuing education. Such deficit thinking is reflected in teachers' indifferent instruction characterized by low expectations and a lack of caring attitudes. As the recipients of indifferent instruction, neither Erika nor Raquel was expected, encouraged, nor guided for higher education, and as a result, they both initially underachieved. Neglected by schools and teachers, gaining the position of paraprofessional was a positive step for both women because it did not require an advanced degree beyond high school.[3] In addition, as Raquel explained, it is reasonable that Latina/o students psychologically disengage themselves from uncaring schooling. Without their conscious knowledge, Latina/os are tracked to detach themselves from schooling, which ultimately has an adverse effect on their motivation to continue education. In summary, schools systemically and systematically funneled both women into a lower-status job track and restricted their upward mobility.

Ironically, both were hired because of their Spanish language skills which their schools had suppressed and ignored. This fact highlights the ethnic and racial dominance of their white teachers and school administrators. They are the one who decided when and how to use the resources the subordinate group, in this case Latinas, possessed. Erika and Raquel both exhibited superb abilities as paraprofessionals, using a combination of sweetness and strictness towards Mexican students. This was similar to the paraprofessionals investigated in other studies (Abbate-Vaughn, 2007; Chopra et al.,

48 *Growing Up, Schooling, and Working*

2004; Ernst-Slavit & Wenger, 2006; Rueda & Monzó, 2002; Wenger et al., 2004). Both women understood where their students were coming from based on their own life experiences, such as doing farmwork. They focused on building personal relationships with the students in a culturally responsive way, which included high expectations and caring attitudes. Although bilingual paraprofessionals are frequently perceived only as language helpers, their popularity among the students did not relate only to the use of the Spanish language. Instead, they were popular because of their caring and demanding attitudes towards the students, and created a family-like school environment for them. Erika and Raquel learned and grew tremendously on the job. By directly working with the students under the supervision of the teachers, they gradually acquired essential knowledge and effective teaching skills to the degree that they were complimented by fellow teachers.

Our ways of thinking are significantly influenced by race, class, gender, and language, and the hierarchical social systems in which they are located (Villegas & Lucas, 2002). Entangled with multiple hierarchies in which they were considered lower in status as racial and ethnic minorities, minority language speakers, and paraprofessionals, Erika and Raquel's sense of social justice was heightened. It was only after they started to work as paraprofessionals that their sense of the need for social change began to grow. Both women witnessed how poorly teachers educated Mexican students. Although they, too, had been the recipients of indifferent education, they were not consciously able to put their own K-12 experiences into context until they became paraprofessionals. Working as paraprofessionals, they saw a parallel between their own K-12 experiences and the mistreatment of Latina/o students they were working with, and could recognize the systemic nature of miseducation of Mexican students. A critical consciousness emerged in them around this time, giving them a clear sense of working towards social justice for their own ethnic group, leading them to decide to become teachers in order to achieve that goal. Some studies suggest that paraprofessionals' decision to become certified teachers was mainly caused by their passionate desire to interrupt and disrupt the systemically oppressive practices they had witnessed at school and provide quality of education to Latina/o students (Flores et al., 2002; Wenger et al., 2004). In the case of Erika and Raquel their consciousness of the reproductive nature of miseducation of Latina/o students, which mirrored their own miseducation, seems to have given them the most incentive. Their experientially derived motivation to "challenge the injustices they experienced in schools and (want to) transform the institutions they viewed as critical to shaping the opportunities available to Latina/o youth" (Irizarry & Donaldson, 2012, p. 168) differs greatly from white teacher candidates.

Raquel's sense of challenging inequality was further enhanced by the sense of powerlessness she felt working as a paraprofessional. Research has documented that paraprofessionals are under-appreciated, under-valued, and marginalized. Weiss (1994) reported that a paraprofessional "might

Growing Up, Schooling, and Working 49

be asked to explain a child's disruptive behavior but is chastised when she volunteered a comment about another child" (p. 339) and concluded that teachers and paraprofessionals are not equals. Ernst-Slavit and Wenger (2006) lament, "Despite their in-depth knowledge of the students' languages and cultures and their pivotal role in educating language minority students, they held marginal positions" (p. 62). For Raquel, her experience of the unequal power relations between her and teachers was the pivotal factor in further motivating her to work toward achieving higher-status, and becoming a teacher. As a teacher, she would gain more power and control to remedy the injustice and a lack of caring experienced by Latina/o students.

Without the experience of working as paraprofessionals, both women may have never entered the teaching field. Without a doubt their experiences as paraprofessionals empowered them personally and professionally. In the next chapter we will examine how Erika and Raquel navigated a teacher education program to become certified teachers. Both felt that they were not academically ready for university studies, but had confidence in their teaching abilities and a strong sense that they must succeed in order to gain new ground for social justice as it relates to the educational success of Latina/o students.

Notes

1 A legal action taken by Chinese students who did not speak English fluently against the San Francisco Unified School District. The plaintiff accused the District for not providing them with necessary language accommodations and for failure to provide them with full access to the district's core curriculum. The Supreme Court unanimously decided that the lack of supplemental language instruction for students with limited English proficiency denied the students a meaningful education, thus, violated the Civil Rights Act of 1964. The ruling mandated additional language instruction mandatory, in other words the ESL services.
2 Dual-language program, a.k.a, two-way bilingual education has been gaining popularity these days. This model serves language minority and majority students simultaneously in one classroom in two different languages: the language minority students' native language and English. This model "encourages socialization between the two groups of students and respect for the others' cultural backgrounds" (Brisk, 2006, p. 37). The advantage of two-way bilingual education is that all students in this model are exposed to two languages: English and the minority language and as a result and get to learn in two languages.
3 The No Child Left Behind Act enacted in 2001 recommends at least two years of study in higher education or an associate degree for paraprofessionals.

References

Abbate-Vaughn, J. (2007). "Para aquí today, para afuera tomorrow": Uncertainty in urban bilingual paraprofessionals' work in the age of NCLB. *The Urban Review*, *39*(5), 567–588.

Brisk, M. E. (2006). *Bilingual education: Froom compensatory to quality schooling* (2nd ed.). Mahwah, NJ: Lawrence Erlbaum Associates.

Bryk, A. S., & Schneider, B. (2002). *Trust in schools: A core resource for school reform*. New York: Russel Sage Foundation.

50   Growing Up, Schooling, and Working

Burbank, M. D., Bates, A. J., & Schrum, L. (2009). Expanding teacher preparation pathway for paraprofessionals: A recruiting seminar series. *Teacher Education Quarterly, 36*(2), 199–216.

Ceja, M. (2004). Chicana college aspirations and the role of parents: Developing educational resiliency. *Journal of Hispanic Higher Education, 3*(4), 338–362.

Chopra, R. V., Sandoval-Lucero, E., Aragon, L., Bernal, C., de Balderas, H. B., & Carroll, A. D. (2004). The paraprofessional role of connector. *Remedial and Special Education, 25*(4), 219–231.

Darling-Hammond, L. (2010). Teacher education and the American future. *Journal of Teacher Education, 61*(1–2), 35–47.

Darling-Hammond, L., & Bransford, J. (Eds.). (2005). *Preparing teachers for a changing world: What teachers should learn and be able to do.* San Francisco: Jossey-Bass.

Dee, T. S. (2004). Teachers, race and student achievement in a randomized experiment. *The Review of Economics and Statistics, 86*(1), 195–210.

Dee, T. S. (2005). A teacher like me: Does race, ethnicity, or gender matter? *The American Economic Review, 95*(2), 158–165.

Downing, J. E., Ryndak, D. L., & Clark, D. (2000). Paraeducators in inclusive classrooms: Their own perceptions. *Remedial and Special Education, 21*(3), 171–181.

Duarte, A. (2000). Wanted: 2 million teachers, especially minorities. *The Education Digest, 66*(4), 19–23.

Ernst-Slavit, G., & Wenger, K. J. (2006). Teaching in the margins: The multifaceted work and struggles of bilingual paraeducators. *Anthropology and Education Quarterly, 37*(1), 62–82.

Fernández, L. (2002). Telling stories about school: Using critical race and Latino critical theories to document Latina/Latino education and resistance. *Qualitative Inquiry, 8*(1), 45–65.

Flores, B. B., Keehn, S., & Pérez, B. (2002). Critical need for bilingual education teachers: The potentiality of *normalistas* and paraprofessionals. *Bilingual Research Journal, 26*(3), 501–524.

Gándara, P., & Contreras, F. (2009). *The Latino education crisis: The consequences of failed social policies.* Cambridge, MA: Harvard University Press.

Giangreco, M. F., Broer, S. M., & Edelman, S. W. (2002). "That was then, this is now!": Paraprofessional supports for students with disabilities in general education classrooms. *Exceptionality, 10*(1), 47–64.

Gibson, M. A., Gándara, P., & Koyama, J. P. (2004). The role of peers in the schooling of U.S. Mexican youth. In M. A. Gibson, P. Gándara, & J. P. Koyama (Eds.), *School connections: U.S. Mexican youth, peers, and school achievement* (pp. 1–17). New York: Teachers College Press.

González, N., Moll, L. C., & Amanti, C. (2005). *Funds of knowledge: Theorizing practices in households, communities, and classrooms.* New York: Routledge.

Gutiérrez, K. D., & Rogoff, B. (2003). Cultural ways of learning: Individual traits or repertoires of practice. *Educational Researcher, 32*(5), 19–25.

Hill, N. E., & Torres, K. (2010). Negotiating the American dream: The paradox of aspirations and achievement among Latino students and engagement between their families and schools. *Journal of Social Issues, 66*(1), 95–112.

Irizarry, J., & Donaldson, M. (2012). Teach for America: The Latinization of U.S. schools and the critical shortage of Latino/a teachers. *American Educational Research Journal, 49*(1), 155–194.

Izquierdo, E. (2011). Two way dual language education. In R. R. Valencia (Ed.), *Chicano school failure and success: Past, present, and future* (3rd ed., pp. 160–172). New York: Routledge.

Kozol, J. (1991). *Savage inequalities.* New York: Basic Books.

Growing Up, Schooling, and Working   51

Ladson-Billings, G. (1994). *The dreamkeepers: Successful teaches of African American children*. San Francisco: Jossey-Bass.

Lareau, A. (1987). Social class differences in in family school relationships: The importance of cultural capital. *Sociology of Education, 60*(2), 73–85.

Lau v. Nichols, 414 U.S. 563 (1974).

Lucero, A. (2010). Dora's program: A constructively marginalized paraeducator and her developmental biliteracy program. *Anthropology and Education Quarterly, 41*(2), 126–143.

Maylor, U. (2009). "They don't relate to Black people like us": Black teachers as role models for Black pupils. *Journal of Education Policy, 24*(1), 1–21.

Milner, H. R. IV. (2008). Critical race theory and interest convergence as analytic tools in teacher education policies and practices. *The Journal of Teacher Education, 59*(4), 332–346.

National Center for Educational Statistics. (2017). *The condition of education.* a) *Public high school graduate rates.* Retrieved from https://nces.ed.gov/programs/coe/indicator_coi.asp b) *Status dropout rates.* Retrieved from https://nces.ed.gov/programs/coe/indicator_coj.asp c) *Immediate college enrollment rate.* Retrieved from https://nces.ed.gov/programs/coe/indicator_cpa.asp d) *Characteristics of postsecondary students.* Retrieved from https://nces.ed.gov/programs/coe/indicator_csb.asp e) Family characteristics of school-age children. Retrieved from https://nces.ed.gov/programs/coe/indicator_cce.asp

National Center for Farmworker Health. (2012). *Farmworker health factsheet.* Retrieved from www.ncfh.org/uploads/3/8/6/8/38685499/fs-migrant_demographics.pdf

National Farm Worker Ministry. (2017). *Farm worker issues: Children in the fields.* Retrieved from http://nfwm-yaya.org/wp-content/uploads/2011/11/pdf_Children-in-the-Fields.pdf

Ochoa, G. L. (2007). *Learning from Latino teachers*. San Francisco: Jossey-Bass.

Orellana, M. F. (2001). The work kids do: Mexican and Central American immigrant children's contributions to households and schools in California. *Harvard Educational Review, 71*(3), 366–390.

Orozco, G. L. (2008). Understanding the culture of low-income immigrant Latino parents: Key to involvement. *The School Community Journal, 18*(1), 21–37.

Ovando, C. J., Collier, V. P., & Combs, M. C. (2011). *Bilingual and ESL classrooms: Teaching in multicultural contexts* (5th ed.). Boston: McGraw Hill.

Palmer, D. (2010). Race, power, and equity in a multiethnic urban elementary school with a dual-language "strand" program. *Anthropology & Education Quarterly, 41*(1), 94–114.

Pérez Huber, L. (2011). Discourses of racist nativism in California public education: English dominance as racist nativist microaggressions. *Educational Studies, 47*(4), 379–401.

Rueda, R., & Monzó, L. D. (2002). Apprenticeship for teaching: Professional development issues surrounding the collaborative relationship between teachers and paraeducators. *Teaching and Teacher Education, 18*(5), 502–521.

Sanderson, L. L., Dukeshire, S. R., Rangel, C., & Garbes, R. (2010). The farm apprentice: Agricultural college students recollections of learning to farm "safely". *Journal of Agricultural Safety and Health, 16*(4), 229–247.

Segura, D. A. (1993). Slipping through the cracks: Dilemmas in Chicana education. In A. de la Torre & B. M. Pesquera (Eds.), *Building with our hands: New directions in Chicanastudies* (pp. 199–216). Berkeley and Los Angeles, CA: University of California Press.

Tovar, J., & Feliciano, C. (2009). "Not Mexican-American, but Mexican": Shifting ethnic self-identification among children of Mexican immigrants. *Latino Studies, 7*(2), 197–221.

52 Growing Up, Schooling, and Working

U.S. Department of Education. (2004). *Title I paraprofessionals: Non-regulatory guidance.* Washington, DC: Author

U.S. Department of Education. (2016). *The state of racial diversity in the educator workforce.* Washington, DC: Author.

Valdés, G. (1997). Dual-language immersion programs: A cautionary note concerning the education of language-minority students. *Harvard Educational Review, 67*(3), 391–429.

Valencia, R. R. (2002). "Mexican Americans don't value education!" On the basis of the myth, mythmaking, and debunking. *Journal of Latinos and Education, 1*(2), 81–103.

Valencia, R. R. (2010). *Dismantling contemporary deficit thinking: Educational thought and practice.* New York: Routledge.

Valencia, R. R. (2011). The plight of Chicano students: An overview of schooling conditions and outcomes. In R. R. Valencia (Eds.), *Chicano school failure and success: Past, present, and future* (3rd ed., pp. 3–41). New York: Routledge.

Valenzuela, A. (1999). *Subtractive schooling: U.S.-Mexican youth and the politics of caring.* Albany, NY: State University of New York Press.

Villegas, A. M., & Lucas, T. (2002). Preparing culturally responsive teachers: Rethinking the curriculum. *Journal of Teacher Education, 53*(1), 20–32.

Weiss, M. S. (1994). Marginality, cultural brokerage, and school aides: A success story in education. *Anthropology & Education Quarterly, 25*(3), 336–346.

Wenger, K. J., Lubbss, T., Lazo, M., Azcarraga, I., Sharp, S., & Ernst-Slavit, G. (2004). Hidden teachers, invisible students: Lessons learned from exemplary bilingual paraprofessionals in secondary schools. *Teacher Education Quarterly, 31*(2), 89–111.

3   Becoming Teachers of Color

Erika and Raquel first met with each other at the teacher education program in Public Northwest University (PNU). PNU is a predominantly white public four-year university with a relatively high percentage of students from families of lower socioeconomic status. Approximately half of first-time, first-year students enrolled in fall 2016 at PNU scored the 500–599 range in critical reading and math and the 400–499 range in writing in SAT. Of all the public universities in the state, it has the lowest academic rank. Reflecting the students' low socioeconomic backgrounds, approximately 65% of all students at PNU are transferring students from the state's community colleges.

At PNU, approximately 14.5% of its students are Latinas/os. The higher rate of Latina/o students compared to other minority groups (Asians—including international students, 4% and African Americans, 3%) reflects where PNU is located, as it is 35 miles away from one of the largest concentrations of Latinas/os in the state. Effectively utilizing this geographical advantage, PNU has successfully established a satellite campus in the community college where Erika and Raquel had obtained associate of arts degrees. Students who are enrolled in the community college can transfer to PNU and obtain a bachelor's degree with a teaching license without commuting to where PNU is located. This situation was ideal for Erika and Raquel who had families to take care of and were restricted in terms of commuting time.

PNU originated in the late 19th century as a teacher preparation college, using the vocabulary of the time, a normal college. It has been known for its strong teacher education program and produces approximately 500 teacher candidates each year. On the satellite campus, the teacher education program offers classes for a teacher license in a cohort model. In the two years Erika and Raquel attended this satellite campus, there were approximately 18 students in the education cohort: three of them were Latinas (including both women) and the others were whites (including two men). Even in the midst of a large Latina/o community on the satellite campus, the students were predominantly white. All the students in the cohort studied for the same major and minor: elementary education and TESL (teaching English

54 *Becoming Teachers of Color*

as a second language). The cohort took classes together in every evening through Mondays and Thursdays for two years, including one summer.

Finding only three Latina/o students in the program, Erika, Raquel, and another Latina (Alma) naturally connected to each other. They sat together in class, studied together, and hung out together on weekends. This newly established friendship in a predominantly white program was relieving, refreshing, and revitalizing for Erika and Raquel. Both were concerned about whether or not they would be able to manage university-level studies. Erika described their friendship as follows:

> We'd get together. We were a little study group. On weekends, we'd have a barbecue and work on our projects together and get through the materials and stuff. We were able to do that. Raquel and I had kids, so we met at our houses, so the kids could play together while we did our work. If she wanted to sleep on the couch, I didn't care. Raquel called me often, too. I felt comfortable.

Erika's sense of camaraderie was shared by Raquel. She stated:

> We and other classmates didn't socialize the same way. There were certain ways we did around and they might have taken it differently. They didn't know the struggles we had gone through. There were things, you know, different topics you ended up talking . . . not the same as when you were with Erika and Alma, and your family.

Several studies found that social support, as a measure of social capital, provided by parents, friend, and teachers is positively associated with school engagement (Brewster & Bowen, 2004; Garcia-Reid, 2007). *Familismo*—"the centrality of family life and its priority over other realities" (Arditti, 2006, p. 246) is a cultural core value of Latinas/os, who have larger family networks, spend more time with family, and rely more on family for instrumental and emotional support relative to non-Latinas/os (Calzada, Tamis-leMonda, & Yoshikawa, 2013; Marin & Gamba, 2003). When *familismo* is extended to include friends at school, school peers can provide emotional, informational, and instrumental support just like real family members. Peers have advantages because they are more likely to be immediately available and to have information relevant to negotiating the college environment (Rodriguez, Mira, Myers, Monis, & Cardoza, 2003). Emotional support from peers is also found to be associated with better social adjustment among Latina/o college students (Schneider & Ward, 2003). Feeling a strong and tight-knit friendship and sharing similar concerns, three women supported each other in anyways they could to succeed in a predominantly white program, which simultaneously strengthened and sharpened their skills to navigate the college life.

Academic Struggles

Academic struggles were something both Erika and Raquel had anticipated. Barely managing to complete an associate degree at the community college, they were both concerned about their academic ability to handle university-level materials. With regard to college readiness, there is a wide gap between white/Asian students and students of color. In 2008, about 27% of white ACT test-takers and 33% of Asians met the college readiness benchmark in all four subjects, while only 10% Latinas/os and 3% of African Americans did (Aud, Fox, & KewalRamani, 2010). In addition, first-generation college students, such as Erika and Raquel, encounter more academic challenges than non-first-generation college students. Bui (2002) contends that the academic difficulties many first-generation university students experience prior to and during university make them vulnerable to lower academic performance. Considering the low-quality education both women received during K-12 and being first-generation college students themselves, it is understandable that Erika and Raquel felt inadequately prepared for university-level studies. They were not alone in the way they felt. Studies have found that Latina/o students in general often graduate from high school without being prepared to enter college (Ocasio, 2014; Vela-Gude, Cavazos, Johnson, Fielding, Campos, & Rodriguez, 2009).

Furthermore, being teacher candidates presented another concern to both women. Given the importance of teacher quality for effective teaching and concomitant student learning, most states require teachers to pass state-mandated licensure tests (Rotherham & Mead, 2004). In the state where PNU is located, all students are required to pass both the entry licensure test before entering a teacher education program and endorsement tests before they can work in the classroom. Unfortunately, teacher candidates of color, in general, suffer from increased standards and competency testing (Torres, Santos, Peck, & Cortes, 2004). Many of them are disproportionately screened out from entering teacher education programs due to low test scores (Goldhaber & Hansen, 2010), which results in a decrease in diversity in the teacher workforce. Latina/o bilingual paraprofessionals attempting to become teachers, in particular, are reported to experience more academic difficulties than most other teacher candidates (Genzuk & Baca, 1998) and have a difficult time passing state-mandated tests for licensure (Waldschmidt, 2002).

Difficulties With Coursework

Ultimately, both women had accurately predicted the level of their academic difficulties. Erika and Raquel passed the state-mandated entry test for the teacher education program which is meant to assess an approximately eighth-grade level of content knowledge in reading, writing, and mathematics. In the writing subset, it took both women a couple of trials to pass. As

56   Becoming Teachers of Color

soon as they started the program at PNU, Erika and Raquel found themselves having serious academic difficulties. They both shrugged their shoulders and sighed: "We saw pretty much the result of the education we had received at K-12." Similar to when they were searching for jobs after high school graduation, both women felt the force of the negative effects of their K-12 schooling again.

Reading had been their weakest point throughout their K-12 schooling. The indifferent instruction both women received had not adequately prepared them to read university-level textbooks and materials. Erika described this situation:

> I realized that my reading level was lower compared to the cohort when I got into PNU. When we'd get little passages to read or an article or something in class, "Read this and discuss with your table," well, I'm like, "I'm not done with reading! Wait!" But, my classmates had already finished reading and were ready to talk.

Lengthy reading assignments became a nightmare.

In order to fully comprehend what is written, a large vocabulary is necessary. Mancilla-Martinez and Lesaux (2010) emphasize the essential role vocabulary plays in reading and assert, "Written text includes vocabulary that is more sophisticated, involves language structures that are more complex, and assumes substantial background knowledge, which is intimately related to vocabulary knowledge" (p. 709). Although fluent in both oral English and Spanish, the Spanish language came much easier to both women partially because they had a larger vocabulary in Spanish than in English. The insufficient amount of vocabulary, particularly academic words in English could be traced to the low-quality education both women received in K-12. Their experiences as paraprofessionals did not help Erika and Raquel improve their level of academic vocabulary in English, as their experiences were limited to the elementary level.

Erika and Raquel did not know many academic words in English, and would be stuck by such words, look them up in a dictionary, and still not comprehend what they meant. When they encountered the words they did not know in English, they frequently checked the meanings in Spanish first and attempted to make sense of them. However, they were not educated in Spanish at school, and their level of academic vocabulary in Spanish also suffered. Erika and Raquel quickly learned that their reading level was much lower than that of their classmates. Both women's struggles with reading comprehension was caused by the fact that they had begun in elementary school to read in English, a language in which they were not fully proficient. The bilingual education Raquel received in K-2 had little positive impact on her development of reading comprehension skills in English. The characteristics of the schools Raquel and Erika attended—poor, Title I, with high concentrations of students of color—are associated with low reading

Becoming Teachers of Color   57

achievement (Lutkus, Grigg, & Donahue, 2007). Their struggles in academics were reflected in the grades they received at PNU.

Another challenge was writing grammatically correct sentences and maintaining a logical flow in assignment papers. Both women were constantly reprimanded by instructors that their written assignments needed to be proofread before submission, because they were full of grammatical and spelling errors. Although hesitant to admit their difficulties in this area, Erika and Raquel had to face their weakness in academic writing in English up front. Raquel was disappointed and discouraged by this:

> My skills were very low, my reading skills, my writing skills, my academics, everything, they were really low. In the fifth year as a para, I took a test and scored very low in reading and writing. I was at remedial reading and writing. But, after 10 years working as a para, my scores were up! That's why I thought I was ready.

Class discussions added another challenge to Raquel. She became silent in class because she could not adequately explain her thoughts on issues in her own words. Soon she found herself using the same strategy that ESL students frequently manifest in mainstream classes. She became quiet and observant in class because of the heavy linguistic demand and social and affective pressures of her classmates and instructors (Campbell, 2007; Duff, 2001). Naturally chatty and outspoken, Erika participated in class discussions much more than Raquel, but even she was intimidated and overwhelmed by her white classmates' ability to express themselves clearly and effectively convince others. Both women, particularly Raquel, began to wonder whether or not she had chosen the right time to go back to school. Raquel was deeply regretful about the inferior K-12 education she had received. She was bombarded with new challenges:

> It was new to me, something that I've never, you know, been used to. Reading a lot of materials and vocabularies that I never heard of. It's right after high school I got a para position, so I didn't do much as far as continuing my education. Whatever I relearned, I learned in classroom along with the kids, learning how to read and write again. I feel so bad about myself!

As a consequence of her educational history and experiences, Raquel began to doubt and blame herself for the difficult situation she found herself in.

The long-term and lingering ill effects of the low-quality education both women received at K-12 could be clearly seen. It reflects what they did or did not study in elementary and secondary school, and poses a dilemma for minority teacher recruitment in general, because many are not adequately prepared academically before entering teacher education. Unfortunately, the low-quality education that many minority and economically disadvantaged

58   Becoming Teachers of Color

students receive in public school continues to block their social mobility and lock them in lower-status. Erika and Raquel felt stuck in this vicious cycle.

Factors other than their own unpreparedness exacerbated both women's struggles in academics. PNU's teacher education program did not offer tutoring services. Furthermore, it did not offer any accommodation to those students who worked full-time despite the fact that half of the cohort members had full-time jobs during the day. These structural restraints had a furthering adverse effects on Erika. As a single-mother with two small children and working as a school substitute every day, Erika clearly saw the advantages of the full-time students who did not work, or held part-time jobs. Many of her cohort members had time during the day and night to study, while Erika did not.

Lack of time to concentrate on studies at PNU produced another adverse effect on Erika:

> It's very, very hard. I worked full-time all day, you know. I subbed during the day, came to school at night, then went home, and put my kids to bed. So, I didn't have the leisure to come and meet with my professors and have a meeting like, "Well, what do you think about this?" It's not fair to say that the professors didn't help me because I also didn't ask for it, but I simply didn't have time. There was no time in the day for me to come and get assistance from them.

Garcia-Reid (2007) contends that social support provided by teachers offer the greatest contribution to school engagement among other supports, such as family and peer. Direct assistance from professors enable struggling students to individually ask questions, seek for constructive feedback, and provide support for raising their level of academic achievement. PNU's "one-size-fits-all" structure impeded students like Erika from receiving much support from her professors.

Coping Strategies

To promote academic success, particularly among former paraprofessionals, Genzuk and Baca (1998) recommend a cohort system because it generates "cooperative, interactive support systems to assist participants" (p. 82). Determined to complete the program and obtain a teaching license, and confronted with the lack of support at the institutional level at PNU, Erika and Raquel proactively took an advantage of the cohort system with regard to academic help. Raquel was particularly motivated and expressed her strong sense of aspiration and navigation: "I'll get through it because I have to. I have to get through it. I don't have a choice. I'm going to get through it. I'm going to finish. It's just part of life."

Crockett, Iturbide, Torres Stone, McGinley, Raffaelli, & Gustavo, (2007) explain that active coping in which the problem is managed cognitively or

Becoming Teachers of Color 59

through action is thought to mitigate the debilitating effects of stress at college. Studies found that active coping methods are linked to better college adjustment for students of color (Zea, Jarama, & Bianchi, 1995) and better psychological well-being among Latinas/os (Gloria, Castellanos, & Orozco, 2005). Instead of ignoring their academic problems, thus increasing their stress level, Erika and Raquel both employed active coping strategy. Seeing each other every evening, both women had some social conversations with several cohort members, particularly nontraditional students who were married, had children, and returned to school after many years. This casual socialization made it easier for them to ask for help. Besides, since PNU did not offer tutoring, if they wanted to complete assignments and projects at a satisfactory level, they desperately needed assistance from the cohort members. Raquel's simple comment reflects the degree of desperation they experienced: "I asked my peers for help instead of Erika and Alma because we struggled with the same things and couldn't rely on each other. We were all struggling." Erika confessed that not every cohort member helped her academically, but the nontraditional students were willing to help. Erika described the help she received:

> There were a lot of collaboration between the peers to help get through, which I thought beautiful. Quite a few people were like, "I'll help you, here you read this, see what you think and maybe it will help you get your train of thought." I thought it was really awesome that we all worked together to get classes done and over with. Yes, I think the cohort helped me, morally, and just, "You can do this, you can do this," and don't stop type of thing.

Their proactive active coping strategy evidences their higher level of maneuvering skills in a situation where the educational system did not accommodate them, but actually impeded their academic achievement.

Erika and Raquel not only helped each other with assignments and projects, Raquel was unofficially tutored by two white female nontraditional students of her same age in the cohort—Sharryn and Megan. Raquel observed what they said, what techniques they used upon reading, and how they discussed in class attentively. She recalled how she learned from these peers:

> Sharryn taught me to talk about what I had read. For example, I used to read and pick things that I thought were important, but usually they were not important. I thought, "What am I doing wrong? Why isn't this coming out?" Sharryn would highlight the main topic and then the main points as she was reading along. By doing so, you would realize they are saying the same thing over and over, but there are certain things that would pop up without realizing they are there. Sharryn will read it, talk about it, then go back and say, "Oh, there's a word I was missing, now I understand." So just watching those little techniques that Sharryn was using and how she was trying to figure it out. I'd think, "Oh, I'll

60   *Becoming Teachers of Color*

have to try that out and see if it works for me." Basically I did what ESL students do—observe, listen, watch, and copy.

The skills Raquel described above should have been already acquired when she was in K-12. The lack of such essential skills even after high school graduation underlines how qualitatively low the education Raquel received was. However, Raquel and Erika's determination to overcome academic difficulties in any means illustrates their high level of what Yosso (2017) calls "aspirational capital"—"the ability to hold onto hope in the face of structured inequality, and often without the means to make such dreams a reality" (p. 122).

According to Staton-Salazar (2004), working-class students most often gain social capital from middle-class peers in institutional contexts, that are rich in social and cultural capital. Studies discovered that low-status students benefit from middle-class peer networks with regard to academic engagement (Stanton-Salazar, 2001; Stanton-Salazar & Dornbusch, 1995). In Erika and Raquel's case, their strong aspirational capital seems to have motivated them to access what Stanton-Salazar (2011) calls "institutional agents"—"high-status, non-kin, agents who occupy relatively high positions in the multiple dimensional stratification system, and who are well positioned to provide key forms of social and institutional support" (p. 1066)—represented by their nontraditional classmates, such as Sharryn. By accessing the capital the nontraditional classmates possessed, Erika and Raquel began to amass the capital necessary to navigate their academic studies at PNU. In this sense, their navigational capital—"skills of maneuvering through social institutions" (Yosso, 2017, p. 124) also intersected to make it through the program.

As a result of the peer social network with the nontraditional cohort members along with their strong aspiration and higher level of navigational skills, both women maintained the required 3.0 GPA throughout PNU's program. Without their proactive stance to overcome their academic difficulties, Erika and Raquel may not have maintained the required GPA since PNU's teacher education program did not offer academic support and left each teacher candidate to figure things out on his or her own.

PNU's laissez-faire structure covertly endorsed the academic gap between those who were prepared and those who were not. This did not facilitate Erika and Raquel's smooth transition from paraprofessionals to teachers. It was fortunate that an informal relationship between the two women and nontraditional women students gave them some of the assistance they needed, rather than a formalized structure created by the university.

Microaggressive Interactions With Peers and Instructors

Hollins and Guzman (2005) state that most teacher candidates of color are prepared in programs where the majority of their peers are white because the number of teacher candidates of color remains disproportionately low (Darling-Hammond, Chung, & Frelow, 2002; Irizarry & Raible, 2011).

Becoming Teachers of Color 61

This learning environment naturally impacts how teacher candidates of color feel: They feel uncomfortable and out of place within the program (Gomez, Rodriguez, & Agosto, 2008; Salinas & Castro, 2010). This is particularly true when white students casually display racial and ethnic stereotyping towards people of color. When students of color are stereotyped by whites, both students and instructors, they are negatively impacted (Lewis, Chesler, & Forman, 2000) and "face charges of unqualified and 'out of place'" (Yosso, Smith, Ceja, & Solórzano, 2009, p. 660). Kornfeld (1999), for example, found that African American teacher candidates often encountered varying levels of racial prejudice and did not feel safe to speak up in predominantly white teacher education programs. Three Latino male teacher candidates in Gomez et al.'s (2008) study interpreted many experiences in the primarily white teacher education program as prejudicial and discriminatory. Sheets and Chew (2002) further revealed that Chinese American teachers in a multicultural education class were disgusted, angered, and disturbed by comments made by white students and eventually became silent under the pressure of majority disapproval. As four teacher candidates of color's experiences in Amos's (2016) study suggests, teacher candidates of color generally have difficulty positioning themselves among the overwhelming silencing power of whiteness in predominantly white teacher preparation programs. All of the findings of the studies above applied to both women.

Microaggressions

The nontraditional cohort members were generally supportive of Erika and Raquel. However, younger cohort members and some instructors were indifferent or openly hostile to the two women. Erika described these peers: "My personality . . . I don't like it when people smile at you when they're with you, but then turn around and speak ill of you. There're lots of that in the cohort."

The hostility was usually expressed in a form of microaggressions or subtle forms of racism. In many cases, these hidden messages may invalidate the group identity or experiential reality of the targeted individuals, demean them on a personal or group level, communicate that they are lesser human beings, and threaten, intimidate or relegate them to inferior status and treatment (Sue, 2010). Microaggressions are often delivered carelessly without thought by well-intentioned people who do not have malice or the intention to harm. People of color describe microaggressions as a pattern of being overlooked, under-respected, and devalued because of their race (Sue & Constantine, 2007). On the surface level, these everyday occurrences may be quite harmless and trivial, but the cumulative effect of them is detrimental to the psychological well-being of the group being marginalized. That is why microaggressions are likened to carbon monoxide, invisible, but potentially lethal (Sue & Sue, 2003). However, to most whites, these microaggressions are invisible, thus go unnoticed.

62 Becoming Teachers of Color

There are three forms of racial microaggressions: 1) microassaults; 2) micro-insults; 3) microinvalidations (Sue, 2010; Sue, Capodilupo, Torino, Bucceri, Holder, Nadal, & Esquilin, 2007). Microassaults are "conscious biased beliefs or attitudes that are held by individuals and intentionally expressed or acted out overtly or covertly toward a marginalized person or socially deval-ued group" (Sue, 2010, p. 8). The perpetrator of this form of microaggres-sions harbors conscious and intentional bias, and expresses overt bigotry and explicitly derogatory verbal and nonverbal attack. Microinsults, on the other hand, occur outside the level of consciousness of the perpetrator. They com-municate "rudeness, insensitivity, slights, and insults that demean a person's racial, gender, sexual orientation, or group identity and heritage" (Sue, 2010, p. 9). Subtle put-downs of someone's racial heritage and identity is an example of microinsults. Microinvalidations also occur unconsciously. However, they are the most insidious, damaging, and harmful because "microinvalidations directly attack or deny the experiential realities of socially devalued groups" (Sue, 2010, p. 10). A belief that visible citizens of racial and ethnic minority groups are foreigners and the myth of meritocracy that asserts race plays a minor role in life success are some examples of microinvalidations.

Microaggressions place the victim in an unenviable position of "(1) try-ing to ascertain the motivations behind the actions of perpetrators and (2) deciding whether or how to respond" (Sue, 2010, p. 16). When they confront the assailants, the victim of microaggressions often expend addi-tional energy and time defending themselves against accusations of being "too sensitive" (Yosso et al., 2009). For this reason, most victims choose or are forced to do nothing in response to their assailants.

Latina/o college students are frequently exposed to insensitive, racist discus-sions and comments in and outside the college classrooms (Robertson, Bravo, & Chaney, 2016; Yosso et al., 2009). The type of microaggressions Latinas/os encounter were categorized by Rivera, Forquer, and Rangel (2010), and they discovered eight domains of microaggressions pertaining to Latinas/os: 1) ascription of lacking intelligence and academic ability—accomplishments and qualifications are questioned and educational success seen as a surprise; 2) second-class citizens—being ignored or excluded; 3) pathologizing cultural values and communication styles; 4) characteristics of speech—demeaning the speaking of Spanish, speaking with a devalued accent, or looking down upon those who speak a dialect of English devalued by the ruling class; 5) alien in own land—U.S. citizenship questioned, negative remarks regarding immi-gration, and assumed undocumented status; 6) criminality—treated like a criminal and received messages implying illegal activity; 7) invalidation of Latina/o experience—verbally dismissed and given excuses for negative treat-ment; 8) other assumed Latina/o attributes—assumed poverty and laziness and generalizations of Latinas/os. Some of the domains seem to be shared by other groups of color as well. Other groups of color also experience their intellects being questioned, for example, which reflects the dominant group's worldview that differences from their race equals inferiority. The "perpetual foreigner" image is also shared by Asians. However, some domains are specifically targeted to Latinas/os. Unlike other marginalized groups, Latinas/os are most likely to be associated with a single common language, Spanish. Rivera et al. (2010) contend, "The possibility of recognizing a language other than English as a commonly used language in the United States appears to cause apprehension for those who have become accustomed to rely solely on English" (p. 76). This type of apprehension, combined with nativism, enables the perpetrator to directly attack Latinas/os regarding the Spanish language, regardless of their ability to speak Spanish.

Throughout the two-year-long teacher education program at PNU, Erika and Raquel became frequent victims of "put-downs." The microaggressions they received ranged from random to specific, and were mostly based on race, ethnicity, gender, language, and social stereotypes.

Counterspaces

An example of random microaggression was the seating arrangement. In class, both women used to sit together with their fellow Latina peer, Alma, while other cohort members also formed several cliques sitting together. In response to the environmental stress coming from their minority status, the three Latinas were able to create a safe ethnic space by sitting next to each other. Raquel adamantly justified this situation: "You know, the first time you go somewhere, you go to people who look like you." What both women did not realize was the degree of irritation Dr. Carpenter, a middle-aged white female professor, felt whenever she saw Erika, Raquel, and Alma sitting together. Her irritation increased gradually, week by week.

First, Dr. Carpenter casually commented to the three women, "Oh, you guys always sit together. You like it, ha?" A following week, seeing Erika, Raquel, and Alma in the same spot together again, Dr. Carpenter raised her voice: "I don't like the fact that you guys always sit together." On the following week, Dr. Carpenter directly ordered that Erika, Raquel, and Alma sit separately and so do other classmates. On the fourth week, she brought a seating chart and demanded that each student sit at a seat assigned by her. In this new seating arrangement, the three Latinas were physically separated, corner to corner in the classroom.

Sue et al. (2007) argue that it is difficult to identify, quantify, and rectify racism if it is subtle, nebulous, and unnamed. Since Dr. Carpenter announced her order to all the members of the class, not only to the three Latinas, her action may have been an unintentional slight. Erika, however, was confident that Dr. Carpenter's action was intentional and targeted to them because Erika had always witnessed Dr. Carpenter rolling her eyes whenever she saw them sitting together. Erika confessed, "It was a direct shot at us. I was comfortable with Raquel and Alma, but she (Dr. Carpenter) didn't like us sticking together. What's wrong with us sitting together? White students sat together, too."

64 *Becoming Teachers of Color*

This type of racial clustering in physical areas is common among people of color at a predominantly white space. Tatum (2003) claims that the ability to see oneself as part of a larger group from which one can draw support is an important coping strategy to avoid considerable social isolation. It offers "counterspaces" where people of color feel that their experiences and racial and ethnic identity can be affirmed and validated because other same-race peers often share similar life experiences and affinity (Carter, 2007; Solórzano, Ceja, & Yosso, 2000; Yosso et al., 2009). In a learning environment where students of color are a demographic minority, these spaces counter "the hegemony of racist and other oppressive ideologies and practices of the institution and its members" (Carter, 2007, p. 543). However, according to Tatum (2003), whites not only want to know why they are sitting together but also what can be done to prevent it. Sometimes, whites even accuse people of color of self-segregating (Villalpando, 2003).

Erika was defiant about this unreasonable imposition by Dr. Carpenter:

> Even though she tried to impose, separate us, it didn't work because when we got together to study, we studied with the same people. So you study and hang out with the same people you feel comfortable with. You get the job done, but at the same time, feel comfortable. What's wrong with feeling comfortable?

Except for Dr. Carpenter's class, the three Latinas continued to sit firmly together, which helped them affirm their ethnic identity, strengthen their peer network, and survive in a predominantly white program. In other words, three Latinas defiantly defended their right to a counterspace and resisted any institutional attempt to dismantle it.

If the purpose of a counterspace is to provide people of color with an invaluable time of the affirmation of their racial and ethnic identity and "self-preservation" (Villalpando, 2003) in the midst of a negative, sometimes hostile college campus climate, it is understandable that in such a space, culturally-oriented ways of speaking and behaving are taken. Whenever Erika, Raquel, and Alma sat together in and outside of the classroom, for instance, they naturally spoke Spanish to each other. It turned out that some of the classmates not only did not like to see them sitting together but also did not like to hear them speaking Spanish in a public space. One day, in the classroom, Stacy, a young white female student, openly displayed her disgust towards their use of Spanish. Stacy asserted, "It's *very* rude to speak the language people don't understand in public!"

Urciuoli (1996) observed that (carefully managed) Spanish is licensed in a public space in contexts like "folk-life festivals," as part of processes of "ethnification" that work to make difference "cultural, neat, and safe" (p. 9). However, other Spanish is condemned by whites as impolite and even dangerous. Urciuoli (1996) reports, for example, "(N)early every Spanish-speaking bilingual I know . . . has experienced complaints about using Spanish in a public place" (p. 35). In this sense, Stacy's condemnation towards the three Latinas' use of Spanish in the classroom buttresses Hill's (2008) statement that whites are not comfortable when marginalized and stigmatized groups "exhibit styles and expressions that are distant from white norms" (p. 23). Whites' explicit display of being uncomfortable when they see visible racial and ethnic communications of people of color reinforces the idea that the public space belongs to the dominant group, to whites and their cultures. Therefore, hearing a marked language—other than English—in what they perceive as their space disrupts the natural state of being where "white virtue" should always prevail. This reinforces "the idea that whites are highest in the hierarchy because their qualities deserve this arrangement" (Hill, 2008, p. 21). Stacy's comment, in reality, revealed the incredible degree of race-entitlement she felt, in which her being comfortable in her English-speaking whiteness was her prime concern, a privileged state that should never be disturbed by people of color who speak a language other than English. Simultaneously, her behavior highlights how many whites consciously or unconsciously promote white supremacy through intense monitoring and surveillance of racialized groups for signs of disorder, abnormality, and markedness. Through the lens of white privilege, white domination is both invisible and normal.

The microaggressions manifested by Dr. Carpenter and Stacy explicitly targeted the rights of both women to associate with fellow Latinas and to speak their native language in a public space. Clearly, the actions of the two Latinas are protected by the Constitution, and Erika and Raquel both felt their rights and well-being were trampled on by Dr. Carpenter and Stacy, behaving in a manner which showed that the two white women felt that both classroom and public spaces were white, English-only spaces.

Presumed Incompetence

Frequently, during their teacher training Erika and Raquel were hurt and offended by microaggressions in which their intelligence and accomplishments were directly questioned without hesitation by the cohort members. The dominant framing of Latinas/os is of a group of people who are in "lower" in the racial hierarchy (Feagin, 2013), less intelligent (Weyant, 2005), and even "dumb" and "lazy" (Solórzano, 1997). Even highly educated Latinas and other women of color who pursue academic careers in higher education struggle daily with internalized presumption that they do not possess the intellectual competence to belong in the rigorous ivory tower of academia, which has traditionally been equated with white male scholars (Gutierrez y Muh, Nieman, Gonzales, & Harris, 2012). There is a clear intersection among race, ethnicity, class, language, and gender in the presumed incompetence of Latinas/os.

Microaggressions do not need to be expressed verbally. Nonverbal behaviors also serve as a form of microaggressions (Sue et al., 2007). Raquel

66   *Becoming Teachers of Color*

certainly sensed an important message in nonverbal behaviors displayed by the peers. According to her, "They'll kind a look at you . . . peers and instructors both. They know right away we're not capable." For example, whenever they made presentations in front of the class, Erika and Raquel witnessed rude behaviors from several peers. These peers frequently and overtly displayed their disapproval by rolling their eyes at what both women said during presentations and sighed with the audible noise, but never uttered anything against them. They would remain silent but not so subtly assault them with a glare. Ladson-Billings (1996) insightfully explains that silence and negative, facial expressions and body language can be used as a means of opposition that shuts down dialogic processes in the classroom.

Some cohort members' silence towards Erika and Raquel clearly accomplished its intention. Both women learned that what was not said was sometimes more important and ominous than what was said.

Sometime certain cohort members talked during Erika and Raquel's presentations. During the interviews Erika complained vigorously about this lack of respect and attempt at disruption of her presentation, behavior which was not censured by the instructor:

> It's just like, show a little bit of respect for us even if we're different. It's really bothering us. Stacy and Marilyn even talk to each other while we talk! If they don't like to see us presenting, just sit there and be quiet! Don't make gestures because it's really obvious. Don't obstruct!

The peers' attitudes seem to have been a reflection of their mindset that "there is nothing of significance to learn from students of color, either intellectually or personally" (DiAngelo, 2006, p. 1993). These nonverbal behaviors gave a clear confirmation to Erika and Raquel that they were perceived as inferior, thus not worth listening to, and in return, reinforced a sense of the white students' intellectual superiority.

In addition, Erika and Raquel's academic standing gave another chance of insult to some cohort members. Raquel herself always perceived that both her cohort members and instructors were doubtful that she was capable of making it through. One day, her perception became real. One of her cohort members, Natalie, asked Raquel if she had passed a difficult class, EDF 311 (classroom management and lesson planning). When Raquel answered positively, Natalie snobbishly replied, "Oh, did you? If you passed, I should have passed that class also. Not to sound like in a mean way, but you know . . ." All Raquel could do at this moment was to silently look at Natalie's face. And to Raquel's disappointment, Natalie had until that point always been friendly to her.

Racial microaggressions cause stress to the victim because they need to decipher the insult and decide what was meant, and whether and how to respond (Yosso et al., 2009; Sue, 2010). Natalie's casual comment was a direct insult to Raquel's intelligence. Since it was uttered as if it were

Becoming Teachers of Color 67

nothing, Raquel needed to ponder for a while what had really happened. A while later, Raquel recovered herself and concluded that the insulting comment uttered by Natalie was based on a stereotype about Mexicans' intelligence. Raquel being a person of color was the reason why Natalie was audacious enough to insult her so overtly. Raquel suspected that if she had not been a person of color, Natalie would have never said such a comment directly to her. It was clear to Raquel that Natalie's comment was uttered because of her racial and ethnic minority status: "It could have been because I'm a Latina, probably yes. And I don't talk a lot in class. So, people don't have expectations of me."

Natalie's casual insult is a manifestation of deficit thinking that contends that "minority cultural values, as transmitted through the family, are dysfunctional, and therefore cause low educational and occupational attainment" (Solórzano & Yosso, 2001, p. 6) and are frequently used to explain students of color's low academic achievement (Valencia, 2010). Raquel's honest feeling at this time was: "I felt like I was not okay. Am I not here? Am I invisible? What's happening?" Sue (2010) contends that the most detrimental impact of microaggressions lies in their invisibility to perpetrators who are usually decent people, and who believe in liberty and justice for all. Invisibility of their own actions, nevertheless, sustains oppressive practices simply because perpetrators do not realize the consequences of their own behavior. Knowing that even nice people with good intentions will still inadvertently act out on their racial biases and prejudices in the form of microaggressions, although they may be unaware, will intensify people of color's "wounds by words" (Solórzano, 1997). Taken aback, and unable to immediately respond directly at the moment to Natalie's profound insult, Raquel was most frustrated with her own inaction and inability to verbally react to the put-down. In other words, Raquel was caught in a vicious trap of racial microinsults, which led to both an erosion of self-confidence and self-image.

In addition, many white cohort members firmly believed that both women had been awarded many scholarships, and had only received them because they were students of color. These cohort members used common terms of white complaint, and called it "reverse discrimination." They felt themselves more worthy, but were denied scholarships because the Latinas and other students of color received unmerited rewards and privileges. Some white peers casually, sarcastically, and directly mentioned the following comment to Erika and Raquel: "You must have lots of scholarships. Nice to be a minority." This type of put-downs is reflective of the white peers' firm belief that students of color receive more than their fair share of scholarships, similar to the anti-affirmative action argument that claims that only individual merit should matter, not race and ethnicity. This type of microinvalidation indicates that the white peers viewed Erika and Raquel as less than competent recipients of undeserved free rides which had only been gained by virtue of their racial and ethnic membership. Lewis, Chesler, and Forman (2000) note, "The racial stereotypes about the academic ability and

68   *Becoming Teachers of Color*

potential of students of color varied for different groups, but the impact of the assumptions were (*sic*) similar in their negative effect" (p. 78).

In reality, neither Erika nor Raquel received scholarships. Raquel contended, "I applied for scholarships, but I was always denied. I never got one." Kantrowitz's (2011) statistical analysis concludes that students of color are less likely to win private scholarships or receive merit-based institutional grants than white students.

Erika remarked in an angry tone:

> It's so hard to be a teacher if you can't accept your peers as colleagues. I am more outspoken than Raquel and Alma. Alma will speak out, but I think she waits longer. I'll speak up, but I don't think about what I say. I don't know if I can finish the quarter without exploding myself.

Clearly, repeated microaggressive comments regarding their intelligence from the cohort members increased both women's feelings of being looked down upon and slighted, and tested their perseverance in a cruel way. Clearly, PNU's teacher education program was not a safe and welcoming place for Erika and Raquel. Ultimately, it was only the three Latinas who felt the pain from the pervasiveness of the white racial frame, not those who acted upon the frame.

Undocumented Immigration

The surrounding communities where PNU's satellite campus is located is heavily populated by Latinas/os, and both Erika and Raquel grew up in these communities. As one of the biggest and most productive agricultural areas in the region, the communities have many farms, orchards, ranches, dairies, and warehouses that employ many Latina/o workers including undocumented workers. Sudden, unannounced raids by agents from Homeland Security are not uncommon at these business locations.

When she was a paraprofessional, Erika was bothered by white teachers' synonymous use of the two terms, "illegal immigrants" and "Mexicans." One day, in a multicultural education class at PNU, she was reminded of how truly pervasive this synonymous use was. The instructor, Dr. Adams, brought up the issue of undocumented children at public school. Erika and Raquel witnessed their cohort members actively expressing their opinions about the issue, which was not unusual at all. Both women also realized that most of the comments of their cohort members were negative stereotypes, such as undocumented immigrants use up the government's money, do not pay taxes, take away jobs, and need to be deported because of the harm they cause society. Since the localities where two women took classes at PNU's satellite campus were heavily populated by Mexicans, the statements made by the cohort members created racist nativist discourses that perpetuate stereotypical beliefs and imagery that falsely portray undocumented

Latina/o immigrants as "'criminal,' 'dangerous,' and a drain of government resources" (Pérez Huber, 2011, pp. 382–383). These comments were racially based political constructs, used to marginalize Latinas/os, and treat them as a threat to the so-called "American way of life."

Dr. Adams's purpose of this discussion was to introduce the U.S. Supreme Court's decision, *Plyler vs. Doe* (1982), which constitutionally guaranteed the rights of undocumented children to free public education. Surprisingly, Erika, though naturally outspoken and argumentative, did not participate in this discussion at all and remained totally silent that evening. Erika, however, was full of anger towards the peers at that time: "It's hilarious! People don't realize that without illegal immigrants, agriculture won't be there. Do they think whites will pick apples and strawberries? Are they naïve? It makes me angry that they think these people so low. What do they know?" As usual, Raquel said nothing during the discussion.

The class discussion on that evening, however, had given Raquel a lingering effect. It reminded of her childhood when she had observed, without understanding, her parents having so much trouble filling out papers that were submitted to government agencies. On the same night, Raquel emailed Dr. Adams (she later explained that she herself did not understand why she decided to email) and confessed that her parents had immigrated to the U.S. undocumented. In the several email exchanges, Raquel honestly expressed how disappointed she felt hearing about insulting comments about Mexicans from her peers, how proud she was to be a Mexican who did agricultural work growing up, and how frightening it was to see her peers becoming teachers while having such prejudicial attitudes towards the Mexican children they would teach. After several email exchanges, Dr. Adams asked if she wanted to talk about her family's experience in class. Raquel thought about this offer overnight, and returned her answer on the following morning: "Yes." Raquel was usually quiet and shy, but bold and courageous as well.

The following week in the same class, Dr. Adams began a class saying, "Raquel has something to say to us." Raquel, gathered all her courage and stood in front of the classroom. Shaking, she quietly said to the class, "My parents were illegal, but they were granted a U.S. citizenship eventually." The class became instantly silent and still. Nevertheless, Raquel continued her talk, sobbing. The cohort members' keen attention never left Raquel. All of sudden, one of the cohort members, Melinda, a middle-aged white woman who was hoping to switch her career from a low-wage job to a teacher, stood up, glared at Raquel angrily for a while, and then left the classroom without saying anything. On the following week, the cohort learned that Melinda had gone to the university administration officer with a stack of printed papers regarding undocumented immigration and fiercely demanded that Raquel be expelled from the program. When PNU's administrators declined her request, Melinda threatened to leave the program. Two weeks after Raquel's presentation, with continued nonaction on her demand

70   *Becoming Teachers of Color*

by PNU's administration, Melinda officially withdrew from the program. In retrospect this was a positive result, as the university administration stood its ground against bigotry (although there was overwhelming legal support for their stand), and a bigoted teacher candidate left the program, sparing, at least for the time being, Latina/o children from being damaged by a profoundly prejudiced white teacher.

Reflecting on this incident, Raquel simply stated, "Yeah, I couldn't believe that. This is the reality. We always get one negative or another. Even though we are U.S. citizens now, people always see us as illegal immigrants." Raquel's comment indicates that she was fully cognizant of racist nativist discourses that racialize all Latinas/os as immigrants, thus nonnative, regardless of actual immigration status. These same discourses confirm formerly undocumented immigrants into the status of permanent criminals even after they become legal U.S. citizens. Witnessing what her brave friend Raquel needed to go through in front of the peers, Erika offered a comfort to Raquel outside the class: "I said to her, 'Raquel, you've got to stick up for yourself!' I'll be with you." Again demonstrating the strong ethnic peer network among the three Latinas at PNU, and the benefit it provided the women when they needed emotional support.

To Erika who was already furious, however, another worry suddenly emerged:

> No matter how hard I try, I'm knowledgeable enough to know there will always be some sort of discrimination, but I have to weigh what I'm looking at. To me that was more of a "Wow, you're at a whole other level!" But at the same time, Melinda really gave me a worry. This town is a Mexican community. How can't people see that? Melinda blew me away. This scared me. If my children were in a class taught by a teacher like Melinda I can see how my children would be treated and possibly discriminated against.

Afterwards, Erika confessed with a grave look, "Besides Raquel, one of us (Alma and Erika) was undocumented, too."

In the current context of the immigration problem, it has been argued that Latinas/os seem to experience biases, prejudices, and stereotypes more overtly in the form of microassaults compared to other marginalized groups (Rivera et al., 2010). Erika's worries indicate that the assaultive nature of the interactions with the peers brought her not only psychological harm and barriers to optimal learning at PNU, but also concerns for the members of the wider community, such as her own children. Having a strong sense of motherhood, after Melinda's incident, both Erika and Raquel became fearful of their children's well-being at school, and began to suspect the possibility that a seemingly decent teacher with no apparent bigotry could be suddenly vicious when it comes to undocumented immigration. To the perpetrators, microassaults are not always intentional, and often they are not conscious of the psychological harm they cause their victims. However, to the victims like Erika and Raquel, microassaults added a layer of fear which was cruelly used by the dominant group for the purpose of manipulation and control of the subordinate groups' thinking and behaviors.

Experiences as Paraprofessionals

As former paraprofessionals, Erika and Raquel were already equipped with the key skills needed for teaching. They already had a clear idea of what they were getting themselves into once they started at PNU. Most importantly, both women had a realistic understanding about how school operated, knowledge they gained through their extensive work experiences at school. When Erika said the comment below, she was with full of pride:

> I have a school experience. I know how a public school works because I worked for seven years in the system. So, I think I have a better understanding of the inner-city schools. I consider my community to be not so much a big city, rural, yeah. I believe I'll be a little more prepared because I'll know what to expect, while maybe my classmates have this perfect picture type of a setup. Their imaginations will be shattered because they think everything's gonna pan out just fine and how they have it envisioned. They haven't realized this yet though. Haha!

Raquel concurred and emphasized the importance of seeing realities: "Other classmates haven't seen much of the real world, seen much of what it's like at school. I was a para, so I saw it exactly. I know what it's like to work with kids."

Academically, both Erika and Raquel could have been behind the peers. However, experience-wise, they felt their edge over other cohort members. In an effort to improve teacher education, well-supervised clinical experiences are considered as a critical element of effective teacher preparation (Darling-Hammond, 2014; Faltis, de Jong, Ramírez, & Okhremtchouk, 2014; Zeichner & Peña-Sandoval, 2015). Both women were cognizant of the market value their experiences as paraprofessionals had brought to them. Erika and Raquel settled in the paraprofessional position simply because the poor-quality of their K-12 education prevented them from moving on to higher education immediately after high school. The position of paraprofessionals was one of the best they could had, as it helped to prepare and qualify them to become full-fledged teachers. It is ironic that both women found their paraprofessional experiences invaluable and advantageous for their desire to be certified teachers, since working as paraprofessionals was the result of the limited employment opportunities they had after high school.

Due to the extensive experiences they had as paraprofessionals, whenever possible, both women tried to connect their school experiences to the class contents. Erika, in particular, was good at this. For example, while listening

72 Becoming Teachers of Color

to professors' lectures, she was easily able to put the discussion topics into a real classroom situation. This was a clear advantage for both Erika and Raquel, compared to other peers who never worked at school. Erika had full of confidence when she remarked:

> I worked as a school secretary. I had perspectives from that angle. I subbed at school. I had the same from that angle, too. So, I already had multiple perspectives of how things all worked together and actually were put at school. When we were talking about certain strategies and things in class, it was easier for me to think of it and to put it to something real because I saw those things at school. Like, "Oh, that kid will benefit from this, you know, this would have worked for that kid." So I had something to attach it to, instead of just reading and not having anything to apply it to.

However, when both women tried to speak based on the practical and experiential knowledge they had gained while being paraprofessionals, most professors would not respond positively. One time, Erika asked a simple question in Dr. Buchanan's class: "I saw this taking place in a classroom when I was a para. How does this relate to what you are talking about now?" She had no intention to challenge Dr. Buchanan. Rather, she wanted to get confirmation from Dr. Buchanan about the point she was making. Dr. Buchanan, however, seems to have felt that Erika's comment was a direct challenge to her knowledge and completely ignored the question. This type of disregard by professors was not uncommon at PNU. Erika and Raquel quickly learned that their practical knowledge and an effort to connect theory to practice were not appreciated by many faculty members who taught in their program.

Erika reflected:

> I didn't know that some professors don't like students talking a lot about their point of view, particularly their experiences at school. So, I kind of put my foot in my mouth because otherwise, I would have said, "I've seen this at this school." I would put practical stuff that was taking place at that time into our class. I didn't realize that was a big no-no for some professors. I think maybe that had a lot to do with why my grade was what it was, but I felt what I was saying in class was important because it was what happened in the classroom at an elementary school in this day and age. So I thought that was very useful.

The outright disregard and hostility towards the knowledge and experiences Erika and Raquel attempted to bring to the classroom at PNU reminds us of Delgado Bernal and Villalpando's (2002) vivid narrative of how the knowledge of faculty of color was discredited and deliberately marginalized by the dominant Eurocentric epistemology in higher

Becoming Teachers of Color 73

education. A normalized view that the experiential knowledge of people of color is not legitimate nor credible because it is too culture specific, thus not universal widely exists in academia. By overlooking and devaluing the knowledge of people of color, Eurocentric knowledge sustains its dominance. Erika and Raquel, in other words, experienced another type of microaggressions—microinvalidation.

Not only was their attempt to express experiential knowledge to expand the existing Eurocentric knowledge in teaching discouraged, their experiences as paraprofessionals were scrutinized and under-valued. Throughout PNU's program, both Erika and Raquel frequently heard several cohort members make insulting comments such as, "Just because you were a para before, that doesn't mean you can teach well." They were further devalued when some professors told students in class, "Paras are good at following whatever directions given to them by teachers. But, can they well teach on their own? Not really." Both cohort members and professors denigrated Erika and Raquel's valuable experiences as paraprofessionals outright, and indicated that such experiences would amount to nothing in the teaching field. Their view sharply contrasts with the findings of several studies that bilingual paraprofessionals have extensive school experiences, cultural insight into the lives of students from minority communities, a learned knowledge of the language system, and a keen awareness of their responsibility as role models (Abbate-Vaughn, 2007; Bonner, Pacino, & Stanford, 2011; Chopra et al., 2004; Lenski, 2007; Rueda, Monzó, & Higaereda, 2004). All of these experiences and qualities contribute to making them strong teachers.

Deficit thinking about Latina/o students was prevalent during Erika and Raquel's K-12 schooling. At PNU, there seems to have been another form of deficit thinking in regards to paraprofessionals. PNU's students and professors both questioned paraprofessionals' capability to become teachers. Although there is convincing evidence that paraprofessionals perform strongly when they become teachers, the peers and professors at PNU assumed that paraprofessionals were not able to teach well. This deficit thinking towards paraprofessionals seems to originate from the perceived lower-status in a hierarchical structure of social relations between teachers and paraprofessionals (Rueda & Monzó, 2002). Trent (2014), for example, observed that teachers consider paraprofessionals as "helpers." This construction naturally creates a hierarchy between them in which paraprofessionals provide teachers with help, and by doing so, they are positioned as subordinates based on an identity of deficit. Weiss (1994) documented one paraprofessional sarcastically expressing her opinion that teachers did not want them to become better educated and improve themselves too much. In this fixed hierarchical structure which rarely allows social mobility, paraprofessionals who aspire to become teachers always carry a subordinate label simply due to their experience as a paraprofessional. Further, the fact that the majority of paraprofessionals are from minority groups seems to aggravate

74 *Becoming Teachers of Color*

this form of deficit thinking, seriously questioning their ability to teach. Clearly, there is an intersection between race, ethnicity. and experiences as paraprofessionals.

During a two-week practicum and a quarter-long student teaching experience, Erika was forced to stop working as a substitute teacher since PNU's teacher education program did not take her experiences into consideration at all. Both Erika and Raquel were simply treated the same as the other students who had no experience at school. Erika felt this "one-size-fits-all" policy to be outrageously discriminatory towards them: "Why did I need to go through the same process as those students who had never worked at public school? Shouldn't my experiences be counted?" Some universities modify the curriculum in order to better prepare paraprofessionals for teaching, such as waiving a field course and shortening student teaching. This was not the case in PNU's program. Despite a growing interest in recruiting paraprofessionals for a teaching force as a way to increase the number of teachers of color, PNU's teacher education program ignored and failed to take into account the knowledge and experience Erika and Raquel brought with them into the program.

Spanish Language Skills

If their legal status was always scrutinized with suspicious eyes, the bilingual ability of these Latinas was also a target of frequent verbal attack. In the community where PNU's satellite campus was located, the Mexican population was slightly higher than the white population. In the small town where Raquel lived, more than 90 percent of the town's population were Mexicans, including the mayor. The increased number of Mexicans in the wider communities brought an increased number of Mexican children into public schools. As a result, there was an increased demand for teachers who could speak both English and Spanish.

Since the beginning of PNU's program, both Erika and Raquel were exposed to explicit jealousy and sarcasm from the cohort members who were monolingual English speakers. Most of the time, the cohort members' jealousy and sarcasm were conveyed in subtle and indirect microinsults, and these insults were not uttered directly at them, either. Rather, Erika and Raquel overheard their white peers making insulting comments in their own conversations but within audible distance of both. Erika had not expected to hear such insults from her peers because they were all mature college students who were pursuing careers in educating children and youth. Before attending PNU, Erika believed those preparing to become teachers had a higher level of common sense and collegiality. However, it soon became a routine to hear comments from classmates such as, "Oh, they don't have to study hard. They're going to get a job anyway because they speak Spanish. They don't have to be smart."

Whenever she heard these comments, Erika immediately and nonchalantly brushed off these demeaning comments with countercomments like, "Yeah, I thank Mom and Dad for that." A bit argumentative and combative in personality to begin with, repeated casual microinsults from the peers made

Erika determined to argue back: "Obviously, I'm not going to sit there and say nothing. I'll defend myself, you know. That's me!" Raquel, on the other hand, always kept it to herself. She did not find any meaning in fighting back against those whom she considered to be "lower." Raquel explained, "Yes, it bothered me, but it was not worth arguing. They were not in my type of moment, you know. So just go with it." Instead, Raquel frequently gave comfort to Erika who was frustrated with the peers' put-downs: "Erika, just calm down. It's okay. Don't pay attention. Don't waste your time with them." Erika and Raquel had very different personalities—argumentative and feisty Erika, and reserved and thoughtful Raquel. These two different personalities seem to have worked perfectly together by pulling each other's strengths and weaknesses towards their shared goal—obtaining a teaching license.

When the economy is not strong and jobs are scarce and competitive, nativist sentiments against immigrants frequently arise, such as fears that immigrants are taking jobs away from Americans. This sentiment aggravates when public schools are desperate to hire Spanish-speaking teachers to accommodate the needs of Spanish-speaking students. In such a climate some whites come to believe that Latina/o Spanish speakers will take over teaching positions and they will be discriminated against as monolingual English speakers. In this context, Spanish-speaking Latinas/os are perceived as a threat to the job security of monolingual Americans, thus racist nativist discourses, which already exist within the culture, emerge even more strongly. Frequently, Erika and Raquel's peers repeated racist nativist discourses in casual and sarcastic microinsults.

The microaggressions regarding paraprofessional experiences and the Spanish language skills Erika and Raquel received are reminiscent of an Aesop's fable, "*The Fox and the Grapes*." Similar to the fox that rationalized that he did not want to eat the grapes he could not reach because they were probably sour, the white cohort members put down both women's experiences and skills because they did not have them. The cohort members' lack of experiences and skills, in other words, their unfavorable condition in comparison to Erika and Raquel, registered as a threat to their positive self-conception as white individuals, prompting them to act in ways to minimize the threat by repressing the undesirable, feared, and repellent ideas, by making self-serving attributions (Campbell & Sedikides, 1999; Levine-Rasky, 2014). The result was disparaging microaggressive comments. The fact that these microaggressions were uttered by several cohort members and sometimes even professors indicates that the threat was perceived not only at the individual level but also collectively at the group level. In the context of obtaining a teaching position, the cohort members as the dominant group responded to the threat to the group's dominant standing with animosity towards those groups seen as threatening their dominant position and the privileged access to economic and material resources (Craig & Richeson, 2014). It underscores the degree of threat the cohort members felt and the degree of effort they thought they would need to defend their group's symbolic position in the racial hierarchy.

76 Becoming Teachers of Color

Simultaneously, the white cohort members' threatened feeling manifested by their destructive comments towards their Latina classmates symbolizes that they psychologically internalized and validated whiteness as "virtuous," "orderly," and "natural," and have naturally assumed that it was their right to dominate other races (Matias, 2016). In the face of others who had superior skills and experience, they were strongly motivated to preserve their fragmented egos. The belief that whites are the one who should be in an advantageous position in the labor market, not nonwhites, appears to have operated to prompt the white cohort members' microaggressive comments. When presented with the uncomfortable reality that nonwhites like Erika and Raquel possessed advantageous experiences and skills which could in turn make the cohort members disadvantaged, the white peers verbally attacked both women. Rustin (1983) calls this form of reaction "projective identification"—"not only a process of fantasy attribution, but also of the actual pushing of feelings onto others, making them experience the aspects of the self in order to relieve the self of mental pain (p. 60)." The white cohort members' pushing of their irrational feelings onto Erika and Raquel in public space indicates their strong desire to control and enforce their dominant positioning.

Bombarded with all of these microaggressions, Erika was flustered, frustrated, irritated, and irate. She described her feeling this way:

> They assume that I am going to get a job because of my Spanish, not because I take the same classes as they do, receive the same grades as they do, and work as hard as they do. Yes, it surely gives me an advantage, but that should not be the only reason why I am going to get a job.

At the same time, this reality hit both Erika and Raquel hard. They realized that their ability to teach would be always judged in conjunction with their ability to speak Spanish and that people would not consider them to be able to teach well without using Spanish. On the contrary, their bilingual ability was also judged with a question mark as was reflected in the white cohort members' frequent comment: "Just because you are bilingual, that doesn't mean you will be a good bilingual teacher." Erika grumbled, "We will need to keep proving ourselves no matter how well we teach. This is not fair. Instead, this is a burden. Catch-22."

Indeed, their bilingual ability turned out to be both advantageous and burdensome. Erika was worried about the very skill that she was dearly proud of:

> My hardest thing would be having my peers and my coworkers say, "She only got the job because she's bilingual." "Oh, you speak Spanish. That's why you have the job." That will be hard for me, having to make sure that I'm always on the ball to prove myself, to prove that I'm worthy of the position that I'm going to get.

As soon as a transition from a paraprofessional to a teacher became a possible reality, their Spanish language skills became their main source of concern. Both women wished to be seen as teachers first and Spanish speakers second. However, in their peers' mind, the ability to teach and the ability to speak Spanish were tightly intersected.

The most harmful aspect of microaggressions is that it is only the victim who gets a consequence, not the perpetrator. Due to the perpetrator having no consequences of their own actions, "their oppressive worldviews that create, foster, and enforce marginalization" (Sue, 2010) prevail. The frequent microaggressions directed at both women made Erika and Raquel feel different, inferior, and marginalized. Erika summarized these feelings: "I can't relate to them (the cohort members)." Meanwhile, the dominant power of the cohort members and some professors who attacked both women in the form of microaggressions was untouched, unchanged, and sustained.

Financial Burdens

Academic struggles and microaggressive interactions with the peers and the professors gave Erika and Raquel torment when they navigated PNU's teacher education program. However, they were also confronted with another problem: how to finance themselves while attending PNU's program.

For many students of color the cost of a college education is prohibitive or will have a debilitating impact on their family's finances (Futrell, 1999), and this is a major reason why the transition from paraprofessionals to teachers becomes difficult. The increased cost of college attendance and the lack of financial assistance hit Latinas/os hard because Latina/o families generally earn well below the white counterparts (Downs, Martin, Fossum, Martinez, Solorio, & Martinez, 2008; Swail, Cabrera, & Lee, 2004). Both Erika and Raquel took student loans from the federal government in order to complete PNU's teacher education program. At the end of the program, they both found themselves in debt for approximately $35,000. Erika, a single-mother, was always worried about her family's finance. She was able to receive a medical coupon for her two children from the state government, while she herself was uninsured throughout the program. The health insurance PNU provided to students was too costly for her. The cost of daycare haunted Erika all the time.

Erika also had a hard time making ends meet during a two-week practicum and a quarter-long student teaching, which PNU did not waive regardless of her experiences. Erika was furious about her impoverished state during this time in her life:

> A lack of accommodations for former paraprofessionals made it really hard because school was not my priority. It comes second to the family, to paying your bills. I needed to feed my children. And you have to have a place to live, especially when you're a parent. Everything else

78   *Becoming Teachers of Color*

came before school. I had to spend time away from subbing to get my classwork done. It made me stressed out on my finances, like "How am I going to pay these bills?" So I wasn't doing as well as I could have in school. I think, you know, it's just a common sense. It's nobody's fault about the choices. But still it makes it a lot harder for people to actually finish school. I can see where some people would just say, "Forget it." And I've had friends who did that. It's really hard. You have to just, you know, struggle through it to make it.

Raquel also struggled: she needed to purchase expensive textbooks, pay full tuition, and worry about gas prices and keeping her car running for the commute while simultaneously relying only on her husband's comparatively low income. She stressed, "There should be some special financial aid for paraprofessionals trying to become teachers, like a credit reduction. We have such a long experience as paras compared to other students. Did we need a classroom management class?"

Furthermore, their poor primary and secondary school training had an adverse effect on their finances as well. Both Erika and Raquel needed to take state-mandated endorsement tests multiple times because they did not pass the tests at the first trial, while most of their white cohort members passed at the first trial with little problem. To make matters worse, every time they retook the tests they had to pay. After a couple of trials, both women eventually passed the elementary education endorsement test, while neither of them passed the bilingual education endorsement test. After failing the bilingual education test twice, both Erika and Raquel simply gave up. After multiple trials and costly fees to retake the tests, Erika was critical about the validity of these tests: "It was just another thing we had to pay. The tests had nothing to do with what we had studied at school. The tests didn't help us in any way to become a good teacher." Clearly, Erika did not see a correlation between the class contents at PNU and the test contents, and was suspicious about the fundamental purpose of the tests mandated by the state. Her priority in teaching was strictly practicality-oriented.

Support Network

Despite their financial burdens, Erika and Raquel were both lucky to have strong family support. None of their parents went to college, nor did they even graduate from high school. Both women were the first generation from their families to go on to higher education. Although they could not expect financial support from family members, their families gave them strong emotional support and encouraged them to finish the program.

Erika's parents attended to her two children while she was taking evening classes. Every day, they picked up Erika's children at school, transported them to a day care, picked them up again, and watched them until Erika stopped by at their house to drive the children back home after the classes at

Becoming Teachers of Color   79

PNU. Raquel's parents, husband, and two children were not only supportive but also pushed her hard because they expected that with a teaching license, she would be able to have a job with which she could earn a decent income. For the last year of her life as a paraprofessional, Raquel made only $12 an hour with no benefits. The goal of obtaining a degree was, in other words, perceived as being synonymous with changing the discourse of poverty (Ocasio, 2014), which illustrates a pragmatic viewpoint of success in life.

Still, balancing between school work and responsibilities at home was difficult. Erika admitted that the situation was really hard on her children. Erika's daughter occasionally complained to her, "You need to stop going to school. You have never been home with us!" It was Erika's parents who comforted her children emotionally. Sometimes, Erika's parents picked up her children on Friday night and brought them back on Sunday night so that Erika could be all by herself to complete her school assignments. Raquel concurred with Erika:

> Balancing family with school. I think that was the hardest thing. Homework, reading, and writing, and doing the family. Taking care of the family. Because you have to cook and you have to clean. Then sometimes you have to tell kids, "No, I can't. Please, I have to finish my homework. I can't play with you today." You feel bad.

The family support both Erika and Raquel received confirms a finding that for Latina/o students attending college full-time, maintaining family relationships is one of most important aspects that facilitates successful outcomes in college (Bernal & Aragon, 2004). Since the entire family placed a pressure on both women to use education as a means of social advancement (Valadez, 2008), attending college and obtaining a degree in essence, constituted as a family endeavor rather than an individual one.

The camaraderie Erika, Raquel, and Alma (who needed to withdraw from the program in the middle of the academic year due to her pregnancy) tremendously helped each other with navigating difficulties. They studied together, gathered at each other's homes on the weekend, and shared their complaints and feelings. Both Erika and Raquel were determined to provide Latina/o students with good education and instill them with the value of education since they knew that they had not experienced it when they were growing up. Because of this strong commitment, they did not let others' negative perceptions towards them disengage them from their ultimate goal: obtaining a teaching license.

The culminating event in any teacher education program is student teaching. Erika and Raquel's student teaching was positively evaluated by both mentor teachers and university field supervisors. Due to her excellent teaching, Erika's mentor teacher nominated her for the student teacher of the year award. Unfortunately, she did not receive the honor: her GPA was not high enough. Not only was their teaching evaluated highly, but also had they

80 Becoming Teachers of Color

wonderful results. Erika received a letter of intent to hire from the school district where she student taught while she was still student teaching. Raquel was encouraged by the building teachers to apply for a position at the elementary school where she student taught while she was still student teaching.

Two and a half months after student teaching, both women started to work as bilingual education teachers without the bilingual education endorsement. Although both women passed the elementary education endorsement test, they never passed the bilingual education endorsement test (Praxis). However, they were excited to begin their new journey as teachers. Erika placed her achievement in gaining a teaching position in context: "I definitely believe that having an experience with Mexican students as a para, and understanding where they are coming from played a much bigger role than the Spanish language ability alone in making me a good teacher."

Teacher Candidates to Teachers

Despite the increasing interest in preparing paraprofessionals, particularly bilingual paraprofessionals of color, for teachers, Erika and Raquel's transition from paraprofessionals to teachers was full of hurdles.

First, PNU's structures were not accommodating for bilingual paraprofessionals transitioning to becoming teachers. One of the greatest concerns for prospective minority and bilingual teachers is the negative impact of high-stakes testing for both entry into the teacher education program and for teacher licensure (Waldschmidt, 2002), which clearly reflects their academic underpreparedness during K-12. The degree of academic struggles Erika and Raquel experienced at PNU's program and their failure to pass the bilingual education Praxis was the direct result of the accumulated negative impact of the low-quality education they had received during K-12. To aggravate the situation, the absence of any kind of academic help in PNU's program was unreasonably punishing them. Providing these students with tutoring and remediation classes in their weakness, and assigning individual academic counselors or mentors to monitor their progress would have surely eased their struggles. However, these proactive actions from the university were nonexistent at PNU.

In addition, PNU's "one-size-fits-all" policy seems to have contributed to escalating Erika and Raquel's financial difficulties and increasing their stress. Despite their rich and practical experiences as paraprofessionals, PNU's program did not wave field-based courses nor shorten their student teaching. This increased the burden on both women. Providing former paraprofessionals who possess extensive experiences at school with curriculum modifications and financial support would have certainly facilitated both women's transition. When universities adopt such policies they will more certainly increase the number of teachers of color. Fundamentally, PNU's strict "one-size-fits-all" model achieved equality among all teacher candidates but neglected equity, and ended up in presenting a prohibitive situation for students like Erika and Raquel. In other words, PNU's teacher education program was not aligned with the societal need for more teachers of color.

Gloria and Rodriguez (2000) contend that major factors affecting the retention of Latina/o students at college are mostly noncognitive. The strongest negative factors to retention are sociocultural in nature. This point is buttressed by Solórzano and Yosso's (2001) finding that Latina/o students experience institutional and interpersonal difficulties which could negatively affect their school outcomes while attending college. In other words, what they experience in school impacts their academic performance and retention. In this sense, the frequent microaggressive comments and behaviors Erika and Raquel encountered in the program were additional barriers placed before, and could have derailed their successful completion of the program. The types of microaggressions Erika and Raquel received were the result of a combination of the pervasiveness of deficit thinking towards Latinas/os and the level of threat the white cohort members felt towards the skills and the experiences both women possessed but they lacked, such as the Spanish language skills and teaching experiences as paraprofessionals. Since Erika and Raquel were deemed inferior to begin with, the white cohort members seem to have vented their frustration and jealousy directly on both women who possessed the advantageous skills and experiences.

The peers' deep resentfulness for the skills and experiences both women possessed and profound competitiveness against those whom they perceived as lower in status and inferior, highlight the white cohort members' fear of losing superiority and dominance. Simply put, they had a binary, loss-gain perspective where they felt "if whites might lose something, then people of color may gain something" (Milner, 2008, p. 335). Fundamentally, the white cohort members simultaneously felt threatened and desired to dominate racial others. Unfortunately, PNU's teacher education program took a laissez-faire approach. They failed to take proactive measures which could have assisted these two former bilingual paraprofessionals, while at the same time, the do nothing stance of the university's program helped to sustain the bigoted attitudes and behavior of many of the of whites in their cohort.

Despite the barriers and active discrimination faced, both Erika and Raquel survived PNU's program and became certified teachers. As they went through the program together, their camaraderie strengthened as their resilient social ties helped them maneuver pass extreme social stress. With strong family support, Erika and Raquel never gave up their aspirations to become teachers who could help Latina/o students. Both women challenged the racism they encountered, both subtle and overt, and overcame it. Their persistence in retaining and promoting the Spanish language, experience in working with the Latina/o students as paraprofessionals, and their understanding of Latina/o students' cultural backgrounds were all weapons they used to successfully obtain teaching licenses and positions.

Yosso (2017) views communities of color as spaces with multiple strengths where resistance to racism and oppression is demonstrated through

82 *Becoming Teachers of Color*

community cultural wealth, although they lack the cultural and economic capital possessed and inherited by privileged groups in society. Erika and Raquel's strong resilience certainly proves this point. In the next chapter, we will examine their experiences as certified teachers. In light of the increasing demand for Spanish-speaking Latina/o teachers in public schools, it is important to understand the challenges Latina/o teachers encounter from school administrators and fellow teachers. Most importantly, it is critical to understand how they survive such challenges, and achieve successful outcomes, both for their students and for their teaching careers.

References

Abbate-Vaughn, J. (2007). "Para aquí today, para afuera tomorrow": Uncertainty in urban bilingual paraprofessionals' work in the age of NCLB. *The Urban Review, 39*(5), 567–588.

Amos, Y. T. (2016). Voices of teacher candidates of color on white evasion: "I worried about my safety!". *International Journal of Qualitative Studies in Education, 29*(8), 1002–1015.

Arditti, J. (2006). Editor's note. *Family Relations, 55*(2), 263–265.

Aud, S., Fox, M., & KewalRamani, A. (2010). *Status and trends in the education of racial and ethnic groups* (NCES 2010–015). U.S. Department of Education, National Center for Education Statistics. Washington, DC: U.S. Government Printing Office.

Bernal, C., & Aragon, L. (2004). Critical factors affecting the success of paraprofessionals in the first two years of career ladder projects in Colorado. *Remedial and Special Education, 25*(4), 205–213.

Bonner, P. J., Pacino, M. A., & Stanford, B. H. (2011). Transition from paraprofessionals to bilingual teachers: Latino voices and experiences in education. *Journal of Hispanic Higher Education, 10*(3), 212–225.

Brewster, A. B., & Bowen, G. L. (2004). Teacher support and the school engagement of Latino middle and high school students at risk of school failure. *Child and Adolescent Social Work Journal, 21*(1), 47–67.

Bui, K. V. T. (2002). First-generation college students at a four-year university: Background characteristics, reasons for pursuing higher education, and first-year experiences. *College Student Journal, 5*(1), 3–11.

Calzada, E. J., & Tamis-leMonda, C. S., & Yoshikawa, H. (2013). Familismo in Mexican and Dominican families from low-income, urban communities. *Journal of Family Issues, 34*(12), 1696–1724.

Campbell, N. (2007). Bringing ESL students out of their shells: Enhancing participation through online discussion. *Business and Professional Communication Quarterly, 70*(1), 37–43.

Campbell, W. K., & Sedikides, C. (1999). Self-threat magnifies the self-serving bias: A meta-analytic integration. *Review of General Psychology, 3*(1), 23–43.

Carter, D. J. (2007). Why the Black kids sit together at the stairs: The role of identity-affirming counter-spaces in a predominantly white high school. *The Journal of Negro Education, 76*(4), 542–554.

Chopra, R. V., Sandoval-Lucero, E., Aragon, L., Bernal, C., Berg de Balderas, H., & Carroll, A. D. (2004). The paraprofessional role of connector. *Remedial and Special Education, 25*(4), 219–231.

Craig, M. A., & Richeson, J. A. (2014). More diverse yet less tolerant?: How the increasingly diverse racial landscape affects white Americans' racial attitudes. *Personality and Social Psychology Bulletin, 40*(6), 750–761.

Crockett, L. J., Iturbide, M. I., Torres Stone, R. A., McGinley, M., Raffaelli, M., & Gustavo, C. (2007). Acculturative stress, social support, and coping: Relations to psychological adjustment among Mexican American college students. *Cultural Diversity and Ethnic Minority Psychology, 13*(4), 347–355.

Darling-Hammond, L. (2014). Strengthening clinical preparation: The holy grail of teacher education. *Peabody Journal of Education, 89*(4), 547–561.

Darling-Hammond, L., Chung, R., & Frelow, F. (2002). Variation in teacher preparation: How well do different pathways prepare teachers to teach? *Journal of Teacher Education, 53*(4), 286–302.

Delgado Bernal, D., & Villalpando, O. (2002). An apartheid of knowledge in academia: The struggle over the "legitimate" knowledge of faculty of color. *Equity and Excellence in Education, 35*(2), 169–180.

DiAngelo, R. J. (2006). The production of whiteness in education: Asian international students in a college classroom. *Teachers College Record, 108*(10), 1983–2000.

Downs, A., Martin, J., Fossum, M., Martinez, S., Solorio, M., & Martinez, H. (2008). Parents teaching parents: A career and college knowledge program for Latino families. *Journal of Latinos and Education, 7*(3), 227–240.

Duff, P. A. (2001). Language, literacy, content, and (pop) culture: Challenges for ESL students in mainstream courses. *The Canadian Modern Language Review, 58*(1), 103–132.

Faltis, C. J., de Jong, D. E., Ramírez, P. C., & Okhremtchouk, I. S. (2014). An introduction to critical issues in teacher education: Building a bridge between teacher education and Latino English language learners in K-12 schools. *Association of Mexican American Educators Journal, 8*(1), 7–9.

Feagin, J. R. (2013). *The white racial frame: Centuries of racial framing and counter-framing* (2nd ed.). New York: Routledge.

Futrell, M. H. (1999). Recruiting minority teachers. *Educational Leadership, 56*(8), 30–33.

Garcia-Reid, P. (2007). Examining social capital as a mechanism for improving school engagement among low income Hispanic girls. *Youth and Society, 39*(2), 164–181.

Genzuk, M., & Baca, R. (1998). The paraeducator-to-teacher pipeline: A 5-year retrospective on an innovative teacher preparation program for Latina(os). *Education and Urban Society, 31*(1), 73–88.

Gloria, A. M., Castellanos, J., & Orozco, V. (2005). Perceived educational barriers, cultural fit, coping responses, and psychological well-being of Latina undergraduates. *Hispanic Journal of Behavioral Sciences, 27*(2), 161–183.

Gloria, A. M, & Rodriguez, E. R. (2000). Counseling Latino university students: Psychosociocultural issues for consideration. *Journal of Counseling and Development, 78*(2), 145–154.

Goldhaber, D., & Hansen, M. (2010). Race, gender, and teacher testing: How informative a tool is teacher licensure testing? *American Educational Research Journal, 47*(1), 218–251.

Gomez, M. L., Rodriguez, T. L., & Agosto, V. (2008). Who are Latino prospective teachers and what do they bring to US schools? *Race, Ethnicity, and Education, 11*(3), 267–283.

Gutierrez y Muh, G., Nieman, Y. F., Gonzales, C. G., & Harris, A. P. (Eds.). (2012). *Presumed incompetent: Intersections of race and class for women in academia.* Boulder, CO: Utah State University Press.

Hill, J. H. (2008). *The everyday language of white racism.* Malden, MA: Wiley-Blackwell.

Hollins, E., & Guzman, M. T. (2005). Research on preparing teachers for diverse populations. In M. Cochran-Smith & K. M. Zeichner (Eds.), *Studying teacher*

84 Becoming Teachers of Color

education: The report of the AERA panel on research and teacher education (pp. 477–548). Mahwah, NJ: Lawrence Erlbaum Associates.

Irizarry, J., & Raible, J. (2011). Beginning with El Barrio: Learning from exemplary teachers of Latino students. *Journal of Latinos and Education, 10*(3), 186–203.

Kantrowitz, M. (2011). *The distribution of grants and scholarships by race.* Retrieved from http://racialequitytools.org/resourcefiles/Distributionracescholarships.pdf

Kornfeld, J. (1999). Sharing stories: A study of African American students in a predominantly white teacher education program. *The Teacher Educator, 35*(1), 19–40.

Ladson-Billings, G. (1996). Silences as weapons: Challenges of a black professor teaching white students. *Theory Into Practice, 35*(2), 79–85.

Lenski, S. D. (2007). Reflections on being biliterate: Lessons from paraprofessionals. *Action in Teacher Education, 28*(4), 104–113.

Levine-Rasky, C. (2014). White fear: Analyzing public objection to Toronto's Afric-entric school. *Race, Ethnicity, and Education, 17*(2), 202–218.

Lewis, A. E., Chesler, M., & Forman, T. A. (2000). The impact of "colorblind" ideologies on students of color: Intergroup relations at a predominantly white university. *The Journal of Negro Education, 69*(1/2), 74–91.

Lutkus, A., Grigg, W., & Donahue, P. (2007). *The nation's report card: Trial urban district assessment reading 2007* (NCES Report No. 2007–455). Washington, DC: National Center for Educational Statistics, Institute for Educational Sciences, U.S. Department of Education.

Mancilla-Martinez, J., & Lesaux, N. K. (2010). Predictors of reading comprehension for struggling readers: The case of Spanish-speaking language minority learners. *Journal of Educational Psychology, 102*(3), 701–711.

Marin, G., & Gamba, R. (2003). Acculturation and changes in cultural values. In K. Chun, P. Balls Organista, & G. Marin (Eds.), *Acculturation: Advances in theory, measurement, and applied research* (pp. 83–93). Washington, DC: American Psychological Association.

Matias, C. E. (2016). *Feeling white: Whiteness, emotionality, and education.* Rotterdam, The Netherlands: Sense Publishers.

Milner, H. R. IV. (2008). Critical race theory and interest convergence as analytic tools in teacher education policies and practices. *The Journal of Teacher Education, 59*(4), 332–346.

Ocasio, K. M. (2014). *Nuestro camino*: A review of literature surrounding the Latino teacher pipeline. *Journal of Latinos and Education, 13*(4), 244–261.

Pérez Huber, L. (2011). Discourses of racist nativism in California public education: English dominance as racist nativist microaggressions. *Educational Studies, 47*(4), 379–401.

Plyler vs. Doe, 457 U.S. 202, 102 S. Ct. 2382 (1982).

Rivera, D. P., Forquer, E. E., & Rangel, R. (2010). Microaggressions and the life experience of Latina/o Americans. In D. W. Sue (Eds.), *Microaggressions and marginality: Manifestation, dynamics, and impact* (pp. 59–82). Hoboken, NJ: Wiley.

Robertson, R. V., Bravo, A., & Chaney, C. (2016). Racism and experiences of Latina/o college students at a PWI (predominantly white institution). *Critical Sociology, 42*(4–5), 715–735.

Rodriguez, N., Mira, C. B., Myers, H. E., Monis, J. K., & Cardoza, D. (2003). Family or friends: Who plays a greater supportive role for Latino college students? *Cultural Diversity and Ethnic Minority Psychology, 9*(3), 236–250.

Rotherham, A. J., & Mead, S. (2004). Back to the future: The history and politics of state teacher licensure and certification. In F. M. Hess, A. J. Rotherham, & K. Walsh (Eds.), *A qualified teacher in every classroom: Appraising old answers and new ideas* (pp. 11–47). Cambridge, MA: Harvard Education Press.

Rueda, R., & Monzó, L. D. (2002). Apprenticeship for teaching: Professional development issues surrounding the collaborative relationship between teachers and paraeducators. *Teaching and Teacher Education, 18*(5), 503–521.

Rueda, R., Monzó, L. D., & Higaereda, I. (2004). Appropriating the sociocultural resources of Latino paraeducators for effective instruction with Latino students: Promise and problems. *Urban Education, 39*(1), 52–90.

Rustin, M. (1983). Kleinian psychoanalysis and the theory of culture. In F. Barker, P. Hulme, M. Iversen, & D. Loxley (Eds.), *The politics of theory* (pp. 57–70). Colchester, UK: University of Essex.

Salinas, C., & Castro, A. J. (2010). Disrupting the official curriculum: Cultural biography and the curriculum decision making of Latino preservice teachers. *Theory and Research in Social Education, 38*(3), 428–463.

Schneider, M. E., & Ward, D. J. (2003). The role of ethnic identification and perceived social support in Latinos' adjustment to college. *Hispanic Journal of Behavioral Sciences, 25*(4), 539–554.

Sheets, R. H., & Chew, L. (2002). Absent from the research, present in our classrooms: Preparing culturally responsive Chinese American teachers. *Journal of Teacher Education, 53*(2), 127–141.

Solórzano, D. G. (1997). Images and words that wound: Critical race theory, racial stereotyping, and teacher education. *Teacher Education Quarterly, 24*(3), 5–19.

Solórzano, D. G., Ceja, M., & Yosso, T. (2000). Critical race theory, racial microaggressions, and campus racial climate: The experiences of African American college students. *The Journal of Negro Education, 69*(1/2), 60–73.

Solórzano, D. G., & Yosso, T. J. (2001). From racial stereotyping and deficit discourse toward a critical race theory. *Multicultural Education, 9*(1), 2–8.

Stanton-Salazar, R. D. (2001). *Manufacturing hope and despair: The school and kin support networks of U.S.-Mexican youth.* New York: Teachers College Press.

Stanton-Salazar, R. D. (2004). Social capital among working-class minority students. In M. A. Gibson, P. Gándara, & J. P. Koyama (Eds.), *School connections: U.S. Mexican youth, peers, and school achievement* (pp. 18–38). New York: Teachers College Press.

Stanton-Salazar, R. D. (2011). A social capital framework for the study of institutional agents and their role in the empowerment of low-status students and youth. *Youth and Society, 43*(3), 1066–1109.

Stanton-Salazar, R. D., & Dornbusch, S. M. (1995). Social capital and the social reproduction of inequality: The formation of informational networks among Mexican-origin high school students. *Sociology of Education, 68*(2), 116–135.

Sue, D. W. (2010). Microaggressions, marginality, and oppression. In D. E. Sue (Eds.), *Microaggressions and marginality: Manifestation, dynamics, and impact* (pp. 3–22). Hoboken, NJ: Wiley.

Sue, D. W., Capodilupo, C. M., Torino, G. C., Bucceri, J. M., Holder, A. M. B., Nadal, K. L., & Esquilin, M. (2007). Racial microaggressions in everyday life. *American Psychologist, 62*(4), 271–286.

Sue, D. W., & Constantine, M. G. (2007). Racial microaggressions as instigators of difficult dialogues on race: Implications for student affairs educators and students. *College Student Affairs Journal, 26*(2), 136–143.

Sue, D. W., & Sue, D. (2003). *Counseling the culturally diverse: Theory and practice.* New York, NY: John Wiley & Sons.

Swail, W. S., Cabrera, A. F., & Lee, C. (2004). *Latino youth and pathway to college.* Washington, DC: Pew Hispanic Center Study.

Tatum, B. D. (2003). *"Why are all the black kids sitting together in the cafeteria?": And other conversations about race* (Revised ed.). New York: Basic Books.

Torres, J., Santos, J., Peck, N., & Cortes, L. (2004). *Minority teacher recruitment, development and retention.* Providence, RI: Education Alliance, Brown University.

86 Becoming Teachers of Color

Trent, J. (2014). "I'm teaching, but I'm not really a teacher": Teaching assistants and the construction of professional identities in Hong Kong schools. *Educational Research*, 56(1), 28–47.

Urciuoli, B. (1996). *Exposing prejudice: Puerto Rican experiences of language, race, and class*. Boulder, CO: Westview Press.

Valadez, J. R. (2008). Shaping the educational decisions of Mexican immigrant high school students. *American Educational Research Journal*, 45(4), 834–860.

Valencia, R. R. (2010). *Dismantling contemporary deficit thinking: Educational thought and practice*. New York: Routledge.

Vela-Gude, L., Cavazos, J., Johnson, M., Fielding, C., Campos, L., & Rodriguez, I. (2009). "My counselors were never there": Perceptions from Latino college students. *Professional School Counseling*, 12(4), 272–279.

Villalpando, O. (2003). Self-segregation or self-preservation?: A critical race theory and Latina/o critical theory analysis of Chicana/o college students. *International Journal of Qualitative Studies in Education*, 16(5), 619–646.

Waldschmidt, E. D. (2002). Bilingual interns' barriers to becoming teachers: At what cost do we diversify the teaching force? *Bilingual Research Journal*, 26(3), 537–561.

Weiss, M. S. (1994). Marginality, cultural brokerage, and school aides: A success story in education. *Anthropology & Education Quarterly*, 25(3), 336–346.

Weyant, J. M. (2005). Implicit stereotyping of Hispanics: Development and validity of a Hispanic version of the implicit association test. *Hispanic Journal of Behavioral Sciences*, 27(3), 355–363.

Yosso, T. J. (2017). Whose culture has capital? A critical race theory discussion of community cultural wealth. In A. D. Dixon, C. K. Rousseau Anderson, & J. K. Donnor (Eds.), *Critical race theory in education: All God's children got a song* (pp. 113–136). New York: Routledge.

Yosso, T. J., Smith, W. A., Ceja, M., & Solórzano, D. G. (2009). Critical race theory, racial microaggressions, and campus racial climate for Latina/o undergraduates. *Harvard Education Review*, 79(4), 659–690.

Zea, M. C., Jarama, L., & Bianchi, F. T. (1995). Social support and psychosocial competence: Explaining the adaptation to college in ethnically diverse students. *American Journal of Community Psychology*, 23(4), 509–531.

Zeichner, K., & Peña-Sandoval, C. (2015). Venture philanthropy and teacher education policy in the U.S.: The role of the new schools venture fund. *Teachers College Record*, 117(6), 1–44.

4 Teaching at Public Schools
Workplace Environment

Spanish-speaking teachers are in demand at public schools these days. From 2003–2013, the number of Latina/o students enrolled in public schools increased from 9.0 million to 12.5 million, and the percentage who were Latinas/os increased from 19% to 25% (U.S. Department of Education, 2016a). This percentage is projected to grow further to 29% in 2025 (U.S. Department of Education, 2016a). Moreover, the percentage of English learners (ELs) also increased to 9.3% in the 2013–14 academic year, and 76.5% (approximately 3.8 million students) of them were Latinas/os (U.S. Department of Education, 2016b). The population of Latina/o English learners represents 7.7% of all public K-12 students (U.S. Department of Education, 2016b).

As the growing number of Spanish-speaking Latina/o students enter public schools, the demand for Spanish-speaking teachers has also grown (Gomez & Rodriguez, 2011; Sakash & Chou, 2007). Schools not only look for teachers who specialize in ESL (English as a second language) to help Spanish-speaking students in the English language development but also teachers who help the students in core subjects in Spanish when they have trouble with math, science, reading, history, and the like. Under the *No Child Left Behind Act* enacted in 2001, English learners in grades three to eight are required to take the same state academic content assessments as all other students in reading/language arts, math, and science (U.S. Department of Education, 2001). In other words, although obviously unfair, English learners are held to the same accountability standards as native English speakers before they master the language of instruction (Short & Boyson, 2012). That is why teachers feel pressured to provide English learners with effective content instructions and opportunities to acquire academic language along with opportunities to develop the English language proficiency.

This is where instruction in the native language of students bear the utmost importance. Theoretically speaking, the importance of students receiving instruction in their native language is well-researched. Research findings suggest that a cognitively and academically beneficial form of bilingualism can be achieved only on the basis of adequately developed first language skills (Cummins, 1979, 1984; Cummins & Swain, 2014). Thomas and

88   Teaching at Public Schools

Collier's (2002) study further supports the importance of native language development among bilingual students as they contend, "The strongest predictor of L2 student achievement is the amount of formal L1 schooling. The more L1 grade-level schooling, the higher L2 achievement (p. 9).[1] When there are many English learners who speak the same language at one school, teachers and administrators have found an instruction in students' native language helpful and effective for facilitating the students' content mastery.

However, hiring teachers who are both orally fluent and literate in Spanish is not easy. White teachers dominate the teaching field in the U.S.—84% (Feistritzer, 2011). The majority of these teachers are monolingual speakers of English who have never seriously experienced learning a second or foreign language (Feistritzer, 2011). In 2011, only 6% of teachers were Latinas/os (Feistritzer, 2011). If such a small number of teachers are Latinas/os, it is reasonable to assume that there are only a small number of teacher candidates who are Latinas/os as well. This reality was symbolically reflected in the student demographics in the teacher education program Erika and Raquel attended: it was predominantly white. In addition, it is important to remember that not all Latina/o teachers and teacher candidates are both fluent and literate in Spanish. According to Taylor, Lopez, Martínez, and Velasco (2012), 82% of Latina/o adults claim oral fluency in Spanish and 78% claim the same for literacy. However, the Spanish proficiency tends to diminish in later generations. The second generation of Latina/o adults' (U.S. born children of immigrants) literacy rate fell to 71%, while the third generations' oral fluency rate in Spanish was 47%, and their ability to read newspapers and books in Spanish was 41% (Taylor et al., 2012). The decline of both oral fluency and literacy rates in Spanish among the third generation of Latinas/os does not contribute to the increased number of Spanish-speaking teachers.

Flores, Keehn, and Pérez (2002) argue that a variety of recruitment methods have been used to compensate for the inadequate supply of Spanish-speaking teachers. For example, some school districts have even begun to recruit Spanish-speaking teachers from other countries, such as Mexico and Spain to compensate for the shortage of Spanish-speaking, bilingual teachers (Fee, 2011; Flores et al., 2002). Tapping into paraprofessional pools is one strategy frequently taken by teacher education programs and school districts to increase the number of bilingual teachers, since a large percentage of paraprofessionals in some urban areas are racial and language minority groups. However, the lack of academic language proficiency in the target language, underpreparation for college-level academic work, and difficulties with fulfilling the requirements for state teacher certification (Flores et al., 2002; Rogers-Ard, Knaus, Epstein, & Mayfield, 2012) frequently hinder schools from hiring teacher candidates of color and in particular, paraprofessionals of color. Fundamentally, the quality of education provided at K-12 to racial and ethnic minorities has an encompassing and determining effect on minority teacher recruitment.

Teaching at Public Schools 89

Luckily, Erika and Raquel's life experiences had well-prepared them for a teaching position. Their family members' insistence on the maintenance of the Spanish language skills had made both women proficient in Spanish, not only orally but also in reading and writing. As former paraprofessionals, both women had added the academic Spanish language skills appropriate to teach at the elementary level. Being able to use academic language in Spanish is an essential qualification that school districts all look for in teaching reading and writing to Latina/o students. Erika and Raquel had been trained in a state-accredited teacher education program and had completed all the certification requirements, despite their struggles in academics, problems financing their education, and having to face the hostile environment created by some of their peers and instructors. Importantly, they were Latinas—under-represented in the field of teaching and perceived as having a much stronger cultural connection to Latina/o students and families, and both had performed extremely well in student teaching. Therefore, it is not surprising that the school districts where they had served as their student teachers were eager to hire Erika and Raquel as soon as they obtained teaching licenses upon graduation from Public Northwest University (PNU). Both women possessed everything schools wanted to better serve Spanish-speaking Latina/o students. Erika and Raquel were, indeed, "wanted."

The "Wanted" Status

As indicated above, Erika and Raquel both completed their student teaching with stellar evaluations. With a strong recommendation to the district by her university supervisor, Erika even received a letter of intent to hire from the district where she student taught while she was student teaching. This letter immediately guaranteed her a job in the district. Erika beamed with pride when she said, "Not everyone got a letter of intent to hire, you know . . ." Furthermore, at the end of her student teaching, Erika was nominated for the student teacher of the year award by PNU's teacher education program, at an awards ceremony hosted each year by her mentor teacher. Erika, unfortunately, did not receive this honor because her GPA was not high enough for the award. Similarly, Raquel was strongly encouraged by the building teachers to apply for a position at the elementary school where she student taught. The teachers made encouraging comments such as, "Where are you going to apply?" and "Do you have your paperwork in? Even if you don't have it ready, just make sure you get it in before you leave."

Immediately after student teaching, both women were offered teaching positions as bilingual education teachers at predominantly Mexican and Mexican American schools. In Erika's case, two schools in the same district both desperately wanted her and fought over her. She eventually settled at "Lincoln Elementary School." The reality that they were sought after by the school districts where they completed their student teaching was clearly transmitted to both women. Erika and Raquel attributed their speedy

90 Teaching at Public Schools

employment to their Spanish language skills. Erika stated, "I definitely think my bilingual ability helped me getting a job. They definitely needed someone who speaks the language." Raquel agreed: "I think my Spanish had a lot to do with it. They were very desperate." Both women further attributed their extensive experiences as paraprofessionals to the direct cause. Smiling with a pride, Raquel said:

> Because I've had some para experience. You know at least if it was management or the way I interact with students. I was able to learn those skills as a para. It does help. I'm glad that I waited so long to become a teacher.

Their life experiences, including persistently maintaining their Spanish language skills in primary and secondary school environments where the use of Spanish language was suppressed and oppressed, and working as paraprofessionals for an extensive period seemed to bear fruit all at once. Ironically, their experiences as paraprofessionals and their Spanish language skills both of which were the target of microaggressive attacks from her cohort members at PNU did indeed contribute to their speedy employment.

Neither, however, passed the state-mandated bilingual education endorsement test (Praxis) although both passed the elementary education endorsement test. Even after their first year of teaching, not only had they not passed the test, they were not even told to retake it by their school districts. Erika said, "I'm teaching a bilingual education class, but they haven't said anything about me not having it (the endorsement)." Raquel concurred: "They haven't pressured me to pass the test yet." The situation where they were permitted to teach bilingual education with only one endorsement made both women keenly aware that the schools were desperate for their language skills and that they were indeed "wanted." With their status as teachers who were "wanted," their future career paths in the profession looked bright. What Erika and Raquel did not realize, at this time, however, was what it meant to be "wanted" differed significantly between the schools and them.

Erika's Story

Erika obtained her first teaching position in the community where 41.3% were Latinas/os and 52.2% were white (U.S. Census, 2010). PNU's satellite campus where Erika and Raquel attended was also in this community. Erika had been living in this community since her childhood except for several years when she lived in a big city. Although almost half and half in population between whites and Latinas/os, schools in this school district were visibly segregated by race, ethnicity, and socioeconomic status. For example, people in the community casually referred to each school as a white school or a Latina/o school. Erika was hired as a third-grade bilingual education

Teaching at Public Schools 91

teacher at Lincoln Elementary School—a majority Mexican and Mexican American school—where 95.3% of students were Latinas/os, 92.4% of students received free or reduced lunches, and 71.9% of students were English learners (OSPI, 2015a). Students at Lincoln Elementary School were the lowest in socioeconomic status and had the lowest academic achievement as measured by the state-mandated standardized tests among the 14 elementary schools in the district.

Although it is generally called "bilingual education," several different models exist. In transitional bilingual education, the native language of English learners is used to teach content subjects while they are learning English. Once students are proficient in English, they are transferred to the mainstream where only English is used to teach content subjects. Thomas and Collier (2002) refer to this model as "remedial bilingual programs" because the English learners' native language is used only to transition to the English-only classes and does not support the development of their native language in the long run. Maintenance bilingual education, on the other hand, seeks to "develop and maintain the native language of the language minority students and develop a positive attitude toward the native culture while also achieving proficiency in English" (Brisk, 2006, p. 38). These two models of bilingual education (transitional and maintenance) are meant strictly for language minority students.

Several studies have found that bilingual education teachers tend to carry heavier workloads compared to monolingual teachers, such as paperwork required by the state and federal government, translation duties, phone calls, etc. (Batt, 2008; Sosa & Gonzales, 2002). These extra duties, in addition to their instructional roles, constrain teaching effectiveness and create much pressure and personal stress. Batt (2008) argues that the responsibilities of English learner education drive some teachers to leave the fields of ESL and bilingual education, which are already the areas of teacher shortage, thus producing a vicious cycle.

Although eagerly hired by the school and excited about her "wanted" status, Erika gradually began to sense that things were not moving into the right direction. Rather than feeling excited, she frequently found herself overwhelmed with the sheer volume of the work she needed to perform and complete as a bilingual education teacher. Erika was stressed out physically and psychologically. Even worse, a sense of being exploited and alienated began to penetrate in her soul.

The Expert Status by Virtue of Being a Bilingual Latina

Erika's school operated as an early exit transitional bilingual education program where students would study all in Spanish from K-2 all day long. English would be gradually introduced at the third-grade, and the instruction would be completely in English from the fourth-grade. The transition to English began at the third-grade because all fourth-graders were

92   Teaching at Public Schools

required to take the state-mandated standardized tests under the *No Child Left Behind Act* (In the year Erika started teaching, the act did not require third-graders to take the test) and the school needed to prepare the students in the bilingual education classes for the tests that were administered only in English.

Two weeks before Erika started her first year of teaching, she was told that it would be a bilingual self-contained class in which she would be responsible for a set of Spanish-speaking Latina/o students in her class only. However, to Erika's bewilderment, this was changed immediately after the school year began. Erika explained with perplexity: "I thought my kids were going to stay in my room all day. But after school started my kids got switched." Erika was told to assume a responsibility of providing another set of pupils from another third-grade class with reading instruction in which the English language was used as the medium of instruction. This situation naturally doubled the number of pupils she was responsible for.

Furthermore, the school ordered Erika to switch the language of instruction on demand. Erika described this teaching condition as below:

> It was first all Spanish, and then they told me to use Spanish with a little English. I would do reading in Spanish, while we did comprehension in English. So we would do different parts in English and different parts in Spanish. My classroom is supposed to be a transitional classroom. So that means they have only had instructions in Spanish, kinder, first, second, and by the end of the third-grade, they are supposed to be all in English. All those studies say it will take seven to ten years to build academic language, but they want it done by the third-grade. And so, you know, the politics of how things get. One day, I was told to switch to all English like, "Oh, you are going back to English." Then around November, they told me to go back to Spanish because the kids were not improving in Spanish. They said that the kids' Spanish was not solid. Of course not! The kids were not given any Spanish instruction. They were getting English instead. I was forced to use English materials as well as Spanish materials. Then, I went to all English again and then I went all the way back to all Spanish.

Erika ended up switching the language of instruction between Spanish and English approximately every three months. This teaching condition continued even in her second-year teaching.

Anyone in this situation would have been upset, as was Erika. However, Erika was keenly aware that she was also a novice teacher without a tenure, and knew how powerless she was. Achinstein and Ogawa (2011) see Erika's dilemma accurately when they stated that novice teachers have little power as they enter a new organization and are vulnerable in regards to their status and job security. Worried that if she complained, she might not be granted a tenure, Erika never complained and kept her anger inside:

Teaching at Public Schools 93

I don't think people recognized how hard it was to switch back and forth. And did I ever complain, did I just do it, and did the best I could? I see a lot of teachers gripe and moan about everything. It doesn't matter what it is. I hope I don't' get that way as years progress because that will be really sad.

Since she was naturally an outspoken individual, holding her complaint inside and not letting it out openly intensified the level of her frustration.

Ladson-Billings (2005) claims that teachers of color are expected to perform at a much higher level than their white counterparts, particularly with regard to teaching students of color. This particular expectation pertaining only to teachers of color seems to originate from several studies on ethnic matching that found that there are academic benefits when students and teachers share the same race and ethnicity (Dee, 2004, 2005; Egalite, Kisida, & Winters, 2015). Therefore, when Erika was hired, it was automatically assumed and expected that she already possessed all the resources to effectively educate Spanish-speaking Latina/o students. This was, in fact, the basis of her "wanted" status. Erika's "wanted" status intensified because not only did she share the same race and ethnicity as her Latina/o students but also she was fluent and literate in Spanish, and had extensive teaching experiences as a paraprofessional. In this sense, Erika was apparently perceived as "the expert" in educating students of her own race and ethnicity. Therefore, the school seems to have simply assumed that "the expert" teacher of Latina/o students would be able to frequently, and effortlessly switch the language of instruction, despite the fact that Erika was still a novice teacher. Erika, in other words, was labeled as the ethnic specialist (Williams & Williams, 2006) who had the expert knowledge and skills of educating Latina/o students simply by virtue of being a Latina herself.

The situation Erika was thrown into by the school is like a miner's canary. Linking a canary to a racially marginalized group, Guiner and Torres (2002) wrote:

Miners often carried a canary into the mine alongside them. The canary's more fragile respiratory system would cause it to collapse from noxious gases long before humans were affected, thus alerting the miners to danger. The canary's distress signaled that it was time to get out of the mine because the air was becoming too poisonous to breathe.

(p. 11)

Just like a canary is sacrificed to benefit humans, the school's unreasonable demands benefited the school, but caused Erika distress. In contrast to the canary that is honored when it sacrifices itself to save the miners, Erika's sacrifices were not honored. When she could not switch back and forth between language programs quickly and effortlessly enough, she was

94   *Teaching at Public Schools*

criticized, and told that she had inadequate skills. When the canary begins gasping for air in a toxic mine environment, the miners save themselves and sometimes the bird by going into the open air. Erika's school administrators never questioned the toxic environment they created for her. If difficulties arose, if Erika could not handle the unreasonable and arbitrary changes that were created by the school for her, it was her fault. As a novice teacher who was concerned about her tenure, Erika was not in a position to complain and fight against the unreasonable conditions and increased demands placed upon her by the school administrators.

In contrast, it is likely that the demands placed on white teachers who teach white students are not as great as the demands placed on bilingual instructors. The curriculum of monolingual white instructors rarely changes as rapidly during the school term as Erika's did, and monolingual instructors do not carry the additional, unrewarded responsibilities of serving as translators and interpreters. Therefore, the unreasonably high expectations set by the school only for Erika when she taught Latina/o students were a discriminatory practice.

Lack of Support

Lapayese (2007) reported that schools serving a higher percentage of students from low-income groups, culturally and linguistically minority communities tend to adopt more prescriptive programs that emphasize direct instruction. Erika's school fell under this category and adopted scripted curricula in math and reading. Some teachers were told by their supervisors that they were not allowed to teach anything beyond the scripts. This type of strict control of what teachers teach by school administrators was evident in Achinstein and Ogawa's (2011) study of novice teachers of color as well. These novice teachers of color wanted more influence over school direction and more autonomy in the classroom to teach what worked, but their voices were hardly heard by administrators. Moreover, some teachers at Erika's school were even told to focus only on students who could improve their test scores. The pressure to increase test scores among English learners both in content areas and English language proficiency seems to be a reality (Lapayese, 2007; Umansky & Reardon, 2014).

Surprisingly, Erika had no objection to using these scripted curricula. She actually thought that they were working perfectly fine with her students. The comment below illustrates the way Erika felt about the scripted curricula:

> It's scripted, like, you say this and this and this. But the way you present is your own style and your own way. What I have to teach, it's already laid out. But you do have to say, "Okay, my goals are for this lesson . . ." You still have to find your goals. I feel my money would have been spent more wisely, during my college years, if PNU had brought these scripted materials or any of other curriculum districts around our communities

Teaching at Public Schools   95

use. We learned all of those theories and logistics at PNU, but until we apply them and actually practice them, we don't even know how they work. I think practical experience is a must.

In the comment above, Erika's emphasis on practicality in teaching and her pragmatic viewpoint is evident. Erika was also critical about the inflexibility of PNU's teacher education program in which it did not incorporate the kind of curricula that were used in the surrounding area schools where many Latina/o students attended. Although strictly following the already laid-out scripts, because of her extensive experiences at school as a paraprofessional, Erika was still able to insert her own personality when she taught with it. Erika was gifted in creativity.

Erika, however, became furious when the school demanded that she repeatedly attend one set of trainings after another so that she would be able to teach the scripted curricula both in Spanish and English, while one third-grade teacher, who was white and competent enough in Spanish, was not asked to do the same because she did not teach in Spanish. On top of these many trainings in both languages, the school rarely provided Erika with instructional support of how to use these scripted curricula effectively. The curricula were indeed scripted, but they were also leveled. Students were supposed to be tested into different levels. Erika's desperate wish for instructional support for how to effectively sort students out into different levels and simultaneously how to deliver differential instructions of the same content according to the different levels was expressed below:

> I get all scores, I put them on a database or an Excel sheet and I say, "Okay, this group I am going to switch all of these kids because they have the same scores and put them in one group." Put these kids in another group. Put these kids in this group. But how am I going to tie each lesson to each of the groups and hit all of them? Just like I made it through college, I will make it through my first year. But, somebody seriously needs to help me. I don't want to keep figuring this out up until two a.m. every day.

To increase retention, Olsen and Anderson (2007) emphasize the utmost importance of teachers of color receiving professional support during their entire careers. However, being a Latina at a school where the teachers are predominantly white seems to have affected the amount of support that Erika received from colleagues and administrators. This does not seem to be an isolated problem, as it appears that compared to white teachers, teachers of color tend to receive little support from white colleagues and administrators. For example, Klassen and Carr (1997) compared teachers of color to white teachers and found that white teachers were rewarded for pursuing antiracist education or equity issues in general, whereas teachers of color pursuing the same issues were made to feel that they were obsessed with

96   *Teaching at Public Schools*

race and racism. Feuerverger's (1997) study in Canada reported that immigrant teachers of color did not feel supported in efforts to create an inclusive, multicultural environment and felt that schools did not recognize their strengths. The five Latina/o teachers in Arce's (2004) study not only felt it was unsafe to share their visions of critical pedagogy with white colleagues, but also believed that their voices were intentionally silenced.

While the multiple trainings to use the scripted curricula both in English and Spanish were useful to better her teaching, they were also disruptive, particularly with the absence of follow-up supports from the school. Erika was angry:

> It's like "Do you speak Spanish? Then, figure it out on your own."
> And there is no one else who does it, so you're going to figure it out by yourself anyway. Other third-grade teachers, all whites, got phenomenal support from the assistant principal (Mrs. Davenport) and even participated in a collaborative decision-making. Why not me?

Within Erika's school, there seems to have been professional divisions by language, similar to the segregation by language she witnessed at her high school. These divisions were apparently created based on the prevalent native vs. nonnative binary where English native speakers were positioned at the top, and nonnative English speakers, especially those languages spoken by nonwhites, at the bottom (Pennycook, 1998; Shuck, 2006). Similar to a racial hierarchy, this hierarchy by language is "reflective and constitutive of power and the underlying power relationships that are normalized in the broader social context and implied as a natural order of things (Liggett, 2010, p. 220)." In this normalized, but prejudiced hierarchy, it is reasonable that those teachers who taught content areas in the language other than English, such as Erika, were perceived as lower, and received less support from administrators. It ultimately indicates that the school's normalized way of thinking is based on the taken-for-granted superior status of the English language. That is, students who study in English and teachers who teach in English should be given priority over students and teachers who study and teach in languages other than English.

Presumed Incompetence and Pressure to Prove Ability

Many paraprofessionals are equipped with the key skills needed for teaching. They have been supervised by teachers in classroom and hold an edge against college graduates with no classroom teaching. Further, most bilingual teachers of color have additional skills which most classroom teachers lack: "linguistic abilities and cultural knowledge which enable them to understand students who are acquiring a second language" (Wenger, Lubbes, Lazo, Azcarraga, Sharp, & Ernst-Slavit, 2004, p. 105). As a former paraprofessional and a bilingual Latina, Erika had plenty of classroom teaching experiences and authentic cultural knowledge. When I observed Erika's teaching at her school, it was obvious from an observer's perspective that she was good at directing and providing models for students. In addition, Erika was creative and invested in hands-on activities for practical learning, reflecting on her emphasis on her pragmatic orientation. For example, she made sandwiches with her students when she taught the concept of fractions. Erika herself was aware how good she was in teaching and attributed her effective teaching skills to her extensive work experiences at school. Comparing her fellow third-grade teachers—all novice—to herself, Erika evaluated her teaching skills objectively and positively as well:

> I do know now after teaching with a couple of people who were team members of mine in third-grade and who had no previous classroom experience, I'm better. It is essential that we get into the classroom a lot more than student teaching only. That does not qualify or substitute for any type of experiences I have had.

Evidently, Erika saw tangible advantages of her experiences as a paraprofessional over other teachers.

Research documents the importance of teacher control—the power to make decisions and influence behaviors or other individuals (Ingersoll, 2003; Priestly, Biesta, & Robinson, 2015). Teacher discretion and agency are particularly significant "given the need to address diverse learners by flexibly drawing on a repertoire of teaching and principles of practice" (Achinstein & Ogawa, 2011, p. 26). Nevertheless, it seems that Erika's specialized knowledge, experiences, and obvious talent in teaching were hardly ever integrated into decisions about language use in her classroom despite that she was treated as the "expert."

Soon after she started her first year, Erika realized that one of her students, Omar, was placed in her Spanish class and he was repeating the third-grade. Furthermore, Erika found out that Omar had never had instructions in Spanish before Erika's class although he spoke Spanish at home. He had received intervention reading instructions all in English before. Realizing Omar's lack of literacy in Spanish and a misplacement, she demanded that Omar be placed in the English class. The response she received from the administrator, however, was to retain Omar in Erika's class and have him keep receiving instruction in Spanish. This type of dismissals of Erika's teacher discretion was not uncommon. Erika sighed, "My opinions are nothing. I don't know the political part. It's really frustrating."

According to Erika, she was excluded from the decision-making process due to her presumed incompetence. Erika described an incident in which Mrs. Davenport at her school once referred to her as a secretary as an example:

> They don't think I'm capable. In Mrs. Davenport's eyes, I'm a secretary. She said to me, "Oh, I could see you as a secretary." When I heard her

98 Teaching at Public Schools

saying that, I didn't say anything because I was just like, I was so taken aback. I went through the same schooling that everybody else did, but it doesn't matter. They still think that I was hired only because of my Spanish skills. Teachers in other grades even urged me to say something to Mrs. Davenport, like, "Why don't you say something to her?"

It seems that Erika was perceived as lacking credibility as a teacher, despite her training and credentials, and her "expert" status.

Erika's feeling resonates with the studies that found that teachers of color's qualifications were frequently met with suspicion and sometimes explicitly challenged. For example, white colleagues in Jones and Maguire's (1998) study constantly made teachers of color to believe that they needed to "prove" their ability in the classroom. In the similar line of research, Bascia (1996a) describes that a white administrator did not seem to respect and recognize an Indian Asian male teacher's expertise and authority because he was "a Black man." This teacher said, when educators "see a Black man, that's not their idea of what a teacher is" (Bascia, 1996a, p. 164). Doubts about the qualifications of people of color also occurs outside of field of teaching. For example, Latina/o professionals frequently receive messages from administrators and colleagues at the workplace that they were only hired because of their race and ethnicity and that their achievements were attributed solely to affirmative action. They were not seen as qualified based on their work experience and accomplishments (Chávez, 2011; Rivera, Forquer, & Rangel, 2010). A teacher having teacher agency means that she/he is permitted to exert higher degrees of professional judgment and discretion within the contexts in which she/he works. This is a key dimension of teachers' professionalism (Priestly et al., 2015). Without being allowed to exert teacher agency and instead feeling her qualifications as a teacher constantly questioned, Erika naturally felt subordinated, marginalized, and deprofessionalized.

Rogers-Ard et al. (2012) contend that racially biased definitions of teacher quality which are strictly based on perceived intelligence and performance on standardized assessment reflect long held beliefs about the inferiority of students and teachers of color, and that such definitions make districts seek out mainly young white teachers. One district administrator, a woman of color, was even reported to mention, "We'd like to have teaches of color—sure! But what (this district) really needs are quality teachers" (Rogers-Ard et al., 2012, p. 462). Rogers-Ard et al. (2012) described another administrator who was told by the district not to send a message to teachers of color that they were looking for quality. These anecdotes clearly suggest a detachment of "teachers of color" from the term "quality teachers." This reflects an underlying assumption that "hiring teachers of color would somehow decrease the quality of teachers" (Rogers-Ard et al., 2012, p. 462). Apparently, many district administrators perceive that teachers of color are not quality teachers compared to white teachers. Unfortunately,

Teaching at Public Schools 99

this erroneous assumption seems to have intensified with Erika because she was mainly perceived to have been hired because of her Spanish language skills. In other words, her "wanted" status was rhetorically used to disqualify her from being perceived and treated as a quality teacher. Doubting her qualifications because of her Spanish language skills but assuming her to be the "expert" because of the same skills is obviously contradictory in nature. The contradictory treatments of the Spanish language skills possessed by Erika—advantageous for teaching but disadvantageous for what it means to be a quality teacher, however, illustrates the dominant group's manipulative power. It underlines the paradox of whiteness in which whites attempt to sustain and perpetuate their advantages while attempting to include others (Chubbuck, 2004; McIntyre, 1997; Yoon, 2012), which in reality results in excluding others.

The skills she was hired for—Spanish language skills—started to haunt Erika. This was just what happened to her in PNU's teacher education program, where her Spanish language skills became the basis for explicit and jealous microaggressive attacks, such as, "Just because you're bilingual, that doesn't mean you'll be a good bilingual teacher." Erika was indeed placed in an impossible situation: if she had failed to effectively teach Latina/o students in Spanish, which would prove to many white administrators and teachers at the school that their assumption that teachers of color are not quality teachers was accurate after all. On the other hand, if she had succeeded to teach Latina/o students effectively, these same white teachers at the school would have attributed her successes only to the fact that she taught using the Spanish language. Erika still would not receive credit for being a quality teacher. It is important to realize that no matter which way Erika had gone, Erika's credibility as a teacher would have been seen with suspicion and doubt. In this situation she could neither demonstrate competence nor receive merit for her achievements. Erika was placed in a situation where she would have never been allowed to prove her quality as a teacher.

What is ironic, however, is the fact that this teaching environment, prejudiced against her, prompted Erika to try even harder to prove herself. The irony was evident when Erika stated, "I always need to prove myself, to prove that I'm worthy of the position that I'm in." Erika was psychologically forced to prove her teaching quality in a work environment that was fundamentally against her no matter how she taught. When strictly and only calculating productivity, however, Erika's determination to prove her skills was beneficial for the school.

A Weak Professional Teacher Network

In addition, Erika found it unfair that she was not able to consult with fellow teachers because she taught in Spanish while others did in English. She began to realize that the amount of feedback any teacher received from other teachers would make a difference in the amount of work they

100   *Teaching at Public Schools*

would end up with. Erika looked lonely when she stated, "Since I taught in Spanish, I didn't have anybody to talk to or share like, 'This is what I'm doing. How can I do better?' That kind of casual conversations would have been nice." Research demonstrates that English learners and bilingual education teachers tend to feel a sense of exclusion from a social fabric of their school communities (Harper, de Jong, & Platt, 2008; Liggett, 2010), and this sense of exclusion keeps marginalizing their experiences as professional teachers. Feeling excluded because of the language of instruction she was told to use by the school, Erika began to understand that professional learning networks—new spaces in which "teachers may learn and grow as professionals with support from a diverse network of people and resources" (Trust, Krutka, & Carpenter, 2016, p. 17)—was not easily accessible for Spanish bilingual education teachers. In contrast, other mainstream teachers tremendously benefited because they were in the system already.

Among the four third-grade teachers (all white women except for Erika) in her first year, Erika was the only one who was required to switch the instruction of language, carry more students from other classrooms, and attend many trainings. This situation inevitably increased Erika's workload while simultaneously creating an unequal workload between Erika and other third-grade teachers. Erika described her workload as:

> There wasn't enough time in a day. A lot of preparation. Using the curriculum that the district wants us to use and making sure we get the state standards, the new one. And making photocopies and sorting all the students into appropriate levels. I was there (at school) every single weekend. There were probably a few weekends when I wasn't at the building, but I was at the building making photocopies, all alone. Doing things like that every single weekend. So, it's just there wasn't time to do the reading analysis.

Unfortunately, the heavy workload made Erika exceptionally busy.

Unable to find time to professionally engage with colleagues during the school hours and to consult with colleagues with regard to the Spanish curriculum, Erika began to feel isolated at school. Feeling no support from her colleagues and being unable to trust Mrs. Davenport who called her a secretary, Erika stated, "By the end of the year I just stayed in my room and did my thing. I didn't even imagine that a workplace will be just like my high school, I mean, segregated by language." Erika was indeed "'locked' in classrooms with little interaction with the rest of the school" (Ada, 1986, p. 390). Erika's social capital as reflected in teacher collaboration, teacher professional network, and mentoring within her school was evidently weaker, compared to other third-grade teachers. It is important to recognize that Erika's weaker professional teacher network was a result of the school's structures in which she was embedded, and not related to her personality at all. By default, any Spanish-speaking bilingual education teacher would have resulted in such a weak professional teacher network because the school structure itself was set up to produce such a result.

Erika's quest to enhance her professional teacher network was further hindered by being assigned to a mentor who did nothing. The district had a first-year teacher assistance program where first-year teachers and their mentor teachers met regularly throughout the school year to discuss topics pertaining to problems and concerns many first-year teachers generally encountered. Out of eight meetings throughout the year, Erika's mentor teacher attended only once. At these meetings, Erika sat alone looking at other first-year teachers sitting next to their mentor teachers. All the others were in deep discussion. Further, her mentor teacher never came to her class to observe her teaching nor gave advice. Erika explained:

> When I had questions, she couldn't help because she was a first-grade teacher and I was a third-grade teacher. She would send me to go somewhere else. And that didn't help me when I had only 15 minutes to get an answer. One time, we needed to come up with a professional development plan. I asked her twice about that. She said, "I don't know anything about that because I didn't have to do that."

Studies have found that it is crucial to match novice bilingual teachers with mentors who model effective practices (Waldschmidt, 2002), have respectful attitudes towards students of color and their parents (Weisman & Hansen, 2008), and teach at the same grade level. Obviously, Erika's mentor did not have these qualities.

Assumed Translation Duty

The volume of her work further increased because of the fact that Erika was bilingual and biliterate. As a bilingual education teacher, Erika translated everything herself—papers that were sent home, phone calls, and conferences with Spanish-speaking parents. Erika, however, soon realized that translation duties were expected only of her, not other third-grade teachers. One of her colleagues, a white third-grade teacher who had an adequate level of the Spanish language skills, was not asked to take up translation duties. Erika expressed her irritation:

> My team members are like, "Oh, I don't have time to send a newsletter because I don't have time to translate it." I make all the phone calls myself in Spanish. But for someone who doesn't speak Spanish? I don't think I had a situation where I had to have the office call for me or do anything like that. So in that part I do think it saves the district time and money. I don't need a translator for conference. I don't need a translator

102   *Teaching at Public Schools*

for any parent communication because I am able to do it myself. From the district's perspective, they are saving time and money because they don't have to hire anybody to help. Yeah, I feel used.

It appears that Erika's linguistic skills presented value to the school not only for instructional purposes but also for financial reasons, and that the school took Erika's skills for granted.

The fact that the school never officially recognized her hard and extra work exacerbated Erika's frustration. Erika was furious about this lack of appreciation:

> I feel no appreciation. Thank you for willingly doing this without any . . . or I want acknowledgement that I've been doing extra things. Just acknowledgements for the things I do. There should be some kind of appreciation. I actually told both the principal (Mr. Logan) and Mrs. Davenport that I didn't get such a thing from them. Of course, I got nothing. Just that small little recognition makes a difference in somebody's life.

Acknowledgement on the side of the school would have meant recognizing and disrupting the normalized practices in which Erika felt overwhelmed with the volume of work, isolated from other teachers, marginalized, and deprofessionalized. Not receiving acknowledgement from any teachers and administrators, Erika finally started to understand what it really meant to be "wanted."

Raquel's Story

Raquel's community was even more heavily populated by Latinas/os: 82.2% were Latinas/os and 15.7% were white (U.S. Census, 2010) with agricultural businesses omnipresent. Immediately after student teaching, Raquel was hired as a kindergarten bilingual education teacher at Washington Elementary School where 93.4% of students were Latinas/os, 85% of students received free or reduced meals, and 49.1% of students were English learners (OSPI, 2015b). It was a poor rural farming town and Raquel herself grew up in this community.

Unfair Workload

Unlike Erika, scripted curricula or trainings were not imposed upon Raquel. She was also fortunate to have a wonderful mentor in her first year. She described her mentor as "terrific" and "a good person." Raquel began teaching bilingual education at a school which adopted a transitional bilingual education model, just like Erika's. According to Thomas and Collier (2002), there are two types of transitional bilingual education. The 90/10

Teaching at Public Schools 103

model starts 90% of the instruction in the minority language, and gradually increases the amount of English until the fifth-grade when students are mainstreamed. On the other hand, in the 50/50 model, students receive equal amount of instructions in English and the minority language for three to four years followed by transition on to the mainstream.

Soon after she started teaching, Raquel found herself overwhelmed with the sheer volume of her workload. Raquel's school adopted the 90/10 model of bilingual education where 90% of the instruction was taught in Spanish, while the rest of 10% was in English. Freeman, Freeman, and Mercuri (2005) explain that many schools have adopted this model with the early emphasis on the non-English language to help compensate for the dominant power of English outside the school context. Raquel explained that the school found the long-term potential of implementing the 90/10 model in kindergarten with the arrival of Raquel, who happened to be hired as a kindergarten teacher and was a native speaker of Spanish. Those Spanish-speaking Latina/o students who were also English learners were assigned to Raquel and her partner teacher's classes. They were taught both in Spanish and English, while another kindergarten teacher taught non-English learner Latina/o and white students in English only. The 90/10 model was implemented only in kindergarten and only at her school.

The most common delivery system for elementary transitional bilingual education is a self-contained classroom taught by one bilingual teacher (Faltis & Hudelson, 1998). However, team teaching where a bilingual teacher works with a group in one classroom, while a mainstream teacher works with a second group in another classroom is not uncommon (Calderón & Minaya-Rowe, 2003). Raquel was told to share two sets of kindergarteners with her partner teacher who was monolingual. Raquel described this team teaching:

> Most of the instructions are done in Spanish. So reading, science, social studies, and writing are all in Spanish, and math is the only thing done in English. My partner takes care of the English part which is math, and I do the Spanish part which is reading, science, social studies, and writing. Then we switch the students.

Raquel ended up teaching 50 students in total in four different subjects, while her partner teacher taught only math to the same population.

Raquel's situation illustrates the reality of the shortage of bilingual education teachers and how schools and districts strategically compensate for it. Sharing two sets of students is often used in areas with limited numbers of bilingual education teachers, because a school can maximize their language resources (Freeman et al., 2005). However, this way of team teaching is usually taken in the 50/50 model where students learn in English 50% of the time and in the other language 50% of the time (Freeman et al., 2005). Sharing the two sets of students in the 90/10

104   *Teaching at Public Schools*

model meant that Raquel's share of the teaching, by default, became considerably larger than that of her partner teacher. Meanwhile, the school could use Raquel's language skills to the greatest extent by maximizing her teaching load.

Witnessing an extremely uneven workload between her partner teacher and herself, Raquel angrily said, "It's a power thing. She has been working longer than I." Although angry, Raquel never let her anger known to other teachers. As a novice teacher, Raquel felt vulnerable in her status and job security and dominated by more senior colleagues who were almost all white and monolingual English speakers. Similar to Erika's school, the linguistic and racial hierarchy were intersected with her novice status and naturally defined Raquel as lower, symbolically a laborer in the hierarchy. In this sense, as Monzó and McLaren (2017) assert that whiteness and capitalism are intertwined, Raquel was embedded in capitalist social relations of production and exploitation where the subordinates—people of color—produce the means of subsistence for the dominant group—whites. Unfortunately, in this hierarchy, Raquel was forced to endure a vastly unfair workload compared to her senior white monolingual partner teacher who took Raquel's labor for granted, and most importantly, tremendously benefited from it.

Lack of Support

Bascia (1996b) reports that teachers who are primarily identified with students in low-status categories tend to receive less administrative support. Unfortunately, there seems to be a mistaken perception of bilingual education teachers as nonprestigious and lower in status. This false perception seems to divide bilingual education teachers and nonbilingual education teachers. Creese (2002), for example, argues that bilingual education teachers are positioned as "helpers" to the subject teachers. This hierarchical thinking appears to allow monolingual teachers to regard bilingual education teachers as nonprestigious (Ada, 1986). Furthermore, Reyes and Kleyn (2010) claim that children in bilingual education are relegated to second-class status because of their race and class backgrounds. Both the hierarchy among subject teachers and the primary identification of bilingual education teachers with students in low-status categories make it difficult for bilingual education teachers to build peer support networks (Ada, 1986). This is exactly what happened to Erika.

Raquel's case supports this situation as well: she felt that she was totally left alone in developing the Spanish curriculum. When she started teaching, Raquel quickly learned that the curriculum the school had been using in the past was too low-level for kindergarteners, lacked solid contents in all subjects, particularly in reading, and had many gaps from the state standards. The curriculum, in other words, was a watered-down curriculum. Because she was determined to educate her students at the grade level,

Teaching at Public Schools 105

Raquel endeavored to develop curriculum by herself. Raquel explained her struggle to create a Spanish curriculum:

> They (the district) know that I work hard. I do a lot of work. I have to. Like if I don't have posters or anything. There is not enough at bookstores. I have to purchase them or I have to create them and that takes times. I created. I bought. I purchased with my own money. My personal money. There's nothing at school. The school doesn't support students. One worksheet and they are done in two seconds. So what else do you do? You need more stuff. You need stuff that is going to help students create those syllables. We need stuff that is going to help them hear those sounds. We need stuff that helps them blend those syllables together. So just a lot of things I've created over time and I purchased from a friend, also created stuff, just stuff that I found out in the bookstores or . . . Just kind of the materials I use. There're not enough materials at the bookstores and I'm having such a hard time finding them online. They (colleagues) always ask me, "Why do you do so much?" They have all these curricula and can go to stores. It's in English!

Raquel showed me the stack of worksheets she had created. They were all colored, full of pictures, in Spanish, and professionally done, and clearly reflected the time she spent creating them. Reyes and Kleyn (2010) note that less educational resources exist in the realm of bilingual education in general and native language instruction in particular. Raquel even asked her mother who was visiting Mexico to purchase workbooks in Spanish.

Raquel's building principal had given her a good evaluation with comments that she interacted with students well and her Spanish was really good (the principal was a monolingual English speaker). However, overwhelmed with the work of creating worksheets all by herself, instead of using the existing ones, which other teachers did, Raquel was desperate for monetary support to purchase more materials. She only received $30 a year as a professional development fund from the district. Raquel wished that she would be given additional as little $15 from the school so that she could purchase more books and workbooks. Raquel uttered in a low voice:

> My personality, naturally, I don't like to be in front of people. I don't go and sell myself, "Hey, this is what I'm doing." It just doesn't make sense to me, bragging my accomplishment to others. You know, just say, "You're doing a good job! Here is a check for it. Go buy some materials you need." That would be super nice.

Her comment above clearly undergirds a lack of communications with and understanding of Raquel's needs by the administrators. The lack of communications and understanding seem to have been caused by a lack of interest in the work Raquel did on the side of the administrators. Furthermore,

106　*Teaching at Public Schools*

unlike the dominant culture in which asserting one's accomplishments and being competitive in public is not only uncommon but also highly recommended as a symbol of strong individualism and personal agency, minority cultures tend to eschew self-promotion and competition as a primary value and favor collaboration (Anastasia & Bridges, 2015; Plaut, 2014). Raquel's personality did not consider such behaviors as commendable and necessary, either. Due to the perceived lower-status that naturally yielded disinterest and the cultural mismatch on how to manifest accomplishments, the work Raquel demonstrated as a bilingual education teacher appears to have been unrecognized. Without recognition, her work was not supported, either. If the school had proactively supported Raquel, and made certain that she was not treated differently from other teachers because she taught in Spanish, this type of miscommunication and misunderstanding could have been prevented. Similar to Erika's case, this presents a paradox in which the school's lack of genuine effort to cultivate the growth of a bilingual education teacher through encouragement and support is contrasted with its effort to include a Spanish-speaking teacher at a workforce.

To my question, "Does anyone guide you through the curriculum development process?," Raquel simply answered, "No." To make matters worse, Raquel found out that her fellow teachers were discussing the possibility of collecting all the worksheets, activities, etc. she and other bilingual education teachers had created in a binder and putting them in a computer file so that any teachers could use them in their classes if necessary. Raquel pondered: "Do we get compensated for our time creating those? Is anyone going to say, 'Thank you!' in a way that is going to benefit us? It's called common sense." Clearly, Raquel's ability to create Spanish curricula all by herself without any monetary and instructional support was convenient to the school and other teachers who lacked such skills. Witnessing her skills being overly exploited by those teachers who lacked such skills, Raquel's sense of being exploited and marginalized as a teacher started to emerge. And the absence of recognitions of her work by the school exacerbated her feeling of exploitation and marginalization.

Translation Duty: In and Outside of School

Translation work further exacerbated Raquel's heavy workload and preparation struggle. Similar to Erika, translation work sneaked into Raquel's job duties. Raquel explained:

> I have to translate. I have to make sure that I translate everything in Spanish and English. Like if I'm sending out any parent notes or anything, I have to translate them if they're in English. If it's something that is very important, I want the parents to read. If I do homework for the kids, I have to put them both in English and Spanish to make sure that every parent can read . . . like in the past I had a parent who was

Teaching at Public Schools 107

only English-speaking, so I had to make sure I translated everything in Spanish and English, so they could follow along. As a para, I translated. I knew translation was going to be part of it (upon becoming a teacher). It takes time. Especially if you want it professionally done, well of course, it's going to take time. Quality work for your time . . . you know?

Writing report cards for all 50 students became a nightmare, which usually consumed her time all weekend and beyond. Raquel wanted to make sure that she made the best decision for each student. She explained, "You can't just say, 'Oh, here is your grade because I just want to give it to you.' You can't do that. You must make sure that the grade I'm giving a child is the grade the child deserves." Constantly looking up each student's test scores and a homework completion rate and analyzing his/her performance in class, Raquel took a significant amount of time writing report cards. On top of this time-consuming effort, Raquel wrote detailed comments in both languages for all the subjects.

Comparing her workload to that of her partner teacher, Raquel lamented:

I just think in dual languages we just work harder. There are more things for us to do. Double classes. Two Spanish classes. I have fifty kids I have to write report cards for in math in Spanish. Mine plus my partner's . . . also in science, also in writing, also in social studies in two languages. And she only has two cohorts only in math and only in English. There is no time to do these report cards, but at home. I spent all Saturday, all Sunday working on report cards and I am still not finished. I still have a couple . . . it takes time. You have to make sure that this is the right grade. Why am I giving this student this grade? I don't have enough time! I'm very much used. They use me quite well! Hiring another teacher? Yes, that would definitely help. Yeah, they do need to hire another bilingual teacher.

Unfortunately, the school did not consider the time she needed to translate at all. It was taken for granted that she provided translation with quality simply because Raquel was literate in Spanish and taught bilingual education. In addition, as a new teacher, Raquel needed to continually translate documents for other teachers who asked her to translate. Raquel said, "Maybe one or two favors, but not 50!"

It seems that Raquel's insistence on perfection in her teaching duties and translation at the professional level was partially a direct reaction to the pressure she felt. Raquel wanted to avoid any situation where her colleagues might describe her as uneducated and incompetent: "I can put up anything just like that, then people will come into the classroom and say, 'Oh, my goodness. She doesn't know how to write (in Spanish).' I don't want to hear anyone saying that kind of thing." Deeply cognizant that she was hired

108   Teaching at Public Schools

primarily because of her Spanish language skills, Raquel felt pressured to prove her language and teaching skills. Raquel was placed in her school's "surveillance structure" (Bushnell, 2003) where novice teachers of color receive intensified scrutiny, are made to feel vulnerable as untenured staff, and are positioned within a climate of fear.

Raquel looked fearful and stressed out when she said:

> I'm in charge of making sure these kids do really well. I feel like I will be blamed if they don't do well. It's a huge pressure. I need to show my best all the time. I knew I was going to be working hard, but I wasn't expecting this hard.

Raquel's feeling of fear resonates with studies' finding that Latina/o bilingual education teachers are under immense pressure to raise test scores of their second language learners and move them into mainstream English-only classrooms as quickly as possible (Lapayese, 2007; Umansky & Reardon, 2014). Raquel was in a similar situation as one bilingual teacher in Valdez's (2001) study who commented, "(B)ilingual teachers seem to be under more scrutiny" (p. 250).

The pressure Raquel felt, however, was due to the exploitive teaching condition she experienced at school. Monolingual white teachers benefited directly from Raquel's language skills without producing their own labor. Raquel, on the other hand, kept providing labor for other teachers because they lacked the language skills. Further, the quantity and quality of the work was closely but covertly monitored by other teachers. Being a novice teacher who were worried about her tenure functioned as a catalyst to increase the demands made upon her. Lipsitz (2006) claims that whiteness does its work as "a structured advantage, as a built-in bias that prevents hard-working people from securing just rewards for their labor and ingenuity. It produces unfair gains and unjust rewards for all whites, although not uniformly and equally" (p. 106). It is obvious that the well-crafted structural advantages were naturally given to white monolingual teachers by default, while the same advantages functioned to disadvantage Raquel.

Raquel encountered another barrier that exacerbated her stress related to translation. In a state where Raquel lived, newly certified teachers were required to change their certificate to the advanced-level certified called a "professional certificate" within five years after they began teaching. In order to do so, novice teachers were required to submit the "proteach portfolio"[2]—an evidence-based assessment designed for teachers seeking the professional certificate. Raquel explained what was required in the portfolio: "We need to prove that we're constantly working with students and communicating with parents. Basically what I've been doing all along, but I need to keep record of all of those in writing." For example, according to Raquel, she was required to write about the lessons she had taught with a thorough analysis of how her students met the targets or not and to reflect

Teaching at Public Schools   109

on how she would improve based on this analysis. Raquel also needed to write about her classroom management strategies based on the voices from her students. In addition, she wrote about how she would involve and incorporate parents and communities to better her teaching.

To facilitate the process of compilation of the proteach portfolio, most novice teachers were enrolled in a professional certificate program operated by a university's teacher education program which usually met after-work or on weekends. Raquel was enrolled in a program which met at an elementary school 10 minutes away from her house. As she began collecting the required documentations for the portfolio, the instructor of the program demanded a full translation of all the documents she had collected from Spanish to English. Raquel explained this dire situation:

> For example, the instructor gave us a student release form for parents to sign. It was not in Spanish. He casually said, "Oh, well, then, you can just translate," as if not a big deal. I said to him back, "Will I get extra credit for this extra work?" He bluntly said, "No." I further asked him, "Most of the work I do is in Spanish. Am I going to translate the work I do with kids in Spanish?" His answer was: "Make it so. That way, I can understand." Almost all of my kids are Mexicans, and I teach in Spanish. He expected me to translate all the documents into English. Basically, this particular translation cost me $1,000!

Evidently, the instructor requested a full translation because it was convenient for him without considering the time Raquel needed to spend. The instructor's unreasonable demand clearly highlights the underlying assumption that any work done in languages other than English needs to be made available, thus translated into English, so that English speakers will understand. This way of thinking is deeply associated with a belief of the superiority of the English language and its speakers. The extra translation work doubled the number of documents Raquel needed to collect and the amount of time she needed to spend as well.

Raquel began to wonder if the advantage of speaking Spanish was really an advantage. She complained: "I'm a bilingual education teacher because of my skills which other teachers don't have. But I'm given extra work because of these skills and I don't even get extra credit. Isn't this unfair?" Possessing skills other teachers lacked—Spanish language skills and knowledge of Mexican culture, Mexican American cultures, and a variety of Latina/o cultures—indeed made Raquel a specialist. However, she was not given any monetary or instructional support for her skills as long as she worked with the title of a bilingual education teacher. In other words, being a specialist did not lead to special and favorable treatment as long as it was within the realm of bilingual education. Lack of recognition of her work as a bilingual education teacher is similar to the absence of knowledge or understanding many teachers and administrators express towards the linguistic skills and

110   *Teaching at Public Schools*

cultural expertise bilingual paraprofessionals possess (Wenger et al., 2004). This situation symbolizes the low-status of bilingual education teachers within the teaching hierarchy. In this sense, the term "bilingual education teachers" seems to be a code term for lower-status teachers.

A Weak Professional Teacher Network

To make matters worse, the skills Raquel possessed began to negatively impact her health. She was stressed out and was seriously thinking about having a doctor prescribe her medicine to reduce her stress. Raquel complained loudly:

> Why all these stresses? Why do I need to pay $1,000 to get stressed out? To show whoever what I'm already doing? Why do I have to do this extra thing which I don't think it's gonna help me become a better teacher? What is gonna help me become a better teacher is me having a desire to be a teacher, a desire to learn, and a desire to try new things.

Similar to Erika, Raquel's teaching philosophy was pragmatic and experience-based, which did not align with the unacknowledged white privileged perspective of her superiors, which deemphasized and delegitimated the experiential knowledge that she had accrued as a person of color.

Eventually, Raquel started to feel disengaged even from the required PLC (professional learning community) meetings with other kindergarten teachers, because meetings that occurred three times weekly and half-an-hour each time consumed too much of her valuable time. Raquel sighed:

> These days I eat lunch in my own classroom. Alone. Only because I have things to do. If I want to go home at a reasonable time, I better get as many things done during my lunch break. I don't have time to even pay attention to who other teachers are and what's going on at my own school! And sadly enough, I don't feel anything bad about that.

With such a busy workload, Raquel was not given luxury to observe other teachers' teaching nor interact with them even during her lunch break.

In studies of teachers of color and social capital, many teachers of color identify with parent and student communities more strongly than with colleagues (Bascia, 1996b; Milner & Hoy, 2003). Being a teacher of color naturally leads to stronger connections with students of color due to shared experiences as people of color. However, Williams and Williams (2006) claim that a feeling of isolation from white colleagues among teachers of color, as was seen in Raquel, could trigger stronger connections with students of color as well, since students of color also often feel isolated. Stronger relationships with minority parents and students are welcome. Weaker collegiality at school, however, is detrimental, particularly to novice teachers because stronger collegiality provides them with stronger social capital which involves a "sense of trust and collaboration, and professional ties to networks and community within and beyond the school" (Achinstein & Ogawa, 2011, p. 25). Since teaching is hardly ever conducted in isolation and always embedded in a larger school context, weaker social capital makes it difficult to access and acquire cultural capital that confers power and status. Without a means to access power and status, novice teachers of color are vulnerable to making themselves marginalized and deprofessionalized.

Workload, Professional Network, and a Feeling of Alienation

The accounts of Erika and Raquel revealed that they felt physically and psychologically alienated from their colleagues at their respective schools. Heavy workloads and concomitant weak professional networks within their schools seems to be the direct cause of this feeling.

Why did they end up carrying such a workload anyway? The fact that both women were keenly aware of their "wanted" status necessarily heightened their consciousness about their teaching skills in Spanish. The fact that both Erika and Raquel felt an imminent need to prove their competence in Spanish is indicative of this consciousness. In return, the schools took advantage of their heightened consciousness. They found effective and efficient use of their language skills, which benefited the school both instructionally and financially. Erika's school invested in her Spanish language skills in the form of scripted curricula and controlled her instruction in language. Raquel's school strategically maximized her language skills in the 90/10 bilingual education model and relied on her ability to develop Spanish curricula. Both schools took for granted that Erika and Raquel provided them with free and quality Spanish translations.

In addition, a carefully monitored surveillance system where Erika and Raquel's credibility as teachers was constantly questioned simply because their employment was understood to be related primarily to their language skills aggravated their heavy workloads. Erika felt that she must show that she could change the language of instruction effortlessly and teach using scripted curricula in both languages, while Raquel felt that she must demonstrate that she could develop Spanish curricula without guidance from the school. Both women made sure that their translations were at the professional level. They were apparently embedded in school structures where they constantly needed to prove and defend the very skills they were hired for—the Spanish language skills. This raised their quality of work but the amount of work as well.

Other school structures further contributed to their heavy workload. The schools did not intervene with the existing practices that added to their workload. Translation was one of them. Wenger et al. (2004) contend that it is not uncommon that Spanish-speaking teachers are called on daily to serve

112 Teaching at Public Schools

as translators without extra pay. This unspoken and unchallenged practice at their schools served as the basis for exploitation; Erika and Raquel were expected to be free and convenient translators. Mentor selection was another problem. A helpful mentor might have advised Erika how to effectively deal with her workload.

It seems that all the factors that contributed to their overwork were made possible because they were placed in unequal power relations at school. Both women consistently held subordinate positions as novice employees, former paraprofessionals, Latinas, Spanish speakers, and bilingual education teachers—all statuses that were low on the power hierarchy of the school systems where they were employed. Achinstein and Ogawa (2011) observed that top-down decision-making approaches in schools reflect problematic power relations between school administrators and novice teachers of color. The lack of voice Erika experienced in directing how her pupils should be educated and the differential treatment between her and other third-grade teachers, who were all whites, are indicative of unequal power relations. This power structure parallels the hierarchical structure of social relations between teachers and paraprofessionals, in which paraprofessionals' inputs are rarely taken into consideration (Rueda & Monzó, 2002; Wenger et al., 2004). Both Raquel and Erika were novices and worried about tenure, in particular. The unequal power relations between Raquel and her partner teacher symbolized the hierarchical power relations between the two individuals—a senior, white English-speaking teacher as dominant vs. a novice Latina Spanish-speaking teacher as subordinate.

It seems that each social category Erika and Raquel held was constructed as lower in status, and this intersection of each category defined them as subordinate. To make matters worse, their subordinate position was taken for granted and treated as a natural condition for novice Latina bilingual education teachers, as was indicated by the schools and the colleagues' lack of support and appreciation of the work both women did. The fact that Erika and Raquel were rarely given the same rights and privileges available to their colleagues was most problematic. Erika's colleagues were given instructional support and participated in decision-making processes, while Raquel's partner teacher enjoyed a substantially lighter teaching load. In this web of unequal power relations, Erika and Raquel were defined as laborers or service providers. The blatantly unequal working conditions between their white colleagues and them, however, ultimately served to privilege their white colleagues, thus worked to protect their property interest in whiteness.

The fact that the schools never attempted to appreciate both women's work seems to have escalated their sense of being "used." Their feeling of "being used" ultimately reveals the schools' disinterest in the well-being of the two women. Both schools were more interested in maximizing Raquel and Erika's skills than caring for their growth as professional teachers. Furthermore, the failure on the part of the schools to simply acknowledge

Erika and Raquel's work sustained the invisible and normalizing workings of racism, while advantaging and elevating the status of their white colleagues. In addition, these mechanisms of exploitation functioned to deprive Erika and Raquel of the opportunities to cultivate professional networks that were essential for their growth as professional teachers over the long-term. In this sense, it was no wonder both women desperately wished that their schools would have at least verbally recognized their work.

Monzó and McLaren (2017) wrote, "(I)nclusion can be conceived as bringing the Other into compliance—coopting or sanitizing our ability to see the world differently and destroying our ability to make whiteness visible and to create structures that decenter it" (p. xv). Hiring more teachers of color is indeed an inclusive practice. However, it could hide traps of whiteness's strong desire to sustain and perpetuate the status quo inside where exclusivity is paradoxically practiced in parallel to inclusion.

Relationships With White Colleagues

To novice teachers, having a solid professional network within their school is one of the most essential elements that determine their success. Naturally, both Erika and Raquel wanted to be included in such a network and receive support from senior teachers and administrators at their schools. However, the heavy workloads both women endured, compared to other teachers, functioned to weaken their professional networks. As a result, they began to feel isolated and marginalized from their colleagues. In addition, the daily lack of warmth Erika and Raquel experienced from their colleagues further removed them from involvement in their schools' professional network and increased their feelings of alienation.

Spanish Language Skills

In PNU's teacher education program, both Erika and Raquel were the frequent victims of microaggressions regarding their Spanish language skills from white monolingual students who were fellow teacher candidates and jealous of the advantageous skills both women held. As the days passed by, the same type of microaggressions began to be expressed by the fellow white monolingual teachers. For example, when Erika explained her disciplinary policy to her fellow teacher, Mrs. Thomasson:

I handle a lot of discipline all by myself. Kids know not to mess around in my class. Whenever I have an issue, whether it be negative or positive, I will tell kids, "Go call your Mom. The phone is right there. Go use it." If it is a bad thing, they will have to explain to their parents what happened. When it is a good thing, they're just so excited. For example, Enrique, a kid who struggled with math. I said to him, "Go

114   *Teaching at Public Schools*

call your Mom and tell her how great you are in math." All of a sudden, I saw his eyes like "boom, boom, boom!" He understood. And he was just so excited and calling, "Mom, Ms. Soto said to call you because I understood math and I'm doing such a great job!" They know I am listening to their conversation. If they don't say what happened and if it was a bad thing, then I would get on the phone and not only would they be in trouble at home, but they also would have consequences at school as well. So it was a double-edge sword for the students and they know they have to walk that fine line because I am listening. Mrs. Thomasson looked at me for a while silent and simply said, "Well, if I spoke Spanish, I would do that." And I just sat back and said to myself, "Did she just miss a whole point?" The point is the kids make a phone call. They do the talking, not I. I was so taken aback.

Erika simply shared her story about the way she disciplined. However, it seems that her intent was not taken as she expected. Erika interpreted Mrs. Thomasson's underlying reaction as jealousy towards her because her colleague was unable to provide such discipline because she did not know Spanish. Erika became disappointed with, frustrated with, and even angry at Mrs. Thomasson's nonacknowledgement of her success and sarcastic response. Erika remarked, "They're not warm. They're so cliquish."

Raquel had a similar experience of microaggressions. In the middle of her first year teaching, all the students in Raquel's kindergarten class reached the benchmark in DIBELS[3] in Spanish. She was pleased with the result and felt good about her teaching and her students. Later, however, Raquel found out that not all the students in her partner teacher's kindergarten class reached the benchmark in English. When her partner teacher, a monolingual white female with a long teaching experience, found out the result, she stated to Raquel with a surprised look:

How can that be possible? You must have overworked them. You drilled them to the test. Teach to the test. Or something's wrong with the Spanish version of the test. Maybe the Spanish test was too easy. Maybe the person giving it to them might have done something to accommodate.

In the incident above, the senior white colleague immediately questioned the reliability of the Spanish test. This blunt put-down seems to illustrate the senior colleague's belief that Latina/o students cannot be higher than white students academically. She was suspicious of the way Raquel taught, instead of congratulating the success of Raquel and her students. Raquel felt that her colleague's suspicion was a camouflage to hide her jealousy. All Raquel could do at this moment was to ponder: "I don't get it because what you teach in reading is basically the same as what kids are tested at. Why this attitude? What's going on?" Her sense of marginalization started to widen further.

Teaching at Public Schools 115

Another incident was even more overtly hostile for Raquel. As a novice teacher, Raquel believed in learning from veteran teachers. Therefore, she did not miss an opportunity of informally observing other teachers whenever possible. Particularly, with her heavy workload, observing other teachers teach was a luxury. One day, Raquel saw a white veteran female teacher teaching a small group of children in the hallway. Raquel quietly began watching the colleague teach from the side of the hallway without disturbing the session. However, the veteran teacher noticed Raquel's presence and said to her with a glare: "Are you gonna take my job away?" Taken aback, Raquel whispered to herself, "You already have the job!" and scurried away from the scene.

In this incident, the veteran white teacher did not hesitate to show her fear and hostility towards Raquel who was novice and had the least seniority at her school. The possibility of her losing her job because of Raquel was certainly remote, but her fear towards this remote possibility seemed to be present and immense. Recalling this incident, Raquel sighed:

> I was both shocked and angry at that time. It bothered me to a point. But, I just kind of let go because it's her business. If she's so worried, she can figure it out on her own. I do what I need to do.

Raquel's comment above symbolizes that a series of incidents of annoying microaggressions and overt aggression from her colleagues made her to believe that collaborative relationships with them were impossible to achieve.

Jealousy and fear seem to be an overarching theme that encompass the incidents above, and this particular theme was also present in PNU's teacher education program where Erika and Raquel received microaggressive assaults with regard to their Spanish language skills. The comments made by their colleagues indicate that Spanish language skills were indeed seen as advantageous in teaching, particularly at majority Mexican and Mexican American schools. The fact that their white colleagues displayed jealousy and fear towards Erika and Raquel underlines another fact that their colleagues firmly believed in their inherent superiority by virtue of being white monolingual speakers. When a white ego encounters a perceived threat from witnessing that people of color possess skills they do not have, fear that people of color might "someday gain the kind of power over whites that whites have long monopolized" (Jensen, 2005, p. 54) emerges. White fear of the skills and abilities of racial and ethnic minorities intensifies and transforms itself into anger and hostility, in order to reestablish their sense of stability and dominance (Britzman, 1998; Levine-Rasky, 2014). The aggressive reactions made by the colleagues of Erika and Raquel clearly indicated their heightened sense of threat to whiteness, their desire to retrieve the normal state of white superiority. The divide between whites vs. nonwhites clearly

116   *Teaching at Public Schools*

existed in the schools in which Erika and Raquel taught, and was systematically maintained and perpetuated in order to sustain the normalcy of white dominance—defining others as a reflection of all that is right with whiteness (Altman, 2006).

The aggression from white administrators and faculty members that both Erika and Raquel experienced made them frustrated and angry. It also heightened their consciousness about what it meant to be a teacher who was of Mexican heritage and fluent in Spanish in a school dominated by white monolingual teachers. Eventually, Erika and Raquel began to feel the existence of the invisible divide between their colleagues and themselves.

Not Supported

Coming from the same ethnic group as their students, both women had a clear understanding of effective methods of teacher-student bonding and discipline, which could not have been easily achieved without sharing the same cultural heritage. Their effective disciplinary skills and concomitant good teaching, however, seem to have created a counter-effective consequence. Erika discussed this problem:

> The assistance a fellow third-grade teacher, a new white female who speaks Spanish, got was phenomenal. There were other teachers in her classroom every day whenever she needed. I mean there were times when we would be in the hallway and she would run into Mrs. Davenport's office, "I don't know what to do," just being hysterical and Mrs. Davenport would go help her. But I, who didn't need to be held by the hand, was just left to float or sink. And she's white and I'm not.

Unlike her colleague above, Erika was left alone with little support. As was described before, she was ordered to use scripted curricula both in Spanish and English, but the instructional support of how to use these curricula was rarely provided. Erika, however, had a strong will and a feeling of competence and independence:

> If they don't want to help me, that's fine. I will make it on my own and I'm not going to beg them for help. That's who I am. I won't ask them for help again. I'll go ask somebody else until I'm able to find a person who is able to help. And so, you know, I'm not gonna sit and say, "Oh, please help. I need help." I'm going crazy. No, I'm not going to do that. I will ask once and if they don't help me, then I find a way to survive. I'm a survivor.

Seeing herself rarely supported, compared to other novice teachers, reinforced her strong mindset.

Teaching at Public Schools 117

Whenever Erika had inquiries regarding the scripted curricula and other teaching related matters, she asked for help via emails and phone calls. They were, however, frequently ignored by other teachers and administrators. Erika narrated one incident as an example:

> At the beginning of the first year, some kids in my reading group were like, "Why are they in Spanish reading? They need to be in English reading." They reached the benchmark in DIBELS both in Spanish and English. They were high enough to be in transition. There was no reason for them to be in my class. Well, I asked Mr. Logan, "What do I do with these kids?" He said, "Get a letter from the parents." So at the conference time, I told the parents. Six of them, I think. I said, "We need it in writing." So those parents sent me notes saying, "I want my kid to be in the English reading group." And I emailed Mr. Logan, "I have the notes. What do we do now?" Waited, waited, and waited. It was March, the end of March, so basically April, when the students got switched to the English reading group. Six months! That's how much my opinion mattered . . .

Erika added with a chuckle: "You email because you get a response. My team members email because they get a response. It became a joke after a while. That's how far it went." Eventually, Erika stopped emailing and asking questions.

The administrators' noncaring attitude towards Erik expanded to her students as well. One time, Erika's class did a fund-raising to go to a field trip to the nearest radio station. One day after school, her students and students' parents all gathered and made homemade tortillas. Despite that tortilla making took place during the day, neither Mr. Logan nor Mrs. Davenport showed up to her class to greet the parents and chat with the students. Erika confessed, "It hurt my feeling a lot. It hurt my students' feeling, too. They didn't come see us. They didn't show us any support." Noticing that "no-response" happened only to her and not feeling supported at all, Erika became increasingly isolated at school, which only served to increased her sense of self-reliance and determination to succeed as a teacher in spite of the absence of assistance and encouragement from the administrators and colleagues.

The administrators' lack of support not only towards Erika but also her Latina/o students support the studies that found that mainstream teachers tend to harbor negative attitudes towards English learners. In Walker, Shafer, and Liams's (2004) study of 422 teachers found that 70% of mainstream teachers were not actively interested in having English learners in their classroom, while 14% adamantly objected to English learners being placed in their classrooms.

Raquel was also not offered help. As we saw in the previous section, not only did she struggle to create the Spanish curriculum by herself, her partner teacher, a senior white female, never offered help. Naturally reserved,

118   *Teaching at Public Schools*

Raquel, never officially complained to the colleagues and the administrators. She stated:

> Even if I express my feeling you get nowhere anyway. I just kind of go to work and teach and do what I've got to do. Kind of stuff like that I don't really think about . . . and I'm not in the lounge often.

Raquel's comment above exemplifies her sense of alienation. Raquel stopped being engaged with the colleagues who not only did not offer assistance, but instead expressed jealousy and fear of her competence with microaggressive statements.

Neither women received even basic guidance from senior teachers and administrators. Studies have found that compared to white teachers, teachers of color tend to receive little support from white colleagues and administrators (Arce, 2004; Ingersoll & Connor, 2009; Ochoa, 2007; Urrieta, 2009), indicating that being a different race and ethnicity seems to directly affect the amount and quality of support they receive from colleagues and administrators. On top of this subtle prejudicial treatment of teachers of color in general, both women were already competent teachers despite their novice status because they had classroom experiences as paraprofessionals and used culturally appropriate teaching. Therefore, the common image of novice teachers being inexperienced did not apply to them. Although already competent, Erika and Raquel would have benefitted from the advice and guidance of senior colleagues and administrators simply because they were still novice in the field of teaching. It would have facilitated their growth as professional teachers, but they were left to figure things out on their own.

Hostile Work Environment

Raquel and Erika were opposite in personality: Raquel was reserved and only acted after thinking about a matter for a long time. In contrast, Erika was naturally strong-minded and assertive by nature. Her assertiveness, however, invited unnecessary and unpleasant scrutiny from one administrator. At the end of the first-year teaching, Erika received an unsatisfactory evaluation from Mrs. Davenport who was the assistant principal, once called her "a secretary," and was charged with evaluating her. Mrs. Davenport's evaluation was followed by detailed negative comments to each observation item. When Erika received it, she became frantic: "Am I doing it right?" At the evaluation feedback meeting, Mrs. Davenport split hairs against Erika's teaching. Erika continued:

> There is always constant noise in my classroom. Radio, TV, the kids, or something. It's part of the Latina/o culture. Most kids love it. When I explained to her (Mrs. Davenport), she said, "Oh, I didn't know that." But continued, "What if there is a kid who has attention deficit? Isn't it

distracting?" I said, "Well, I will turn it off if it's distracting." I always ask my kids if the music bothers them, they need to let me know, then I will turn it off. Most kids love it. Around nine to ten a.m., a local radio station has a children's program and they play children's music.

Erika went on complaining further:

> Then she asked why I had my kids grouped at a high, medium, and low separately. She asked, "Don't you ever mix them up? Sometimes pairing up with students in a different ability group?" I said, "Yes, but for this activity, it was a group project, so it was the right way to group." They had to present to the whole class, so they were working on a visual presentation at that time. I always have my lower students do an oral presentation. That way, I can make sure if they've understood everything they're supposed to do. If they can verbalize, then they have comprehended. Well, she questioned about the validity of that method, too. On and on and on.

Erika felt that Mrs. Davenport kept questioning on trivial matters regarding her teaching. Erika also sensed that Mrs. Davenport was unfamiliar with Latina/o cultures. Furthermore, growing up upper-middle-class made it harder for Mrs. Davenport to relate to the students who were poor, Erika thought. Mrs. Davenport's negative perspectives and lack of interest in understanding Erika's teaching style from a cultural perspective made Erika feel not only unsupported and but also under attack.

One of many difficulties people of color face every day in a predominantly white workplace which systemically advantages whites, and in return, disadvantages them, is the burden of proving the perpetrators' discriminatory intentions. From Erika's perspective, Mrs. Davenport's excessively and unreasonably detailed and negative evaluations of Erika's teaching, for example, were perceived as an act of bullying based on her being a racial and linguistic minority. However, proving that Mrs. Davenport's act of bullying was racially motivated, was an impossible task for Erika. Without a proof, Erika's complaining about the incident would have resulted in her being perceived as obsessed with racial discrimination, and meanwhile Mrs. Davenport would have received no consequences for her bullying. This is the unfortunate result of the persistent casual deployment of microaggressions. The cumulative effect of daily microaggressions is a major part of the hostile professional environment people of color face and are forced to endure.

When I met with Erika at the end of her first-year teaching, she said with a fearful look, "I don't think I have a good relationship with the school administrators. Her (Mrs. Davenport) calling me a secretary was an offense. That put a real wrench in our relationship. Anyway, I haven't been fired yet, at least." As a novice teacher without a tenure at that time, it seems that the term "being fired" was constantly in Erika's mind. The direct and explicit

120   *Teaching at Public Schools*

exercise of power and the enticement of fear are effective and historically proven ways to control and manipulate people of color's behaviors, and to sustain the superior positioning of whites. Erika was afraid of the power Mrs. Davenport presided over her because she was the one who had the decision-making authority and power for her tenure.

Moreover, Erika perceived that she was kept under surveillance. This perception is not uncommon among teachers of color who frequently feel they are treated in an over-judgmental and over-scrutinized manner, compared to their white counterparts (Jones & Maguire, 1998; Kohli, 2016; Matias, 2016). Erika explained her feelings:

> One week, we needed to change a schedule. I switched my calendar on Sunday night and emailed it to the administrators. When she (Mrs. Davenport) came in (without notice), we were doing social studies. I was reading about Cesar Chavez in Spanish. She left. When she came back again, the kids were watching a video clip on Chavez. She left and came in again when we were ready to go to a PE or a library. Then she asked, "Aren't you supposed to be doing math?" I said, "No, I changed the schedule as I was told. You came in three times within a short period of time and my kids are really distracted. I really don't appreciate that." She said, "I thought you were doing math." I said to her, "Do you check up on other teachers to make sure they follow the right schedule?" Then she said, "Oh, no, I wasn't double-checking. I just wanted to see you teaching math, didn't see it, so I came back."

This incident occurred when Erika started her third-year teaching and Erika was the only teacher who experienced this type of surveillance.

Erika did not let this incident go easily. By the third year, her frustration with Mrs. Davenport's continuous deployment of microaggressive comments and repressive, distracting action had reached the boiling point. One the day when the incident above happened, Erika confronted Mrs. Davenport and demanded that she recognize Erika's effort: "I just switched my schedule. Recognize that I did it without complaining. Many other teachers would complain and throw a fit about!"

Reflecting on Mrs. Davenport's treatment of her, Erika stated:

> Mrs. Davenport had doubts about my teaching. In the first year, she said, "I see you as a secretary." I said to myself, "Wait a minute! You don't think I'm capable?" I didn't say anything to her because I was just like, I was so taken aback. I was really offended. Afterwards, when I needed something, I never went to her. Never. In her mind, a new person should be more needy. For some reason, I am able to establish a rapport with the troubled kids. So there are kids who are now fifth-graders who sneak into the building to come to my room. And I'm like, "Guys, you are going to get me in trouble." What I see is that those

Teaching at Public Schools 121

people who are really clingy and needy, need her (Mrs. Davenport) to guide them. What I have noticed is that she really thrives on that. Because I don't go to her, I think that might be part of the issue. I handle a lot of discipline myself. I just don't send kids to go get disciplined or I handle it in my classroom. What I think it is, she expects me to be more needy. Needing her guidance more. Because I don't do that, I think she is checking up more, but I could be wrong. But I think that's what it is. That I don't go to her as much as she would like me to do. But I don't have time during the day to go chit-chat with her.

In Mrs. Davenport's mind, Erika symbolized someone who should be incompetent and needy. Instead, Mrs. Davenport observed Erika as an independent and skillful teacher. This unexpected reality caused Erika to be seen as a competitor and a threat, and stirred resentment in Mrs. Davenport. Whites resenting people of color's accomplishments is well-documented in Chávez (2011) who investigated Latina/o professionals at predominantly white workplaces. One Latina lawyer in the study revealed that her white colleagues were condescending while simultaneously being resentful of her accomplishments. Another Latina lawyer in her study found herself being taught her proper "place" and being reminded of the inherent inferiority of her ethnic group. It is a way in which whites project their fear on people of color when the latter thwart expectations of presumed incompetence. Erika concluded that a sense of threat seems to have triggered a series of emotionally-driven and unprofessional treatment by Mrs. Davenport who was in a position of structural dominance over her. Because of her dominant position in the administration, Mrs. Davenport's display of crude, demeaning words and behavior towards Erika was not condemned at all.

Fighting Back

Even after she voiced her anger directly to Mrs. Davenport, Erika continued to experience hostile treatment at the hands of this administrator. After receiving another negative evaluation, Erika finally initiated a series of meetings with Mr. Logan (the principal), Mrs. Davenport, and a union representative. Before the first meeting, she wrote down her responses on sticky notes to every single question Mrs. Davenport made on the evaluation forms and rehearsed what she was going to say and how she was going to say it. During the meeting, Erika professionally presented her observations about the unfair treatment and negative evaluations she had received from Mrs. Davenport:

I went in heads up. I honestly said I felt being attacked. Then continued, "I know what students need because I am with them all day long. You come in for five minutes, you don't know the children, and you don't know why I am doing it. I know my students and know what they

122   Teaching at Public Schools

need." I said, "You question me, that is fine, but at the same time you need to do it in a way that is not sounding like as if I'm not doing my job correctly. You know, this is my third year and I don't like working like this. I've never been at a place where I felt like this mad, you know. I see you treat everybody the same on my floor. I don't appreciate this. You need to know I have gone through the same schooling that everybody else did, and obviously I put in my time. I'm learning, this is my third year, I'm learning still and I will always be learning, but just to understand where I'm coming from and how I feel. You talk about being positive with the kids, but you don't do that to me." And she just said "Okay." I continued, "So really in order for me to believe what you say, to believe what your purpose is here, I need to see you doing what you preach to us to do."

What Erika said directly to Mrs. Davenport could have been interpreted as defiant and disrespectful to the authority and to the immediate supervisor, particularly as she was a novice teacher. However, she felt she needed to do so in order to save her integrity as a teacher and more importantly, as a human being. After the meetings, Mr. Logan became in charge of Erika's evaluation. Erika welcomed this change: "Mr. Logan has been coming to my classroom more often, which is super nice. I feel like he's just giving fresh perspectives. So, he's trying to alleviate the horrible tension I felt whenever Mrs. Davenport came in." Luckily, Erika possessed a strong will, and could challenge the hostile practices of her superior in the school hierarchy, whose discriminatory intentions were difficult to prove.

The key to Erika's successful negotiation with the administrators derived from the ways she presented herself. Erika had learned from her confrontation with Mrs. Davenport the importance of professionally expressing one's grievances in an objective and professional manner, and not engaging herself in the same manner Mrs. Davenport deployed. Erika summarized her current situation:

> I feel a lot of tension still, but not as bad as before. I have learned how to . . . just express myself. When I don't feel something is right, I have learned how to express that in a positive way. I still need a lot of work expressing, "I didn't like how you said that." And say it in a way where I am not offensive and not sounding like accusing someone. I do need a lot of growth in that area. More professional and more self-confident in what I do and say.

It is clear that Erika had learned to resist her oppressor's emotional violence and to present herself with calmness and rationality against an irrational and aggressive perpetrator. This tactic naturally highlighted Mrs. Davenport's irrationality in favor of Erika, who was incredibly clever when navigating against someone more powerful than herself.

Teaching at Public Schools   123

After this encounter against Mrs. Davenport, Erika began proactively speaking up at faculty meetings. As Erika attempted to present herself more professionally against the discriminatory behavior against her, she began to notice a change in Mr. Logan as well. At meetings Mr. Logan began to recognize Erika's work and occasionally mentioned, "Yeah, just like what Erika said before . . ." Erika was elated: "I feel like he now listens to what I say, and furthermore agrees with what I say. That's so good! I'm finally getting some type of feedback." When Mr. Logan praised her discipline skills, Erika felt honored and simultaneously appreciated him acknowledging her work.

The series of meetings that Erika initiated with the administrators in her school resulted in giving her an opportunity to raise her complaints, and in a highly professional manner describe the indignities that she received at the hands of administrators and colleagues. These meetings began to bear fruit. At one faculty meeting, Mrs. Davenport, for the first time, responded positively to a comment Erika made. Erika said with a smile, "I was like, 'Wow!' Just the fact that Mrs. Davenport said something positive about what I said was shocking. This is my third year. It took a while, but I think it's finally getting better."

Splits in Collegiality

Bascia (1996b) states that many teachers of color are engaged in special advocacy roles with respect to racial minority students. In particular, Weisman (2001) found that Latina/o teachers have a strong identification with Latina/o culture, have political consciousness, and value the Spanish language as a means of affirming the cultural identities of Latina/o students. Coming from similar backgrounds as the students in their own classrooms, both women were passionate about advocating for Latina/o students academically, socially, and personally. Their advocacy, however, sometimes caused conflicts with their colleagues. These conflicts are not uncommon. Several studies found that teachers of color's passionate attempt to create more equitable educational environments tend to cause conflicts with white colleagues and administrators (Achinstein & Ogawa, 2011). The Latina/o teachers interviewed by Ochoa (2007), for example, fought against their schools' tracking and unequal distribution of resources, but the result of their actions was unfavorable. Chicano/a activist teachers in Urrieta's (2009) study attempted to challenge oppressive educational practices, while understanding that their activism was always partly compromised at predominantly white schools. Frequently, bilingual education teachers have to deal with the insensitive, disrespectful and racist comments and behaviors of white teachers who have negative attitudes towards both bilingual education and English learners. One bilingual education teacher of Cuban origin in Cahnmann-Taylor, Wooten, Souto-Manning, and Dice's (2009) study, for example, mentioned that she wanted to choke her dismissive colleagues because they undermined advocacy practice for English learners.

124   *Teaching at Public Schools*

Erika perfectly understood her confrontational personality and did not feel like apologizing for it, either, if being so was related to her advocacy for Latina/o students. Erika was nonapologetic at all when she commented:

> I think I probably attract confrontations because when I see something I don't like, I will say something. Some people don't like that. I see myself as sticking up for the underdogs basically. The kids who are in trouble. I will say, "No! Don't judge them. Look for what he did do. Look for the positives." I'm always defending the kid who is always in trouble.

Erika delineated a conflict she had with a white third-grade teacher, Mrs. Smith, over a boy who accidentally shaved his eyebrow while playing at home:

> When he came to school, I made a joke about it like, "Oh, you look really cute!" When he went to the English reading class in the afternoon, his teacher, Mrs. Smith, sent him to the principal's office. Mrs. Smith said to him that he was involved in a gang. At recess at two o'clock p.m., the boy came to me with tears in his eyes and with a band-aid, a brown ugly band-aid. Not even a clear one, a brown ugly band-aid on his eye! I said to him, "What's that on your eye?" Tears in his eyes and his head down, he said, "Mrs. Smith said I had to put it on my eyebrow." How humiliating for him to walk around the school with a big brown band-aid on his eyebrow! So, I went to Mrs. Smith and said, "Why did you send him to the principal's office?" Mrs. Smith said, "That's gang-related." I said, "No, it has to be one versus three or four for it to be gang-related. You have to get a right combination. Also, it is crooked first of all. Second of all, it is only one eyebrow." Mrs. Smith said, "How do you know that?" I said, "I know. I lost my friends from gang-related accidents. Look at the environment you teach at. You have to realize where you're teaching." After that, I removed a band-aid and penciled his eyebrow with an eyeliner.

The difference between Erika and Mrs. Smith's handling of the boy illustrates that Erika had a stronger connection to the Latina/o students because her own life experiences gave her better understandings of the students' cultures and social environment. Erika felt that Mrs. Smith, lacking such cultural and social understanding, interpreted the incident with her own cultural and social framework and disciplined the student in haste and in a punitive way.

Erika's caring attitudes towards her students was evident in the incident below where she and another white third-grade teacher, Mrs. Brown, had an argument over a boy who was in Erika's homeroom, but went to Mrs. Brown's classroom for a reading lesson:

Mrs. Brown: "Jorge doesn't have his glasses yet."
Erika: "Yes, I know that."

Teaching at Public Schools 125

Mrs. Brown:	"Are you going to report to CPS?"
Erika:	"No, you don't do a CPS report for that."
Mrs. Brown:	"It's been five days already since they were notified. This is neglect. His parents are neglecting to provide him with glasses."
Erika:	"If you want, you do a report. Have you thought about them making an appointment with the eye doctor? How long does it take to get glasses? Do they have insurance to pay for those glasses? How do they pay? Have you thought about all those? I have already phoned Jorge's mother. They already made an appointment with the doctor and are in the process of getting him a pair of glasses."

Similar to Mrs. Smith, Erika interpreted that Mrs. Brown failed to understand Jorge's cultural and social background and hastily judged that the student's parents needed to be disciplined by state authority.

Both Mrs. Smith and Mrs. Brown's disciplinary approach shows their negative and dysfunctional image towards students of color and their families, which seems to be deeply ingrained in their consciousness. Repeated confrontations with her white colleagues regarding their inappropriate actions towards Latina/o students and their families necessarily created a psychological distance between Erika and her colleagues.

Raquel, naturally reserved, generally did not confront fellow teachers up front. She rarely had lunch in a teachers' lounge and rarely interacted with fellow teachers simply because she wanted to complete her work as soon as possible. However, she sometimes argued against colleagues when she thought that it was necessary to advocate for her students. The fact that she completed PNU's teacher education program and obtained a bachelors' degree had boosted Raquel's feeling of confidence in counter-arguing her colleagues. For example, Raquel fiercely fought for the adoption of a new curriculum. Raquel's rationale was if she had a brand new set of curriculum, she would be able to double-up activities along with the existing curriculum. That way, she would not have to make activities all by herself. Desperate for more activities she could do in Spanish, Raquel was determined to translate the activities in the new curriculum in Spanish. However, her idea did not make any sense to her fellow teachers: first of all, it cost additional money. Second, they all had a plenty of materials and activities in the existing curriculum since they taught in English.

Whenever Raquel defended and advocated for Latina/o students against her colleagues, she met with disapproval and hostility. Raquel explained her feeling:

I like working with children more than adults. Some teachers are so set in certain ways and don't want to try new things, don't take criticisms, and don't listen to me. You have to be open-minded, try to understand the best you can, and listen and see how you can work things through.

126 *Teaching at Public Schools*

You know, try to find the best for students, which is a lot of work. But, most teachers don't.

Trying to maintain amicable relationships with her colleagues, Raquel chose to self-isolate from her school's professional teacher network to avoid further confrontations. Raquel concluded, "They see us as 'Well, what are you doing here? You need to go back to where you came from.' We get that attitude. I always feel negative perceptions in one way or another." Having a difficult time professionally engaging with her colleagues, by the end of the first year, Raquel's feeling of psychological distance between her and her colleagues had widened.

Collegiality and a Feeling of Alienation

Erika and Raquel's discussion of their relationships with white colleagues and administrators seem to support existing studies that documented how alienated teachers of color are at the workplace. Both women did not form amicable and collaborative relationships with their white colleagues. They effectively handled their jobs with little guidance, and even challenged their colleagues for their use of unnecessarily punitive actions against Latina/o students and their families. However, both women felt that the attributes that contributed to their competence—Spanish language skills, cultural knowledge, and understanding of effective methods of teacher-student bonding and discipline—were not recognized nor appreciated by the colleagues. In their minds, their colleagues displayed irrational jealousy and fear. The emotionally-driven responses of their colleagues towards Erika and Raquel suggest that their competence was on some levels actually recognized by their colleagues, and that is why it was threatening.

Lin, Kwan, Cheung, and Fiske (2005) explain that an out-group's competence could engender group competition, given the tendency for positive attributes to be appreciated as assets only when they reflect well on the ingroup. Both women's positive attributes to their successful teaching were something their colleagues did not possess. Therefore, it is possible that these attributes possessed by the out-group—Erika and Raquel—were perceived to be *too* good among the in-group, and stirred feelings of jealousy and envy by white colleagues. They were seen as competitors who might threaten the status quo.

Both Erika and Raquel felt that their colleagues reacted towards their competence with emotionally-driven responses, such as microaggressions and overt aggression, repressive surveillance, and the like. This suggests how imminent the feeling of threat was to the dominant group. It illustrates their fear of losing superiority in the loss-gain binary perspective and highlights that whiteness is a valued property (Harris, 1993; Ladson-Billings & Tate, 2017) that must be protected. It appears that the skillful and successful teaching of these Latina teachers challenged the white teachers and school administrators who for years used rigid, inappropriate educational methods, and punitive behavioral controls with Latina/o students and their families. Evidently, the colleagues' fear of losing the edge over Erika and Raquel prevented the former from openly and collegially acknowledging the latter's accomplishments. The existence of strong loss-gain perspective further prevented any power-sharing between both women and their colleagues. This is evident because no colleagues expressed any interest in expanding their instructional knowledge by learning from them. The end result was an unwelcoming and hostile work environment experienced by Erika and Raquel.

Since the white racial frame is systemically embedded in the ways the dominant group feels, thinks, and behaves, it is as invisible, prevalent but as important as oxygen in the air (Scheurich, 2000). It is both consciously and unconsciously "produced and reproduced in communicative interaction" (Bonilla-Silva, 2010, p. 11). Therefore, although both women perceived that their colleagues reacted negatively towards their competence, their white colleagues had difficulty in understanding and acknowledging any negative behavior on their part. The problem of this both overtly conscious and unconscious racial frame is the fact that it not only gave Erika and Raquel ill-feelings but also gave them a negative material effect at the institutional level. It caused them to gradually withdraw from their school's professional network of teachers and administrators, which could have been useful if the colleagues had been more supportive and collaborative. Meanwhile, their colleagues continued to express their jealousy and fear without reflecting on their own lack of cross-cultural teaching ability. Ultimately, to Erika and Raquel, their white colleagues reified racism. Their colleagues' attempts to sustain their dominant positioning in the face of their cross-cultural incompetence negatively affected not only Erika and Raquel, but all the Latina/o students and families associated with the school.

Teachers need professional communities in which they share an orientation toward teaching and the groups to whom they turn for moral support, practical aid, and intellectual stimulation (Bascia, 1996b). Achinstein and Ogawa (2011) claim that professional communities help articulate what it means to be a professional, teach one's subject, engage with other professionals, and define practice and values in relation to students. It is within the context of such professional communities that teachers make meaning of their professional lives (McLaughlin & Talbert, 2001). There are clear costs when there is an absence of professional community *within* schools (Bascia, 1996b), particularly to novice teachers.

It seems that Erika and Raquel were entangled with the systemic and institutionalized racism of their schools, and that the unexamined white racial frame omnipresent throughout their schools systemically marginalized them. They wanted to cultivate strong and positive professional networks but were instead discouraged from doing so. The aggression towards and neglect of both women operated to disadvantage their growth as professional teachers. In this sense, just hiring more teachers of color does not

128   *Teaching at Public Schools*

transform their marginalized condition. Rather, it keeps reproducing marginalization, thus serving the interest of the dominant group.

Teachers of Color At a Workplace

Teachers of color are desperately needed. However, they are hard to get and hard to keep. Ingersoll and May (2010) warn that once they become teachers, teachers of color change schools and abandon the profession at higher rates than whites. There could be a number of reasons why teachers of color leave the profession. However, Achinstein and Ogawa (2011) conclude that stressful work conditions teachers of color frequently experience seem to be the main reason for their low retention rate. Studies found that stress indeed impacts persistence attitude significantly (Johnson, Wasserman, Yildirim, & Yonai, 2014; Wei, Ku, & Liao, 2011).

As was evident in Erika and Raquel's accounts of their experiences, their feelings of alienation from white colleagues and administrators made their work more stressful. Alienation was caused by several factors. In contrast to their white counterparts, Erika and Raquel frequently felt that they were treated in an over-judgmental and over-scrutinized manner. Schools did not recognize their strengths and good teaching. The clash of values between them and the white teachers and administrators was problematic. They believed in equity and social justice, while the white teachers and administrators wanted maintenance of the status quo. In line with Ingersoll and Connor's (2009) analyses, both women lacked administrative support. Furthermore, their colleagues openly expressed sarcasm and jealousy towards Erika and Raquel. Unfavorable and discriminatory treatment, microaggressive statements and aggressive incidents, conflicts with colleagues and administrators, and lack of administrative support all lead to a sense of marginalization and alienation among both women. As Vance, Miller, Humphreys, and Reynolds (1989) argued in their study, teachers of color in predominantly white schools experience stress and conflict. One consequence of this conflict in predominantly white schools, is that professionals of color leave teaching in great numbers.

In the next chapter, we will examine how the unfair workload and the cold relationships with the colleagues affected Erika and Raquel's motivation to teach Latina/o students. Although experiencing toxic work environments, both women were strong-minded, and determined to make their Latina/o students successful at school.

Notes

1 L1 refers to native language, while L2 refers to second language.
2 The proteach portfolio evaluates teachers on their ability to impact student learning in three standards—the knowledge and skills for effective teaching that ensure student learning, the knowledge and skills for professional development, and professional contributions to the improvement of the school, the community, and the profession. Under all standards, there are 12 criteria in total.

Teaching at Public Schools 129

3 DIBELS stands for Dynamic Indicators of Basic Early Literacy Skills and it is a test invented by the University of Oregon and assesses the acquisition of early literacy skills from kindergarten through sixth-grade.

References

Achinstein, B., & Ogawa, R. (2011). *Change(d) agents: New teacher of color in urban schools.* New York: Teachers College Press.

Ada, F. (1986). Creative education for bilingual teachers. *Harvard Educational Review, 56*(4), 386–395.

Altman, N. (2006). Whiteness. *Psychoanalytic Quarterly, 75*(1), 45–73.

Anastasia, E. A., & Bridges, A. J. (2015). Understanding service utilization disparities and depression in Latinos: The role of fatalismo. *Journal of Immigrant and Minority Health, 17*(6), 1758–1764.

Arce, J. (2004). Latino bilingual teachers: The struggle to sustain an emancipatory pedagogy in public schools. *International Journal of Qualitative Studies in Education, 17*(2), 227–246.

Bascia, N. (1996a). Teacher leadership: Contending with adversity. *Canadian Journal of Education, 21*(2), 155–169.

Bascia, N. (1996b). Inside and outside: Minority immigrant teachers in Canadian schools. *International Journal of Qualitative Studies in Education, 9*(2), 151–165.

Batt, E. G. (2008). Teachers' perceptions of ELL education: Potential solutions to overcome the greatest challenges. *Multicultural Education, 15*(3), 39–43.

Bonilla-Silva, E. (2010). *Racism without racists: Color-blind racism and racial inequality in contemporary America.* Lanham, MD: Rowman and Littlefield.

Brisk, M. E. (2006). *Bilingual education: Froom compensatory to quality schooling* (2nd ed.). Mahwah, NJ: Lawrence Erlbaum Associates.

Britzman, D. P. (1998). *Lost subjects, contested objects: Toward a psychoanalytic inquiry of learning.* Albany, NY: State University of New York Press.

Bushnell, M. (2003). Teachers in the schoolhouse panopticon: Complicity and resistance. *Education and Urban Society, 25*(3), 251–272.

Cahnmann-Taylor, M., Wooten, J., Souto-Manning, M., & Dice, J. L. (2009). The art and science of educational inquiry: Analysis of performance-based focus groups with novice bilingual teachers. *Teachers College Record, 11*(1), 2535–2559.

Calderón, M. E., & Minaya-Rowe, L. (2003). *Designing and implementing two-way bilingual programs: A step-by-step guide for administrators, teachers, and parents.* Thousand Oaks, CA: Corwin Press.

Chávez, M. (2011). *Everyday injustice: Latino professionals and racism.* Lanham, MA: Rowman & Littlefield.

Chubbuck, S. M. (2004). Whiteness enacted, whiteness disrupted: The complexity of personal congruence. *American Educational Research Journal, 41*(2), 301–333.

Creese, A. (2002). The discursive construction of power in teacher partnerships: Language and subject specialists in mainstream schools. *TESOL Quarterly, 36*(4), 597–616.

Cummins, J. (1979). Linguistic interdependence and the educational development of bilingual children. *Review of Educational Research, 49*(2), 222–251.

Cummins, J. (1984). Wanted: A theoretical framework for relating language proficiency to academic achievement among bilingual students. In C. Rivera (Ed.), Language proficiency and academic achievement (pp. 2–19). Clevedon, North Somerset, UK: Multilingual Matters.

Cummins, J., & Swain, M. (2014). *Bilingualism in education: Aspects of theory, research and practice.* New York: Routledge.

130 Teaching at Public Schools

Dee, T. S. (2004). Teachers, race and student achievement in a randomized experiment. *The Review of Economics and Statistics, 86*(1), 195–210.

Dee, T. S. (2005). A teacher like me: Does race, ethnicity, or gender matter? *The American Economic Review, 95*(2), 158–165.

Egalite, A. J., Kisida, B., & Winters, M. A. (2015). Representation in the classroom: The effect of own-race teachers on student achievement. *Economics of Education Review, 45,* 44–52.

Faltis, C. J., & Hudelson, S. J. (1998). *Bilingual education in elementary and secondary school communities: Toward understanding and caring.* Boston: Allyn and Bacon.

Fee, J. F. (2011). Latino immigrant and guest bilingual teachers: Overcoming personal, professional, and academic culture shock. *Urban Education, 46*(3), 390–407.

Feistritzer, C. E. (2011). *Profile of teachers in the U.S. 2011.* Washington, DC: National Center for Education Information.

Feuerverger, G. (1997). "On the edges of the map": A study of heritage language teachers in Toronto. *Teaching and Teacher Education, 13*(1), 39–53.

Flores, B. B., Keehn, S., & Pérez, B. (2002). Critical need for bilingual education teachers: The potentiality of *normalistas* and paraprofessionals. *Bilingual Research Journal, 26*(3), 501–524.

Freeman, Y. S., Freeman, D. E., & Mercuri, S. P. (2005). *Dual language essentials for teachers and administrators.* Portsmouth, NH: Heinemann.

Gomez, M. L., & Rodriguez, T. L. (2011). Imagining the knowledge, strength and skills of a Latina prospective teacher. *Teacher Education Quarterly, 38*(1), 127–146.

Guiner, L., & Torres, G. (2002). *The miner's canary: Enlisting race, resisting power, transforming democracy.* Cambridge, MA: Harvard University Press.

Harper, C. A., de Jong, E. J., & Platt, E. J. (2008). Marginalizing English as a second language teacher expertise: The exclusionary consequence of *No Child Left Behind. Language Policy, 7*(3), 267–284.

Harris, C. I. (1993). Whiteness as property. *Harvard Law Review, 106*(8), 1707–1791.

Ingersoll, R. M. (2003). *Who controls teachers' work: Power and accountability in America's school.* Cambridge, MA: Harvard University Press.

Ingersoll, R. M., & Connor, R. (2009, April). *What the national data tell us about minority and Black teacher turnover.* Paper presented at the annual meeting of the American Educational Research Association, San Diego, CA.

Ingersoll, R., & May, H. (2010). *Recruitment, retention, and the minority teacher shortage.* Philadelphia, PA: Consortium for Policy Research in Education, University of Pennsylvania and the Center for Research in the Interest of Underserved Students, University of California, Santa Cruz.

Jensen, R. (2005). *The heart of whiteness: Confronting race, racism, and white privilege.* San Francisco: City Lights Books.

Johnson, D. R., Wasserman, T. H., Yildirim, N., & Yonai, B. A. (2014). Examining the effects of stress and campus climate on the persistence of students of color and white students: An application of Bean and Eaton's psychological model of retention. *Research in Higher Education, 55*(1), 75–100.

Jones, C., & Maguire, M. (1998). Needed and wanted? The school experiences of some minority ethnic trainee teachers in the UK. *European Journal of Intercultural Studies, 9*(1), 79–91.

Klassen, T. R., & Carr, P. R. (1997). Different perceptions of race in education: Racial minority and white teachers. *Canadian Journal of Education, 22*(1), 67–81.

Kohli, R. (2016). Behind school doors: The impact of hostile racial climates on urban teachers of color. *Urban Education.* Advance online publication. doi:10.1177/0042085916636653.

Teaching at Public Schools   131

Ladson-Billings, G. (2005). *Beyond the big house: African American educators on teacher education*. New York: Teachers College Press.

Ladson-Billings, G., & Tate, W. F. (2017). Toward a critical race theory of education. In A. D. Dixon, C. K. Rousseau, & J. K. Donnor (Eds.), *Critical race theory in education: All God's children got a song* (pp. 11–31). New York: Routledge.

Lapayese, Y. V. (2007). Understanding and undermining the racio-economic agenda of No Child Left Behind: Using critical race methodology to investigate the labor of bilingual teachers. *Race, Ethnicity and Education*, 10(3), 309–321.

Levine-Rasky, C. (2014). White fear: Analyzing public objection to Toronto's Africentric school. *Race, Ethnicity, and Education*, 17(2), 202–218.

Liggett, T. (2010). "A little bit marginalized": The structural marginalization of English language teachers in urban and rural public schools. *Teaching Education*, 21(3), 217–232.

Lin, M. H., Kwan, V. S. Y., Cheung, A., & Fiske, S. T. (2005). Stereotype content model explains prejudice for an envied outgroup: Scale of anti-Asian American stereotypes. *PSPB*, 31(1), 34–37.

Lipsitz, G. (2006). *The possessive investment in whiteness: How white people profit from identity politics* (Revised & Expanded Ed.). Philadelphia: Temple University Press.

Matias, C. E. (2016). *Feeling white: Whiteness, emotionality, and education*. Rotterdam, The Netherlands: Sense Publishers.

McIntyre, A. (1997). *Making meaning of whiteness: Exploring racial identity with white teachers*. Albany, NY: State University of New York Press.

McLaughlin, M. W., & Talbert, J. E. (2001). *Professional communities and the work of high school teaching*. Chicago: University of Chicago Press.

Milner, H. R. IV, & Hoy, A. W. (2003). A case study of an African American teacher's self-efficacy, stereotype threat, and persistence. *Teaching and Teacher Education*, 19(2), 263–276.

Monzó, L. D., & McLaren, P. (2017). Foreword: Unleashed—whiteness as predatory culture. In T. M. Kennedy, J. I. Middleton, & K. Ratcliffe (Eds.), *Rhetorics of whiteness: Postracial hauntings in popular culture, social media, and education* (pp. xiii–xvii). Carbondale, IL: Southern Illinois University Press.

Ochoa, G. L. (2007). *Learning from Latino teachers*. San Francisco: Jossey-Bass.

Office of Superintendent of Public Instruction (OSPI). (2015). *Washington state report card*. a) Retrieved from http://reportcard.ospi.k12.wa.us/Summary.aspx?schoolId=3166&reportLevel=School&ye ar=2015-16&yrs=2015-16 b) Retrieved from http://reportcard.ospi.k12.wa.us/Summary.aspx?schoolId=3211&reportLevel=School&ye ar=2015-16&yrs=2015-16

Olsen, B., & Anderson, L. (2007). Courses of action: A qualitative investigation into urban teacher retention and career development. *Urban Education*, 42(1), 5–29.

Pennycook, A. (1998). *English and the discourses of colonialism*. New York: Routledge.

Plaut, V. C. (2014). Models of success in the academy. In S. A. Fryberg & E. J. Martínez (Eds.), *The truly diverse faculty: New dialogues in American higher education* (pp. 35–60). New York: Palgrave Macmillan.

Priestly, M. R., Biesta, G., & Robinson, S. (2015). *Teacher agency: An ecological approach*. New York: Bloomsbury Publishing.

Reyes, S. A., & Kleyn, T. (2010). *Teaching in 2 languages: A guide for K-12 bilingual educators*. Thousand Oaks, CA: Corwin.

Rivera, D. P., Forquer, E. E., & Rangel, R. (2010). Microaggressions and the life experience of Latina/o Americans. In D. W. Sue (Ed.), *Microaggressions and marginality: Manifestation, dynamics, and impact* (pp. 59–83). Hoboken, NJ: Wiley.

132   Teaching at Public Schools

Rogers-Ard, R., Knaus, C. B., Epstein, K. K., & Mayfield, K. (2012). Racial diversity sounds nice; Systems transformation? Not so much: Developing urban teachers of color. *Urban Education, 48*(3), 451–479.

Rueda, R., & Monzó, L. D. (2002). Apprenticeship for teaching: Professional development issues surrounding the collaborative relationship between teachers and paraeducators. *Teaching and Teacher Education, 18*(5), 502–521.

Sakash, K., & Chou, V. (2007). Increasing the supply of Latino bilingual teachers for the Chicago public schools. *Teacher Education Quarterly, 34*(4), 41–52.

Scheurich, J. (2000). *White antiracist scholarship within/against the academy.* Paper presented at the annual meeting of the American Education Research Association, New Orleans, LA.

Short, D. J., & Boyson, B. A. (2012). *Helping newcomer students succeed in secondary schools and beyond.* Washington, DC: Center for Applied Linguistics.

Shuck, G. (2006). Racializing the nonnative English speaker. *Journal of Language, Identity, and Education, 5*(4), 259–276.

Sosa, A. S., & Gonzales, F. (2002). *Teachers need teachers: An induction program for first year bilingual teachers.* Paper presented at the Annual Meeting of the National Association for Bilingual Education.

Taylor, P., Lopez, M. H., Martínez, J., & Velasco, G. (2012). *Language use among Latinos.* Pew Research Center. Retrieved from www.pewhispanic.org/2012/04/04/iv-language-use-among-latinos/

Thomas, W. P., & Collier, V. P. (2002). *A national study of school effectiveness for language minority students' long-term academic achievement.* Santa Cruz, CA: Center for Research on Education, Diversity and Excellence.

Trust, T., Krutka, D. G., & Carpenter, J. P. (2016). "Together we are better": Professional learning networks for teachers. *Computers and Education, 102*, 15–34.

Umansky, I. M., & Reardon, S. F. (2014). Reclassification patterns among Latino English learner students in bilingual, dual immersion, and English immersion classrooms. *American Educational Research Journal, 51*(5), 879–912.

Urrieta, L. Jr. (2009). *Working from within: Chicana and Chicano activist educators in whitestream schools.* Tucson, AZ: The University of Arizona Press.

U.S. Census. (2010). *State and county quick facts.* Retrieved from http://quickfacts.census.gov/qfd/states

U.S. Department of Education. (2001). *PL 107–110 The no child behind act of 2001.* Washington, DC: Author.

U.S. Department of Education. (2016). a) *Racial/ethnic enrollment in public schools.* Retrieved from https://nces.ed.gov/programs/coe/indicator_cge.asp b) *English language learners in public school.* Retrieved from https://nces.ed.gov/programs/coe/indicator_cgf.asp

Valdez, E. O. (2001). Winning the battle, losing the war: Bilingual teachers and post-proposition 227. *The Urban Review, 33*(3), 237–253.

Vance, B., Miller, S., Humphreys, S., & Reynolds, F. (1989). Sources and manifestations of occupational stress as reported by full-time teachers working in a BIA school. *Journal of American Indian Education, 28*(January), 21–31.

Waldschmidt, E. D. (2002). Bilingual interns' barriers to becoming teachers: At what cost do we diversify the teaching force? *Bilingual Research Journal, 26*(3), 537–561.

Walker, A., Shafer, J., & Liams, M. (2004). "Not in my classroom": Teachers' attitudes towards English language learners in the mainstream classroom. *NABE Journal of Research and Practice, 2*(1), 130–160.

Wei, M., Ku, T.-Y., & Liao, K. Y.-H. (2011). Minority stress and college persistence attitudes among African American, Asian American, and Latino students: Perception of university environment as a mediator. *Cultural Diversity & Ethnic Minority Psychology, 17*(2), 195–203.

Weisman, E. M. (2001). Bicultural identity and language attitudes: Perspectives of four Latina teachers. *Urban Education, 36*(2), 203–225.

Weisman, E. M., & Hansen, L. E. (2008). Student teaching in urban and suburban schools: Perspectives of Latino preservice teachers. *Urban Education, 43*(6), 653–670.

Wenger, K. J., Lubbes, T., Lazo, M., Azcarraga, I., Sharp, S., & Ernst-Slavit, G. (2004). Hidden teachers, invisible students: Lessons learned from exemplary bilingual paraprofessionals in secondary schools. *Teacher Education Quarterly, 31*(2), 89–111.

Williams, B., & Williams, S. (2006). Perceptions of African American male junior faculty on promotion and tenure: Implications for community building and social capital. *Teachers College Record, 108*(2), 287–315.

Yoon, I. H. (2012). The paradoxical nature of whiteness-at-work in the daily life of schools and teacher communities. *Race, Ethnicity and Education, 15*(5), 587–613.

5 Teaching With Warmth and Demands

Conventional wisdom says that students of color are more likely to excel academically when matched with teachers who share their race and ethnicity. This conventional wisdom is mainly based on the assumption that teachers of color bring to teaching first-hand knowledge about minority cultures and personal experiences as members of marginalized groups in the U.S., and that the shared knowledge and experiences enable them to build rapport between home and schools for students from marginalized communities (Cochran-Smith & Villegas, 2016). Furthermore, teachers of color are presumed to be able to function as role models for students of color and to monitor the latter's academic progress more appropriately due to their familiarity with the students' cultures (Bone & Slate, 2011; Maylor, 2009; Villegas & Irvine, 2010). These assumptions and presumptions are frequently used to promote the idea of recruiting more teachers of color to increase students of color's overall school engagement and academic achievement. Since the demographic mismatch, where white teachers consistently dominate the teaching force while the student population has become increasingly diverse, is unlikely to change, recruiting more teachers of color and providing students of color with cultural affirmation at school is not only reasonable but also imperative. It benefits students of color in all domains: academically, psychologically, and socially (Irizarry & Donaldson, 2012; Villegas & Lucas, 2004).

Effects of Teacher Demographics on Students' Learning

Broadly speaking, there are two ways to explain how teachers' demographics influence students' educational outcomes: passive and active. The passive teacher effects are "simply triggered by a teacher's racial, ethnic, or gender identity, not by explicit teacher behaviors" (Dee, 2005, p. 159). "Role model" effects are one of the passive teacher effects. Students may respond better to demographically similar role models by raising their motivations and expectations. It is possible that the mere presence of a teacher of color who shares the same-race and ethnicity background could excite and motivate students of color who are accustomed to being taught by white teachers,

Teaching With Warmth and Demands   135

and might yield positive effects. Passive teacher effects also include reducing race and ethnicity based "stereotype threats" (Steele, 1997; Steele & Aronson, 1995). When students perceive that they could be viewed through negative stereotypes by teachers, they may inadvertently lower their academic engagement and performance as well. On the other hand, when students are taught by a teacher who shares their race and ethnicity, they may not experience stereotype threats, and this situation is conducive to achievement.

Active teacher effects are unintended biases or different expectations for students with different demographic traits from their own (Cahnmann & Remillard, 2002; Ferguson, 2003). In general, teachers seem to evaluate students who share the same race and ethnicity more favorably than other students (Ehrenberg, Goldhaber, & Brewer, 1995; McGrady & Reynods, 2013; Ouazad, 2014), which directly influence students' perception of teachers and their level of academic engagement. Teachers' subjectivity affects a specific student-teacher relationship. Teachers of color may evaluate the behaviors and the ability of the same-race and ethnicity students more favorably than or differently from other teachers who do not share such traits. However, these positive effects are simultaneously constrained to a specific teacher in a specific class or grade, thus producing a temporary effect, compared to objective measures such as test scores which are the result of cumulative learning experiences that are influenced by several teachers in earlier grades.

In reality, the effect of the same-race and ethnicity teachers on the academic achievement of students of color is more complicated than the assumptions of the above premise. Whether or not the assignment of the teachers of same race and ethnicity yields to better student achievement has produced ambiguous and mixed evidence so far and yet to be fully resolved (Driessen, 2015; Egalite, Kisida, & Winters, 2015; Fryer & Levitt, 2004; Howsen & Trawick, 2007; McGrady & Reynolds, 2013). Given teachers of similar quality, students of color seem to benefit from teachers who share their race and ethnicity, but this conclusion is hardly decisive. Positive effects seem to affect low-performing students and students with lower-socioeconomic status the most (Dee, 2005; Egalite et al., 2015), but this occurs regardless of students' race and ethnicity. It was also found that a result of the effects of ethnic matching may vary depending on how we define "student success." The positive effects apply to subjective teacher evaluations of students to a much greater extent than objective achievement outcome measures, such as test scores (Driessen, 2015). Buttressing this point, Wright, Gottfried, and Le (2017) revealed that Latina/o English learners demonstrated much lower externalizing behaviors (e.g., aggressive, noncompliant, and/or oppositional) when paired with Latina/o teachers, but the same effect was not observed when paired with Spanish-speaking non-Latina/o teachers. Thus, the effect of the ethnic matching seems to have a bigger impact on students' socioemotional skills development than academic achievement.

136 Teaching With Warmth and Demands

Given the mixed evidence of the effects, it is fallacious to conclude that students of color automatically and always achieve better outcomes when taught by teachers of color who share the same race and ethnicity. And this hasty conclusion conveniently ignores the fact that teacher quality is the most determinant factor of student achievement. Even if taught by the same-race and ethnicity teacher, if the teacher is unable to produce quality teaching, there will be no learning, thus little achievement. I have seen teachers of color who taught the same-race and ethnicity students but lacked quality of teaching. As a result, these teachers did not produce desirable outcomes. Teachers who successfully taught African American students, according to Ladson-Billings (2009), included not only African American teachers but also teachers who did not share the same race and ethnicity with the students. A key to successful teaching was the use of culturally relevant pedagogy, which indicates high levels of teacher quality regardless of race and ethnicity.

At the same time, what is important to understand, and often missing in the discussions on the effects of ethnic matching, are the kinds of educational opportunities teachers of color provide to students of color of the same race and ethnicity. One must also analyze the kinds of classroom environments, and how these factors are driven by "systematic differences in teacher quality" (Egalite et al., 2015, p. 50) that may include factors that vary by a teacher's race and ethnicity. For example, teachers of color may create a different classroom environment from white teachers because of their past experiences, beliefs, and motivations that are fundamentally different from white teachers. In other words, we should pay more attention to how teachers of color perceive, teach, and interact with students of color who share the same-race and ethnicity background in a classroom and how their perceptions, teaching styles, and interactions are culturally specific and affirming, and different from those of teachers who do not share the same traits. With this in mind, this chapter assesses Erika and Raquel's quality of teaching. How did they teach Latina/o students? What is it that makes a difference in Latina/o students' school engagement?

Trusting Relationships

How teachers interact with students can influence students' attitudes, and subsequently their behaviors (Blad, 2017). Growing research shows that social trust, mutual expectations, and shared values among teachers, students, and parents is critically related to student achievement (Dika & Singh, 2002; Shoji, Haskin, Rangel, & Sorensen, 2014). For example, when students believe that teachers are neither fair nor caring, students disengage themselves from learning when taught by such teachers. On the contrary, when interactions between students and teachers show that the latter hold higher expectations of the former and firmly believe in the former's potential, better outcomes occurs because a trusting relationship has been

Teaching With Warmth and Demands 137

established. Student trust in teachers has positive effects on students, while mistrust in teachers has lasting negative effects.

Schools are comprised of multiple stakeholders. Each stakeholder, students, teachers, administrators, and parents, socially interacts with the others at an organization called school. During these interactions distinct relationships emerge: teachers with students, teachers with colleagues and administrators, teachers and parents, students with other students, and the like. Each party in a relationship maintains an understanding of his or her role obligations and holds some expectations about the obligations of the other parties. Bryk and Schneider (2002) call this mutually dependent obligatory relationship "relational trust" and contend that for a school community to work well, it must achieve agreement in each role relationship in terms of the understanding held about these personal obligations and expectations of others. It is predicted that the higher the level of relational trust between teachers and students and parents, the higher the level of student engagement in school. In sum, how teachers relate to students influences and shapes the latter's educational opportunities and growth.

It appears that Erika and Raquel effectively taught their Latina/o students primarily because both were able to establish trusting relationships with their students and parents. How did both women build trusting relationships and what were they like? These questions will be addressed in what follows below.

Cultivating a Family-Like Relationship

With regard to academic aspirations and achievement of Latina/o students, it is imperative to consider the impact of families. Studies consistently found that Latina/o parents and their expectations and *consejos*, or advice, are an important bridge to student success in education and beyond (Behnke, Piercy, & Diversi, 2004; Gomez, 2010; Ocasio, 2014). The emphasis on strong ties among families—*familismo*—is an essential cultural value among Latinas/os (Villalba, 2007), but these ties are not necessarily confined to immediate family only, such as parents and siblings. They usually include "extended families" which may include aunts, uncles, grandparents, and friends who are considered to be part of one's *familia*. These tight kinship provides Latina/o children with the importance of maintaining a healthy connection to the community and its resources (Yosso, 2017). It also provides Latina/o children with opportunities to adopt social behaviors, linguistic conventions, and cognitive proficiencies that reflect the norms of their culture (Sameroff & Fiese, 2000). In this sense, it seems that both Erika and Raquel saw themselves as being part of their students' extended family. Their teaching philosophy certainly reflects this point. Erika mentioned, "My teaching philosophy is to build personal relationships with students." Raquel concurred: "It's important to build a family-like relationship with students."

138   *Teaching With Warmth and Demands*

The establishment of a family-like relationship with students is indeed one of the key interactional styles that Latina/o paraprofessionals and teachers frequently exhibit. For example, Rueda, Monzó, and Higareda (2004) found that Latina bilingual paraprofessionals interacted with students with *cariño* (caring) by using Spanish language terms of endearment. Erika and Raquel's use of the Spanish terms "*mi hijo*" and "*mi hija*" in classroom was culturally effective, and helped to create a family-like classroom environment. Raquel rationalized her use of the terms:

> I would say, "*mi hijo* or *mi hija*" which means my child. It just comes out. "You know, *mi hijo*, you have to do it this way." It's what my Mom and Dad always call us. My students have called me Mom. And grandma. I am a father, too. I've been called Papa. Very close, like a family.

Erika was the same. When I observed Erika's teaching one day, her class was learning a concept of fraction. Erika first made jam and butter sandwiches in front of the class, then began cutting the sandwiches into pieces. Then she explained, "If I cut this way, we can make two half sandwiches. Oh, *mi hijo*, could you please hand me the plastic knife over there?" Immediately, the boy who was referred to handed out the knife to Erika as told. From the observer's viewpoint, this interaction was nothing but natural.

The effectiveness of the Spanish terms "*mi hijo*" and "*mi hija*" lie not so much on the terms themselves but on Erika and Raquel's students beginning to see their teachers as approachable like their own parents. In a culture where adults and professionals in the community are highly esteemed—*respeto* (Villalba, 2007), having a teacher who conveys the warmness of their parents instead of a teacher who demonstrates intimidation and distance bolsters psychological closeness between the teacher and the students. Feeling close to the teacher, the students feel at ease and increase the interactions with the teacher by asking questions, discussing, and just casually chatting. These friendly interactions, when repeated, ultimately lead to deep trust towards each other. The deep trust Erika and Raquel gained from their students can be symbolized by a comment Raquel made: "You're going to make me cry . . . They bring me bouquets of flowers. You know, my room has three or four bouquets of flowers. I'm like, 'What's this?' 'Just for you, *maestra*.' I'm like, 'Oh, my goodness.' That's what you call valuing."

Erika disclosed the level of her approachability among her students. As seen in Chapter Four, in the first year of her teaching, Erika used to spend every weekend at her school trying to manage the heavy preparation that was necessary in order to make her classes successful. Every time she entered her empty classroom the first thing she noticed was a red blinking light on the desk telephone. Erika recalled:

> I go into the room. There are messages on my phone. They're from the kids: "Ms. Soto, are you in the classroom?" Or a phone rings as soon

as I enter. The kids around the neighborhood call me, "Can we see you? Can we help you?" I say, "No guys, it's Saturday. You can't come in the classroom and help me." But they're playing on the playground. They see me enter the building and want to come help. They would go home, call me in the classroom and ask. That makes me feel that I'm so special to them. Whether Mrs. Davenport (assistant principal) thinks I'm a secretary or not, I don't care!

The perceived approachability of Erika among her students helped the students interact with her more, and as a result made her feel that she was valued. It is reasonable to assume that she could have been wanting to hear someone recognize her hard work in the absence of such recognition and appreciation by the school administrators and her colleagues in particular. Students regarding their teacher as approachable and the teacher feeling valued by students naturally enhances a sense of connection and reciprocity between them. In Erika's case, this sense must have intensified because of the lack of support she experienced at her school.

It is obvious that both women juxtaposed their students' trust towards them with their colleagues and administrators' lack of support. Their administrators and colleagues were generally disrespectful towards both women, rarely acknowledging their successes with students, which was the fruit born of extremely hard work, preparation, knowledge, and caring. Raquel, for example, described trusting relationships with her students as "Look at me crying! Look at the tears in my eyes! It's wonderful. When I'm talking about my students, I'm happy." However, immediately after saying this, she did not forget to add: "When I'm talking about the politics that go on at school, I don't like it at all." Feeling their skills being explicitly exploited but never receiving recognition and acknowledgement they thought they deserved at school, it seems that both Erika and Raquel were unofficially comforted by their students' deep trust towards them. In other words, a feeling of trust was reciprocal between Erika and Raquel and their students. Erika bitterly described her tribulations and lack of recognition:

> I had to go through so many trainings (to transition from Spanish to English). But no acknowledgement like "You're doing things that have not been asked. Thank you." I told both Mr. Logan (the principal) and Mrs. Davenport that I don't get that from them. But when I have parents from two years ago who still call me, talk to me, or just to say "hello," I said to them (Mr. Logan and Mrs. Davenport) that that is a validation for me and that that is more important to me . . .

Clearly, both women perceived their students and their families as their strong allies while simultaneously positioning their colleagues and administrators at school at the opposite end (but not necessarily as enemies). This

140   Teaching With Warmth and Demands

juxtaposed positioning is similar to the counterspace Erika and Raquel harbored during teacher education as opposed to the not-so-friendly, in fact sometimes openly hostile, peers and instructors. In their own counterspace they were able to unite and affirm their cultures and resist oppressive practices.

Empowering and Being Empowered by Parents

A much stronger counterspace, not in a sense of a physical space but in a psychological sense, naturally emerged with the cooperation of the students' parents. As part of the extended families of the students, both women frequently communicated with the students' parents. And their Spanish proficiency was maximized for this purpose. Erika explained: "I communicate with their parents in Spanish. I was very fortunate this year because I had an awesome group of parents. I got close to a lot of them and really got to know their families." Raquel followed:

> I have a good relationship with my students and their parents. We have good communication, almost every day in Spanish. I let them know what's going on, so they are aware. Like, "How is your child doing at home? Is she enjoying school? Are there any problems?" So they don't get surprises later.

These frequent communications with the parents were effective in building deep, trusting relationships. Low-income, Latina/o, and immigrant parents often experience cultural dissonance and discomfort in interacting with their children's teachers and schools (Ramirez, 2003; Stanton-Salazar, 2001). The parents may not be fluent in English, may not be knowledgeable about how schools operate in the U.S., and may feel uncomfortable and afraid when discussing their children's education issues to those whom they perceive have authority over them. Strained parent-teacher and school interactions disadvantage students since they lack strong school-based relationships (Stanton-Salazar, 2001; Valenzuela, 1999), while strong school-based relationships facilitate the success of students within academic contexts.

Erika and Raquel seem to have become institutional agents who could serve as a bridge between parents and schools. The effectiveness of their frequent communications with parents seems to be characterized by their responsiveness and reciprocity. Shoji et al. (2014) define responsive communication as communication in which the listener(s) react readily and with interest and enthusiasm. This type of communication is vital in building a trusting relationship. Raquel, for example, explained how she managed to connect with Jorge's family by being responsive to his problems. Further, his parents advised her as well.

Jorge was a chaotic boy. He would go anywhere to anybody, climb on top of the chair, take down his pants and his underwear, and start mooning

Teaching With Warmth and Demands   141

his classmates and teachers at school. He would further poke other students and even kicked a teacher one time. Raquel called Jorge's parents almost every day and with them tried to figure out his problems:

> I would call them up and tell them, "Okay, this is what happened today. Has his diet changed recently? Is he having a trouble reading? Why does he do that? Some boys like to do that." Then, we talked to each other trying to figure out. His parents gave me lots of ideas to deal with the problem, too.

With daily correspondence with his parents, Jorge's mooning behavior gradually disappeared and he started to behave at school. Raquel recalls:

> One day, I called the parents and said to them, "He did really good today." Then, I heard Jorge telling his parents, "Momma, I did good today!" I said to the parents, "Thank you for supporting Jorge and working on disciplining him at home. I needed your help." It was nice of his parents to respond back to me, "Thank you for taking care of Jorge." Super nice feeling!

In dealing with Jorge's case, Raquel not only responded to his problem with a sincere concern as his teacher but also proactively sought for his parents' assistance with which his parents willingly provided. In a culture where parents consider teachers as having authority and respecting the authority of a school to do its work while they maintain active involvement in the life of a child at home is a norm (Delgado-Gaitan, 2004), the act of advising and making suggestions to a teacher could have been viewed as interfering with the work of a school (Walker, Ice, Hoover-Dempsey, & Sandler, 2011). However, by intentionally including Jorge's parents and allowing them to advise her, Raquel made it possible for the reciprocal social exchanges to take place, and these exchanges were equal in status. This social environment seems to have enhanced a sense of connection and shared identity among Jorge's parents. Raquel reinforced the recognition of the important role each person played (Bryk & Schneider, 2002), thus deepened their trust with each other.

Erika and Raquel were both particularly appreciative of the explicit recognition and acknowledgements they received from their students' parents. It certainly left them feeling emotionally supported and affirmed as teachers. More importantly, it gave them a stronger incentive to navigate the sometimes hostile work environment towards their most important goal: educating Latina/o students. Erika detailed one incident where she felt her work was validated by a student's parents:

> I had frequent phone conversations with one boy's mom. She even needed to talk to a counselor at school regarding his behavior problems.

142　*Teaching With Warmth and Demands*

She said to me, "When he doesn't behave on weekends, I will tell him, 'I'll call Ms. Soto on Monday morning.' Then, he would straighten himself up." They actually moved out of our school area during Christmas and he was supposed to go to the new school in a new area. According to the Mom, he said, "No, I'm not going. Ms. Soto won't be my teacher." So the Mom drove him across town every day because he wanted to be in my class. If she had switched him, he would have gotten suspended and I'm sure he would have. He could do it. He would have got himself into trouble. He himself said he was going to. I had that type of relationship. The Mom said, "Hugo has two Moms, me and Ms. Soto." I was like, I don't care about the rest. This is what I'm here for. I'm here for the families and I'm here for the kids.

Being explicitly appreciated for the work she did for Hugo, Erika felt flattered. It is important to understand that the compliment of Hugo's mom was rooted in the mutual trust Erika and Hugo's family had gradually cultivated. Otherwise, no parent would call their child's teacher another mom. It is also essential to notice that the compliment of Hugo's mom affected Erika to the degree that she was determined to better educate Latina/o students. In this sense, both sides—the teacher and the parents—mutually empowered each other and were empowered by each, and created a dependent social environment. This type of social environment fostered trustworthiness, mutual expectations and obligations, and positive affective feelings toward others in the group (Molm, 2010). It certainly cultivated a strong sense of a learning community with the students and the parents and strengthened family-school relations, which has been identified as a way to close demographic gaps in achievement and maximize students' potential (Hill & Tyson, 2009).

A juxtaposition between the absence of recognition and appreciation of their work at school and on the contrary, the vivid presence of affirmation and trust from the parents, may have aggravated Erika and Raquel's feeling of isolation from her administrators and colleagues at school. It could have also escalated both women's feeling of resentment towards their colleagues and administrators, in particular since the obvious differences were visibly sharp. For example, while discussing how much her students trusted her, Raquel's wish for recognition from the school was evident:

One parent said to me the other day, "I just want to let you know, Mrs. Ortiz, that my child is not afraid to come and talk to you. I always encourage my child to talk to you first when something happens. And she says, 'Okay, I will talk to Mrs. Ortiz myself.' My daughter has so much confidence in you." This tells me a lot and makes me trust them as well. Parents value me and students value me. But the school has no idea of what I do. I don't hate my school. We're so busy every day. But, people skills, like how to communicate and how to listen is important.

Teaching With Warmth and Demands 143

Raquel's reflection above indicates that she appreciated the fact that her student had high confidence in her and that this feeling was sharply contrasted with the school's nonappreciation of her work. Because the students and their families highly valued her teaching, Raquel's wish for more communication with the school escalated and, which increased her feelings of isolation.

Erika shared a similar feeling of disappointment and isolation. As was briefly mentioned in Chapter Four, in preparation for a field trip to a nearby radio station in town, Erika planned a fund-raising event in which her students made homemade tortillas and sold them. The students, their parents, their siblings, and cousins all gathered in her classroom after school one day and made tortillas together. The classroom was so packed with people that they could barely move. Despite the fact that Erika announced this event to both Mr. Logan and Mrs. Davenport beforehand, to her disappointment, neither of them showed up to greet the participants on the day of the event. Erika's comment below underlined her frustration and disappointment:

> They didn't even come to support us! It was after school and on a school day. There were so many people in my classroom. Our presence was so visible on the school premises. Everyone at school knew what was happening. They surely could have come. That hurt my feeling deeply.

Erika's presence at the event went beyond her official obligations, which the parents must have interpreted as a sign of genuine caring. Expressions of personal regard such as this promote a sense of self-worth in others and mitigate uncertainty and dependence inherent in social exchange, thus foster reciprocation and trust between actors (Blau, 2002; Bryk & Schneider, 2002). On the other hand, the absence of such personal regard can exacerbate negativity in a relationship. Although there could have been multiple reasons why neither Mr. Logan nor Mrs. Davenport attended the event to show support, from Erika's perspective, their physical absence was a clear indication that the school did not care about her students, their family members, and her efforts as a teacher at the school. Since Erika made an effort to connect the parents to the school as an institutional agent who possessed "knowledge of, access to, or control over institutional resources" (Shoji et al., 2014, p. 607), the event provided the parents with a perfect opportunity to enhance their social networks at school by interacting with the school administrators. It certainly solidified the trusting relationships between Erika and the students and their families. However, it also demonstrated how difficult it was for Erika to penetrate the school's social network without minimal collaboration, even a brief moment of greeting of visitors and acknowledgment of Erika's efforts, from the administrators.

144   *Teaching With Warmth and Demands*

Discipline and Excellence

While cultivating a family-like relationship with the students and extending a trusting network with parents, both Erika and Raquel had one goal in mind: students' academic engagement and achievement. In order to achieve this goal both women focused heavily on establishing and enforcing rules in the classroom and delivering higher-academic standards in their lessons.

Discipline

Novice teachers are frequently overwhelmed with classroom discipline. For example, Onafowora (2005) found that novice teachers portray themselves as being challenged by discipline issues which overshadow instruction issues. However, this was not the case for Erika and Raquel. First of all, both women had already had considerable classroom teaching experiences as former paraprofessionals. Their classroom teaching had also been consistently supervised by teachers. Therefore, both Erika and Raquel were already equipped with the key skills needed for teaching and disciplining students. Furthermore, they had additional skills which monolingual white classroom teachers lacked: linguistic abilities and cultural knowledge. Both their teaching experiences and the shared cultural and linguistic knowledge they had with the students were advantageous with regard to classroom management. In addition, the lack of discipline problems in both women's classrooms had more to do with Erika and Raquel's establishment of fair and reasonable rules and consistent enforcement of such rules. Their efforts in this regard were all made possible because of the trusting relationships they cultivated with the students.

Raquel

In terms of teaching styles, Raquel could be characterized as being rule-governed and strict. Raquel rationalized her strictness in the following way:

> I guess I do some of the things that Mom might do like giving the look, you know. Yeah, we Latinas are very strict. Very, very strict, you know . . . in academics. We want to make rules. We want them to succeed. I want them to succeed. When they are in my classroom, I put them to work. "You get to work. You're going to do this."

According to Raquel, her strictness mirrored how she was brought up. She was brought up to aim for the best and never give up by the parents who emphasized the importance of education and work ethic. Raquel reflected:

> It's just the way I was brought up. You got to do work. You got to do your work well. You know it's just my background. If I'm going to do a job, I have to do it well. You know I just can't leave it there. Half way . . . no. I'm

Teaching With Warmth and Demands 145

going to get it done and I'm going to do it well. That is how I was brought up. You have to do the best you can.

The goal-oriented stance her parents instilled in her while growing up was part of the familial capital that surrounded her. Raquel benefitted from the goal-orientation that her parents brought her up with, and tried to instill the same in her students through her teaching.

Interestingly, Raquel's strict, "work-hard" stance was met without resistance. In fact, it was well-received by her Latina/o students. Referring to Jorge who used to be disruptive, moon, and be chaotic, Raquel described how he behaved in her classroom:

When he was in my classroom, I put him to work saying in Spanish, "Jorge, you need to work. You're going to do this." And he did. When he was in my classroom, he never did mooning or being chaotic at all. He's a pretty busy, pretty hyper boy. If I keep him busy, he'll be fine. I don't know if he's bored in other classrooms. He knows he has to behave in my class.

Jorge responded to Raquel's commands with obedience, while he continued to act in disruptive ways in the classrooms of other teachers.

It is important to differentiate the term, "authoritarian" from another term, "authoritative." An authoritarian teacher demands total obedience even if she/he is unfair or obeying the teacher could cause a loss of personal freedom. On the other hand, an authoritative teacher expects respect and demands total obedience from students, but students can trust the teacher as legitimate, fair, and doing right things for the students. Past research has shown that when students perceive a teacher to be fair, the former tend to be less defiant and more cooperative, and that African American, Latina/o, and Native American students in particular respond well to an authoritative approach (Gregory & Thompson, 2010; Turiel, 2005). Ultimately, authoritative approaches to classroom management can nurture trusting and positive student-teacher interactions in many classrooms (Gregory, Clawson, Davis, & Gerewitz, 2016), especially at the elementary school level. Fundamentally, whether or not students and teachers have trusting relationships influence whether or not an authoritative approach becomes successful. In this sense, the trusting relationships Raquel cultivated with her students worked in her favor. The trust between Raquel and her students and their families was mutual, reciprocal, and dependent, and the students treated her like their extended family member. This made it possible for students to naturally trust Raquel's authority and obey whatever she demanded as a matter of respect.

Erika

At Erika's school, teachers usually sent misbehaving students to the principal's office. However, Erika resisted to this school-wide practice and avoided sending the students to the principal's office as much as possible. Instead,

146 *Teaching With Warmth and Demands*

she had deep one-on-one discussions with the students whenever disciplinary issues occurred. Erika rationalized her action and proudly declared:

> I didn't have one kid suspended in my class at the end of the first year. I didn't have any behavioral issues in my class. Teachers need to know that we can have a good relationship with even the hardest kids. Sending them to the principal's office just ruins that relationship.

She was proud to run a disciplined classroom with no disruption where her students were respectful, worked hard, but also had fun.

Students immediately sense whether or not they can trust their teachers or not, and act accordingly. The aforementioned example of Jorge in Raquel's class exemplifies the level of trust he felt towards Raquel and the lack of it towards other teachers. That is why Jorge misbehaved badly in other teachers' classes, while the opposite was true with Raquel. When Erika was teaching in the third year, her reputation as a disciplinarian was finally recognized by the administrator. One day, Erika received a call from Mr. Logan.

Mr. L.: "Didn't you have Jose and Juan in your class three years ago? Were they ever disrespectful to you?"

Erika: "Are you really going to make me answer that? The answer is No."

Mr. L.: "Did they ever get in trouble with you?"

Erika: "No."

Mr. L.: "Okay, that's strange. They're having huge troubles with their new teacher."

Erika: "Well, the only teacher they got in trouble with was the music teacher. But others all got in trouble with him, too."

Mr. L.: "When Jose and Juan were in your class, they knew how to behave, right?"

Erika: "Yes. Are Jose and Juan with you now? Then, can you send them up to my room and we will have a talk."

Mr. L.: "Okay. I will."

With a smile, Jose and Juan immediately rushed up to Erika's room expecting to be disciplined by her. After successfully convincing them to behave in their teacher's classroom, Erika let both students return to their homeroom and reported it to Mr. Logan. Erika confessed that at that time she felt nothing but honored: "Mr. Logan acknowledged the fact that I run a very disciplined classroom." Although the recognition of her disciplinary skills did not come directly with words of compliments from Mr. Logan, the reality that he made Erika handle disciplining Jose and Juan spoke the truth. It was two and a half years after Erika began teaching at Lincoln Elementary School.

Erika certainly knew that her trust in students was the key to her and their successes in her class and this perception was obviously shared

Teaching With Warmth and Demands   147

by her students. She declared in a firm voice, "The kids learned really fast in my class that they had my trust to begin with and if they lost it, it was not a good thing. They learned that within the first two weeks."

One of Erika's effective discipline strategies was that the students made a call home themselves while Erika listened attentively. This strategy naturally required the students to remember their home phone numbers. Erika continued:

> My kids entered my classroom not knowing their phone numbers. Third-grade. How do they not know their phone number? But by the end of the year, every single one of them had it written down in their notebooks or knew it by memory because they knew on any given day, it was their turn to call whether it would be good or bad. Most of the time it was good, but once in a while, not. So, the kids learned not only that I celebrate when they do something good, but also the important fact that they need to know their home phone numbers or the parents' cell phone numbers. When they tried not to tell the truth to their parents, they were reminded that I was ear to ear on what they said. And they knew after school I would be making phone calls to their parents anyway.

From the students' perspective, maintaining a trusting relationship with Erika required them to keep up with the expectations set by her. They must follow whatever she demanded without questioning her motives. The students knew that as long as they trusted Erika and followed her commands, they would be fine at school. Since Erika's demands were sometimes challenging, the students must have felt some tension in the classroom. However, the tension was most likely followed with the students feeling confident in and comfortable with whatever expectations Erika set for them. Apparently, Erika's students regarded her as a "warm demander" (Kleinfeld, 1975) who provided "a tough-minded, no-nonsense, structured, and disciplined classroom environment" (Irvine & Fraser, 1998, p. 1), and this approach was culturally congruent with her students.

Excellence

Since neither woman needed to spend much time disciplining students compared to other novice teachers who usually needed to spare the majority of their instructional time for discipline, they were able to concentrate on teaching.

Simply put, both women's academic expectations were high. The following comment made by Raquel symbolizes their high expectations: "I need to teach these kids. These kids need to learn. I need to teach them. I need to show them how to achieve their goals. They need to feel their success at school." Raquel's rather urgent-sounding high expectations were reciprocated by her students. Her partner teacher frequently remarked with awe,

148   *Teaching With Warmth and Demands*

"I don't know what you do with these kids, but they just want to learn." Students in general favor teachers who believe in their ability to succeed (Curwin, 2013). Because their students respected and trusted their authority as teachers, it seems that the students took Erika and Raquel's high expectations of them for granted.

Raquel

Raquel's high expectations were most vividly demonstrated through her teaching of higher-than-grade-level content to her kindergarteners. Raquel was seriously concerned about the level of the kindergarten curriculum at Washington Elementary School. Besides lacking teaching materials in Spanish, Raquel found the expectations set by the school for kindergarteners were both too low and unacceptable. She explained:

> It's very difficult because the expectations are very low for the kinder. What I'm wanting is for the kinders to do is basically the first-grade. I feel like I'm teaching the first grade to the kinders. Because you know, we do a lot of work. They have to meet their expectations. I don't know if it's just me who wants them to learn and wants them to know or if that's part of the pressure we have from school. "Oh, they need to meet this by the end of the year or we need to see these improvements." But I do feel it's more like the first-grade than kindergarten.

As we saw in Chapter Four, Raquel felt pressured to prove her teaching talents to her colleagues and the administrators simply because her hiring was perceived to be the result of her Spanish language skills.

> On one morning in late April, Raquel's lesson began with carpet time. Twenty-five Latina/o kindergarteners sat all together in front of Raquel. Raquel held up one card of alphabet letter after another and pronounced the sound of each letter in Spanish followed by the students spontaneously blurting out the words that began with the particular sounds of the letters. After this review session was complete, all the cards were distributed to the students with each student holding one card. Then Raquel pronounced one consonant sound, "n." Immediately, one student who held the corresponding letter card stood up. Next, she pronounced a vowel sound, "a." Again, immediately another student who held the corresponding vowel sound stood up. Raquel then told the class to pay attention to the student pairs. The two students began physically getting closer and finally stood side by side. Raquel then said to the class, "'N' and 'a' together make a 'na.'" The class continued with other consonant-vowel combinations, such as "re," "mu," and "du." The students genuinely looked like they were having fun with this activity.

Then, the students were broken into three groups: worksheets, writing, and reading. One group filled out the worksheets on letters and sounds. In another group, a paraprofessional read out words in Spanish and the students were instructed to write down the first letter of the words. The final group was taught by Raquel. She read a short book out loud with a focus on particular sounds and letters for the day. While reading out loud, Raquel constantly instructed the students with a stern look, "Look! Point!" After rotating each of the mini-group lessons, the students were told to begin observing how their caterpillars had changed from the day before and to record their shape in a drawing. Each student had his/her own caterpillar in the classroom. They were all on task: watching and drawing the caterpillars. This science activity, however, took only five minutes, and Raquel did not specifically instruct the students what to look for in their caterpillars. Finally, Raquel instructed all students to gather and began reading a book to them. While reading out loud, Raquel gave the students one comprehension question after another.

It took approximately one and a half hours to complete all of these activities with each activity seamlessly proceeding. Most of the time, the students received direct instructions except for the time when they freely observed and drew caterpillars. Each activity was structured and sufficiently short so that the kindergarteners would not get bored. And none of them looked restless and bored. The students were all engaged in the tasks and did not chit-chat, either. In the end, the students all lined up neatly at the entrance door of the classroom for a recess. From a strictly outsider's perspective, while observing, I jotted down a following comment in my fieldwork notebook: "The curriculum level is high in her class. Is this really a kindergarten class?"

The pressure she felt from school would certainly have influenced her high expectations. It was also her genuine desire to give Latina/o students vigorous learning experiences because she did not experience this during her K-12 education. These experiences propelled Raquel to teach with higher-academic standards. Raquel passionately stated:

If somebody had taught me better, maybe I would have done better in kinder. I kind of look at it as "Okay, what did I miss out? Why didn't I have the same success?" You know, what can I do to make this better for these students so that they can have success, even the students who struggle. Because I struggled as well, I always think what I can do to help them out.

Deeply resentful of the inferior education she received at K-12, Raquel was determined to give the best to her own Latina/o students. Her desire not to duplicate the negative experiences she had as a student, hoping to create a

150   *Teaching With Warmth and Demands*

better future for the students she taught, resonates with other Latina/o teachers. Irizarry and Donaldson (2012), for example, observed that Latina/o teachers "consistently described teaching as a platform from which to challenge the injustices they experienced in schools [and want to] transform the institutions they viewed as critical to shaping the opportunities available to Latina/o youth" (p. 168).

Raquel's high expectations are based on her subjective experiences. In the lesson above, Raquel focused on teaching literacy because her own K-12 schooling reminded her of the utmost importance of developing higher-level literacy skills in any content areas for academic achievement at school. Raquel's science lesson (drawing caterpillars), for example, was criticized by an exchange student from Japan who accompanied me for this observation on that day. This student remarked, "What's the point of drawing a caterpillar without teaching the basics of how it morphs and why it morphs? Just drawing a caterpillar without such fundamental knowledge doesn't constitute real learning." From an educator's perspective, this Japanese student's statement has a merit. Although free observation and drawing could lead to students' making their own discovery, this possibility could also mean that the students may miss out important information about caterpillars if observed with less attention. Raquel did not dwell on the science activity on that day. Moreover, the students did not seem to have been given any direct instruction about caterpillars and butterflies previously. Jerald (2006) contends, "Cutting too deeply into social studies, science, and the arts imposes significant long-term costs on students, hampers reading comprehension and thinking skills, increases inequity, and makes the job of secondary level teachers that much harder" (p. 5). Raquel's teaching was indeed heavily and intentionally focused on literacy, which resulted in the students' spending only a little time on science without much direct instruction on science content.

Raquel's intentional focus on literacy lessons at the expense of a science lesson, however, were based on her subjective professional judgment in which she felt that she needed to expand her instructional time in reading to enable students to become fluent readers (Jerald, 2006). Her judgment most likely originated from her deep knowledge about the students' social and academic backgrounds. As a former second language learner who received inadequate and ineffective instruction in her native language, Spanish, Raquel knew how important it was for her Latina/o students to develop solid literary skills in Spanish so that such skills would transfer to those in English later. She lamented, "I know what it's like to suffer as a second language learner."

Apparently, her expectations were directly and indirectly communicated to the students and affected the students' own beliefs and attitudes towards learning as was evident in their test scores in DIBELS described in Chapter Four. Because Raquel regularly and frequently provided the students with the same high expectations of literacy development, it appears that the

Teaching With Warmth and Demands 151

students incorporated her expectations into their own beliefs, and gradually formed their attitudes towards educational attainment. Raquel's firm belief about her students' potential was clearly expressed:

> They can be whatever they want to be. They know that they can also go to college. They know they can learn to read and write, you know. Have the same opportunities that I have. They know that if they stay in school and if they work hard they'll be able to have their own cars or houses someday.

Teachers likely play an important role in shaping students' beliefs about their academic prospects (Burgess & Greaves, 2013; Dee, 2015). This is particularly true for relatively disadvantaged students who rarely interact with college-educated adults outside of school settings (Jussim & Harber, 2005; Lareau, 2011). In this sense, Raquel's high expectations, particularly with regard to literacy development, protected against or counteracted negative expectations created by a lack of access to educationally-successful role models (Gregory & Huang, 2013).

Erika

Erika displayed her high expectation for students with a strict, no-nonsense stance in an extremely structured lesson using the scripted curricula mandated by her school. According to Ede (2006), Title 1 schools, those in high-poverty and low-income communities adopt scripted curriculum more often than those in more affluent communities. Further, urban schools that often serve predominantly African American and Latina/o students, and those with high levels of poverty and those whose native language is not English are heavily populated with underprepared teachers. Consequently, students in these environments experience scripted curriculum (Darling-Hammond, 2010) because all these teachers have to do is to just follow the script while teaching, which is helpful if one is an underprepared or untalented teacher. Scripted curriculum is defined as "scripting the conduct of teachers" (Hlebowitsh, 2007, p. 28). Teachers follow a detailed and prescribed instructional model that requires them to teach with faithful attention to the script for each lesson (Carl, 2014).

Criticisms towards scripted curriculum are abundant. It deprives students from the exposure to a more sophisticated, complex curriculum (Milner, 2014). The materials are unappealing to students and the flow and the organization of the curriculum is incoherent and unresponsive to the particular idiosyncrasies of student needs (Smagorinsky, Lakly, & Johnson, 2002). Teachers feel deprived of autonomy (Carl, 2014) and feel they are turning into mechanical delivery systems (Kohl, 2009) or being reduced to actors on a stage reading a script (Sawyer, 2004). Despite these downsides of scripted curriculum, Erika felt that the scripted curricula she was required to use in

152 Teaching With Warmth and Demands

reading and math actually maximized her students' learning opportunities and effectively served its purpose at the lowest-performing school in the district where she taught.

On a sunny February afternoon at 12:30 p.m. right after lunch, Erika began her literacy lesson to the third-graders in English. The students were divided into four groups with each of three paraprofessionals being responsible for a group along with Erika. As a former paraprofessional herself, Erika's handling of them was succinct and efficient. After briefly instructing the paraprofessionals, letting them know what to do, Erika took one of the four groups. Her group consisted of five Latinos and three Latinas and they began a literacy lesson using the scripted curriculum. From the beginning, Erika was strict, with no smiles. Before the lesson began, she bluntly said to the students, "You guys were not nice to the substitute yesterday. What you guys did is not acceptable." Then, continued looking at each student's face, "Am I clear?"

The literacy lesson in English that consisted nothing but drills finally began. It was like a military cadet being trained by a Sargent. The students yelled out each letter of the word in chorus with perfect harmony, "often" as "O-F-T-E-N!" and recited, "Often!" immediately afterwards. The students repeated this recitation several times, then moved on to the new words, such as "slow," "rabbit," and "flat." Meanwhile, Erika strictly followed the script. While the students were reciting, she yelled out at Mateo, "Sit down properly!" Obviously the students had learned how to sound out the words before and looked relieved when Erika said, "Much better than before." Still no smile appeared from Erika.

After the above brief review of the words, the students were told to read aloud a paragraph in the text that contained the words they had just reviewed. Diego began reading aloud to the group. Then, Erika immediately stopped him and warned, "Take off your jacket. Breathe deeply. You guys need to pronounce the ending of each word clearly. Slow down." Diego took off his jacket and began reading aloud again, this time exaggerating the endings. Meanwhile, the rest of the group diligently followed Diego's oral reading with their fingers. When Diego couldn't pronounce certain words, Erika immediately asked the rest of the group, "What's that?" In immediate return, the group yelled out the words in chorus, again with perfect harmony.

Erika proclaimed, "So much better than before. Good. Now, I want you to read more fluently." (At this moment, my immediate reaction was, "Not again!") The students began reading aloud the same paragraph in chorus while Erika attentively listened to how each student pronounced the words and checked the individual student's pronunciation mistakes. In the middle, Erika stopped the students and asserted, "Don't guess! You guys are doing lots of guessing! You are making too many mistakes. You guys should know all of these words already."

Teaching With Warmth and Demands   153

Then, Erika gave each student a brief comment: "Camila reads pretty loud, good. Alejandro, you're not listening well. Angel, you can read loud, but you're choosing not to do so. You're wasting our time. Now, Isabella, begin! Continue!" This activity that contained nothing but repetitive and dull drills lasted for 50 minutes, non-stop.

Erika moved on giving comprehension questions about the paragraph they had just read out loud. By that time, having just finished lunch, the students began yawning one after another. Erika completely ignored their desire for a break which was displayed by yawning, and showered one comprehension question after another mercilessly: "Everybody, look at me! Follow me!" As soon as the questions were uttered, the students shouted out the answers in chorus. It was obvious that the students were drowsy. Ignoring their sleepy faces, Erika frequently and silently instructed each student to pay attention to the task on hand through gestures, such as pointing her finger at them, glaring at them, and using a more dramatic tone of voice, which effectively woke up the students. This comprehension check time lasted for 15 minutes.

At 1:35 p.m., Erika put the textbook down and sharply instructed: "You will do page 118 now!" With this concise direction only, the students knew what to do. They immediately opened the page in the textbook (there was one short paragraph followed by five comprehension questions) as told and began writing down the answers to the comprehension questions in their notebooks. While the students were silently writing down the answers, Erika began pulling one student after another for testing—another fluency reading. Erika pulled out Maria first and remarked encouraging comments before she started reading out loud a paragraph to Erika. After Maria was complete, another student was pulled until all the students were tested. Meanwhile, the rest of the group was engaged in the comprehension questions quietly and completely on task. Except for the student reading out loud one by one, the group was not making a sound.

By 1:50 p.m., the testing was complete for her group. Erika casually asked, "What page did I tell you to do? Continue!" She then began testing the students in other groups, while the students in her group kept doing the comprehension work. It was obvious that while testing other students in a different area of the classroom from where her group members were sitting, Erika's eyes were still on her group. She occasionally yelled out to her group: "Santiago, do you work!" Santiago replied, "Yes, *maestra*."

At 2:05 p.m., Erika returned to her own group. The group began another drill again, but this time the students in her group were more eager to participate. Holding the textbook in her left hand and snapping her fingers with her right hand, Erika began another drill with a rhythm.

154 Teaching With Warmth and Demands

Erika: "What's the first letter in 'combination?'"
Ss: "C!"
Erika: "What's the last letter in 'combination?'"
Ss: "N!"
Erika: "Spell 'wrong!'"
Ss: "W-R-O-N-G! Wrong!"
Erika: "Say Word 1! Say Word 2! Say Word 3! Get ready! Santiago, participate!"

When the bell rang indicating that it was time for a recess, the students looked physically exhausted and drained, but also relieved. Erika, however, was still in a strong spirit. As soon as the bell finished ringing, she commanded pointing at several students including her own group members, "You guys stay here! No recess!" Several students immediately came towards her with their heads deeply down. Erika continued with a warmer tone but with a serious look, "You guys are going to have a test on Friday, and you're way behind. I know you can do it when you set your mind properly. Let's continue while others are out." The students in her group (turned out to be the lowest-performing group) and some from other groups started another drills again. Deprived of a recess, the students genuinely looked sad and disappointed. However, none of them complained nor resisted Erika. On the contrary, they all began another drill with Erika's remark, "Let's begin!" It was 2:16 p.m.

Teaching using scripted curriculum could be dull, unappealing, uncreative, and mechanical. Teachers criticize scripted curriculum mainly because it deprives them of their autonomy and ability to exercise professional judgment (Carl, 2014). Unlike many teachers who lament that they want their more autonomy when they are required to use scripted curriculum, Erika did not seem much concerned about the lack of autonomy the scripted curriculum forced upon her. She was rather mostly concerned about her students' not meeting the academic standards. Erika claimed, "Socially I do a really good job for the students. But, academically, they still struggle." Teaching students at the lowest-performing school at the district where each year, two thirds of students (in some years, three fourths) do not meet the standards in literacy and math, Erika was seriously concerned about increasing the students' basic skills in reading and math rather than whether or not she was able to exercise greater autonomy in teaching. For this reason, Erika, although admitting that she did not agree with the nature of the scripted curriculum, felt that it served its purpose by allowing the students to spend expanded time in building up the basic skills because it narrowly and intentionally focused on developing discrete skills, such as phonemic awareness, phonics, fluency, vocabulary, and comprehension (Duncan-Owens, 2009; Yatvin, Weaver, & Garan, 2003) with which the students all struggled. Phonological and alphabetic skills will foster reading development (Vadasy &

Teaching With Warmth and Demands 155

Sanders, 2008). Without fully developed basic literacy skills, students will struggle with building fluency, which in turn will negatively impact comprehension (Torgesen, 2004).

Putting aside the argument of whether or not scripted curriculum produces a desirable educational outcome, it is evident that Erika's instructional skills were of high quality. Erika was indeed an authoritative teacher who showered one command after another in an assertive voice with no mercy, made blunt and sometimes harsh comments, rarely showed smiles, and kept the no-nonsense stance throughout. Her students, however, were never defiant nor resistant. Three reasons seem to explain the students' compliance.

First, Erika's teaching was highly structured, and the use of the scripted curricula in literacy and math actually enhanced the degree of its structure. This highly structured classroom environment Erika created was conducive to her Latina/o students' learning, in particular when accompanied with Erika's insistence on high expectations without excuses (Bondy & Ross, 2008; Ware, 2006). Due to the structure, Erika's students were able to concentrate on her explicit directions and anticipate what to do next, which eliminated uncertainties, reduced the students' anxiety level, and increased their engagement on the tasks with a feeling that they were learning. A highly structured learning environment is indeed found to be beneficial for struggling readers (Lindo, Weiser, Cheatham, & Allor, 2017).

Second, the teacher-centered direct discourse style Erika employed, which included expressing assertive and blunt comments, repeating requests, and delivering warnings and consequences (Bondy, Ross, Gallingane, & Hambacher, 2007; Brown, 2004; Ware, 2006) was culturally more synchronous with the style with which the students were familiar at home. Studies found that Latina/o parents use authoritative practices more frequently than white parents (Varela et al., 2004), such as conversing with children about choices and consequences (Calzada, Huang, Anicama, Fernandez, & Brotman, 2012). An important Latina/o cultural orientation, *familismo*, has been said to promote more controlling dispositions among Latina/o parents (Halgunseth, Ispa, & Rudy, 2006). Erika's discourse style made a direct contrast to that of most white female teachers who prefer the soft-spoken, nonconfrontational, indirect speech (Brantlinger, Morton, & Washburn, 1999; Thompson, 2004). Although her direct discourse style could be perceived as mean and harsh to cultural others, it is apparent that Erika's students did not perceive so. That is why they did not even make an attempt to complain when their recess was taken away by Erika.

Third, the students regarded Erika's authority as absolute, which made only possible because of the shared culture, history, and a frame of reference in the trusting relationship. As Ford and Sassi (2014) claim, race and ethnicity play a critical role in building teacher-student authority relationships in classrooms. And Erika certainly knew that her sharing the same race and ethnicity with the students made a difference in how they perceived her authority as a teacher. She explained, "I really think my race and ethnicity

156 *Teaching With Warmth and Demands*

do matter. To my students, it meant a lot because I would always say, 'You can do this. Don't use anything as an excuse. If I can do it, you can do it.'" However, the students' admiration for Erika seems to have been triggered not only from the passive "role model" teacher effects but also the active teacher effects that included how Erika subjectively interacted with the students.

Erika being the same race and ethnicity with the students was certainly a precursor for gaining trust from the students. As the trusting relationship grew among Erika, the students, and the parents, her reputation as a teacher who held effective balance of discipline and care, held high expectations, and used a culturally congruent interactional and discourse style also grew. Furthermore, Erika explicitly and intentionally made her belief open to the students and the parents that she shared the same knowledge, customs, values, languages, norms, behaviors, and experiences with the students and the parents. For example, Erika proclaimed:

> I think just my little experience of working in the field and working in the warehouse, also listening to those in my community, my aunts, my family friends that work in that environment have made me more realistic. Sensitivity, well I don't think that other teachers will understand what it's like to go work in the field. They won't understand the labor. Sure we want parent participation and parent this and parent that, but yet we also have to understand that at our local warehouses, parents start at six in the morning and during harvest they don't get home until nine or ten p.m. So they're gone during harvest in the warehouses, in the fields they may start at five a.m., get off at two or three p.m. depending on which orchards they work for, yet the labor is so intense for that. I don't think that they understand what it is to work out in the fields and to be part of the migrants. My relationship with the kids was developed to another level because I also understood where they were coming from. I understand working in a warehouse. I understand working on the fields. I understand those things. So, I think having that perspective helped me gain the respect of the parents as well as, the children's.

Erika's deep understanding of the collective struggles Latinas/os in her community encountered must have been perceived as a sign of displaying authentic care. Caring relationships Erika cultivated with the students and the parents based on her subjective experiences made it possible for Erika to solidify her authority in her classroom.

The following episode illustrates what caring meant to Erika:

> I had two kids in my Spanish class. Neither of them, one of them in particular, was in retention. He had never had Spanish instruction before, even though he spoke Spanish. He should have never been placed in my class. So that was really hard. He wrote me a note, "Mrs. Soto, I want

Teaching With Warmth and Demands 157

you to teach me in Spanish." He wanted to be in my classroom badly. I said, "*Mi hijo*, you don't need to be in the Spanish class. You need to be in English and you need to show them what you can do." He crossed his arms and pouted. But I insisted he be in English.

Despite the student's desperate plea to be taught by her, Erika mercilessly sent the student to the English class because he had not developed literacy skills in Spanish before. Her decision was based on her professional judgment that he would not need instruction in Spanish and should continue to be taught in English. To Erika, "caring" meant acting for the academic benefit of students even if it may cause dissatisfaction among students.

Teacher Quality

Teachers of high quality produce favorable educational outcomes. Traditionally, research on teacher quality has been focused on two factors: the cognitive capabilities with which people enter the teaching career (Kennedy, Ahn, & Choi, 2008), and professional knowledge acquired during teacher education (Shulman, 1987). The former posits that bright and well-educated people make the best teachers because they are smart and thoughtful enough to figure out the nuances of teaching in the process of doing it (Kennedy et al., 2008). The study conducted by Kunter et al(2013), however, refutes this hypothesis and found that teachers' general academic ability did not affect their instruction. Professional knowledge seems to be more important than general attributes to make one a good teacher. Knowledge alone, however, does not sufficiently explain differences in teachers' behaviors and success, either (Kunter et al., 2013). For example, even if teachers possess similar professional knowledge through similar teacher preparation programs, how they instruct and how much their instruction contributes to students' success are all different.

Key aspects that determine teacher quality appear to be professional competence which includes beliefs, motivations, and self-regulation, in addition to field-specific knowledge (Kunter et al., 2013). Teachers' beliefs influence their behavior in the classroom and affect the ways they teach and the kinds of learning environment they create (Guskey, 2002; Palak & Walls, 2009). Teachers who have a passion for teaching and regard teaching as intrinsically rewarding provide students with more support, which in turn has a favorable impact on their students' motivation (Frenzel, Goetz, Lüdtke, Pekrun, & Sutton, 2009; Roth, Assor, Kanat-Maymon, & Kaplan, 2007). In addition, successful teachers learn to cope with the constant demand of their work and stressful situations while regulating their engagement (Klusmann, Kunter, Trautwein, Lüdtke, & Baumert, 2008).

Both Erika and Raquel firmly believed that Latina/o students were capable academically, thus had high expectations for their students. Unlike

158 *Teaching With Warmth and Demands*

other novice teachers who tend to be altruistically motivated, and want to contribute to the development of young people (van Uden, Ritzen, & Pieters, 2014) in a sincere but vague sense, the two women's motivation to become teachers were intrinsically driven—experienced-based, concrete, and focused. Both women were determined to have their Latina/o students feel success at school and contribute to the betterment of not only their students but also the wider Latina/o communities. Their determination was not weakened even by the microaggressions, harassment, and discrimination they experienced at work. On the contrary, these obstacles made their determination stronger and solid. The two women's belief, motivation, and level of self-regulation were all grounded in their experiences as Latina bilingual students, paraprofessionals, and teachers. The fact that they demonstrated high quality teaching buttresses the argument that teachers of color who share the same race and ethnicity with students have the potential to make superb teachers.

Teacher belief and motivation are indeed important because they dictate how they teach. Yet, how one teaches is subjective, requires each individual teacher's creativity and skills, thus is different from one teacher to another. Even if teachers share the same belief, how they demonstrate the belief in their instructions is different. The differences highlight each teacher's quality of teaching. What is important to consider is what kinds of instructions Erika and Raquel made and as a result, what kinds of learning environment, conducive to better student engagement, their instructions created.

The two women were strongly motivated to provide their Latina/o students with quality instruction and have the latter experience academic success because they both believed in their student's academic ability. As was noted above, the two women were able to achieve their teaching objectives by cultivating family-like relationships with the students and the parents, and this helped them establish a strong school-based parent network. Latinas/os and working-class or poor families typically have stronger familial ties, but tend to be more isolated from schools (Gamoran, López Turley, Turner, & Fish, 2012). Even in communities with strong Latina/o roots, Latina/o parents could feel ignored by schools or "abandoned and helpless when trying to gain information regarding their children's education" (Ramirez, 2003, p. 93). Both women made efforts to connect the two disconnected worlds—school and the Latina/o parents and their communities—by strengthening the relationships, which in return increased the students and the parents' access to valuable school resources. The demands of these two women for discipline and excellence within their classrooms were built upon the foundation of trusting relationships. Without deep trusting relationships between themselves, students and their families, neither their authoritative disciplinary approach nor their strictly academically-focused instruction would have been accepted by students and parents. The result of establishing positive social bonds with students and families was a higher level of student engagement.

Teaching With Warmth and Demands   159

Being Latinas with high expectations for their students, Erika and Raquel positively affected the ways the students responded to their teaching. However, the ways their students were engaged in the classroom and at school indicates that both women, in addition to being teachers with backgrounds similar to their students, were gifted teachers. The students' trust of Erika and Raquel's authority as manifested by their compliance and engagement on academic tasks would not have been possible if the two women had not taught well. Since white teachers' success with white students are not usually attributed to their being white, attributing the teaching successes of Erika and Raquel only to their race and ethnicity excludes them from the category of professional teachers.

Two cautionary notes need to be added with regard to teacher quality. Student engagement is an important precursor for learning (van Uden et al., 2013). High levels of engagement have been shown to be related to better achievement at school (Archambault, Janosz, Fallu, & Pagani, 2009; Reschly & Christenson, 2006). Although the students in both Erika and Raquel's classrooms were highly engaged, Raquel's kindergarteners made better academic achievement than Erika's third-graders, at least as measured by their test scores. In fact, Erika stated that her students still struggled academically. Thus, if measured strictly by the degree of academic achievement the students made, Erika would not be called a successful teacher. This reality suggests that we may be missing out important aspects of student engagement if we define teacher success strictly on student achievement. Teacher quality should include long-term noncognitive student outcomes, such as their socioemotional skills development.

Another caution is the impact teachers in earlier grades make. Students' academic achievement is a result of cumulative learning experiences taught by different teachers who all possess different instructional styles. In this sense, Erika's students being third-graders and having been not taught by Erika for the previous three years could have explained the level of achievement her students were able to make. Erika's students were already below the grade level before being taught by her, and it can be hypothesized that her students would have performed better on the standardized tests if they had been taught by classroom instructors who were more similar to her. Raquel, being a kindergarten teacher, had an advantage over Erika. Her students had not been taught by any teachers previously and they had not been affected by negative influences from other teachers. The difference in test scores between Erika and Raquel's students suggests that it may be important for students of color to always be taught by teachers who deeply understands their cultural background, who display culturally responsive and caring teaching with a focus on academic achievement. In this regard, teachers of color of high quality are, of course, the most suitable candidates.

160 *Teaching With Warmth and Demands*

References

Archambault, I., Janosz, M., Fallu, J., & Pagani, L. S. (2009). Student engagement and its relationship with early high school dropout. *Journal of Adolescence, 32*(3), 651–670.

Behnke, A. O., Piercy, K. W., & Diversi, M. (2004). Educational and occupational aspirations of Latino youth and their parents. *Hispanic Journal of Behavioral Sciences, 26*(1), 16–35.

Blad, E. (2017, February 14). Mistrust in school can have lasting negative effects: Teachers' "wise feedback" to students may help. *Education Week, 36*(21), 9–11.

Blau, P. M. (2002). Reflections on a career as a theorist. In J. Berger & M. Zelditch Jr. (Eds.), *New directions in contemporary sociological theory* (pp. 345–357). Lanham, MD: Rowman and Littlefield.

Bondy, E., & Ross, D. D. (2008). The teacher as warm demander. *Educational Leadership, 66*(1), 54–58.

Bondy, E., Ross, D. D., Gallingane, C., & Hambacher, E. (2007). Creating environments of success and resilience: Culturally responsive classroom management and more. *Urban Education, 42*(4), 326–348.Bone, J., & Slate, J. (2011). Student ethnicity, teacher ethnicity, and student achievement: On the need for a more diverse teacher workforce. *The Journal of Multiculturalism in Education, 7*(1), 1–22.

Brantlinger, E., Morton, M. L., & Washburn, S. (1999). Teachers' moral authority in classrooms: (Re)Structuring social interactions and gendered power. *Elementary School Journal, 99*(5), 491–504.

Brown, D. F. (2004). Urban teachers' professed classroom management strategies. *Urban Education, 39*(3), 266–289.

Bryk, A. S., & Schneider, B. (2002). *Trust in schools: A core resource for improvement.* New York: Russell Sage Foundation.

Burgess, S., & Greaves, E. (2013). Test scores, subjective assessment, and stereotyping of ethnic minorities. *Journal of Labor Economics, 31*(3), 535–576.

Cahnmann, M. S., & Remillard, J. T. (2002). What counts and how: Mathematics teaching in culturally, linguistically, and socioeconomically diverse urban settings. *The Urban Review, 34*(3), 179–204.

Calzada, E. J., Huang, K. Y., Anicama, C., Fernandez, Y., & Brotman, L. M. (2012). Test of a cultural framework of parenting with Latino families of young children. *Cultural Diversity and Ethnic Minority Psychology, 18*(3), 285–296.

Carl, N. M. (2014). Reacting to the script: Teach for America teachers' experiences with scripted curricula. *Teacher Education Quarterly, 41*(2), 29–50.

Cochran-Smith, M., & Villegas, A. M. (2016). Preparing teachers for diversity and high-poverty schools: A research-based perspective. In J. Lampert & B. Burnett (Eds.), *Teacher education for high poverty schools* (pp. 9–31). Cham, Switzerland: Springer International Publishing AG.

Curwin, R. (2013). Believing in students: The power to make a difference. *Reclaiming Children and Youth, 22*(2), 38–39.

Darling-Hammond, L. (2010). *The flat world and education: How America's commitment to equity will determine our future.* New York: Teachers College Press.

Dee, T. S. (2005). A teacher like me: Does race, ethnicity, or gender matter? *The American Economic Review, 95*(2), 158–165.

Dee, T. S. (2015). Social identity and achievement gaps: Evidence from an affirmation intervention. *Journal of Research on Educational Effectiveness, 8*(2), 149–168.

Delgado-Gaitan, C. (2004). *Involving Latino families in schools: Raising student achievement through home school partnerships.* Thousand Oaks, CA: Corwin Press.

Dika, S. L., & Singh, K. (2002). Applications of social capital in educational literature: A critical synthesis. *Review of Educational Research, 72*(1), 31–60.

Teaching With Warmth and Demands   161

Driessen, G. (2015). Teacher ethnicity, student ethnicity, and student outcomes. *Intercultural Education, 26*(3), 171–191.

Duncan-Owens, D. (2009). Scripted reading programs: Fishing for success. *Principal, 88*(3), 26–29.

Ede, A. (2006). Scripted curriculum: Is it a prescription for success? *Childhood Education, 83*(1), 29–32.

Egalite, A. J., Kisida, B., & Winters, M. A. (2015). Representation in the classroom: The effect of own-race teachers on student achievement. *Economics of Education Review, 45*, 44–52.

Ehrenberg, R. G., Goldhaber, D. D., & Brewer, D. J. (1995). Do teachers' race, gender and ethnicity matter? Evidence from the National Educational Longitudinal Study of 1988. *Industrial and Labor Relations Review, 48*(3), 547–561.

Ferguson, R. F. (2003). Teachers' perceptions and expectations and the black-white test score gap. *Urban Education, 38*(4), 460–507.

Ford, A. C., & Sassi, K. (2014). Authority in cross-racial teaching and learning: (Re) considering the transferability of warm demander approaches. *Urban Education, 49*(1), 39–74.

Frenzel, A. C., Goetz, T., Lüdtke, O., Pekrun, R., & Sutton, R. E. (2009). Emotional transmission in the classroom: Exploring the relationship between teacher and student enjoyment. *Journal of Educational Psychology, 101*(3), 705–716.

Fryer, R., & Levitt, S. (2004). Understanding the black–white test score gap in the first two years of school. *Review of Economics and Statistics, 86*(2), 447–464.

Gamoran, A., López Turley, R. N., Turner, A., & Fish, R. (2012). Differences between Hispanic and non-Hispanic families in social capital and child development: First-year findings from an experimental study. *Research in Social Stratification and Mobility, 30*(1), 97–112.

Gomez, M. L. (2010). Talking about ourselves, talking about our mothers: Latina prospective teachers narrate their life experiences. *The Urban Review, 42*(2), 81–101.

Gregory, A., Clawson, K., Davis, A., & Gerewitz, J. (2016). The promise of restorative practices to transform teacher-student relationships and achieve equity in school discipline. *Journal of Educational and Psychological Consultation, 26*(4), 325–353.Gregory, A., & Huang, F. (2013). It takes a village: The effects of 10th grade college-going expectations of students, parents, and teachers four years later. *American Journal of Community Psychology, 52*(1–2), 41–55.

Gregory, A., & Thompson, A. R. (2010). African American high school students and variability in behavior across classrooms. *Journal of Community Psychology, 38*(3), 386–402.Guskey, T. R. (2002). Professional development and teacher change. *Teachers and Teaching, 8*(3), 381–391.

Halgunseth, L. C., Ispa, J. M., & Rudy, D. (2006). Parental control in Latino families: An integrated review of the literature. *Child Development, 77*(5), 1282–1297.

Hill, N. E., & Tyson, D. F. (2009). Parental involvement in middle school: A meta-analytic assessment of the strategies that promote achievement. *Developmental Psychology, 45*(3), 740–763.

Hlebowitsh, P. (2007). First, do no harm. *Education Week, 27*(11), 28.

Howsen, R., & Trawick, M. (2007). Teachers, race and student achievement revisited. *Applied Economics Letters, 14*(14), 1023–1027.

Irizarry, J., & Donaldson, M. (2012). Teach for America: The Latinization of U.S. schools and the critical shortage of Latino/a teachers. *American Educational Research Journal, 49*(1), 155–194.

Irvine, J. J., & Fraser, J. W. (1998, May). Warm demanders: Do national certification standards leave room for the culturally responsive pedagogy of African American teachers? *Education Week*, 41–42.

162   Teaching With Warmth and Demands

Jerald, C. D. (2006). *The hidden costs of curriculum narrowing*. Washington, DC: The Center for Comprehensive School Reform and Improvement.

Jussim, L., & Harber, K. D. (2005). Teacher expectations and self-fulfilling prophecies: Knowns and unknowns, resolved and unresolved controversies. *Personality and Social Psychology Review, 9*(2), 131–155.

Kennedy, M. M., Ahn, S., & Choi, J. (2008). The value added by teacher education. In M. Cochran-Smith, S. Feiman-Nemser, D. J. McIntyre, & K. E. Demers (Eds.), *Handbook of research on teacher education* (3rd ed., pp. 1249–1273). New York: Routledge.

Kleinfeld, J. (1975). Effective teachers of Eskimo and Indian students. *School Review, 83*(2), 301–344.

Klusmann, U., Kunter, M., Trautwein, U., Lüdtke, O., & Baumert, J. (2008). Teachers' occupational well-being and quality of instruction: The important role of self-regulatory patterns. *Journal of Educational Psychology, 100*(3), 702–715.

Kohl, H. (2009, January 8). *The educational panopticon*. Teachers College Record. Retrieved from www.tcrecord.org, ID Number: 15477.

Kunter, M., Klusmann, U., Baumert, J., Richter, D., Voss, T., & Hachfeld, A. (2013). Professional competence of teachers: Effects on instructional quality and student development. *Journal of Educational Psychology, 105*(3), 805–820.

Ladson-Billings, G. (2009). *The dreamkeepers: Successful teachers of African American students* (2nd ed). San Francisco: Jossey-Bass.

Lareau, A. (2011). *Unequal childhoods: Class, race, and family life*. Oakland, CA: University of California Press.

Lindo, E. J., Weiser, B., Cheatham, J. P, & Allor, J. H. (2017). Benefits of structured after-school literacy tutoring by university students for struggling elementary readers. *Reading and Writing Quarterly*. Advance online publication. doi:10.1080/10573569.2017.1357156.

Maylor, U. (2009). "They do not relate to Black people like us": Black teachers as role models for Black pupils. *Journal of Education Policy, 24*(1), 1–21.

McGrady, P. B., & Reynolds, J. R. (2013). Racial mismatch in the classroom: Beyond black-white differences. *Sociology of Education, 86*(1), 3–17.

Milner, H. R. (2014). Scripted and narrowed curriculum reform in urban schools. *Urban Education, 49*(7), 743–749.

Molm, L. D. (2010). The structure of reciprocity. *Social Psychology Quarterly, 73*(2), 119–131.

Ocasio, K. M. (2014). Nuestro camino: A review of literature surrounding the Latino teacher pipeline. *Journal of Latinos and Education, 13*(4), 244–261.

Onafowora, L. L. (2005). Teacher efficacy issues in the practice of novice teachers. *Educational Research Quarterly, 28*(4), 34–32.

Ouazad, A. (2014). Assessed by a teacher like me: Race and teacher assessments. *Education Finance and Policy, 9*(3), 334–372.

Palak, D., & Walls, R. T. (2009). Teachers' beliefs and technology practices: A mixed methods approach. *Journal of Research on Technology in Education, 41*(4), 417–441.

Ramirez, A. Y. F. (2003). Dismay and disappointment: Parental involvement of Latino immigrant parents. *The Urban Review, 35*(2), 93–110.

Reschly, A. L., & Christenson, S. L. (2006). Prediction of dropout among students with mild disabilities. *Remedial and Special Education, 27*(5), 276–292.

Roth, G., Assor, A., Kanat-Maymon, Y., & Kaplan, H. (2007). Autonomous motivation for teaching: How self-determined teaching may lead to self-determined learning. *Journal of Educational Psychology, 99*(4), 761–774.

Rueda, R., Monzó, L. D., & Higareda, I. (2004). Appropriating the sociocultural resources of Latino paraeducators for effective instruction with Latino students: Promise and problems. *Urban Education, 39*(1), 52–90.

Teaching With Warmth and Demands 163

Sameroff, A. J., & Fiese, B. H. (2000). Transactional regulation: The developmental ecology of early intervention. In S. Meisels & J. Shonkoff (Eds.), *Early intervention: A handbook of theory, practice, and analysis* (pp. 135–159). New York: Cambridge University Press.

Sawyer, K. (2004). Creative teaching: Collaborative discussion as disciplined improvisation. *Educational Researcher, 33*(2), 12–20.

Shoji, M. N., Haskin, A. R., Rangel, D. E., & Sorensen, K. N. (2014). The emergence of social capital in low-income Latino elementary schools. *Early Childhood Research Quarterly, 29*(4), 600–613.

Shulman, L. S. (1987). Knowledge and teaching: Foundations of the new reform. *Harvard Educational Review, 57*(1),1–22.

Smagorinsky, P., Lakly, A., & Johnson, T. S. (2002). Acquiescence, accommodation, and resistance in learning to teach within a prescribed curriculum. *English Education, 34*(3), 187–211.

Stanton-Salazar, R. D. (2001). *Manufacturing hope and despair: The school and kin support networks of U.S.-Mexican youth.* New York: Teachers College Press.

Steele, C. M. (1997). A threat in the air: How stereotypes shape intellectual identity and performance. *American Psychologist, 52*(6), 613–629.

Steele, C. M., & Aronson, J. (1995). Stereotype threat and the intellectual test-performance of African Americans. *Journal of Personality and Social Psychology, 69*(5), 797–811.

Thompson, G. L. (2004). *Through ebony eyes: What teachers need to know but are afraid to ask about African American students.* San Francisco: Josey-Bass.

Torgesen, J. K. (2004). Avoiding the devastating downward spiral: The evidence that early intervention prevents reading failure. *American Educator, 28*(3), 6–47.

Turiel, E. (2005). The many faces of parenting. In J. G. Smetana (Ed.), *Changing boundaries of parental authority during adolescence: New directions in child and adolescent development* (pp. 79–88). San Francisco: Jossey-Bass.

Vadasy, P. F., & Sanders, E. A. (2008). Benefits of repeated reading intervention for low achieving fourth-and fifth grade students. *Remedial and Special Education, 29*(4), 235–249.

Valenzuela, A. (1999). *Subtractive schooling: U.S.-Mexican youth and the politics of caring.* Albany, NY: SUNY Press.

van Uden, J. M., Ritzen, H., & Pieters, J. M. (2014). Engaging students: The role of teacher beliefs and interpersonal teacher behavior in fostering student engagement in vocational education. *Teaching and Teacher Education, 37*, 21–32.

Varela, R. E., Vernberg, E. M., Sanchez-Sosa, J. J., Riveros, A., Mitchell, M., & Mashunkashey, J. (2004). Parenting style of Mexican, Mexican American, and Caucasian-non-Hispanic families: Social context and cultural influences. *Journal of Family Psychology, 18*(4), 651–657.

Villalba, J. A., Jr. (2007). Culture-specific assets to consider when counseling Latina/o children and adolescents. *Journal of Multicultural Counseling and Development, 35*(1), 15–25.

Villegas, A., & Irvine, J. (2010). Diversifying the teaching force: An examination of major arguments. *The Urban Review, 42*(3), 175–192.

Villegas, A., & Lucas, T. (2004). Diversifying the teacher workforce: A retrospective and prospective analysis. In M. A. Smylie & D. Miretzky (Eds.), *Developing the teacher workforce: 103rd yearbook of the national society for the study of education* (pp. 70–104). Chicago, IL: University of Chicago Press.Yatvin, J., Weaver, C., & Garan, E. M. (2003). Reading first: Cautions and recommendations. *Language Arts, 81*(1), 28–33.

Walker, J. M. T., Ice, C. L., Hoover-Dempsey, K. V., & Sandler, H. M. (2011). Latino parents' motivations for involvement in their children's schooling: An exploratory study. *TheElementary School Journal, 111*(3), 409–429.

164 Teaching With Warmth and Demands

Ware, F. (2006). Warm demander pedagogy: Culturally responsive teaching that supports a culture of achievement for African American students. *Urban Education*, *41*(4), 427–456.

Wright, A., Gottfried, M. A., & Le, V-N. (2017). A kindergarten teacher like me: The role of student-teacher race in social-emotional development. *American Educational Research Journal*, *54*(1S), 78S–101S.

Yosso, T. J. (2017). Whose culture has capital? A critical race theory discussion of community cultural wealth. In A. D. Dixson, C. K. Rousseau Anderson, & J. K. Donnor (Eds.), *Critical race theory in education: All God's children got a song* (2nd ed., pp. 113–136). New York: Routledge.

6 Conclusion

Latina/o Teachers Advancing in the Profession

Teachers impart knowledge necessary for the young generation to survive, prosper, and improve in our society. Perhaps, this task could be done by robots, but machines are not capable of attending to and nurturing children's physical, emotional, and social development. Only human teachers can do that. Thus, teaching goes beyond disseminating knowledge and involves all dimensions of human development. Whether or not our society advances in the future depends on what the young generation does and how they behave. The influences of teachers are significant, which is why teaching is called "a noble profession."

Latinas/os and the Teaching Profession

Although noble and highly regarded in some countries, such as Finland and Japan, teaching is not a higher status occupation in the U.S. (Hargreaves, 2009) mainly because it is a profession closely associated with women—lower in status and consequently lower in salary. Teaching is viewed as a culturally feminized occupation (Flores & Hondagneu-Sotelo, 2014) because it is a profession historically dominated by females. Currently, approximately 84% of all public school teachers are females nationwide (Feistritzer, 2011). Like nursing and social work, teaching is perceived as a quintessentially feminine "helping" profession, one associated with motherhood, nurturance, and caring for others (Acker, 1989). In reality, the overwhelming dominance of women in the teaching field has little to do with characteristics commonly attributed to women. Rather, it is a legacy of historical and structural constraints in a job market: women were simply prohibited from entering workplaces dominated by men, teaching was one of few professions women were allowed to enter, and men could obtain higher-paying jobs other than teaching. Whites dominate the teaching profession at the rate of 84% (Feistritzer, 2011). This is also a result of a historical discrimination at the institutional level.

It is well documented that Latinas/os accord great respect to the authority of teachers and want their children to be well-educated and well-behaved (Galindo, 1996; Gordon, 2002). For this reason, although the salary may

166 *Conclusion*

not be great, the teaching profession is perceived as being much more desirable than blue-collar professions, such as a farmwork, gardening, or restaurant work, or pink-collar professions, such as domestic work, day-care, and clerical work, where Latinas/os tend to be hyper-segregated. Gordon (2002) argues that for many working-class and racial and ethnic minority groups, teaching has served as a platform for promoting not only individual mobility but also community mobility and "racial uplift." That is why many Latinas/os see teaching as a noble service, one that will advance students of color who have been devalued and disadvantaged by poverty and racism (Suárez, 2002; Ochoa, 2007). Teaching, in their eyes, is a community endeavor.

These days, teaching is the most pursued career, in particular for college-educated Latinas, and they are entering the teaching profession at greater rates than other minority groups (Feistritzer, 2011). Two factors seem to be accelerating this phenomenon. First, compared to other professions, such as law or medicine that require longer courses of preparation, the teaching profession requires a relatively short course of study in higher education, usually four or five years. Latinas/os who are from working-class or poor backgrounds with limited financial resources are able to "take advantage of socioeconomic constraints which limit access to college education" (Lortie, 2002, p. 48). In addition to its financial feasibility, the teaching position is readily available most of the time, which gives another incentive to those who are financially constrained. A more important factor is the demand for Spanish-speaking teachers fueled by the increase of Latina/o student population at public school nationwide. Since the 1990s, the labor queue shifted as school districts sought teachers who could communicate with Spanish-speaking Latina/o parents and children, and thus bilingual Spanish-speaking teachers (mostly women) have been increasingly recruited (Flores & Hondagneu-Sotelo, 2014).

What Erika and Raquel Taught Us

What should we take from the stories Erika and Raquel narrated? Despite the favorable factors that attract Latinas/os into the teaching profession, Erika and Raquel's odysseys to becoming and being teachers have been full of obstacles. It suggests that they have needed to pay a price in their attempts to penetrate a traditionally white female, white-collar occupation. This section will summarize the penalties Erika and Raquel paid from critical lenses.

A Fundamental Problem of Minority Teacher Recruitment

At a conference I attended several years ago, Harvard professor Lani Guinier (2010) told the audience a following anecdote. In a party children played a game. Girls kept winning, while boys kept losing. Carefully watching them play and its result,—constantly the same—Guinier's mother grinned: "Look

Conclusion 167

who made the rules." The rules had been made entirely by the girls. Not realizing that they were never meant to win a game whose rules were systemically against them, the boys kept losing and the girls kept winning.

Erika and Raquel encountered many obstacles as they went through each phase of their life. Their K-12 education phase, however, was the cruelest and the most unjust. It was cruel because the education both women received was not only low in quality but almost condemned them to careers of menial jobs. Similar to the boys in the above anecdote, Erika and Raquel were never meant to succeed at school. Rather, their failure was almost guaranteed through a series of practices involving negligence, low expectations, deficit thinking, and language suppression. As a consequence, both women were left with self-doubts about their academic capabilities, rather than blaming school practices that made them view themselves as inferior. It was unjust because the education they received sorted them into blue- and pink-collar jobs after high school and blocked their upward mobility through higher education and beyond. Even when Erika and Raquel entered college and university after years of absence, the lingering ill-effects of the indifferent education they received continued to haunt them. This was evidenced by their multiple trials to pass the entry level test for the teacher education program, their academic struggles as teacher candidates, and their failure to pass the bilingual education Praxis test.

We lament how low a number of teachers of color are in the profession. Thus, we make an effort to vigorously recruit students of color into teacher education and the profession. However, Erika and Raquel's academic struggles in college clearly suggests that the cause of their problems began while they were at K-12 rather than at the college level and needs to be traced to "what these students did or did not study in elementary and secondary school" (Futrell, 1999, p. 31). Erika and Raquel's academic problems indicate what needs to be addressed more vigorously, particularly for what they did not learn at the K-12 level and how little they were prepared for tertiary education. Trying to fix the problem only during teacher education by, for example, providing tutoring and requiring remedial courses, although crucial for their academic success, will not address the fundamental reason why many teacher candidates of color struggle academically and why there are so few Latinas/os in four-year-colleges to begin with.

The low-quality education that many students of color and economically disadvantaged students receive in public school continues to block their social mobility and lock them into lower status. To the dominant group who benefits from this reproductive situation, providing the subordinate group with high quality education is dangerous since it may shake the foundation of their privilege and dominance. That is why, historically, people of color were denied access to education. Under current legislation, depriving anyone of education is not only strongly condemned as ethically wrong but is also illegal. However, providing inferior or ineffective education still occurs with little condemnation when it comes to minority populations.

168 Conclusion

Erika and Raquel's story convinces us that the problem of minority teacher recruitment cannot be solved without addressing the fundamental problems of minority educational preparation. If we are serious about increasing the number of teachers of color, it is imperative that we improve K-12 education for students of color because this is a problem that begins early. For example, low-income Latina/o kindergarteners are already behind both their white and African American peers in reading and white peers in math. This group difference persists and does not narrow as they move on to a next grade (Padilla, Cabrera, & West, 2017). In this sense, Erika and Raquel's overwhelming emphasis on literacy and math in their teaching was a sound professional judgement for the students they served. In order to improve K-12 education, particularly in earlier grades, more teachers of color who demonstrate high quality teaching should be produced and hired. And in order to produce and hire more teachers of color, the ideologies of racism and classism that have made indifferent instruction possible should be widely disseminated, discussed, and addressed with funding for effective programs for students of color, including programs that provide culturally appropriate and effective educational resources. This would allow the improvement of education for students of color to begin.

Without receiving quality education at public schools, Latinas/os will not be able to take advantage of the favorable labor market and become teachers. In this sense, by not explicitly linking K-12 education to minority teacher recruitment, the teaching field has been systemically preventing Latinas/os from entering the profession, and by doing so protecting and perpetuating the dominance of a white female workforce.

Collegiality and Fear

As a predominantly single race and gender occupation, the dominant group has established rules, customs, and practices that benefit them the most and exclude others. Once these system-wide practices are set in place and kept being practiced, they become normalized. These systemic and normalized practices are hard to detect if one benefits from them. It is when others appear, raise concerns about, complain about, and try to change such practices that do not benefit them that the dominant group becomes aware of the existence of such practices. Even with the awareness, the dominant group is often reluctant to admit the existence of unfair practices. When others are perceived to possess advantageous skills the dominant group lacks, the degree of resistance becomes fierce because the dominant group perceives others as being in a direct competition with them, thus intruding into a profession they have long dominated and owned. In a way, Erika and Raquel were thrown into a tug-of-war with their white college peers and teacher colleagues who wanted the teaching profession to remain predominantly white.

From the perspective of the dominant group who did not possess much knowledge about the cultures of the students they taught and lacked Spanish

Conclusion 169

language skills, Erika and Raquel were perceived as a threat to the system that had been traditionally occupied by monolingual white women. The white peers and colleagues of Erika and Raquel were fearful of the two women's advancement in the teaching profession. In both the teacher preparation program at college and when they began work as certified teachers, white female peers openly expressed their fears through various types of microaggressions, offering neither support nor collaboration with them, sometimes intentionally sabotaging their efforts. All these acts contributed to a hostile environment at college and work for both women.

White colleagues were certainly jealous and threatened by Erika and Raquel's Spanish language skills. However, the degree of their fear implies that they were fearful of the two women's overall teaching skills that they perceived were at least equal or even superior to theirs, not necessarily only their Spanish language skills. If the fear was thoroughly driven by their lack of Spanish language skills, it is difficult to imagine why their colleagues, who were already tenured and senior to the two women were so threatened by them. After all, teaching in Spanish was not a requirement for everyone at the school. If the fear was caused by the quality of teaching Erika and Raquel demonstrated, it is understandable that the professional teachers who had been teaching longer than the two women felt resentful and threatened by the newcomers' talents. Even if one is tenured, not possessing quality teaching skills is indeed demoralizing. The fact that the hostile work environment was collectively created by white colleagues throughout their schools underlines the degree of perceived threat caused by the arrival of the skillful Latina newcomers. Whenever the white colleagues talked to Erika and Raquel they suggested that only their Spanish language skills gave them an advantage, camouflaging their fear with regard to their teaching quality.

Because it was inherently a teacher quality issue rather than the Spanish language skills, it was vexing and abhorrent for many of their white colleagues to envision sharing their power and privilege with Erika and Raquel. Latinas had long been perceived by them as outsiders to the teaching profession. Thinking of them as equal and *real* colleagues was not possible for those with this mindset. To most of their white colleagues, being a colleague to Erika and Raquel was impossible as long as the latter were perceived as "bodies out of place" (Puar, 2004). That is, where their capabilities were viewed with suspicion, where they were automatically thought of as less capable and inferior to whites.

The level of fear the white colleagues displayed when presented with evidence that Erika and Raquel were competent, even superior teachers, implies that their feelings of superiority towards people of color had been unpleasantly challenged, causing them great discomfort. The white colleagues' emotional responses through various microaggressive comments, aggressive actions, including outright harassment, highlight that they were indeed "disgusted" with the two women's superb teaching skills. In the white colleagues' mind, Erika and Raquel threatened to violate their space and purity

170   *Conclusion*

(Matias, 2014) and injure their privilege (Leonardo & Zembylas, 2013). Linking the history of normative whiteness for instance as fear, desire, terror, and fantasy, Yancy (2008) argues that whites relegate people of color to marginalized and segregated spaces that restrict them from "'disturbing' the tranquility of white life, white comfort, white embodiment and white being" (p. xvi). In this way of thinking, it was Erika and Raquel who must be blamed for invoking feelings of terror and disgust, causing disturbance among them, inconveniencing them, thus needing them to be disciplined. A reflective moment about their own teaching skills did not seem to exist. By prioritizing their own emotions of distress and disgust at the expense of their own shortcomings and projecting such emotions directly and sarcastically onto racial others, the white colleagues successfully reified whiteness through the use of entitlement. This was indeed "bullying" and this was also a norm at the schools at which Erika and Raquel taught.

Resilience and Experiential Knowledge

Despite suffering from numerous practices at school that disadvantaged them, Erika and Raquel displayed superb resilience. Resilience is defined as the ability to experience normal developmental achievement in the face of social and environmental adversities (Carver, 1998). In particular, resilience in their maintenance of the Spanish language skills was formidable against the language suppression policies that attempted to eliminate the use of Spanish, segregate Latina/o students by language, and break their social network at school and present them with token, low-quality bilingual education. Far from abandoning the Spanish language, both women, in fact, took a full advantage of their experiences while they were paraprofessionals, and sharpened and polished the literacy domain of the Spanish language as they learned academic language in Spanish. Their persistence in maintaining and improving their Spanish skills resulted in their rapid hiring and employment as bilingual education teachers at schools. Erika and Raquel were both skillful in navigating against adversities at the schools where they taught and eventually turning their opponents' bigotries against them into advantage. They did not waste their experiences at all. Rather, they thrived on them.

The effects of the hostility Erika and Raquel faced at their schools was compounded by their weak professional networks where they worked. However, the more hostile their work environments became, the stronger their resilience grew. In response, the two women strengthened their relationships with their students and extended it to parents. Eventually, the parents became their most trusting allies. No matter what the colleagues came up with to professionally disadvantage and weaken the two women, Erika and Raquel always effectively counter-reacted and overcame the negative expectations that they encountered. This was made possible because of the strong *familismo* both women cultivated with the students, the parents, and with one another. Students' academic resilience is heavily influenced by

Conclusion 171

the relationships teachers build with them (Sosa & Gomez, 2012). In return, teachers are also influenced by students. In a sense, Erika and Raquel's superb resilience was heavily supported by the students and the communities, forming a constellation of the Latina/o community power.

If "collaborative relations of power" (Cummins, 2001) had existed between Erika and Raquel and their colleagues, and the two women had been easily conferred with the same power, status, and autonomy enjoyed by their colleagues, their resilience would not have shown as strongly. In this sense, it is the white colleagues who inadvertently strengthened the degree of the two women's resilience. It is ironic that the proactive and vigorous defense of white teaching dominance by white monolinguals contributed to strengthening the resilience of Erika, Raquel, and the Latina/o communities at their schools. What the white colleagues miscalculated was the fact that the bullying took place at predominantly Mexican schools, in predominantly Latina/o communities where Erika and Raquel were able to cultivate and deepen their trusting relationships with students and parents. In so doing they empowered themselves as professional teachers. Thus, the physical and cultural location of the tug-of-war was favorable to Erika and Raquel. Their white colleagues underestimated the positive effect that the Latina/o communities' hidden resilience and strength had on the two women.

Inclusion and Exclusion

Dominance is the essential ingredient for sustaining whiteness. In order to dominate others, others need to be excluded. Fine (1997), for example, contends, "Whiteness was produced through the exclusion and the denial of opportunity to people of color . . ." (p. 60). A result is a social system in which resources and rights are favorably given to whites and unjustly denied to people of color. If whiteness is an exclusive club whose membership is closely and grudgingly guarded (Harris, 1993) and members are afforded with privilege, power, and high status, how can diversifying a predominantly white teaching force and recruiting more teachers of color do any good to the sustenance of whiteness? How do exclusion and inclusion go together so peacefully?

The power of whiteness lies in its function of advantaging whites without them feeling advantaged (Johnson, 2012; Rothenberg, 2012). After all, the advantages they receive are taken-for-granted, everyday occurrences, and embedded in the normal daily practices of society. Since the advantages they receive are invisible and the system works for them, most whites are eager to give voice to the thought that they promote democratic and egalitarian practices, and work to create a fairer and more equal society for all, including themselves. This would be commendable thinking if it wasn't actually an illusion. Whiteness can claim to embrace diversity-related policies and practices and sing their praises with tagged words like multicultural and diversity (Castagno, 2014). The underlying thinking is that as long as we

172   *Conclusion*

avoid having racist attitudes, we can succeed in avoiding racist practices. By focusing on fairness and equality, whiteness encourages sameness and neutrality. However, in a system where one group receives built-in advantages by default and is already advantaged, the emphasis on equality, neutrality, and sameness, no matter how genuine, accentuates the already-existent advantages and furthers the gap between the dominant group and the subordinate groups with regard to the benefits they receive. Whiteness, in other words, both perpetuates and advances by verbally challenging itself.

During the PNU's teacher education program, Erika and Raquel complained that they did not receive any treatment different from their white peers, though they thought their experiences as paraprofessionals should have been recognized and taken into account. Having taught as paraprofessionals before, the two women felt that field-based courses should have been waved and their student-teaching should have been shortened. This special accommodation would have certainly helped them financially. Nevertheless, PNU strictly adhered to a "one-size-fits-all" policy and treated every teacher candidate exactly the same. After they were hired as classroom teachers, Erika and Raquel were also not given differential treatment from monolingual white teachers. The end result of this same treatment was a vastly different workload from other teachers. The same treatment both women received under their schools' neutral treatment of all doctrine did not produce fair and equal outcomes for all. Rather, this doctrine ironically increased the two women's workload from which the white colleagues directly benefited. It is important to note that the neutral policy negatively affected Erika and Raquel, while it had no effect on their white peers at PNU and gave direct advantages to the white colleagues at work. Invisibility of whiteness and its unmarked category persisted since they were not affected negatively at all.

Given the unfair disadvantages Erika and Raquel encountered in the attempt to be included in the teaching profession, it is no wonder that Waldschmidt (2002) questions "the motives of those of us who would 'diversify' the teaching force" (p. 556). Even if "included," teachers of color run the risk of being constrained and forced to be compliant and become "changed agents" (Achinstein & Ogawa, 2011) by the very education system they desire to change. This will remain true as long as normative and invisible practices of whiteness remain the same and are not challenged and disrupted. Hiring teachers of color is a welcome step for greater equity because it diversifies the predominantly white teaching force. However, if inclusion requires teachers of color to be "docile and useful" (Foucault, 2003), it prevents them from effectively assisting the very student population they are recruited to serve.

What Erika and Raquel Advise Us

Listening to Erika and Raquel's stories remind us that the teaching profession needs to do a lot more work if it is to become serious about increasing the number of teachers of color. This section will list several recommendations.

Conclusion 173

Paraprofessionals in Teacher Education

In order to transition paraprofessionals of color into teaching more smoothly, teacher education programs need to change structurally. First, an explicit effort to raise the status of paraprofessionals needs to be taken. One way is to offer curriculum modifications, particularly in field-based courses, to seasoned paraprofessionals instead of offering a one-size-fit-all type of curriculum. Ignoring their rich experiences as paraprofessionals, neither Erika nor Raquel received modifications in their curriculum. Since their experience was practical, waiving a field-based course and shortening their student teaching would not have affected the quality of Erika and Raquel's teaching. They were already strong, experienced teachers at the beginning of their teacher education program. The experiences of paraprofessionals should be treated with respect and should be appreciated by both instructors and other teacher candidates. Teacher education instructors should be especially cognizant of the practical experience paraprofessionals bring to their programs, because education instructors have the greatest influence on the ways teacher candidates think and behave. In Rintell and Pierce's (2003) study, the respect Latina/o paraprofessionals received from professors tremendously facilitated their success. Erika and Raquel's effort to connect theories to practice should have taken with appreciation, been discussed openly in class, and used as part of the curriculum.

As for the cost, Futrell (1999) warns that for many students of color the cost of a college education is prohibitive and can have a debilitating impact on family finances. Several successful paraprofessional-to-teacher programs have provided teacher candidates with financial support in the form of tuition and books, scholarships, and stipends. As Raquel stated, financial support should be available for seasoned paraprofessionals even in traditional teacher education programs.

Academic difficulties Erika and Raquel experienced during teacher education could have been eased if the professors in their teacher education program had had more caring attitudes. Becket (1998) found that the success of Latina/o paraprofessionals aspiring to become certified teachers depended on whether or not they could connect coursework to their ongoing experiences in classrooms and to programs' value-added model, which "includes elements that add value to the program such as cooperative learning, cohorts, special sessions and interventions, and careful counseling and nurturing from program administrators and mentors" (Becket, 1998, p. 203). Since passing required teacher exams is unavoidable for certification, each program needs to be structured so that it will provide paraprofessionals with more caring, nurturing, and flexible academic environments.

The most important structural change teacher education programs need to make is to intentionally and directly incorporate the issues of race and culture into their required curriculum. Sue, Capodilupo, Torino, Bucceri, Holder, Nadal, & Esquilin, (2007) contend that the prerequisite for cultural competence is always racial self-awareness. Are teacher candidates as well

174　*Conclusion*

as teacher educators "challenged to explore their own racial identities and their feelings about other racial groups" (Sue et al., 2007, p. 283)? Do they experience a process of learning and critical self-examination of racism and its impact of one's life and the lives of others (Thompson & Neville, 1999)? These questions need to be systematically addressed and openly discussed with everyone involved in teacher preparation, from teacher candidates to teacher educators to administrators, to create a collegiate and collaborative academic environment for prospective teachers of color.

Microaggressions

According to Sue et al. (2007), "Racial microaggressions are potentially present whenever human interactions involve participants who differ in race and culture (p. 284) and the cumulative effects of racial microaggressions could be devastating. To relieve paraprofessionals, teacher candidates, and teachers of color from enduringly stressful and sometimes hostile classroom and work environments, social, academic, and professional counterspaces are recommended. Solórzano and Villalpando (1998) argue that counterspaces allow people of color to foster their own learning and to nurture a supportive environment wherein their experiences are validated and viewed as important knowledge. Erika and Raquel created a counterspace by and for themselves and with the help of the Latina/o students and the parents, which tremendously helped their journey through the teacher preparation program and a workplace. As part of creating a more caring and nurturing academic and professional environment, teacher candidates and teachers of color should be not only encouraged to create their own counterspaces but also such an opportunity should be proactively provided.

At a Workplace

Because it is part of our everyday reality, we tend to view the public education space as racially neutral. This tendency fosters an inability to critically understand the role of race and racism in public education. Thus, we need to analyze whether or not everyday policies and practices we take for granted at public schools are really race neutral and who benefits the most from school policies and practices.

For example, is the fact that translation duty is frequently assigned to Spanish speakers race neutral? Since Spanish speakers could be any race, it sounds colorblind, but who is actually mostly assigned this task? Why do we assume that Spanish native speakers can create Spanish curriculum without assistance? If monolingual English teachers benefit from Latina/o bilingual teachers' cultural resources, what do they offer to Latina/o bilingual teachers in return? Questioning and discussing these unchallenged everyday policies and practices is the first step towards working for equal power relations.

Conclusion 175

Erika and Raquel's stories revealed that the mere presence of teachers of color does not necessarily guarantee their engagement in the professional communities of their schools. Because the white racial frame is dominant, pervasive, and institutionally embedded in school cultures and structures, it affects challenges to reform it. Attempts by teachers of color to use their cultural knowledge and collaborate with white colleagues in creating a more positive educational environment for students of color are undermined by their schools' white supremacist framework. Further, teachers of color frequently perceive that their white colleagues are playing a zero-sum game. That is, they believe their white colleagues feel if students and teachers of color make gains, white teachers will lose power, authority, and prestige.

First, courageous leadership that is willing to transform school cultures is needed. Administrators and teachers of all color should take responsibility to educate themselves regarding how the white racial frame works in the U.S. in general and at their schools in particular, and how it leads to structural racism. Without deep understanding of the pervasive and systemic nature of white racism, the positive influence of school administrators and teachers at their schools will be minimal. School administrators can emphasize collaboration, not competition among white teachers and teachers of color, but all teachers and administrators must reflect on how the white racial frame's loss-gain perspective operates within U.S. schools. Without the positive reflection of teachers and administrators, exhortations for collaboration between white teachers and teachers of color are merely rhetorical. Since administrators can have a profound impact on teachers at the professional level, their ability to train their fellow teachers becomes essential.

Second, schools need to take affirmative steps to make sure teachers of color, particularly novice teachers of color, are engaged in a positive collegial network. As Erika and Raquel perceived, novice teachers are, in general, overwhelmed at their work. Novice Latina/o bilingual education teachers, however, carry an extra burden that creates unnecessary stress when colleagues and administrators are unaware or refuse to be aware of the double demands placed upon them. On top of this, teachers of color frequently need to deal with their white colleagues' feelings of jealousy and fear. Knowing the importance of a positive professional community within schools that provides access to multiple opportunities and support networks (Quiñones, 2016), administrators need to proactively intervene with the existing professional network of their schools to insure that teachers of color participate in a constructive manner. Those who are vulnerable for marginalization because of the tenacious and systemic nature of the white racial frame and the concomitant system of structural and institutional racism need to be protected, cultivated, and affirmatively encouraged. Without the schools' conscious effort to facilitate their growth as professional teachers, Latina/o bilingual education teachers are vulnerable to the revolving door syndrome in which they come and leave. This situation negatively affects the students of color's success at school.

176 *Conclusion*

Conclusion

Listening to Erika and Raque's stories gives us hope that strongly motivated and diligent teachers of color will continue to succeed in navigating through minefields of racial and ethnic aggression. At the same time, it reminds us that the teaching profession has been historically dominated by whites, and as a result, collegiality with teachers of color is saturated as contemptuous tolerance (Tate, 2013). Inclusion of teachers of color merely signifies their marked status so that those left unmarked—white teachers—are seen as able, intelligent, proficient, and having the temperament for success (Puar, 2004).

Latina/o students' success at school is largely dependent on whether or not they are taught by Latina/o teachers who possess high quality teaching skills. Therefore, we must seriously reflect upon the challenges Latina/o paraprofessionals, teacher candidates, and teachers are forced to endure and make appropriate programmatic changes in order to facilitate the smooth transition and successful retention of these populations in the profession. Student populations continually grow more diverse at public schools. If we are determined that all students must succeed academically, socially and psychologically, we cannot afford not to produce and not to retain teachers of color any longer.

References

Achinstein, B., & Ogawa, R. (2011). *Change(d) agents: New teacher of color in urban schools*. New York: Teachers College Press.

Acker, S. (1989). *Teachers, gender and careers*. London: Taylor and Francis.

Becket, D. (1998). Increasing the number of Latino and Navajo teachers in hard-to staff schools. *Journal of Teacher Education, 49*(3), 196–205.

Carver, C. S. (1998). Resilience and thriving: Issues, models, and linkages. *Journal of Social Issues, 54*(2), 245–266.

Castagno, A. E. (2014). *Educated in whiteness: Good intentions and diversity in schools*. Minneapolis, MN: University of Minnesota Press.

Cummins, J. (2001). *Negotiating identities: Education for empowerment in a diverse society* (2nd ed.). Covina, CA: California Association for Bilingual Education.

Feistritzer, C. E. (2011). *Profile of teachers in the U.S. 2011*. Washington, DC: National Center for Education Information.

Fine, M. (1997). Witnessing whiteness. In M. Fine, L. C. Powell, L. Weis, & L. Mun Wong (Eds.), *Off white: Readings on race, power, and society* (pp. 57–65). New York: Routledge.

Flores, G. M., & Hondagneu-Sotelo, P. (2014). The social dynamics of channelling Latina college graduates into the teaching profession. *Gender, Work, and Organization, 21*(6), 491–515.

Foucault, M. (2003). *Power/knowledge: Selected interviews and other writings, 1972–1977* (C. Gordon, Ed.). New York: Pantheon Books.

Futrell, M. H. (1999). Recruiting minority teachers. *Educational Leadership, 56*(8), 30–33.

Galindo, R. (1996) Reframing the past in the present: Chicana teacher role identity as bridging identity. *Education and Urban Society, 29*(1), 85–102.

Gordon, J. (2002) *The color of teaching*. London: Routledge.

Guinier, L. (2010, October). *Rethinking race and class within the context of our crisis in education*. Speech presented at the Race and Pedagogy National Conference, the University of Puget Sound, Tacoma, WA.

Conclusion 177

Hargreaves, L. (2009). The status and prestige of teachers and teaching. In L. J. Saha & A. Dworkin (Eds.), *International handbook of research on teachers and teaching* (pp. 217–29). New York: Springer.

Harris, C. I. (1993). Whiteness as property. *Harvard Law Review*, 106(8), 1707–1791.

Johnson, A. G. (2012). Privilege as paradox. In P. S. Rothenberg (Ed.), *White privilege: Essential readings on the other side of racism* (pp. 115–119). New York: Worth Publishers.

Leonardo, Z., & Zembylas, M. (2013). Whiteness as technology of affect: Implications for educational theory and praxis. *Equity and Excellence in Education*, 46(1), 150–165.

Lortie, D. C. (2002). *Schoolteacher: A sociological study*. Chicago, IL: University of Chicago Press.

Matias, C. E. (2014). *Feeling white: Whiteness, emotionality, and education*. Rotterdam, The Netherland: Sense Publishers.

Ochoa, G. (2007). *Learning from Latino teachers*. Hoboken, NJ: Jossey-Bass.

Padilla, C. M., Cabrera, N., & West, J. (2017). *The development and home environment of low-income Hispanic children: Kindergarten to third grade*. Bethesda, MD: National Research Center on Hispanic Children and Families.

Puar, N. (2004). *Space invaders: Race, gender and bodies out of place*. Oxford: Berg Publishers.

Quiñones, S. (2016). "I get to give back to the community that put me where I am": Examining the experiences and perspectives of Puerto Rican teachers in western New York. *Urban Education*. Advance online publication. doi:10.1177/0042085915623336.

Rintell, E. M., & Pierce, M. (2003). Becoming maestra: Latina paraprofessionals as teacher candidates in bilingual education. *Journal of Hispanic Higher Education*, 2(1), 5–14.

Rothenberg, P. S. (2012). Introduction. In P. S. Rothenberg (Ed.), *White privilege: Essential readings on the other side of racism* (pp. 1–5). New York: Worth Publishers.

Solórzano, D. G., & Villalpando, O. (1998). Critical race theory, marginality, and the experiences of students of color in higher education. In C. A. Torres & T. R. Mitchell (Eds.), *Sociology of education: Emerging perspectives* (pp. 211–224). Albany, NY: State University of New York Press.

Sosa, T., & Gomez, K. (2012). Connecting teacher efficacy beliefs in promoting resilience to support of Latino students. *Urban Education*, 47(5), 876–909.

Suárez, E. (2002). *A calling of the heart: A comparative study of meanings and motivations of Chicana and Mexican American teachers*. PhD dissertation, Department of Education, Claremont Graduate University.

Sue, D. W., Capodilupo, C. M., Torino, G. C., Bucceri, J. M., Holder, A. M. B., Nadal, K. L., & Esquilin, M. (2007). Racial microaggressions in everyday life. *American Psychologist*, 62(4), 271–286.

Tate, S. A. (2013). Racial affective economies, disalienation and "race made ordinary". *Ethnic and Racial Studies*, 37(13), 2475–2490.

Thompson, C. E., & Neville, H. A. (1999). Racism, mental health, and mental health practice. *Counseling Psychologist*, 27(2), 155–223.

Yancy, G. (2008). *Black bodies, white gazes: The continuing significance of race*. Lanham, MD: Rowman and Littlefield.

Waldschmidt, E. D. (2002). Bilingual interns' barriers to becoming teachers: At what cost do we diversify the teaching force? *Bilingual Research Journal*, 26(3), 537–561.

Index

academic difficulties 28, 55, 56, 60, 173
academic standards 144, 149, 154
accommodations: lack of 77
approachability 138, 139
aspirational capital 60
authoritarian 34, 145
authoritative 145, 155, 158

bilingual education 7, 9, 10, 30,
37, 38, 56, 91, 92, 105, 170; and
endorsement 78, 80, 90, 167;
maintenance 91; transitional 91, 102,
103; two-way 38; 90/10 model 102,
103, 111
bilingual education teachers 7, 9, 14, 123;
and nonprestigious status 104, 110,
112; shortage of 103; stress of 175

caring 34, 36, 38, 48, 124, 138, 139,
143, 156, 157, 159, 165, 173, 174;
and uncaring 29, 34, 47, 49, 112,
117, 136
Catch-22 76
collegiality 74, 110, 111, 123, 126,
168, 176
counterspaces 63, 64, 174
critical race theory 9, 12; and Latino
critical race theory (LatCrit) 14
cultural capital 5, 60, 111
curriculum: rigor of 154

deficit thinking xi, 28, 30, 39, 47, 67,
73, 74, 81, 167
discipline 18, 34, 41, 113, 114, 121,
123–126, 144, 146, 147, 156, 158

empowerment 43
ethnic matching 93, 135, 136

ethnic specialists 91, 93, 96, 98, 99
excellence 144, 147, 158
exclusion 12, 100, 171
expectations 33, 37, 134–137, 142; and
colleagues 67; parents 137; teachers
(high) viiii, x, 2, 34, 36, 48, 147,
148, 149, 150, 151, 155, 156, 157,
159; teachers (low) 3, 28, 29, 41, 47,
167, 170; unreasonable 93, 94
experiential knowledge 14, 16, 72, 73,
110, 170
expert see ethnic specialists

familial capital 34, 54, 137, 145, 155, 170
familismo see familial capital
farmwork 25, 26, 27, 36, 37, 43, 48, 166
fear 5, 36, 70, 71, 75, 108, 119, 120,
169: whites 10, 11, 43, 81, 115, 118,
121, 126, 127, 169, 170, 175
financial burden 77, 78, 173

hostility 4, 11, 31, 61, 72, 115, 125, 170

immigration xi, 13, 14, 62, 70: and
undocumented 26, 37, 62, 68, 69, 70
inclusion i, 96, 113, 171, 172, 176
indifferent instruction 28, 29, 32, 47,
56, 168
institutional agents 60, 140,
intra-ethnic hostilities 31
isolation xiiii, 64, 110, 111, 142, 143

juxtaposition 139, 140, 142

laborer 104, 112
Latino critical race theory (LatCrit) see
critical race theory
linguistic capital xi, 32

marginalization viiii, 8, 18, 31, 77, 128, 175; and bilingual education teachers 100, 101, 104, 106, 114

mentoring 80, 100, 101; and importance of 101, 112, 173; lack of 11

mentors see Mentoring

microaggressions xi, 9, 11, 18, 61, 62, 63, 65, 66, 73, 75, 76, 81, 126, 158, 169, 174; and definition 11; effects 12, 62, 66, 77, 115, 119; invisibility to perpetrators 61, 67, 77; pertaining to Latinas/os 62; Spanish language skills related 113, 114; types 62

miner's canary 93

minority teacher recruitment 57, 88, 166, 168

motherhood 35, 70, 165

noble profession 165

"no-nonsense" stance 147, 151, 155

one-size-fits-all 58, 74, 80, 172

overwork 112, 114

paraprofessionals 5, 8, 15, 17, 18, 33–36, 81, 88, 89, 118, 138; and advantages 5, 6, 90, 96, 110, 144, 152, 158, 170, 172, 173, 174, 176; cultural connectors 6; definition 32; hierarchical relations with teachers 45, 112; lack of training 38; recruiting for teaching 5; under-appreciation 6

peer pressure 26, 27

peer support 7, 104

peer tutoring 59–60

pressure to prove: bilingual education teachers 108

presumed incompetence xiii, 11, 65, 96, 97, 121

professional network 100, 111, 113, 127, 170, 175

racial uplift 166

recognition: absence of 102, 106, 109, 139, 142; from administrators 146; from students and parents 141, 142

relational trust 33, 137; and family-like relationship 137, 138, 144, 158

resilience xiiii, xi, 41, 82, 170, 171

resistant capital 31, 37

role model ix, 5, 33, 35, 73, 134, 151, 156

schooling: ill-effect 36, 167

scripted curriculum 94, 95, 96, 111, 116, 117, 151, 152, 155; criticism 151, 154

segregation by language: at K-12 31; at work 96

self-doubt 40, 167

self-image: negative 29, 67

social capital 27, 54, 60, 100, 110, 111

social support 54, 58

Spanish language skills: advantages 18, 32, 90, 109; apprehension 63; connection to students 48, 123, 138; contradictions 42, 47, 99; exploitation 18, 111; in demand 5; lower in status 30; maintenance 26, 36, 56, 81, 89, 170; microaggressions 18, 74–77, 81, 99, 113; pressure 108, 111–115, 148; suppression 31; unrecognized 126

stereotype threats 135

student engagement 137, 158, 159

support: lack of 3, 8, 58, 94–96, 109, 112, 117, 139

surveillance 65, 108, 111, 120, 126

teachers of color: changed agents 172; expectations for 93; marginalization 4, 95, 96, 98, 108, 111, 118, 126; shortage of ix, 2, 46; strengths x, 2, 123, 134; subjectivity 136; surveillance 120; work environment 6–8, 128

teachers: authority 40, 98, 120, 140, 141, 145, 148, 155, 156, 159, 165, 175; attitudes 7, 11, 28, 29, 44, 47, 69, 117, 123; control 97; network 99–101, 110–111, 126; novice 15, 92, 94, 108, 109, 111, 112, 113, 116, 118, 127, 144, 147, 158, 175; quality 55, 98, 136, 157–159, 169; subjectivity 135

threat (Threatening) 6, 10, 11, 17, 18, 69, 75, 76, 81, 115, 121, 126, 169

translation duty 33, 91, 101–102, 106–110, 111, 174

uncaring see caring

"used" 102, 107, 112

"wanted" status 89–90, 91, 93, 99, 111

warm demander xiiii, 147

180 Index

white racial frame xi, 9–11, 68, 127, 175
whiteness 9, 12–13, 61, 65, 76, 108, 112, 113, 115, 116, 126, 170, 171, 172; and capitalism 104; invisibility 172; paradox 99

"work hard" stance 145
workload 18, 91, 100, 102–104, 106, 107, 110, 111, 112, 113, 115, 128, 172
workplace 6, 7, 8, 11, 98, 100, 119, 121, 126, 165, 174